Springer Series in Computational Mechanics

Edited by S. N. Atluri

S. N. Atluri, S. G. Sampath, P. Tong (Eds.)

Structural Integrity of Aging Airplanes

With 319 Figures

Springer-Verlag
Berlin Heidelberg New York
London Paris Tokyo
Hong Kong Barcelona Budapest

Editor of the series:

Prof. S. N. Atluri

Center for Computational Mechanics
Georgia Institute of Technology
Atlanta, GA 30332-0356, USA

Editors of this volume:

Prof. S. N. Atluri

J. C. Hunsaker Visiting Professor
Aeronautics & Astronautics, Rm. 33-313
M.I.T., Cambridge, MA 01239, USA

Dr. S. G. Sampath
Dr. Pin Tong

Transportation Systems Center
Structures and Dynamics Division
Kendall Square
Cambridge, MA 02142

ISBN-13: 978-3-642-84366-2 e-ISBN-13: 978-3-642-84364-8
DOI: 10.1007/978-3-642-84364-8

Library of Congress Cataloging-in-Publication Data
Structural integrity of aging airplanes / [edited by] S. N. Atluri, S. G. Sampath, and P. Tong.
(Springer series in computational mechanics)
Papers presented at the International Symposium on Structural Integrity of Aging Airplanes, held Mar. 20-22, 1990, in Atlanta, Ga., and sponsored by the Georgia Institute of Technology, Center for Computational Mechanics and the Transportation Systems Center of the U.S. Dept. of Transportation.

ISBN 978-3-642-84366-2

1. Airplanes--Reliability--Congresses. 2. Airframes--Fatigue--Congresses.
3. Aeronautics--Safety measures--Congresses.
I. Atluri, Satya N. II. Sampath, S. G. III. Tong, Pin.
IV. International Symposium on Structural Integrity of Aging Airplanes (1990 : Atlanta, Ga.)
V. Georgia Institute of Technology. Center for Computational Mechanics.
VI. Transportation Systems Center. VII. Series.
TL671.S58 1991
629.134'31--dc20 90-26077

Softcover reprint of the hardcover 1st edition 1991

61/3020-543210 – Printed on acid-free paper

Preface

The emergence of civil aviation as a means of mass transportation is primarily due to the large-scale construction of jet airplanes in the past 30 years or so. A large number of these jet airplanes is currently operating at or beyond their designed fatigue lives. Thus, the structural integrity of these aging airplanes has become an issue of major concern to all nations of the world.

To bring the needed technical and research focus on the issues involved in the life-enhancement and safety-assurance of aging airplanes, the Federal Aviation Administration sponsored a symposium in Atlanta, GA, USA, during 20-22 March 1990. This symposium, under the title "International Symposium on Structural Integrity of Aging Airplanes" was organized jointly be the Georgia Institute of Technology (Center for Computational Mechanics) and the Transportation Systems Center (Cambridge, MA) of the U.S. Department of Transportation. Industrial and academic experts from several countries in North America, Europe and Asia, were invited to discuss their experiences and proposed solutions. This monograph contains the original papers that represent the expanded and edited versions of the talks presented at this symposium.

This book aims to bring the collective experience, from across the world, with problems related to the structural integrity of aging airplanes to the attention of the professional and research community at large - in the hope that it may stimulate further fruitful research on this important topic of global concern.

A synopsis of the contents of this monograph is as follows: In Chapter 1, Achenbach and Thompson summarize the principal nondestructive evaluation techniques, their applications, advantages and disadvantages. They also briefly discuss four major components of a quantitative nondestructive evaluation program for aging aircraft. In Chapter 2, Atluri and Tong review some simplified computational strategies for the integrity analysis of fuselage panels in aging airplanes. Such topics as an alternating technique for multiple site damage near a row of fasteners, bulging of cracks, analysis of cracks with bonded repair patches, are discussed. In Chapter 3, Berens, Burns and Rudd discuss a computer based risk analysis tool for the decision making process of scheduling inspection, repairs, replacement or retirement maintenance actions on aging aircraft fleets.

In Chapter 4, Bristow discusses, from a regulatory viewpoint, such issues of aging aircraft, as: Proof pressure testing, inspection reliability, effects of corrosion and definition of corrosion inhibiting fluids, and implementation management of aging aircraft programs. In Chapter 5, Broek presents a brief review of the damage tolerance requirements for civil aircraft, and of their impact and significance for both newly developed and aging aircraft. In Chapter 6, Chubb, Morad, Hockenhull and Bristow discuss the effects of prior exfoliation corrosion on the fatigue behavior of two specific aluminum alloys - 7178-T6 and 2024-T351.

In Chapter 7, Cowie discusses, in depth, a damage tolerance approach for the management of aging gas turbine engines. In Chapter 8, Dawicke, Poe, Newman and Harris present a generic analytical and experimental investigation related to the pressure proof test concept for a 2024-T3 aluminum alloy sheet. In Chapter 9, Goranson and Miller focus their attention on the principles of durability and damage tolerance technology standards suitable for large teams of structural engineers; and provide some preventive maintenance recommendations that will permit continued safe operation of aging jet transport until their retirement from service for economic reasons.

In Chapter 10, Harris discusses the combination of a fatigue crack growth prediction methodology and a nondestructive examination technology with the focus on multi-site damage at riveted connections. In Chapter 11, Hendricks discusses the factors relating to the Aloha Airlines accident which occurred on 28 April 1988 in the Hawaiian Islands. In Chapter 12, Hoggard presents a manufacturer's approach to the design of longitudinal splices in a fuselage.

In Chapter 13, Ingraffea, Grigoriu and Swenson discuss issues related to the representation of a number of arbitrarily growing cracks in a finite element model with codes built on a topological data structure and employing automatics remeshing. In Chapter 14, Jones, Bridgeford, Wallace and Molent discuss the topic of repairs to cracked fuselages using bonded composites, and discuss the results of a test program which confirm the effectiveness of such a repair. In Chapter 15, Kanninen, O'Donoghue, Green, Leung, Roy and Burnside focus on the conditions in which inelastic-dynamic effects occur in the linking up of local damage in a lap splice to form a major crack, and on the large-scale fuselage failure by a rapidly occurring fluid/structure interaction process.

In Chapter 16, Kosai and Kobayashi discuss a model of axial rupture of an aircraft fuselage based on the mechanisms of rapid crack propagation, crack curving, and arrest in a pressurized, thin-walled ductile steel tube. In Chapter 17, Mar provides an interesting personal perspective on structural integrity of aging airplanes, based on his many years of experience in various high-level technical committees of the United States Government. In Chapter 18, Mayville and Warren discuss a laboratory study of fracture in the presence of lap splice multiple site damage.

In Chapter 19, Orringer presents a simple conceptual model for estimating the risk of multiple site damage. In Chapter 20, Pelloux, Warren and O'Grady present a fractographic analysis of initiation and growth of fatigue cracks at rivet holes. In Chapter 21, Roll, van Dalen and Jongebreur review the philosophy developed at Fokker, with regard to structural maintenance, in conjunction with Piedmont and USAir.

In Chapter 22, Samavedam, Hoadley and Davin describe a new test facility for evaluating the fatigue and fracture strength of stiffened and jointed aircraft fuselage panels. In Chapter 23, Sampath and Broek address issues concerning the probability of detection of multi-site damage in fuselage lap-splices in aging airplanes. Their results indicate that reduction in the mandated period between inspections should be considered. In Chapter 24, Schmidt presents a summary of the evaluation methodologies applied for pressurized fuselages of the Airbus.

In Chapter 25, Selikhov, Raikher, Leibov and Nesterenko present the Soviet experience in specifying and prolonging the airframe time limit; and provide an insight into the Soviet approach wherein the airframe is considered as an aging one from the beginning of the aircraft fleet service. In Chapter 26, Smith presents an overview of the optical methods of stress analysis applied to cracked components. In Chapter 27, Swift presents a very comprehensive summary of the procedures of repairs to damage tolerant aircraft. He presents the results of a displacement compatibility analysis representing a variety of repair doubler and lap splice configurations, and illustrates how structural repairs can degrade the fatigue initiation life and damage tolerance capability of the primary transport aircraft structure. In Chapter 28, Terada and Ohtsuka present a case study, a simulation of fatigue damage and damage tolerance evaluation of aging aircraft, based on a review of research in Japan.

The editors thank all the authors for their timely cooperation in preparing their respective chapters in this volume. The editors also thank Ron Allen (Delta Airlines), Ben Cosgrove (Boeing Aircraft Co.), and Nelson Miller (Federal Aviation Administration, Technical Center) for their contributions to the organization of the symposium, on which this volume is based. Finally, it is a pleasure to acknowledge the untiring efforts of Ms. Brenda Bruce, of the Computational Mechanics Center at Georgia Tech, in not only organizing the symposium, but also in putting this volume together.

July 1990
Atlanta, Cambridge and Hong Kong

Satya N. Atluri
Sam G. Sampath
Pin Tong

Professor James W. Mar

A Dedication

Professor James W. Mar was born in Oakland, California, on 10 March 1920. He received his S.B. (1941), S.M. (1947) and Sc.D. (1949), all from the Massachusetts Institute of Technology. He was head of the Structural Test Section at Curtiss Wright Corp. during 1941-44; and has been on the faculty of Aeronautics and Astronautics at M.I.T. since 1950. In 1990 he retired from M.I.T. as the Jerome Clarke Hunsaker Professor of Aerospace Education.

He was Chief Scientist of the U.S. Air Force during 1971-72; a Member of the U.S. National Academy of Engineering, and heads the Task Group on Aging Aircraft for FAA. He received the Structures, Structural Dynamics and Materials Award of the American Institute of Aeronautics & Astronautics in 1987.

In recognition of his many contributions to the technology of structural design for longevity of airplanes, this volume is dedicated to Professor James W. Mar.

Satya N. Atluri
Sam G. Sampath
Cambridge, MA

Pin Tong
Hong Kong

Sept. 1990

Table of Contents

1. Towards Quantitative Non-Destructive Evaluation of Aging Aircraft

J. D. ACHENBACH

Center for Quality Engineering and Failure Prevention
Northwestern University, Evanston

and

D. O. THOMPSON
Center for Non-Destructive Evaluation
Iowa State University, Ames

Summary

Non-destructive testing techniques, as they are practiced in the field of quantitative non-destructive evaluation, are at the basis of a comprehensive approach to secure the safety of aging aircraft. The applications, advantages and disadvantages of the principal NDE techniques are summarized in this paper. It is discussed that measurement models for these techniques, in conjunction with the POD concept, scanning plans and methods of graphical display, facilitate the selection of optimal procedures for specific inspection problems. These models also suggest NDE standards and calibration techniques and they can be an important part of inspection system validation and operator training. Four major components of a comprehensive QNDE program for aging aircraft are identified and briefly discussed.

Introduction

The media and the public became aware of aging aircraft in April 1988 when an Aloha Airlines B737 lost part of its fuselage, twenty thousand feet above the Pacific. For some years the aircraft manufacturers and the operators had known that many aircraft were exceeding their original design lives, and a number of actions had been taken, particularly with regard to aircraft inspection and maintenance procedures, see e.g. Ref.[1]. To the general public the advanced age of many airplanes came, however, as a surprise, and April 1988 marks a quantum jump of public interest in the airworthiness of the commercial aircraft fleet.

The Aloha accident caused the Aircraft manufacturers, the Airlines and the FAA to reevaluate their efforts. In June 1988, the FAA convened its first conference on aging aircraft. Experts from all over the world in the fields of aircraft structures, non-destructive inspection, human factors and propulsion systems attended. One of the products of this meeting were eight recommendations by a Committee of representatives of the Airlines and the Manufacturers. These recommendations were subsequently accepted by the FAA as a constructive outcome of the conference. The recommendations were

1. Continue to use the present system of maintenance and inspection with diligence and thoroughness. Find out why a single aircraft suffered major

Springer Series in Computational Mechanics
Atluri, Sampath, Tong (Eds.)
Structural Integrity of Aging Airplanes

structural failure and adjust the system as necessary.

2. Initiate research to find better ways to assess structural condition and detect structural problems.

3. Continue to pursue the concept of teardown of the oldest aircraft to determine structural condition, and conduct fatigue tests of older airplanes.

4. Pursue transfer of the currently available body of knowledge of NDT and its application to aircraft inspection.

5. Put R&D money into improving NDT techniques and methods.

6. Examine all aspects of human factors involved, including training and qualification of airline inspectors.

7. Ensure that the communications systems between airlines, manufacturers, and the FAA are adequate.

8. Establish task forces from the airlines, manufacturing industry, FAA, and NASA to continue the work begun in this workshop.

Since that first meeting numerous committee meetings have been held and other International and National Conferences have been organized. The Aviation Safety Act (Public Law 100-591) has been passed and other legislative proposals are being considered by Congress. It is not the purpose of this paper to present a historical account. It should, however, be noted that the recommendations of the 1988 Meeting specifically recognized the need for extensive efforts in the general area of nondestructive inspection, and that, so far, direct action in this all-important area has been very limited. Some organizations have offered proposals to develop non-destructive inspection programs. A proposal for a broad-based approach anticipates the formation of a Center for Aviation Systems Reliability, but as of the writing of this paper a comprehensive program of research and development on non-destructive inspection of aircraft had not yet been initiated.

Most structural parts of modern transport aircraft are designed to satisfy damage tolerant requirements. This does not mean that aging aircraft should be assumed to necessarily suffer structural damage such as fatigue cracking. It does, however, mean that when structural damage does occur it should be detected and characterized by appropriate inspection procedures and repaired through routine maintenance. Efficient and reliable inspection methods play a crucial role in the detection of unforeseen defective conditions before strength is degraded to an unacceptable level. The longer an aging aircraft or any other load-bearing structure is operated, the more likely it is to develop structural damage which may lead to a need for unscheduled repair. It is in the long term economic interest of air fleet operators to avoid the additional costs which result from unscheduled repairs. This can be done by a systematic approach to the overall inspection requirements of the airframe, the engine and other components. This approach should take into account the statistics of the occurrence of damage, various non-destructive detection techniques and the probabilities of detection of certain classes of defects, loads and damage deterioration properties, inspection intervals, human factors, costs benefits and repair methodology.

Fatigue and corrosion, and to a lesser extent improper maintenance, have been recognized as the main potential causes of structural failure of aging aircraft. Although the failure mechanisms of fatigue and corrosion are quite different, and require their own specific approaches to prevention and detection, they are particularly harmful when occurring together. This occurred in the Aloha Airlines Boeing 737 which lost the top half of its forward fuselage. As pointed out by several authors, see e.g. Bobo [2], a comprehensive approach to secure the safety of aging aircraft should have as a principal component a very strong program of non-destructive evaluation. In this paper we will discuss some essential parts of such a program.

Current NDE technology is not well positioned to efficiently deal with the problem of aging aircraft. A principal reason is that NDE is not currently practiced as a quantitative engineering discipline but as a piecemeal, empirical technology. At the present time NDE systems are not engineered with detailed quantitative design guides and specifications regarding flaw detectability, system validation, performance reliability, and other useful specifications for particular structures; rather, they are assembled and validated on a one-by-one, completely heuristic and empirical basis, usually after the problem has arisen. It has become apparent that the practice of NDE must be raised to that of an engineering based discipline if it is to cope effectively with problems of aging aircraft and other structures. It is also apparent that new ways must be found for an engineering based NDE to enter at the design phase of aircraft, to assure integrity and efficient inspection procedures in service, and to eliminate or at least considerably reduce the aging aircraft dilemma in the future.

Fatigue Damage

Fatigue is probably the most intensely studied damage mechanism in aircraft. In the original design of an aircraft its usage is estimated based on a 20 year economic life and the customer's anticipated use of the aircraft. Sometimes these estimates do not completely take account of the ultimate use of the aircraft in service. Hence, the original fatigue goals based on these initial estimates and the fatigue tests conducted to verify the design, may at some point not adequately cover the aircraft. Then new inspection procedures and new fatigue tests of the structure are appropriate. This was, for example, the case when multiple site damage (MSD) was uncovered as a new and widespread case of fatigue damage. In general it should however, be possible to anticipate the occurrence of fatigue in the design stage, and to plan, at that stage, appropriate inspection programs. Ways to deal with the fatigue problem in aging aircraft have been discussed by Hagemaier et al. [1] and Broek [3].

Corrosion

Much attention has been paid to the development of a baseline corrosion inspection and control programs. Corrosion is neither new nor limited to aging aircraft. In 1965 two Handley Page Herald aircraft, each only three years old, were lost through fuselage lap joint corrosion. Corrosion is highly operator dependent and is widespread in aircraft which have not had the benefit of a program for inhibiting corrosion from early in service life. Thresholds for the introduction of such a program should be well below and typically half of the design life. The program should require periodic access to and inspection of all areas where there is any history of corrosion, and selective inspection of those areas where corrosion has not been found before. Immediate corrosion removal and reprotection should be required. Any structure found corroded beyond acceptable limits should be repaired immediately. In practice, corrosion inspection may actually determine maintenance scheduling. Aircraft corrosion and detection methods have been discussed by Hagemaier et al.[4].

Improper Maintenance and Repairs

Improper maintenance and repairs can be the cause of serious accidents. An example is provided by the loss of a Japan Air Lines 747-SR. The rear pressure bulkhead of this aircraft had been repaired by a partial replacement after a landing accident. The execution of the repair was not in accordance with the drawing. Loads in the bulkhead due to pressurization were effectively being transmitted through only one row of rivets over a length of some three feet. An added stiffener forced the fatigue crack which developed to run across the circumferential tear stops which were designed to stop and turn it. Cracking would have been indicated by difficulty in pressurizing the aircraft, but in this case the crack was held closed until the bulkhead collapsed. Air surged into the fin and destroyed it together with the rudder attachments. When the rudder separated from the aircraft all the hydraulic services were severed.

Quantitative Non-Destructive Evaluation

Fracture mechanics and, in a more general sense, failure mechanics have made great strides in the prediction of the integrity of structural components. For a component made of a material of known properties subjected to a given set of loads, it is possible to calculate the critical size of a crack at a specified location. The component is judged to be safe if the crack is smaller than a critical size and when it may be expected that it will not grow to that size during the service life or prior to the next inspection. Naturally, to tolerate cracks of subcritical size, reliable methods must be available to detect and characterize cracks. Quantitative non-destructive evaluation provides the techniques to detect a crack (or more generally a flaw), and to determine its location, size, shape and orientation. The role of QNDE, particularly ultrasonic methods in strength evaluation has been discussed by Achenbach [5].

A single crack is only one kind of material discontinuity. Cracks are particularly objectionable since they are very obvious causes of catastrophic failure, but voids, cavities, inclusions, interfaces, distribution of cracks, or in general terms damaged regions of a material, may have equally deleterious effects on the strength of components. By the use of appropriate nondestructive evaluation methods it should be possible to discriminate between a broad spectrum of flaws and to determine the relevant characteristics of each kind.

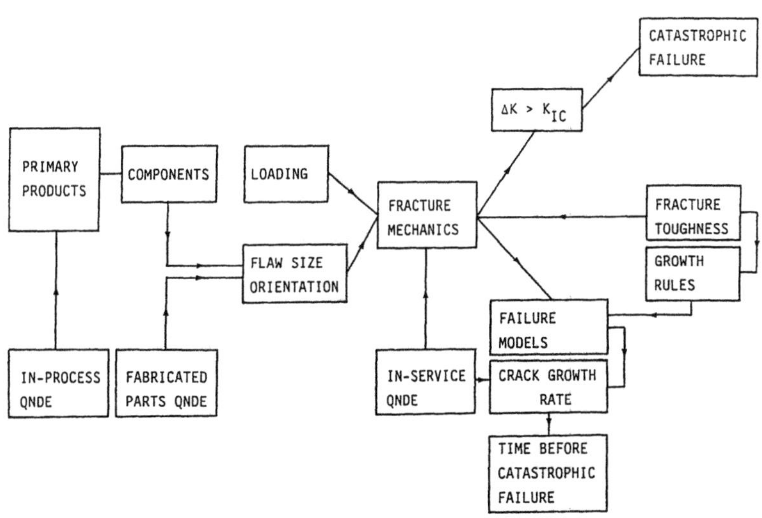

Fig. 1 Role of Quantitative NDE in the life cycle of a component.

The role of quantitative non-destructive evaluation during the various stages of the life cycle of a structural component is illustrated in Fig. 1.

The schematic depiction applies to metal components which are subjected to cyclic loading, and hence may sustain metal fatigue. As indicated in Fig. 1, Quantitative NDE methods should enter in the material processing stage to play a role in maintaining material quality of primary products. At this stage QNDE methods ensure that primary products such as plates, sheets and strips do not contain cracks or other flaws whose dimensions exceed a certain specified level. In the next stage, QNDE methods should be applied to characterize flaws that have been induced in the process of fabricating components. The third and most important application of QNDE techniques is to detect and characterize in-service cracks. The maximum dimensions of such cracks, in conjunction with the magnitude of the cyclic load, can be used to calculate the maximum values of corresponding stress intensity factors. Naturally these maximum values, ΔK, should be less than the fracture toughness. If this is indeed the case, then a crack may still propagate, but at a controlled rate which in principle is predictable. Hence within the framework of the damage tolerant philosophy, a part containing a macroscopic flaw is acceptable if it can be shown that at the predicted stress levels, the flaw will not grow to critical size during a specified time. Reliable quantitative methods of non-destructive in-service inspection of parts, in conjunction with the concept of probability of detection, are clearly essential for a successful implementation of the damage tolerant philosophy. Flaws which at the time of in-service inspection are greater in size than consistent with the design lifetime, must be detected, characterized, and the part must be repaired. On the other hand, the QNDE procedure should not reject components that contain only smaller size flaws. A part should be returned to service if no flaws are found, or if it can be shown that the size of a detected flaw is small enough that it will propagate to failure only over a period substantially larger than the next inspection interval. This approach has been successfully implemented in the retirement for cause procedures of turbine disks [6].

Inspection Techniques

It has been reported that about eighty percent of all inspections of the aircraft structure is carried out by visual-optical techniques, i.e., by an experienced inspector who is equipped with a spotlight and a mirror. Penetrants for the detection of surface-breaking flaws are also extensively used. The remaining inspections are carried out by the use of more sophisticated techniques such as high and low frequency eddy current methods, ultrasonic techniques, x-ray radiography, gamma-ray radiography and magnetic particle methods. Other techniques that are being considered are neutron radiography and thermal wave imaging.

The applications, advantages and disadvantages of the most useful non-destructive inspection techniques are well known to experts in the field. For a broader audience they may be summarized as follows.

VISUAL INSPECTION

<u>Application</u>: Detection of surface defects and surface damage.

<u>Advantages</u>: Simplicity. Optical aids can be used for visual enhancement.

<u>Disadvantages</u>: Reliability is strongly inspector dependent. Direct accessibility is required.

PENETRANT

<u>Application</u>: Detection of surface-breaking cracks.

<u>Advantages</u>: Simplicity, reliability, speed.

<u>Disadvantages</u>: Defect must be surface breaking, open and accessible. Part must be cleaned before application.

HIGH-FREQUENCY EDDY CURRENT

<u>Application</u>: Detection of surface defects in metal components: cracks, pits, intergranular corrosion. Cracks near fasteners.

<u>Advantages</u>: Can be used when defects cannot be detected visually or by penetrants. Fast, sensitive, portable; works for tight fatigue cracks.

<u>Disadvantages</u>: Requires a trained operator and special probes for specific applications. Current applications require reference standards. "Lift-off" effects affect accuracy.

LOW-FREQUENCY EDDY CURRENT

<u>Applications</u>: Detection of sub-surface defects in metal components, corrosion damage.

<u>Advantages</u>: Can be used for local inspection of complicated structural assemblies. Deeper penetration than with high-frequency techniques.

<u>Disadvantages</u>: Same as for high-frequency eddy current methods.

SONIC

<u>Application</u>: Can be used to obtain indications of delaminations, disbonds, voids, crushed core in composite and honeycomb materials.

<u>Advantages</u>: One-sided approach. Does not require surface preparation.

<u>Disadvantages</u>: Minimal sensitivity.

ULTRASONIC

<u>Application</u>: Detection and characterization of surface-breaking and sub-surface defects. Thickness gaging. Material characterization.

<u>Advantages</u>: Easy to operate, fast, reliable. Highly sensitive and accurate, portable.

<u>Disadvantages</u>: Requires a trained operator. Some a-priori information of the type of defect is required for appropriate selection of probe and wave modality. Current applications require reference standards.

X-RAY RADIOGRAPHY

Application: Detection of damage such as internal flaws (cracks, voids, inclusions), corrosion and thickness variations.

Advantages: Can be used for assembled structures, has high sensitivity and provides a permanent record on film. Real-time observations are feasible.

Disadvantages: Radiation hazard. Trained operators and film processing equipment required. X-ray beams must be parallel to crack plane.

NEUTRON RADIOGRAPHY

Application: Detection of corrosion and bond integrity.

Advantages: Corrosion products have an order of magnitude higher absorption coefficients than aluminum.

Disadvantages: Moveable thermal neutron sources are low flux and require long exposure times. Hydrogen compounds such as trapped moisture, fuel or sealing materials can mask corrosion products and reduce detectability.

MAGNETIC PARTICLE METHODS

Application: Detection of surface-breaking defects in ferromagnetic materials.

Advantages: Conceptually simple, easily applicable and fast. Portable equipment.

Disadvantages: Trained operator required. Parts must be demagnitized after testing. Magnetic flux must be normal to plane of defects.

LASER-BASED OPTICAL METHODS

Application: Non-contact generation of ultrasound and non-contact inspection of surfaces.

Advantages: Full field inspection with rapid scanning for shearography and holography, but low resolution. Higher resolution with interferometry for point probing.

Disadvantages: Susceptible to extraneous noise.

X-RAY DIFFRACTION

Application: Detection and characterization of near-surface conditions, particularly residual stress states.

Advantages: Accuracy.

Disadvantages: Slow procedure, requiring a trained operator. Not suitable for large-scale testing.

THERMAL WAVE IMAGING

Application: Detection of anomalies in thin metal skins and composites.

Advantages: Full-field images.

Disadvantages: Coating required for satisfactory absorption of heat in aluminum parts.

ACOUSTIC EMISSION

Application: Detection of flaw initiation and propagation.

Advantages: Only requires receiving transducers.

Disadvantages: Difficulties in separating flaw-generated signals from noise, and in identifing source characteristics from the data.

Measurement Models

A fundamental approach to NDE must be based on quantitative models of the measurement processes, i.e. a "measurement" model for each of the various inspection techniques. The model's principal purpose is to predict, from first principles, the measurement system's response to specific anomalies in a given material or structure, (e.g. cracks, voids, corrosion, deviations in material properties from specifications, and others). Thus, a measurement model must include the geometric details of the component being inspected, i.e. the configuration of probe and part. It must also include a description of the generation, propagation, and reception of the interrogating energy. In the ultrasonic case, this description requires knowledge of the transducer radiation pattern, refraction of the beam at the parts' surface, the beam profile and propagation characteristics in the host material including effects of material anisotropy, attenuation, diffraction losses, etc. Information on critical flaws that is usually available from materials engineering (size, orientation, deviation in material properties) must also be considered. Detailed models of the field-flaw interactions which generate the measurement system's response function should also be included, as well as knowledge of material and other conditions that produce noise and add an uncertainty to the measurement results. Thus, a measurement model must be able to predict specific instrumental responses to anomalies in complex materials and structures as well as to "standard" flaws placed in various calibration blocks.

A number of measurement models have been formulated in the past several years for different inspection technologies. In principle, the formulation of such models is straight forward. In practice, the challenge lies in making approximations that permit the computations to be tractable while retaining sufficient accuracy so that the engineering applications are not compromised. The status of models for ultrasonics, eddy current methods and radiographic techniques has recently been discussed by Gray, Gray, Nakagawa and Thompson [7].

The implementation of a measurement model should be coupled to the probability of detection (POD) concept. This is a statistical representation of the probability that a given measurement system will be able to detect a specific flaw (or condition) in a given material/structure.

It incorporates knowledge of the signal detected by the measurement system together with statistical information concerning flaw distributions and instrumental noise/thresholding levels. An idealized POD curve is shown in Fig. 2 in which the probability of a flaw's detection is shown as a function of flaw size for both an ideal and a real inspection technique. For the ideal technique, the POD of flaws smaller than a critical size (as determined by performance requirements and material properties) is zero whereas the POD for any flaw greater than this size is unity. In this case, there are neither false rejects (FR) of good parts nor false accepts (FA) of defective ones. However, real NDE Techniques are never as sharp and as discriminatory as indicated by the ideal curve, with the result that there are regions of uncertainty shown by the false reject and false accept areas. This representation of the capabilities of a given NDE Technology includes the necessary information from which various NDE engineering applications, including economic benefits, can be designed with predictable performance characteristics. The importance of measurement models has been discussed in greater detail by Thompson [8].

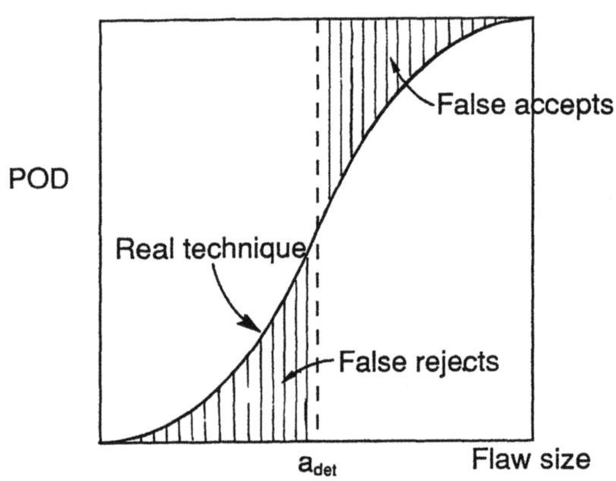

Fig. 2 Probability of detection versus flaw size

The POD concept has been discussed by several authors, see Refs.[2] and [8].

A productive approach towards an inspection problem combines the measurement model, enriched by the POD concept, with a scanning plan and with graphical techniques for comprehensive display. By employing a solids modeling software package, simulated results of inspection procedures can be displayed. In such a display, different colors on different shades of gray correspond to a range of POD values from zero to unity. Examples of displays and a more detailed discussion have been presented by Gray, et al. [7].

The approach described above, makes it possible to select the optimal procedure for a specific inspection problem. Performance results for different postulated measurement modalities and test procedures can be determined in terms of POD's for critical "test" flaws (as defined by past performance profiles, material properties, and fracture mechanics) to select an optimum NDE system. The same set of procedures can also be used to predict the quantitative performance of the selected system in other inspection conditions in which the geometry or critical test flaw conditions are changed. None of these predictive capabilities are possible with the current empirical technology. Cost alone would prohibit the empirical approach from ever producing an optimal design solution.

The beginning of a new class of NDE standards and calibration techniques is also conceptually included in this approach, and can be developed for special cases with many technical and cost advantages. For example, standard blocks with various specific flaws can be used in the model simulation for all the measurement models.

A third advantage of the approach concerns NDE system validation. NDE system validation is currently a costly, completely empirical procedure that usually requires the preparation of a statistically significant number of test specimens. The extensive number of samples required for this procedure could be significantly reduced if coupled with the predictability of the POD model approach. A verification of the model predictions at a few data points coupled with predictions for simulated cases should serve to validate the approach at considerably reduced cost. With the models the validation can also be extended to other cases.

The approach may also serve to separate equipment measurement capability from operator effects. The skills of operators are usually quantified empirically by determining how successfully various flaws are detected, using a given test technique. It is implicitly assumed in these various operator qualification procedures that the POD of the various test flaws is unity. The fallacy of that assumption should be evident. Some of the poor showings that have been registered for operator performance may have been, at least in part, due to limited measurement capabilities of the NDE system as defined by the POD characteristics. The POD provides a way of separating the physical measurement capabilities of the NDE inspection system from operator capabilities. The best operator in the world is not going to do any better than this limitation. A proper account of the POD concept provides a way to assess an operator's training needs and to separate those from the performance limits of the NDE system.

Towards QNDE of Aging Aircraft

A comprehensive approach to secure the safety of aging aircraft should be an internationally focused, joint government/industry/university effort. It should include technology applications and implementation, as well as engineering research and development, education and training, and economics and management of new technology. Targets within these areas should include:

TECHNOLOGY APPLICATIONS, IMPLEMENTATION AND TRANSFER

- Continuous review, evaluation, and selection of research advances for reduction to practice and transfer to field problems.
- Validation of inspection schemes and instrumentation using real components and model-based POD approaches.
- Hands-on training for NDE engineers and technicians in education programs, the commercial aviation industry, and FAA.
- Establishment of proficiency standards and evaluation programs for NDE engineers.
- Development of a "Simulator" for a number of uses including validation of inspection schemes and hardware; establishment of improved inspection schemes and hardware with quantified reliability; training of inspectors through programs using human factors inputs and proficiency standards.

ENGINEERING RESEARCH AND DEVELOPMENT

- Improved and quantified inspection reliability.
- Advanced quantitative NDE techniques.
- Development of new in-service air-safety monitoring sensors, such as acoustic emission devices, embedded corrosion sensors, and sensors to measure stress, strain and cumulative fatigue.
- Inspection and process-control techniques for materials and processes, particularly in adhesively bonded structures and composite bonds.
- NDE and failure modes of emerging aerospace materials.
- Systematic life extension techniques.
- Integrated design, NDE and manufacturing for new aircraft including design for inspectability and design for pre-programmed inspection intervals.

EDUCATION AND TRAINING

- Extended NDE curriculum programs.
- Two-year technician programs at junior colleges linked to programs at research universities.
- Continuing education programs.

•Human factors concerns incorporated into education/training curricula.

ECONOMICS AND MANAGEMENT OF NEW TECHNOLOGY

•Evaluation of the cost benefits of new NDE technologies.

•Development of optimum managerial approaches to enhanced reliability testing and evaluation.

•Examination of expected benefits from integrating NDE inspectability into aircraft design for improved reliability.

•Evaluation and optimization of man-machine interfaces in NDE systems.

Concluding Comment

A comprehensive approach to the quantitative assessment of the safety of aging aircraft is proposed. Non-destructive testing techniques, as they are practiced in the field of quantitative non-destructive evaluation, are at the basis of the approach. As discussed in this paper, important advances can be achieved by the use of measurement models for NDI techniques, the POD concept, scanning plans and graphical display techniques. The four principal components of the overall approach have been summarized.

References

1. Hagemaier, D.J., Abelkis, P.R. and Harmon, M.B., Supplemental inspections of aging aircraft, Materials Evaluation 44 (1986) 989-997.

2. Bobo, S.N., The aging aircraft fleet - a challenge for nondestructive inspection, in Review of Progress in Quantitative Nondestructive Evaluation, Vol. 9B, 2097-2111, D. O. Thompson and D. E. Chimenti (eds), Plenum Press 1990.

3. Broek, D., The civil damage tolerance requirements in theory and practice. This volume.

4. Hagemaier, D.J., Wendelbo, A.H. and Bar-Cohen, Y., Aircraft corrosion and detection methods, Materials Evaluation 43 (1985) 426-437.

5. Achenbach, J.D., Strength Evaluation by ultrasonic methods, in Proc. ICF7, Advances in Fracture Research Vol 5, 3099-3119, K. Salama, K. Ravi-Chandar, D.M.R. Taplin and P. Rama Rao (eds), Pergamon Press 1989.

6. Cowie, W. D., Fracture control philosophy, in Nondestructive Evaluation and Control, Metals Handbook, Volume 17, ASM International 1989.

7. Gray, J.N., Gray, T.A., Nakagawa, N., and Thompson, R.B., Models for predicting NDE reliability, in Nondestructive Evaluation and Control, Metals Handbook, Volume 17, ASM International 1989.

8. Thompson, D.O., Tools for an engineering approach to NDE., in Flight-Vehicle Materials, Structures and Dynamics Technology. Assessment and Future Directions, to appear.

9. Berens, A., Burns, J. and Rudd, J., Risk analysis of aging aircraft fleets. This volume.

2. Computational Schemes for Integrity Analyses of Fuselage Panels in Aging Airplanes

Satya N. Atluri
Center for Computational Mechanics
Georgia Institute of Technology
Atlanta, GA 30332-0355, USA

Pin Tong
Transportation Systems Center
Kendall Square
Cambridge, MA 02142, USA

Abstract

This paper reviews some simplified computational strategies for the integrity analyses of fuselage panels in aging airplanes. The topics covered include: (i) a "direct stiffness" method for stiffened panels (with fastener flexibility being accounted for) with multiple cracks, using the alternating technique, (ii) MSD near a row of fastener holes, (iii) analysis of cracks with bonded repair patches, (iv) weight functions for multiple cracks, and (v) bulging near crack-tips in pressurized fuselages, and equivalent domain integral methods for computing fracture parameters for bulged cracks.

Introduction

Elastic and elastic-plastic fracture mechanics methodologies, that are easy to apply, are mandatory in a strategy for assessing the structural integrity of curved, stiffened, fuselage panels with multiple-site-damage (MSD). One common occurance of such MSD is in a row of fastener holes in a lap joint in a fuselage structure. Straight-forward finite element/boundary-element analyses of such MSD problems, with detailed numerical modeling of crack-tip singularities, would be prohibitively expensive, even in the linear elastic regime. Furthermore, such straight-forward numerical analyses are useful for a single configuration of crack(s) and loads; and whenever the crack-configuration or the load-configuration changes, the entire analysis must be repeated. Thus, alternate strategies for efficient analyses of structural integrity in the presence of MSD are needed. This paper reviews some such possible strategies.

The topics that are covered in this brief overview include: (i) alternating methods for determining stress-intensity factors for multiple cracks in stiffened fuselage panels; (ii) extension of these techniques for moderate-scale yielding plasticity; (iii) weight functions for multiple cracks in fuselage panels; (iv) effects of bulging near crack-tips in pressurized curved panels; and (v) equivalent domain integral methods for the evaluation of fracture parameters for bulged cracks, with large deformation and plasticity effects near the crack-tips. This overview is necessarily cursory in nature; and additional background material may be found in the recent comprehensive summaries: Atluri (1986), and Atluri and Nishioka (1989).

Springer Series in Computational Mechanics
Atluri, Sampath, Tong (Eds.)
Structural Integrity of Aging Airplanes

- **Alternating Technique for Multiple 2D or 3D Embedded/Surface Cracks**

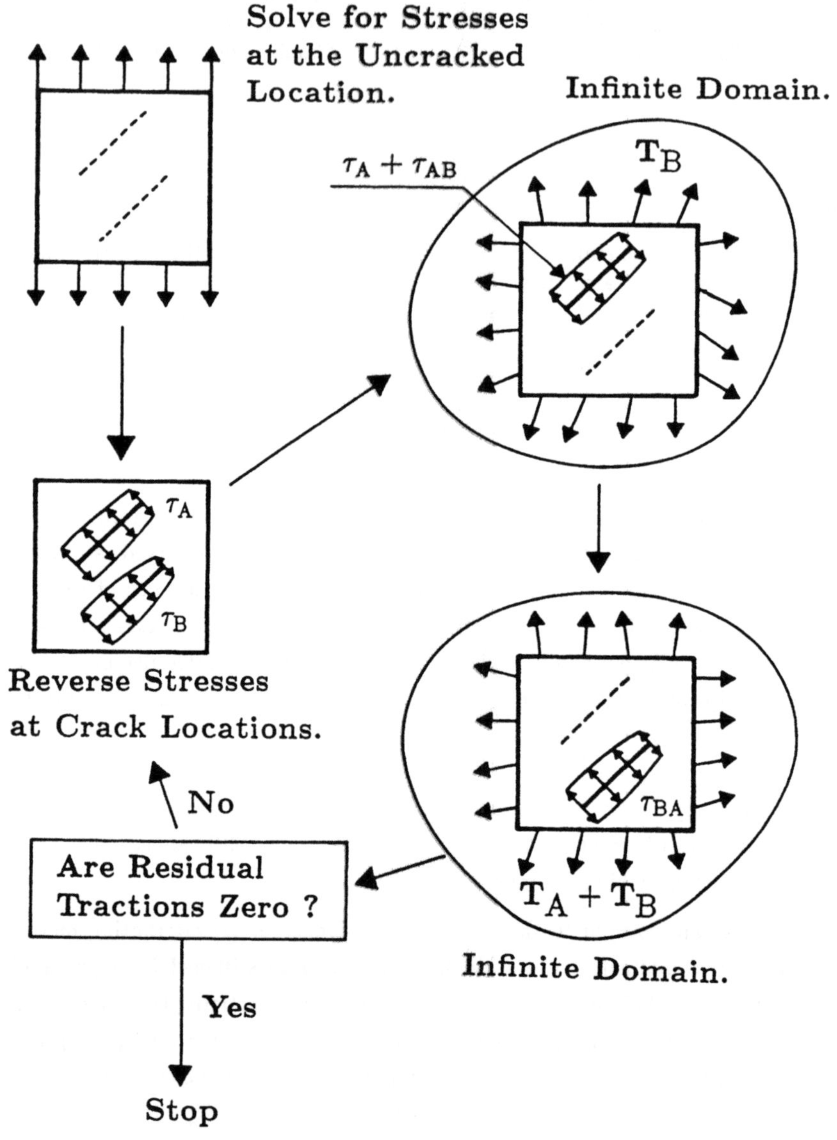

Fig. 1

2. The Alternating Technique for Multiple 2 or 3 Dimensional Embedded/Surface Cracks

Consider a 2 or 3 dimensional homogeneous solid containing multiple cracks, as shown in Fig. 1, and assume that the crack-faces are traction-free. In the alternating technique [O'Donoghue, Nishioka and Atluri (1984), and Chen and Atluri (1988)], the stresses in the *uncracked* body are first analyzed, by using a numerical/analytical method such as the finite element/boundary element method, for the given system of external loading. In order to assess the effect of the crack, the tractions at the locations of the cracks in the otherwise uncracked body must be erased. Assume for simplicity, for now, that there is a single crack. Thus, one has to deal with the problem of a *finite body* with a crack, the faces of which are subject to arbitrary tractions, and the outer boundaries are traction free. Denoting by S_c the crack-face, and by S_σ the external boundary of the *finite-body* V, this problem may be posed as:

$$Solve: \; L(u) = 0 \; in \; V; \quad B(u) = \bar{t} \; at \; s_c; \quad B(u) = 0 \; at \; S_\sigma \tag{1}$$

where u are the displacements, and L and B are the appropriate differential operators. To solve problem (1) by using the alternating method, the finite body is replaced by an *infinite body*, with stresses going to zero at infinity. The problem of an infinite homogeneous body containing a crack, the faces of which are subject to tractions $\bar{t}$ as in (1), does have an *analytical* solution often times, and hence need not be solved numerically. The far-field stresses from this *infinite body* solution do not satisfy the condition $B(u) = 0$ at S_σ. Thus, the residual tractions at the boundaries of the finite-body are erased, by first solving an *uncracked* body with these residual boundary tractions, and then erasing the tractions at the location of the crack in the uncracked body. This last problem of erasing crack-face tractions is similar to problem (1) above. This iterative loop is continued until the analytical solution for the infinite body satisfies also the zero-traction condition, $B(u) = 0$ at S_σ, of the finite body. Thus, in the alternating method, problem (1) is recast as:

$$L(u) = 0 \; in \; V_\infty; \quad B(u) = \bar{t}^{(i)} \; at \; S_c; \; and \; B(u) = \bar{t}_\sigma^{(i)} \; at \; S_\sigma \tag{2}$$

where i denotes the ith iteration, and V_∞ is the infinite domain. The iteration continues until $\bar{t}_\sigma^{(i)} = 0$. Let $\bar{t}^{(i)}$ be the converged value at S_c, corresponding to which the stress-intensity factors are determined from the analytical infinite-body solution [Chen and Alturi (1988); Atluri(1986)].

For multiple cracks, the alternating method is analogous; and is represented schematically in Figs. 1 and 2. In a variety of practical cases, this alternating method has been established to be a very simple and cost-effective tool of analysis [Atluri and Nishioka (1989)].

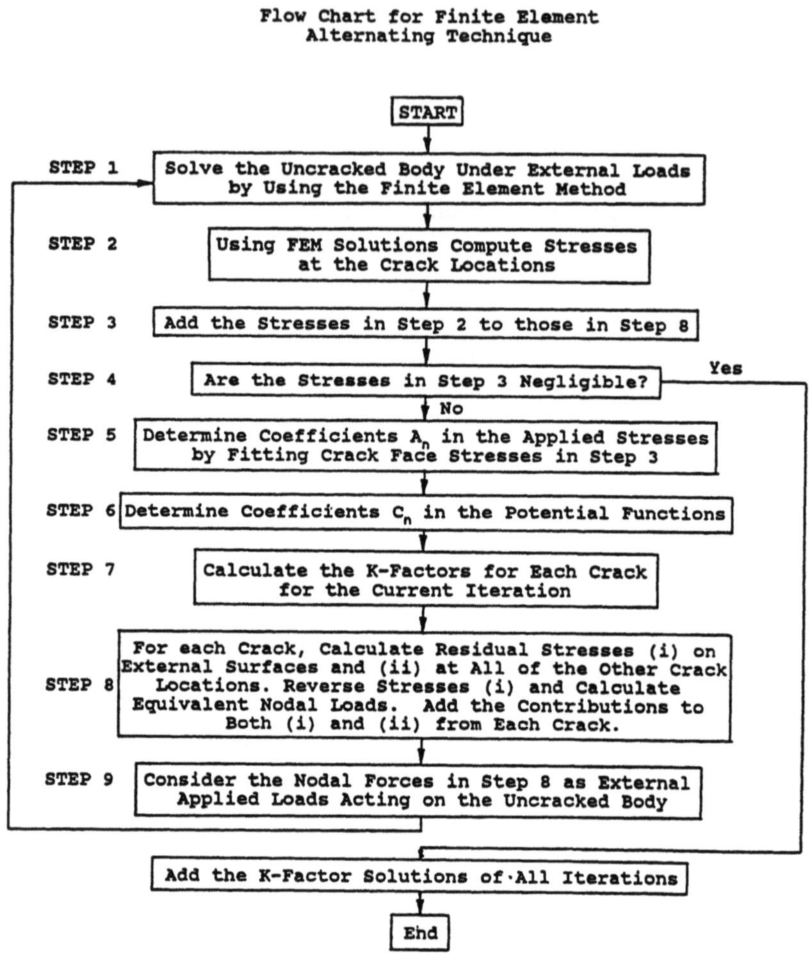
Flow Chart for Finite Element
Alternating Technique
START
STEP 1
Solve the Uncracked Body Under External Loads by Using the Finite Element Method
STEP 2
Using FEM Solutions Compute Stresses at the Crack Locations
STEP 3
Add the Stresses in Step 2 to those in Step 8
STEP 4
Are the Stresses in Step 3 Negligible?
Yes
No
STEP 5
Determine Coefficients A_n in the Applied Stresses by Fitting Crack Face Stresses in Step 3
STEP 6
Determine Coefficients C_n in the Potential Functions
STEP 7
Calculate the K-Factors for Each Crack for the Current Iteration
STEP 8
For each Crack, Calculate Residual Stresses (i) on External Surfaces and (ii) at All of the Other Crack Locations. Reverse Stresses (i) and Calculate Equivalent Nodal Loads. Add the Contributions to Both (i) and (ii) from Each Crack.
STEP 9
Consider the Nodal Forces in Step 8 as External Applied Loads Acting on the Uncracked Body
Add the K-Factor Solutions of All Iterations
End

Fig. 2

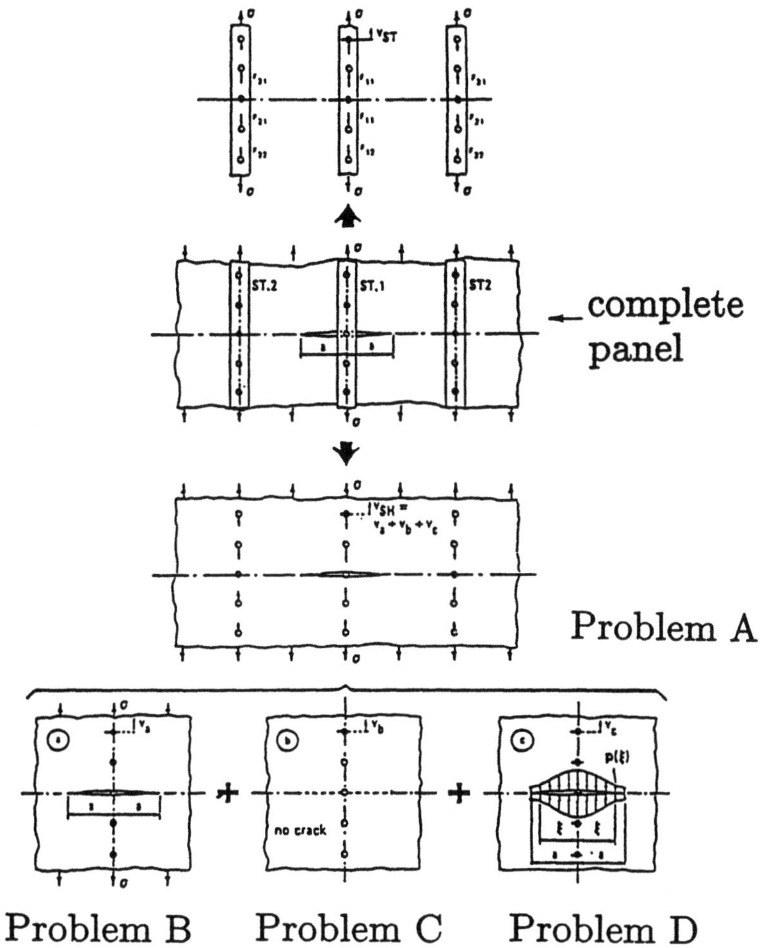

DISPLACEMENT COMPATIBILITY METHOD

Fig. 3

3. The Current State-of-the-Art: "The displacement Compatibility Approach" for Cracked, Stiffened Fuselage Panels

The methods that are currently used for the fracture analysis of stiffened panels are based on the so-called "displacement-compatibility approach" as developed by Swift (1970, 1984), and others [see, for instance, Vlieger, 1988]. The "displacement-compatibility" approach is schematically illustrated in Fig. 3. This method can be classified as a general "flexibility matrix" approach in structural mechanics. In this approach, the effects of the fasteners on the sheet are represented by a series of concentrated forces on the sheet, which may contain multiple cracks, and the sheet is subjected, in addition, to the far-field hoop stress [labelled henceforth as problem A]. The stringers, in turn, are acted upon by a series of concentrated forces, equal in magnitude and opposite in direction to those in problem A. The dislacement compatibility between the sheet and the stiffener, at each fastener location, is enforced, by taking into account the fastener flexibility, as:

$$v_{sheet} = v_{stiffener} + v_{fastener} \tag{3}$$

(at each fastener location)

Problem A above is again broken-up, by the superposition principle, into 3 problems, labelled here as problems B,C and D, respectively. Problem B is that of a sheet, subjected to hoop membrane stress, and containing multiple cracks. Problem C is that of the uncracked sheet subjected to a series of concentrated forces as in Problem A, at the fastener locations. Problem D is that of a sheet containing multiple cracks, each of which is subjected to tractions that are equal in magnitude, but opposite in direction, to the residual tractions existing at the locations of the cracks in an otherwise uncracked sheet, as in Problem C.

$$v_{sheet} = v_B + v_C + v_D \tag{4}$$

(at each fastener location)

or

$$v_b + v_C + v_D = V_{stiffener} + v_{fastener} \tag{5}$$

(at each fastener location)

In the methods used by Swift (1970, 1984) and Vlieger (1988), one needs: (i) analytical solutions for displacements in the stringer at a given fastener location due to concentrated forces at any other fastener location; (ii) analytical solutions for displacements in the sheet, at a given fastener location, for the Problem B; (iii) analytical solution for displacements, in Problem C, at any given fastener location, due to a set concentrated forces at any other fastener location; and (iv) analytical soltions for displacements in the sheet, at a given fastener location, for Problem D. It should be noted that not all of these analytical solutions are readily available in literature; and, when available, they are not suitable for the multiple crack case. when the displacement solutions (i) through (iv) above are used in Equation (5), one obtains the linear system of equations:

$$\boldsymbol{C} \cdot \underline{\boldsymbol{F}} = \underline{\boldsymbol{A}} \tag{6}$$

where $\boldsymbol{C}$ is the "flexibility matrix" as assembled from solutions (i) through (iv) above, $\underline{\mathrm{F}}$ is the vector of fastener forces (at all fastener locations) acting on the sheet, and $\underline{\mathrm{A}}$ is the given vector that accounts for fastener flexibility, etc. In equation (6) the vector $\underline{\mathrm{F}}$ is the unknown.

Once Eq. (6) is solved for the vector $\underline{\mathrm{F}}$, the problem of the fracture of the sheet alone, i.e., Problem A above, is well-posed [i.e., the far-field hoop stress, as well as the fastener reaction forces, are known]. This Problem A is broken up into Problems B, C, and D as above. The stress singualrities at the crack-tips arise only in *Problems B and D.* Thus, the stress-intensity factor at a crack-tip can be written as:

$$K = K_B + K_D \quad (\ at\ \ each\ \ crack - tip\ ) \tag{7}$$

It should be recognized that the key element in the approach described above, is the assembly of the matrix $\boldsymbol{C}$ in Eq. (6) from the various analytical solutions as described under (i) through (iv) earlier. Thus, the approach is by means a trivial one for adaptation on a modern personal digital computer. Further, in writing Eq. (7), Problems B and D are assumed to correspond to *infinite* domains in Swift (1984) and Vlieger (1988). Thus, finite-dimension correction factors are ignored, which is not necessarily a conservative approximation.

4. A Proposal for a Simple Finite Element Alternating Method for Stiffened Panels with Multiple Cracks

A straight-forward finite element modeling (including the crack-tip singularities) of a cracked, stiffened, fuselage panel, especially when used for a parametric study during the service/design phase of the aircraft, is prohibitively expensive [Swift, 1970]. In order to circumvent this, displacement compatibility methods wherein the panel is assumed to be infinite, have been developed by Swift (1970, 1984), and Vlieger (1988), among others.

In the following, a very simple alternative to the current state-of-the-art displacement compatibility approach, is presented. The proposed method essentially involves two steps: (i) evaluation of the fastener reaction forces on the sheet through a very course finite element model, wherein the details of the crack-tip singularities are not modeled; and (ii) the application of the finite-element-alternating method, employing the same finite element mesh as in stage (i) for the *unstiffened, uncracked* panel, with the applied membrane hoop stress, and the fastener reaction forces as solved in stage (i), to determine the fracture parameters. The present method may be classified as the now-standard finite element stiffness (or displacement) method.

Stage (i) Solution

The skin is descritized using a very coarse finite element mesh, such that the fastener locations are taken to be the nodes of the finite element mesh. The axial deformation of the stiffener is modeled by using the conventional "truss-type" elements. Since the fastener shear forces are usually offset from the stiffener neutral axis, an out-of-plane bending is also induced in the stiffeners. The out-of-plane bending deformation of the stiffener, between two fasteners, is given by the elementary beam theory:

$$\delta_b = C\left(\frac{ML}{E_{St}I}\right) = a_b F \tag{8}$$

where:
C = distance from the neutral axis of the stiffener to the point of action of the fastener shear force
I = stiffener cross-sectional inertia
L = distance between two fasteners, or the length of the stiffener "truss" element
E = Young's modulus of stiffener material
M = F · C ; where F is the force in the truss element

Note that δ_b is in the same direction, as the stiffener axial force, F. From (8), it is seen that:

$$a_b = \left(\frac{C^2 L}{E_{St}I}\right) \tag{9}$$

The axial stretch of the stringer is given by:

$$\delta_a = \frac{F}{E_{St}A} \equiv (a_S)F \quad ; \quad a_S = \frac{1}{E_{St}A} \tag{10}$$

We use the total axial deformations at the ends of the stringer, at the points where the stringer is attached to the skin, as the generalized degrees of freedom for the stringer. For these degrees of freedom, the stiffness matrix of the stringer element is given by:

$$\boldsymbol{K}^e_{St} = \begin{bmatrix} \dfrac{1}{(a_b + a_s)} & \dfrac{-1}{(a_b + a_s)} \\[2ex] \dfrac{-1}{(a_b + a_s)} & \dfrac{1}{(a_b + a_s)} \end{bmatrix} \tag{11}$$

The flexibility of the fasteners has been found [Swift, 1984] to be an important factor that influences the stress-intensity factors for a crack in the stiffened skin. If Q is the shear force acting on the fastener, the shear deformation of the fastener can be represented by the empirical relation [Swift, 1984].

$$\delta_F = \frac{1}{E_{Sh}D}\left[A + C\left(\frac{D}{B_1} + \frac{D}{B_2}\right)\right] F \equiv a_F Q \tag{12}$$

where

E_{Sh} = modulus of sheet material

D = rivet diameter

B_1 and B_2 = thicknesses of joined sheets

$A = 5.0$ for Al rivets and 1.66 for steel fasteners

$C = 0.8$ for Al rivets and 0.86 for steel fasteners

The "stiffness" of the rivet in shear is thus given by:

$$K_F = \frac{E_{Sh}D}{\left[A + C(\frac{D}{B_1} + \frac{D}{B_2})\right]} \tag{13}$$

consider, for simplicity (but without any loss of generality), that the skin is descritized into finite elements, with nodes being only at the locations of the fasteners; and likewise, the stringers are descritized into finite elements with nodes being only at the fastener locations. Let the number of fasteners be N. Let the number of stringer elements be N_{St}; and the number of sheet elements be N_{Sk}. Let the generalized displacements of the skin at the nodes of the finite element mesh be denoted as q_{sk}; and those of the stringers at the nodes be denoted by q_{st}. Then the total strain energy of the stiffened fuselage skin, with flexible fasteners, is given by:

$$W = \sum_{ele=N_{Sk}} \frac{1}{2} q_{Sk}^{t_e} K_{Sk}^{e} q_{Sk}^{e} + \sum_{N_{St}} \frac{1}{2} q_{St}^{t_e} K_{St}^{e} q_{St}^{e} + \sum_{N} \frac{1}{2} K_F (q_{Sk} - q_{St})^2 \tag{14}$$

where $\boldsymbol{q}_{Sk}^{e}$ is the vector of nodal displacements of a skin element; and $\boldsymbol{q}_{st}^{e}$ is the vector of nodal displacements of a stringer (stiffener) element.

Let $\boldsymbol{q}_{Sk}$ be the master vector of nodal displacements of the skin; $\boldsymbol{K}_{Sk}$ the assembled nodal stiffness matrix of the skin; $\boldsymbol{q}_{St}$ be the master vector of nodal displacements of the stringers; and $\boldsymbol{K}_{St}$ the assembled nodal stiffness matrix of the stringers. Let $\boldsymbol{K}_F$ be the "assembled" (diagonal) stiffness matrix of the fasteners; ie.,

$$\boldsymbol{K_F} = \begin{bmatrix} K_F & 0 & 0 & 0 \cdots 0 \\ 0 & K_F & 0 & 0 \cdots 0 \\ 0 & & \ddots & 0 \\ \vdots & & & \vdots \\ 0 & 0 & \cdots\cdots & K_F \end{bmatrix} \tag{15}$$

$$\mathrm{N} \times \mathrm{N}$$

Then, W of Eq. (14) can be written as:

$$W = \frac{1}{2} q_{Sk}^t (\boldsymbol{K}_{Sk} + \boldsymbol{K}_F) \boldsymbol{q}_{Sk} + \frac{1}{2} \boldsymbol{q}_{St}^t (\boldsymbol{K}_{St} + \boldsymbol{K}_F) \boldsymbol{q}_{St} - \boldsymbol{q}_{Sk}^t \boldsymbol{K}_F \boldsymbol{q}_{St} \tag{16}$$

If the fastener flexibility is ignored, then $\boldsymbol{q}_{Sk} \equiv \boldsymbol{q}_{St}$; and equation (16) reduces to:

$$W = \frac{1}{2} q_{Sk}^t (\boldsymbol{K}_{Sk} + \boldsymbol{K}_{St}) \boldsymbol{q}_{Sk} \tag{17}$$

The potential of the external forces (the hoop stress in the fuselage) may be represented, in general, as:

$$U = (\boldsymbol{q}_{Sk}^t \boldsymbol{Q}_{Sk} + \boldsymbol{q}_{St}^t \boldsymbol{Q}_{St}) \tag{18}$$

Let $\pi = W - U$. The finite element equations that arise from the vanishing of the variation of $\delta\pi$ (ie., $\delta\pi = 0$) are given by:

$$(\boldsymbol{K}_{Sk} + \boldsymbol{K}_F) \boldsymbol{q}_{Sk} - \boldsymbol{K}_F \boldsymbol{q}_{St} = \boldsymbol{Q}_{Sk} \tag{19a}$$

and

$$(\boldsymbol{K}_{St} + \boldsymbol{K}_F) \boldsymbol{q}_{St} - \boldsymbol{K}_F \boldsymbol{q}_{Sk} = \boldsymbol{Q}_{St} \tag{19b}$$

for the case of flexible fasteners. Eqs. (19) may be rearranged as:

$$\begin{bmatrix} (\boldsymbol{K}_{Sk} + \boldsymbol{K}_F) & (-\boldsymbol{K}_F) \\ -(\boldsymbol{K}_F) & \boldsymbol{K}_{St} + \boldsymbol{K}_F \end{bmatrix} \begin{Bmatrix} \boldsymbol{q}_{Sk} \\ \boldsymbol{q}_{St} \end{Bmatrix} = \begin{Bmatrix} \boldsymbol{Q}_{Sk} \\ \boldsymbol{Q}_{St} \end{Bmatrix} \tag{20}$$

After the imposition of appropriate boundary conditions, $\boldsymbol{q}_{Sk}$ and $\boldsymbol{q}_{St}$ can be solved for, from Eq. (20). Once $\boldsymbol{q}_{Sk}$ and $\boldsymbol{q}_{St}$ are solved for, the *reactions* of the *stiffeners* on the skin, at the locations of the fasteners, can be easily calculated as:

$$F_{stiffener} = K_F(q_{St} - q_{Sk}) \equiv Q_{St} - K_{St}q_{St} \tag{21}$$

with care being exercised to determine the direction of these reactive forces.

Once the effects of the stringer (with flexible rivets) on the skin are determined, one can consider the free-body diagram of the *cracked skin alone* [See Fig. 4]; the skin being subject ot the far-field hoop stresses, and the stringer reaction forces. The stress-intensity factors for the multiple cracks in the skin, subjected to these forces, may now be determined in the stage (ii) alternating solution.

Stage (ii) solution

In the finite element alternating method to determine the k-factors for multiple cracks in the skin [after the effects of the stiffeners and fasteners are isolated as in stage (i) above], one may employ the same finite element mesh as in stage (i), to model the *uncracked, unstiffened* free-body of the skin alone. In stage (ii), as explained in Section 2 of this paper, one needs to know: (a) the tractions to be erased at the locations of the cracks in an othewise uncracked skin; and (b) the analytical solution for a crack subjected to arbitrary tractions in an infinite body. Solution (b) is included below for completeness.

ii.1 Analytical Solution for an Infinite Plane with an Embedded Line Crack

The basic equations of plane elasticity have the general solution [England, 1971].

$$\begin{aligned} 2\mu(u+iv) &= \kappa\Omega(z) - z\overline{\Omega'(z)} - \overline{\omega(z)} \\ \tau_{xx} + \tau_{yy} &= 2\Omega'(z) + 2\overline{\Omega'(z)} \\ \tau_{yy} - i\tau_{xy} &= \Omega'(z) + \overline{\Omega'(z)} + z\overline{\Omega''(z)} + \overline{\omega'(z)} \end{aligned} \tag{22}$$

where the complex potentials $\Omega'(z), \omega'(z)$ are holomorphic functions in the region occupied by the body, and μ, κ are elastic constants. Moreover, u, v are displacements and τ_{xx}, τ_{yy}, τ_{xy} are stresses at the point (x, y), $z = x + iy$.

Suppose a line crack on $y = 0$, $|\, x \,| \leq a$ in an infinite plane is inflated by equal and opposite tractions over the faces of the crack given by

$$\tau_{yy} - i\tau_{xy} = -[p(t) + is(t)]\ ; \qquad |\, t \,| \leq a \tag{23}$$

with zero tractions at infinity. This problem may be solved by choosing $\omega(z)$ so that [England, 1971, §3.10]

$$\omega(z) = \overline{\Omega(\overline{z})} - x\Omega'(z) \tag{24}$$

CONSIDER SKIN ONLY:

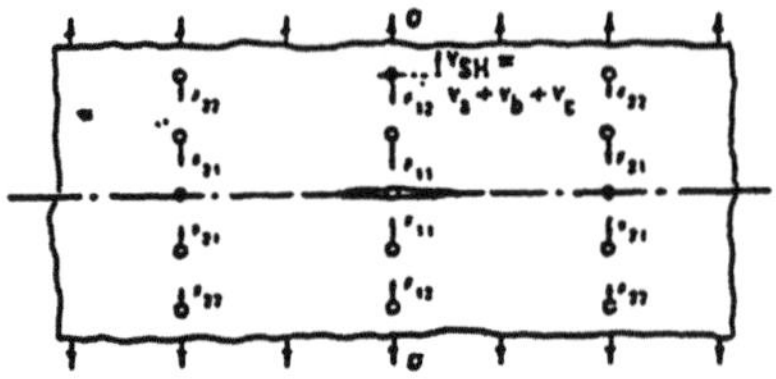

F_{ij}: Fastener forces on sheet.

ASSUME NO CRACK:

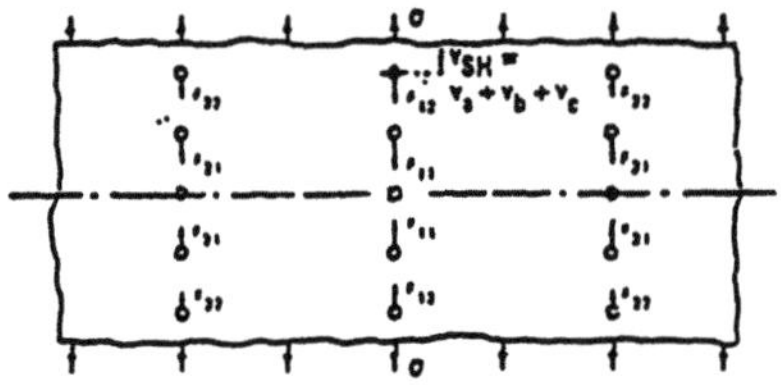

Use coarse mesh FEM.

[$\mathbb{K}_{sheet}$ is already formed once before.
$\underset{\sim}{F}_{stiff}$ is known] $\Rightarrow \mathbb{K}_{sheet}\underset{\sim}{q} = \underset{\sim}{F}_{stiff} + F_{\sigma}$ (Uncracked sheet)

Solve the uncracked sheet using very coarse FEM.

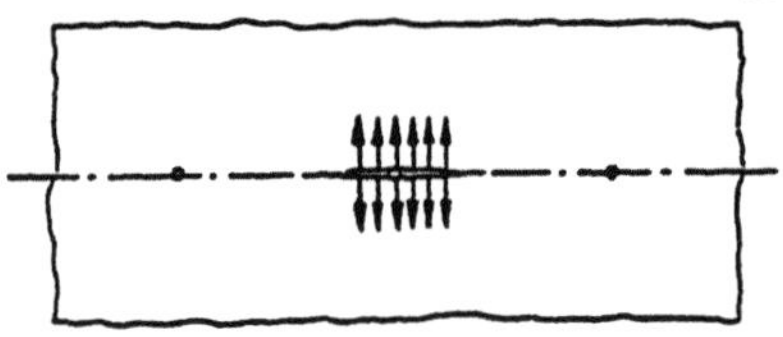

[Pressurized crack only]

Fig. 4

The problem then reduces to determining the potential $\Omega(z)$, holomorphic in the plane cut along $y = 0, |x| \leq a$ across which it satisfies

$$\Omega'^{+}(t) + \Omega'^{-}(t) = -[p(t) + is(t)] \; ; \qquad |t| \leq a \tag{25}$$

where (+) implies the upper crack face and (-) implies the lower crack face. Approximate the applied tractions in the form

$$p(t) + is(t) = -\sum_{n=1}^{N} b_n U_{n-1}(t) \; ; \qquad |t| \leq a \tag{26}$$

where $U_{n-1}(t)$ is the Chebyshev polynomials of the second kind and is defined as

$$U_n(t) = \sin[(n+1)\theta]/\sin\theta, \qquad t = a\cos\theta \tag{27}$$

It could be easily shown that

$$2\Omega' = \sum_{n=1}^{N} b_n G_{n-1}(z), \qquad 2\Omega' = \sum_{n=1}^{N} b_n \frac{R_n(z)}{n} \tag{28}$$

where $R_n(z)$ and $G_{n-1}(z)$ are defined as follows:

$$R_n(z) = a\{z_1 - (z_1^2 - 1)^{\frac{1}{2}}\}^n \qquad G_{n-1}(z) = -(z^2 - a^2)^{-\frac{1}{2}} R_n(z) \tag{29}$$

where $z_1 = z/a$.

The stresses on $y = 0, |x| \geq a$ are given by

$$\tau_{yy} - i\tau_{xy} = -sgn(x)(x^2 - a^2)^{-1/2} \sum_{n=1}^{N} b_n R_n(x) \tag{30}$$

The stress-intensity factor K_I and K_{II} at $x = +a$ are defined as

$$K_I = \lim_{x \to a} \left[\sqrt{2\pi(x-a)} \;\; \tau_{yy}\right] \qquad K_{II} = \lim_{x \to a} \left[\sqrt{2\pi(x-a)} \;\; \tau_{xy}\right] \tag{31}$$

making use of Eq. (9) in Eq. (10) we get

$$K_I - iK_{II} = -\sqrt{\pi a} \sum_{n=1}^{N} b_n \tag{32}$$

Similarly at $x = -a$ the stress intensity factors are

$$K_I - iK_{II} = \sqrt{\pi a} \sum_{n=1}^{N} (-1)^n b_n \tag{33}$$

Hence the complete stress and displacement fields and the stress-intensity factors at the crack tip have been determined.

ii.2 Finite Element Alternating Method

The present finite element alternating method requires the following steps:

(1) Solve the uncracked panel under the given external loads and fastener reactions by using the finite element method. The uncracked body has the same geometry as in the given problem *except the crack.*

(2) Using the finite element solution, we compute the stresses in the uncracked panel at the location of the original crack.

(3) To satisfy the stress-free crack face boundary conditions, reverse the residual stress in step (2). Then determine the coefficients b_n of Eq. (5) by using the orthogonality of Chebychev polynomials.

(4) Calculate the stress intensity factors for the current iteration by substituting coefficients b_n in Eqs. (11-12). Compare them with the previous iteration S.I.F. If the differences are less than some permissible magnitude, stop the iteration procedure.

(5) Calculate the residual stresses on external boundaries of the body due to the loads in step (3). To satisfy the stress condition, reverse the residual stresses and calculate equivalent nodal forces.

(6) Consider the nodal forces in step (5) as external applied loads acting on the uncracked body. Repeat all steps in the iteration process until the criterion in step (4) is satisfied. To obtain the final solution, add the stress intensity factors of all iterations.

ii.3 Solution Process for an Edge Crack

In this case, because the crack is only a half of the embedded crack in the imaginary infinite body, a specific procedure must be followed. It is necessary to define stresses over the entire crack length including the portion of the crack which lies outside of the finite body. According to Nishioka and Atluri (1983), if we choose the stresses in the fictitious part to be the same as those at the free end of the edge crack (Type C of Fig.13 in Nishioka and Atluri), we can obtain monotonic convergence. Moreover, this type of fictitious stress is the easiest way to code. Therefore these types of fictitious stresses are used in this analysis.

After the stresses on the full crack are determined, the rest of the algorithm remains the same as before.

ii.4 Solution Procedure for Multiple Cracks

Without loss of generality, we only discuss the case of two cracks in the panel. Because there is interaction between the two cracks, the following procedure is adopted.

(1) Solve the uncracked body under the given external loads by F.E.M. as in the first step of the "single-crack" case.

(2) Using F.E.M., calculate the stresses in the uncracked body at the locations of the two given cracks.

(3) Reverse the residual stress at the first crack. Determine the coefficient, b_n. Calculate the S.I.F. at the first crack.

(4) Calculate the residual stress on boundaries due to the loads in the first crack. Reverse them and calculate equivalent nodal forces, T_1.

(5) Calculate the residual stresses at the location of the second crack due to the load in the first crack.

(6) Add the stresses from step (2) and step (5) of the second crack together; they are the residual stresses at crack 2. Reverse them. Determine b_n. Calculate the S.I.F. at the second crack.

(7) Calculate the residual stress on boundaries due to the load in the second crack. The equivalent nodal forces are called T_2.

(8) Calculate the residual stresses at the location of the first crack due to the load in the second crack.

(9) Sum up T_1 (step 4) and T_2 (step 7). Now these nodal forces are the external loads in step (1).

(10) After evaluating the residual stresses at crack 1 due to $T_1 + T_2$, add these stresses and the stresses of step (8) together. Now, these are the residual stresses at crack 1. Go to step (3) and repeat step (3)-step (10) until a convergence is achieved (ex. S.I.F. convergence).

4.1 An Example Problem Analyzed by the Alternating Method

To illustrate the application of the proposed "direct-stiffness" finite element alternating methodology, for analyzing cracked stiffened fuselage panels, the example problem as shown in Fig. 5 has been analyzed. The problem involves a stiffened panel, with a central crack and a broken stiffener, with the crack being symmetrically located with respect to the broken stiffener. For the purpose of illustration of the types of finite element meshes that may be sufficient in the present methodology, the fastener flexibilty is ignored in the analysis. A typical finite element mesh that is used is shown in the inset of Fig. 5. The stress-intensity magnification factors for various crack lengths are shown in Fig. 6, which also shows the convergence of the results with mesh refinement. It is seen that the results are insensitive to the finite element mesh size, with acceptable results being obtained for a (16 × 3) mesh. It is also seen that the stress-intensity magnification factor decreases as the crack-tip approaches the stiffener.

5. MSD Near a Row of Fastener Holes

A typical multiple site damage near a row of fastener holes is illustrated schematically in Figs. 7 and 8. In example 4.2, the crack was supposed to be much larger in size as compared to the rivet diameter; and thus, the detailed distribution of stresses, on the hole-surface in the skin, due to fastener reaction forces, was not a factor in determining the fracture parameters near the crack-tip. In fact, in Section 4, these reaction forces were simply treated as concentrated forces. However, in a typical MSD situation as depicted in Figs. 7 and 8, the fastener interaction stresses on the hole-surface in the skin are likely to play a significant role in the fracture parameters for the cracks emanating from fastener holes in the skin.

The fastener reaction forces can still be determined for the lap-splice joint configuration of Figs. 7 and 8, through the direct-stiffness finite element method presented above in Section 4. Once these reaction forces, treated as concentrated forces in Section 4, are determined, one may use the known elasticity solution, to approximate the detailed stress-field on the hole-surface in the skin, that is equivalent to these concentrated forces. Under the action of these fastener interaction stress-fields, the stress-intensity factors for MSD near fastener holes can be determined using the alternating method described earlier. Such methods are currently being developed at Georgia Tech.

6. Analysis of Cracks with Bonded Repair Patches

Repair of cracked structures with bonded composite laminate patches appears to be a cost-effective and reliable way of enhancing the fatigue life of fuselage panels [Jones and Callinan (1984)]. The load-transfer to the composite patch is the primary cause for lessening the stress- intensity near the crack-tip in the main panel, and thus improves its fatigue life. This reduction in the stress-intensity for a crack in the main panel depends on the laminate properties, its thickness, the properties of the adhesive material, and the thickness of the adhesive layer. Analytical solutions have been obtained for: (i) the case of an infinite isotropic material panel with a crack, bonded to an infinite orthotropic panel, and (ii) the case of an infinite isotropic panel with a crack emanating from a hole, bonded to an infinite orthotropic panel. These analytical solutions have been repeated for orthotropic patches of inifinite width and finite height (perpendicular to the crack-line in the isotropic panel), with the edges of the patch being parallel to the crack-axes.

These analytical solutions, when implemented in the finite-element alternating method, would provide useful design tools for *designing composite patches* of *arbitrary shape* to arrest further growth of a crack emanating from stress-concentrations in the main panel. Such a design tool, for implementation on a personal computer, is being developed at Georgia Tech.

7. Weight Functions for Multiple Cracks in Fuselage Panels

The concept of weight functions [Atluri (1989)] is quite useful in linear fracture mechanics. The weight-functions are the appropriately normalized first-order variations in displacements (say along the crack-surface), for a reference state of loading, due to variations in the characteristic dimensions of the crack. Once these weight-functions are evaluated, for a body of given finite-dimensions with a given crack-shape, for a reference state of loading, then the stress-intensity factors for any other state of loading can be determined

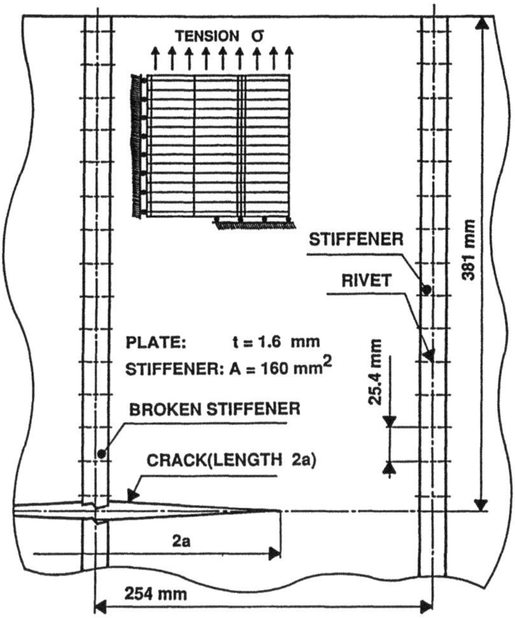

TENSION σ
STIFFENER
RIVET
381 mm
PLATE: t = 1.6 mm
STIFFENER: A = 160 mm2
25.4 mm
BROKEN STIFFENER
CRACK(LENGTH 2a)
2a
254 mm

Fig. 5

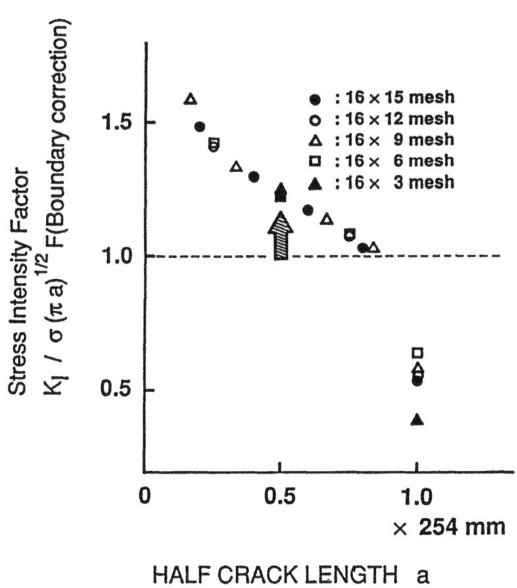

Stress Intensity Factor
KI / σ (π a)1/2 F(Boundary correction)
: 16 × 15 mesh
: 16 × 12 mesh
: 16 × 9 mesh
: 16 × 6 mesh
: 16 × 3 mesh
1.5
1.0
0.5
0
0.5
1.0
× 254 mm
HALF CRACK LENGTH a

Fig. 6

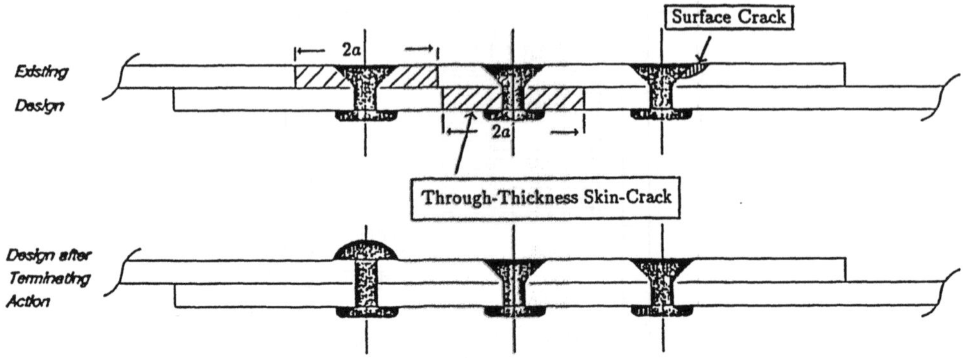

Fig. 7

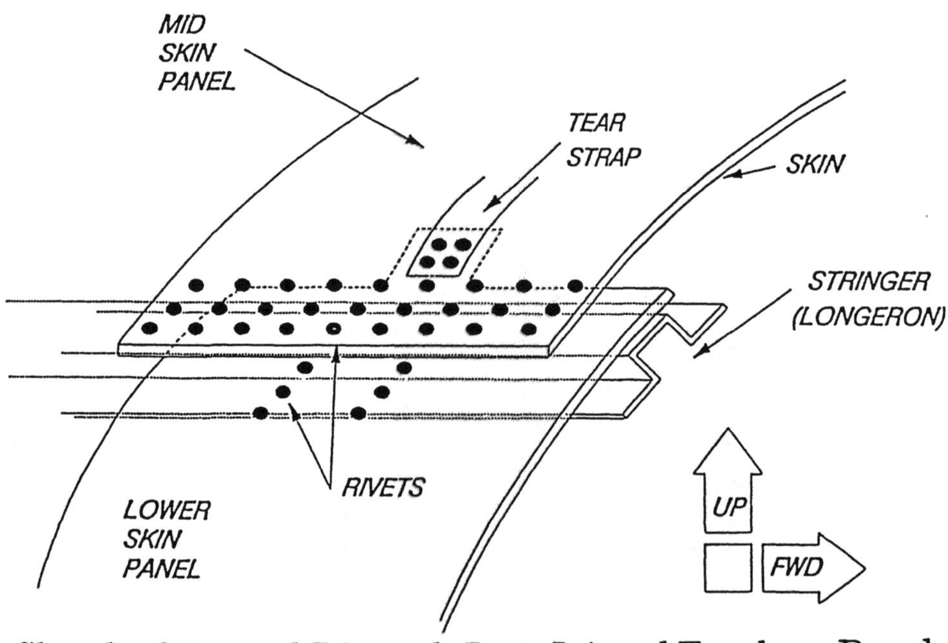

Sketch of Typical Riveted, Lap-Jointed Fuselage Panels

Fig. 8

by a work-like integral of the product of the wight-function and the applied traction at a point on the crack-surface. The weight- functions along the crack-surface, for a reference state of loading on the crack-face in a fuselage panel, can be determined once and for all, and stored in a data base, for subsequent analysis of different load-cases.

For a panel with multiple site damage, the weight functions for each crack may be computed for a *constant normal-pressure case, considering the sheet alone*, ie., ignoring the stringers and fasteners. The detailed procedures for computing these weight-functions are discussed in Chen and Alturi (1988), and Rajiyah and Atluri (1989). For the problem of the stiffened, pressurized, fuselage panel, the tractions to be erased from the location of the crack are determined as in steps (1) and (2) given in the subsection, ii.2 Finite Element Alternating Method, earlier in this paper. While using the weight-function method, steps (3) to (6) of subsection (ii.2 Finite Element Alternating Method) above, are no longer necessary. Once the crack-face tractions are determined as per step (2) of subsection (ii.2), the k-factors can be simply determined by evaluating a work-like integral of a product of these tractions and the predetermined weight functions. This procedure is particularly convenient in the case of cylindrically curved fuselage panels. In this case, the analytical solution for an axial crack, in a cylindrical shell, that is subject to constant crack-face pressure is readily available; while that for arbitrary crack-face traction is not.

8. Bulging Near Crack-Tips in Pressurized Fuselages; and Equivalent Domain Integral Methods for Evaluation of Fracture Parameters

For longitudinal cracks in pressurized fuselages, the bulging of the area near the crack-tips plays a significant role. For long cracks, the stress on a longitudinal plane (perpendicular to the crack axis), induced by bulging, may be substantial, and may cause crack-flapping. Thus, a detailed analysis of crack-bulging is of importance. Crack-bulging is essentially a finite-deformation phenomenon, confined, however, to a region close to the crack-tip. Plasticity also plays a significant role. In addition to an accurate determination of longitudinal stresses due to crack-flapping, the computation of an appropriate fracture parameter in the presence of large- deformations and plasticity is an important issue.

In the presence of large deformations and plasticity, one of the fracture parameters that is valid is the so-called T^* (or J_ε) integral [Atluri (1986)]:

$$T^* \equiv J_\varepsilon = \int_{\Gamma_\epsilon} \left(W n_1 - t_i \frac{\partial u_i}{\partial x_1} \right) dA \tag{34}$$

where t_i are tractions per unit area of the *undeformed* shell, $t_i = n_i t_{ij}$; t_{ij} is the nominal (first Piola) stress tensor, and W is the strain energy density per unit volume of the undeformed shell. For the finitely deformed shell, the stress-state can be characterized by the nominal (first Piola) stress-resultants and stress-couples; and the strain-energy can be expressed in terms of these nominal resultants, and the rotations [Atluri (1984)]. In Eq. (34), Γ_ϵ is a cylindrical surface that is asymptorically close to the crack-tip.

Since the stress-field or the strain-energy density-field cannot be computed highly accurately vary near to the crack-front, the computation of T^* of Eq. (34) is often made convenient through the so-called "equivalent-domain-integral [EDI]" method [Nishioka and Atluri (1987)]. In this approach, one defines an arbitrary function S such that $S = 1$ at Γ_ϵ and $S = 0$ at a far-field cylindrical surface Γ. Thus,

$$T^* = \int_{\Gamma_\epsilon} \left[W n_1 - t_i \frac{\partial u_i}{\partial x_1} \right] S dA - \int_{\Gamma} \left[W n_1 - t_i \frac{\partial U_i}{\partial x_1} \right] S dA \tag{35}$$

$$= - \int_{V-V_\epsilon} \left[\frac{\partial}{\partial x_1}(WS) - \frac{\partial}{\partial x_j} \left(t_{ji} \frac{\partial U_i}{\partial x_1} S \right) \right] dV \tag{36}$$

In the presence of large-deformations and plasticity, the volume integral in Eq. (36) can be computed directly from the obtained finite element solution. Dynamic fracture calculations based on a crack-velocity dependent T^* criterion are feasible [Yoshimura, Yagawa and Atluri (1989)].

Acknowledgements

These exploratory studies were preformed with the financial support of the Transportation Systems Center, Cambridge, MA. The useful discussions with S.G. Sampath are thankfully acknowledged. The assistance of Ms. Brenda Bruce in preparing this manuscript is noted with appreciation.

References

1. Atluri, S.N. : Alternate Stress and Conjugate Strain Measures, and Mixed Variational Formulations Involving Rigid Rotations, for Computational Analysis of Finitely Deformed Solids, with Applications to Plates and Shells. Com. & Struc. 18 (1984) 93-116.

2. Atluri, S.N.: Computational Methods in the Mechanics of Fracture. Amsterdam: North-Holland Publishing Co. 1989.

3. Atluri, S.N.: Advances in Computation of Weight Functions. 5th Int. Conf. Num. Meth. Frac. Mech. 1990.

4. Atluri, S.N.; Nishioka, T.: On Some Recent Advances in Computational Methods in the Mechanics of Fracture. Adv. Frac. Res. 3 (1989) 1923-1989.

5. Chen, K-L.; Atluri, S.N.: Weight Functions for Mixed-Mode Multiple Cracks in Anisotropic Media by VCE and FEM Alternating Methods. Eng. Frac. Mech. (to appear).

6. England, A.H.: Complex Variable Methods in Elasticity. New York: Wiley-Interscience 1971.

7. Jones R.; Callinan, R.J.: Bonded Repairs to Surface Flaws. Theor. Appl. Frac. Mech. 2-1 (1984) 17-26.

8. Nikishkov, G.P.; Atluri, S.N.: Calculation of Fracture Parameters for an Arbitrary 3-D Crack by the EDI Method. Int. Jnl. Num. Meth. Engg. 24-9 (1987) 1801-1822.

9. Nishioka, T.; Atluri, S.N.: Analytical Solution for Embedded Elliptical Cracks and Finite Element Alternating Method for Elliptical Surface Cracks, Subjected to Arbitrary Loadings. Engg. Frac. Mech. 17 (1983) 247-268.

10. O'Donoghue, P.E.; Nishioka, T.; Atluri, S.N.: Analysis of Interaction Behavior of Surface Flaws in Pressure Vessels. Jnl. Pres. Ves. Tech. 108-5 (1985) 24-32.

11. Swift, T.: The Effects of Fastener Flexibility and Stiffener Geometry on the Stress Intensity in Stiffened Cracked Sheet. Prospects of Fracture Mechanics. Leydon, Netherlands: Noordhoff International 1974.

12. Swift, T.: Fracture Analysis of Stiffened Structures. Damage Tolerance of Metallic Structures. ASTM STP 842 (1984) 69-107.

13. Vlieger, H.: Damage Tolerance of Stiffened - Skin Structures: Prediction and Experimental Verification. ASTM STP 969 (1988) 169-219.

14. Yoshimura, S.; Yagawa, G.; Atluri, S.N.: Generation and Propagation Analysis of High-Strain-Rate Dynamic Crack Propagation in a Viscoplastic Solid. Nuc. Engg. Des. III (1989) 273-289.

3. Risk Analysis for Aging Aircraft Fleets

A. P. Berens

University of Dayton Research Institute
Dayton, OH 46469

J. G. Burns and J. L. Rudd

Wright Research and Development Center
Wright-Patterson Air Force Base, OH 45433-6553

INTRODUCTION

To date, the Air Force has applied the durability and damage tolerance requirements of MIL-STD-1530A in three areas: a) designing new aircraft; b) evaluating the durability and damage tolerance characteristics of aircraft which were designed prior to the current requirements of MIL-STD-1530A; and, c) evaluating structural repairs and modifications. These applications have used fracture mechanics principles in a deterministic manner. That is, flaw growth was predicted using a fixed potential flaw size, a fixed da/dn vs ΔK relationship, and a stress spectrum derived from a predicted average usage. While it was realized that there are many stochastic elements in the initiation and growth of cracks, the applied process was considered conservative. Initial flaw size assumptions were generally severe, tracking programs accounted for variations in usage severity, and inspections (if necessary) were scheduled at half the time required for the specified initial flaws to grow to a critical size. At the time of the assessments, there was a high (but unquantified) degree of assurance that fatigue failures and widespread cracking would not occur within the design operational lifetime.

The realized life of individual airframes is seldom equal to the design life planned for a fleet. The life of an aircraft fleet is

Springer Series in Computational Mechanics
Atluri, Sampath, Tong (Eds.)
Structural Integrity of Aging Airplanes

determined more by its inherent operational capability and maintenance costs than by the number of flight hours specified at the design stage. As the population of airframes ages, fatigue cracks will initiate and grow and all of the stochastic elements which can influence the fatigue process will be active. The exact status of fatigue cracks in any specific airframe is somewhat unknown even after inspections due to the uncertainties of inspection processes. Regardless, in managing aging fleets, decisions must be made concerning the timing and extent of inspections, repairs, modifications and life extension options. Since the readiness and cost ramifications of these decisions are extremely important, every possible tool that can assist in making cost effective decisions should be used.

The objective of this Air Force program on structural risk analysis is to stochastically assess structural integrity in terms of both safety (as quantified by the probability of fracture of a population of structural details) and durability (as quantified by the expected number and sizes of the cracks requiring repair that will be detected at an inspection). The characterization of structural integrity can then be applied as an additional tool in making decisions concerning the timing of inspection, replacement, and retirement maintenance actions. This paper describes the methodology that has been implemented as a computer program, PRobability Of Fracture (PROF), for performing the risk analyses and presents one example of its potential application to a representative scenario.

AIR FORCE DATA BASE

Implementing structural risk analyses involves compromises between the ability to model reality and the data that is available to feed analytical models. In general, the more detail that is required by the model, the less reliable the available data is. Because of the Aircraft Structural Integrity Program (ASIP) requirements of MIL-STD-1530A [1], the Air Force has an extensive data base on each system for the deterministic evaluation of structural integrity. Of particular application to risk analysis are the data associated with the damage tolerance [2,3] and durability [2,4] analyses that are performed for all potential airframe cracking sites and the data associated with the force management tasks of ASIP [5].

<u>Damage Tolerance Analysis</u>: For every fracture critical location in an airframe, analyses have been performed for deterministically predicting the growth of assumed initial fatigue cracks at these locations. These analyses require: a) representative stress spectra from which flight by flight stress sequences can be derived; b) stress intensity factor coefficients for accounting for differences in detail geometries; and, c) crack growth rates as a function of stress intensity factor. These data items can be used to predict crack growth versus time (spectrum hours) relationships in the usage environment. These relationships when coupled with the probability of detection (POD) curve for quantifying inspection capability, produces the curves from which the Air Force damage tolerance approach to establishing inspection intervals has been formulated.

<u>Durability Analysis</u>: Durability is normally defined in terms of the number of flight hours until cracking in the airframe impairs the function of the structure. In the past, durability has been deterministically demonstrated by showing that a crack representative of the initial quality of the structure (e.g., 0.005 or 0.010 inch) will not grow to a functional impairment size (e.g., through the skin thickness or to an uneconomical repair size) in two design lifetimes. A schematic of the deterministic durability approach is shown in Figure 1.

A stochastic approach to durability has recently been developed [4]. In the stochastic approach, the initial quality of a structural element is characterized in terms of an Equivalent Initial Flaw Size Distribution (EIFSD). The EIFSD is obtained by conducting tests, collecting data and determining the distribution of times for a crack to reach a given reference size in a structural element subjected to a specified loading history. This distribution of times is known as the Time To Crack Initiation (TTCI) distribution. The TTCI distribution is then stochastically "grown" backward in time to obtain the flaw size distribution at time zero (EIFSD). The EIFSD is then grown forward in time (using the same stochastic model) to determine the flaw size distribution at any service time. Durability is then based on the number of spectrum hours for low percentiles of the flaw size distribution to reach the functional impairment size. A schematic of the stochastic durability approach is shown in Figure 2.

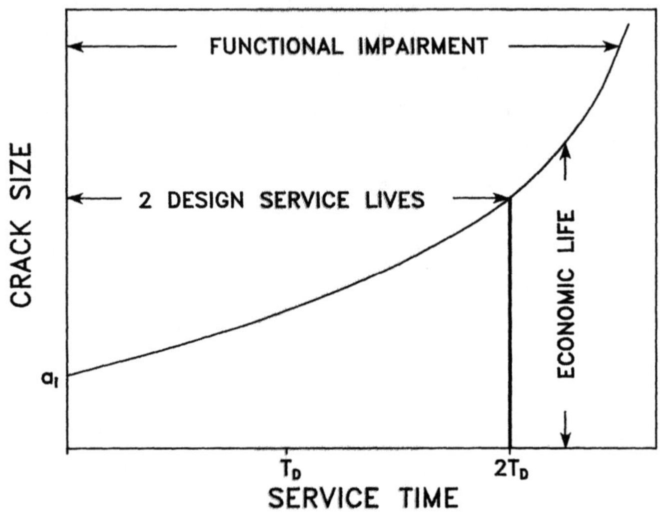

Figure 1. Deterministic durability approach

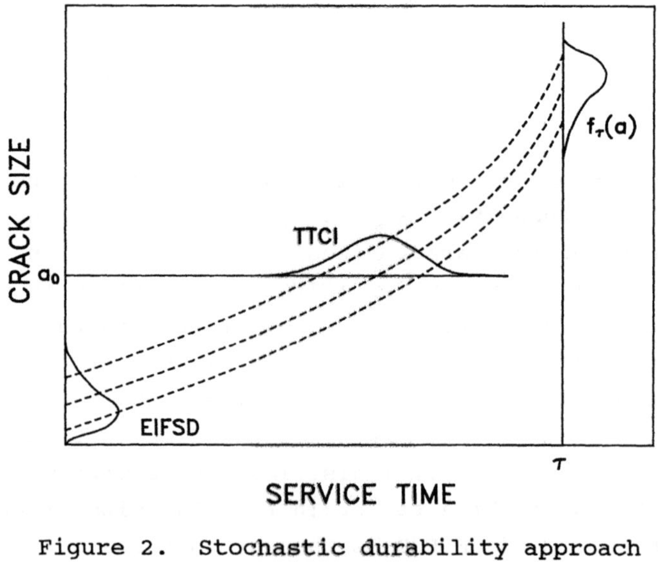

Figure 2. Stochastic durability approach

Further research is currently being performed on the stochastic durability approach by the Flight Dynamics Laboratory of the Wright Research and Development Center in conjunction with Alcoa. In this research, two different qualities of 7050 aluminum plate are being investigated to determine the role material quality plays in the initial quality (EIFSD) and thus the durability of a structural element. At the same time, the stochastic durability approach is being evaluated and any needed improvements will be identified.

Force Management: The Air Force collects two types of data within the Force Management ASIP requirements. These are individual aircraft tracking (IAT) and loads/environment spectra survey (L/ESS) data. The IAT data are potentially useful in risk analyses because they provide a method for correlating the results of inspection feedback between different aircraft of a fleet which may have experienced different usage. The L/ESS data can be used to verify and update the data on the magnitude of the stresses that the aircraft is expected to encounter.

RISK ANALYSIS METHODOLOGY

The risk analyses model (PROF) is applicable to a population of structural elements which is defined in terms of details which experience essentially equivalent stress histories. Because of the available data, the model was constructed around the growth of a distribution of cracks as illustrated in Figure 3 [6,7]. The distribution of crack sizes to initiate the analysis would be estimated from the best available data. These data could result from routine inspections, teardown inspections, equivalent initial crack size distributions obtained in a manner previously described, or engineering judgement. Given an initial distribution at some reference time, the program estimates the distribution at a later time by projecting the percentiles using the crack growth versus time curve from the damage tolerance analyses.

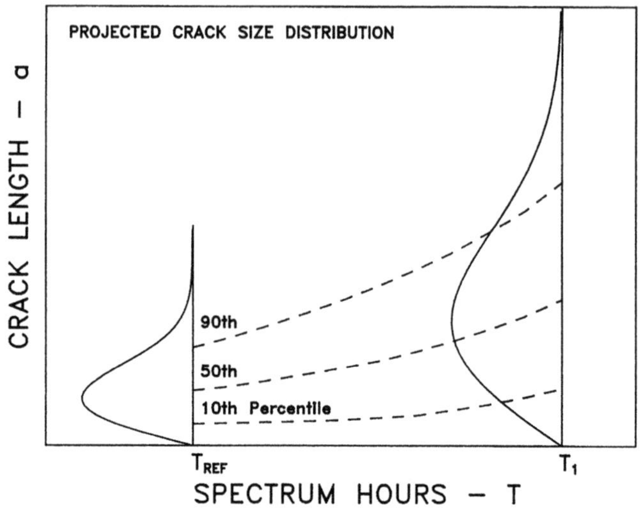

Figure 3. Schematic of projectd crack size distribution

At a maintenance action, the population of details are inspected and all detected cracks are repaired. This maintenance action will change the crack size distribution and the change is a function of the inspection capability and the quality of repair. Inspection capability is modeled in terms of the probability of detection as a function of crack size, POD(a). Repair quality is expressed in terms of an equivalent repair crack size distribution, $f_R(a)$, which is analogous to an equivalent initial flaw size distribution. If $f_{before}(a)$ and $f_{after}(a)$ represent the density function of crack sizes in the population of structural details before and after a maintenance action, then

$$f_{after}(a) = P \cdot f_R(a) + [1-POD(a)] \cdot f_{before}(a) \tag{1}$$

where P is the percentage of cracks that will be detected

$$P = \int_0^\infty POD(a) \cdot f_{before}(a) \cdot da \tag{2}$$

The post maintenance crack size distribution is then projected forward for the next interval of uninspected usage. The process is continued for as many inspection intervals as desired.

Safety - Probability of Fracture: Safety is quantified in terms of the probability of fracture (POF) due to the maximum stress encountered in a flight. POF is calculated as the probability that the maximum stress encountered in a flight will produce a stress intensity factor that exceeds the critical stress intensity factor for a structural detail. This calculation is performed in two contexts. The single flight POF is the probability of fracture in the flight given that the detail has not fractured previously. This number can be compared to other single event types of risks, such as the risk of death in an automobile accident in an hour of driving. The interval probability is the probability of fracture at any time between the start of an analysis (reference time of zero or after a maintenance action) and the number of spectrum hours, T. This POF is useful in predicting the expected fractures in a fleet of aircraft. The basic algorithm for calculating POF is

$$POF(T) = \int_0^\infty f(a) \int_0^\infty g(K) \cdot PF(a,K) \cdot dK\, da \qquad (3)$$

where f(a) = probability density function of crack sizes at start of analysis interval,

g(K) = density function of critical stress intensity factors for the structural detail,

$$PF(a,K) = 1 - \prod_{i=1}^{m} H_{\Delta T}(\sigma_{cr}(a(T_i),K)$$

$H_{\Delta T}(\sigma_{cr}(a(T_i),K)$ = probability of maximum stress intensity factor in ΔT flights exceeding the critical stress intensity factor

$$T_i = i \cdot \Delta T, \qquad i = 1,\ldots,m$$

For single flight POF, m = 1.

Expected Maintenance Costs: Given the predicted crack size distribution at the time, T_j, of an inspect/repair maintenance action and the POD(a) function, the expected number and sizes of the cracks that will be detected can be calculated. The probability of detecting a crack between a_i and a_{i+1} is given by

$$P(\Delta a_i) = \int_{a_i}^{a_{i+1}} POD(a) \cdot f_{before}(a) \cdot da \tag{4}$$

If the total population being modeled comprises k details in each of N airframes, then the expected number of cracks to be repaired at T_j between sizes a_i and a_{i+1} is $k \cdot N \cdot P(\Delta a_i)$. If C_i represents the cost of repairing a crack in size range i, C_F represents the cost of a fracture, and I represents the cost of inspecting each detail, then the expected costs of fracture and repairs in the usage interval are given by

$$E_j(C) = POF(T_j) \cdot N \cdot C_F + k \cdot N \cdot [\ I + \sum_i P(\Delta a_i) \cdot C_i] \tag{5}$$

Summing over usage intervals (maintenance periods) yields the total expected maintenance costs. Figure 4 is a schematic illustrating expected maintenance costs over 4000 flight hours as a function of inspection interval.

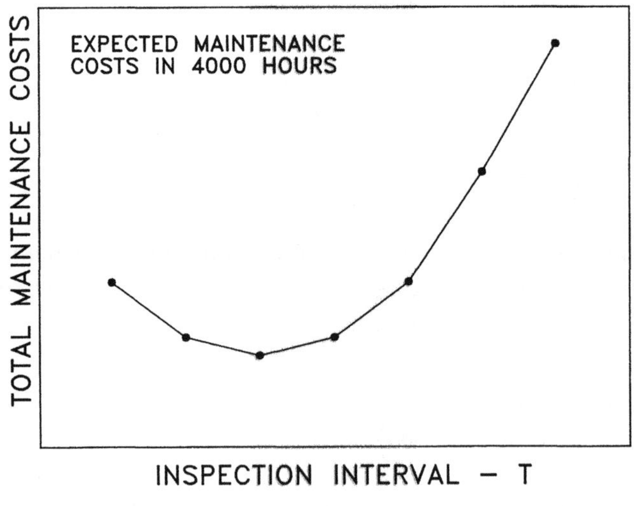

Figure 4. Schematic of maintenance costs vs inspection interval

Input Requirements: Table 1 summarizes the input data required by the risk analysis program. Crack size distributions are output for immediately before and after each maintenance action and for the cracks which were detected and repaired at the maintenance action. These crack size distributions can be coupled with cost data to obtain expected maintenance cost information.

Table 1 Input Data for PRobability Of Fracture (PROF)

DATA	FORMAT	SOURCE/COMMENTS
DETERMINISTIC		
K/sig vs a	Table	DTA Analysis
a vs T	Table	DTA Analysis
T1,...,Tm	Table	Spectrum hours at inspections -user defined
STOCHASTIC		
g(KIC)	mu,sigma	Normal distribution of critical stress intensity factor, Ref. 8
h(sig)	A,B	Gumbel distribution of max sigma per flight. Derived from L/ESS data or stress sequences of DTA analysis
f0(a)	Table	Crack size distribution at start of analysis
fR(a)	Table	Repair crack size distribution. Can be entered to remove repaired cracks from further analysis.
POD(a)	mu,sigma	Cumulative log normal POD function frpm evaluation of NDE system. exp(mu) = 50% detectable size

EXAMPLE

As an example application of the risk analysis program, representative data from an attack/fighter/trainer type of aircraft are used to compare the risks of fracture associated with two approaches to scheduling inspection/repair actions. In this example all input except the inspection intervals are fixed. The stress spectrum is representative of severe A/F/T usage. The analyses are started at zero spectrum hours (new) with an equivalent initial flaw size distribution that is log normal (median = 0.0008 in., σ = 0.62). The equivalent repair flaw size

distribution is assumed to be exponential with the probability of exceeding 0.050 in. equal to 0.0001. Inspection capability is characterized by a cumulative log normal model [9] with median detectability of 0.030 in. and σ = 0.5. This capability is representative of a good (automated) eddy current system for which the probability of missing a crack larger than 0.100 in. (length or depth since the aspect ratio of the detail is one) is 0.008.
It is further assumed that the population of details of the example comprise fatigue cracks growing from countersunk rivet holes, Figure 5, and there are six such details on each airframe in the fleet. For the population of 7075-T7351 aluminum details it was assumed that the critical stress intensity factors were normally distributed with a mean of 30 ksi$\sqrt{\text{in}}$. and a standard deviation of 1.5 [10].

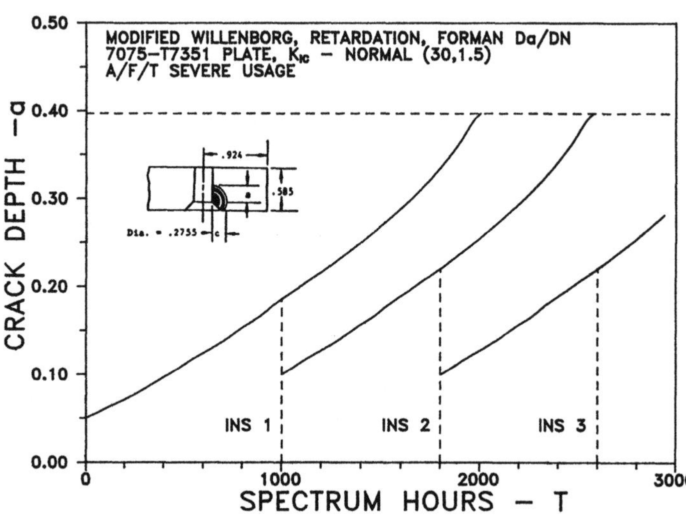

Figure 5 DTA inspection schedule

Figure 5 displays the crack size versus spectrum hours curve which was calculated from the severe A/F/T spectrum. The DTA inspection intervals are also shown on Figure 5. The first inspection occurs at 1000 hours, one half the time required for a 0.050 in. crack to grow to a critical size (0.4 in.). After an inspection, the reset crack length is 0.100 in. Subsequent inspections are scheduled at 800 hour increments or one half the time for a 0.100 in. crack to grow to a critical size.

The probability of fracture (POF) as a function of spectrum hours for the DTA inspection schedule is shown in Figure 6. The bottom (solid) curve presents the POF in a single flight at T spectrum hours while the top (dashed) curve presents the POF anywhere in the interval between 0 and T spectrum hours. Note that the first inspection was scheduled at 1000 hours and had minimal effect on the probability of fracture. This suggests that the likelihood of finding and repairing cracks at the first inspection was small. Note also that the single flight POF's stabilize after the second inspection, implying that the inspections are effectively finding the large cracks.

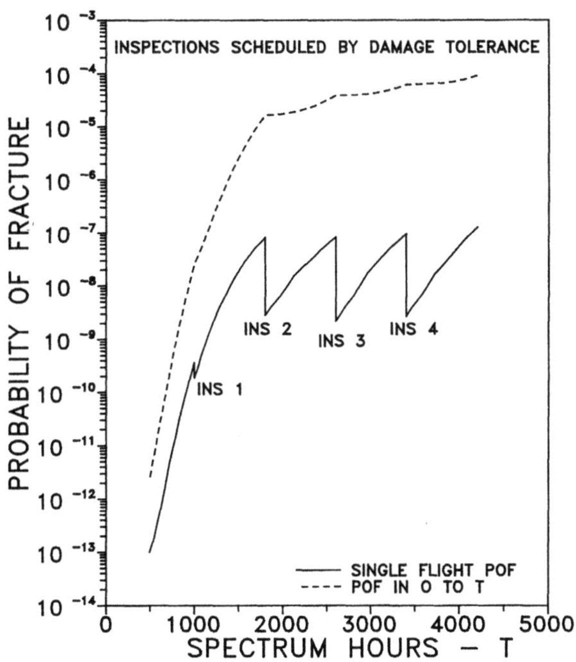

Figure 6. Fracture risks from DTA inspection schedule

Since the first inspection from the DTA inspection schedule resulted in an inspection which had little effect on the POF, an alternative "durability" schedule was devised. The durability schedule called for inspections at times when one crack would be expected to be found on each airframe, i.e., at the number of spectrum hours for which a crack could be expected to be found at one of the six countersunk holes in the zone of equivalent stresses and geometries on each airframe in the fleet. This interval depends on the crack size distribution at the beginning of the usage period, the crack growth curve, and the POD(a) function. Figure 7 illustrates the calculation of the first inspection interval. The axis on the top of the figure shows the probability of detecting cracks of the indicated depth. Since cracks of all sizes might be detected, the time for the 1 in 6 size equivalent initial crack (0.0016 in., as calculated from the equivalent initial crack size distribution) to grow to the median detectable crack size (0.030 in.) was calculated as the time to the first inspection. Under this scheme the first inspection was scheduled at 2000 spectrum hours. At 2000 hours, the program predicted that cracks would be found in 14 percent of the details as opposed to the 17 percent as planned. This difference was ignored. The process was repeated using the crack size distribution after the first inspection to schedule the second inspection, and so on. The second interval was calculated to be 1200 hours with essentially equal subsequent intervals of 700 hours between inspections.

Figure 8 compares the single flight and interval fracture probabilities for the DTA and durability inspection schedules. The durability schedule has regions of higher single flight fracture probabilities prior to 3200 spectrum hours but the single flight risks are essentially equivalent after the third DTA inspection at 3400 spectrum hours. As noted earlier, the inspection process is controlling the risks after the first two inspections. The durability schedule has higher interval fracture probabilities after the first DTA inspection at 1000 spectrum hours. The costs of inspections, repairs, and fractures would be required to further evaluate the trade-off between the increased fracture risks from the durability schedule and the added costs of the additional inspections from the DTA schedule. Introducing cost data is beyond the scope of this paper.

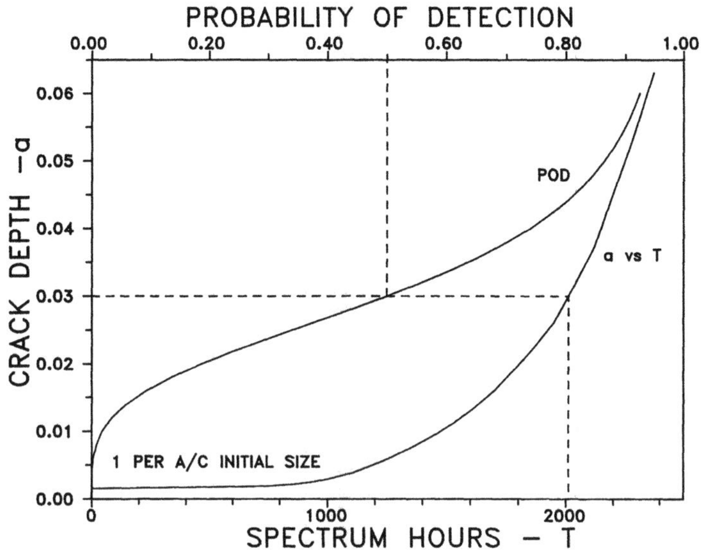

Figure 7. Durability inpsection schedule

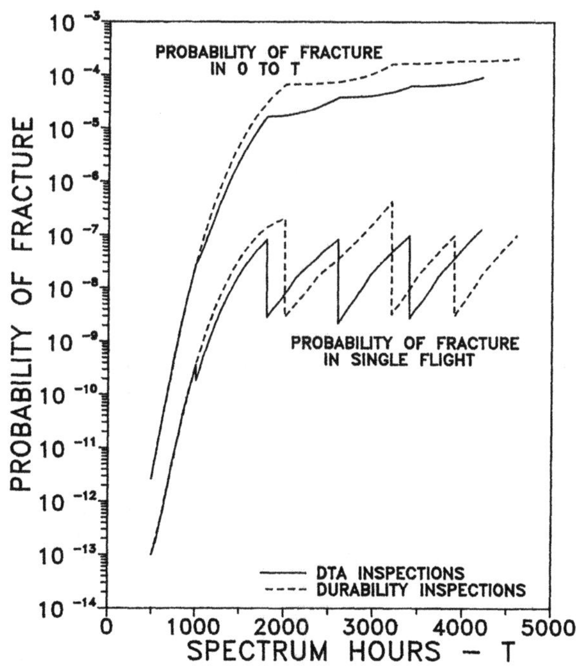

Figure 8. Fracture risks from durability and DTA inspection schedules

SUMMARY

The computer program, PROF, has been written to provide the Air Force with a risk analysis tool for the decision making process of scheduling inspection, repairs, replacement, or retirement maintenance actions on aging aircraft fleets. The program is based on a growing population of fatigue cracks for zones of equivalent stress histories. The data required for the model is generally compatible with the ASIP data that is or is expected to be available on Air Force aircraft. The output provides information that is applicable to both risk evaluation and risk management. Risk management is expressed in terms of the probability of fracture in both single flights and intervals of usage. Risk management information is obtained from the distributions of crack sizes before and after a maintenance action and the distribution of the sizes of the cracks that would be expected to be detected during an inspection. An example was presented which illustrates a potential application of the program using representative data for an attack/fighter/trainer class of aircraft.

ACKNOWLEDGEMENT

This work was sponsored under Contract No. F33615-87-C-3215 by the Flight Dynamics Laboratory, Air Force Wright Aeronautical Laboratories, Aeronautical Systems Division (AFSC), United States Air Force, Wright-Patterson Air Force Base, Ohio, 45433-6553.

REFERENCES

1. Military Standard, "Aircraft Structural Integrity Program, Airplane Requirements," MIL-STD-1530A (USAF), December 1975.

2. Military Specification, "General Specification for Aircraft Structures," MIL-A-87221, February 1985.

3. Gallagher, J.P., Giessler, F.J., and Berens, A.P., "USAF Damage Tolerant Design Handbook: Guidelines for the Analysis and Design of Damage Tolerant Structures," Air Force Wright Aeronautical Laboratories, AFWAL-TR-82-3073, May 1984.

4. Military Specification, "Airplane Strength, Rigidity, and Reliability Requirements, Repeated Loads, and Fatigue," MIL-A-8866B, August 1975.

5. Manning S.D. and Yang, J.N., "USAF Durability Design Handbook: Guidelines for the Analysis and Design of Durable Aircraft

Structures," Air Force Wright Aeronautical Laboratories, AFWAL-TR-83-3027, January 1984.

6. Berens, A.P., et al., "Handbook of Force Management Methods," Air Force Wright Aeronautical Laboratories, AFWAL-TR-81-3079, April 1981.

7. Yang, J.N., "Statistical Estimation of Service Cracks and Maintenance Cost for Aircraft Structures," J. of Aircraft, Vol. 13, No. 12, December 1976.

8. Lincoln, J.W., "Risk Assessment of an Aging Military Aircraft," J. of Aircraft, Vol. 22, No. 8, August 1985.

9. Berens, A.P., "NDE Reliability Data Analysis," Metals Handbook, Volume 17, 9th Edition: Nondestructive Evaluation and Quality Control, 1988.

10. Stumpf, P.L., et al., "Damage Tolerant Design Data Handbook," Air Force Wright Aeronautical Laboratories, AFWAL-TR-83-4144, December 1983.

4. Aspects of Aging Aircraft: A TransAtlantic View

John W Bristow

Safety Regulation Group
Civil Aviation Authority
Gatwick Airport. U.K.

Summary

This paper will discuss from a regulatory viewpoint some of the key technical matters in the ageing aircraft issue. Attention is focused primarily on research topics of interest, for an overview of the CAA approach to ageing aircraft in general the reader is referred to Williams [1] and James [2]. The topics covered here are

Experience with proof pressure testing.
Inspection reliability in relation to the task
Potential effects of corrosion on fatigue crack growth and toughness
Definition of corrosion inhibiting fluids
Implementation management of ageing aircraft programmes.

Experience with Proof Pressure Testing

For many years in pressure vessels, such as steam boilers, successful application of a pressure above working pressure has been taken as a demonstration of structural integrity the so called "proof" test. In fact following initial construction of an aircraft fuselage and before flight it is normal to apply 1.33 times the maximum cabin differential pressure. Recently the concept has been revived as a demonstration of structural integrity for pressure fuselages of high time aircraft [3]. A significant amount of analytical work has been done [4], [5] on the crack growth following application of such proof pressure loading. From this work it is possible to derive an interval at which the proof pressure application must be repeated to ensure that any failure will occur during proof pressure application which is on ground and therefore "safe". Such intervals work out to be of the order of 500 flights for cracking in lap splice joints at typical working stresses. It is interesting to compare these very recent analytic results with the actual experience gained over 35 years ago with the Comet I aircraft, shown in fig 1 below.

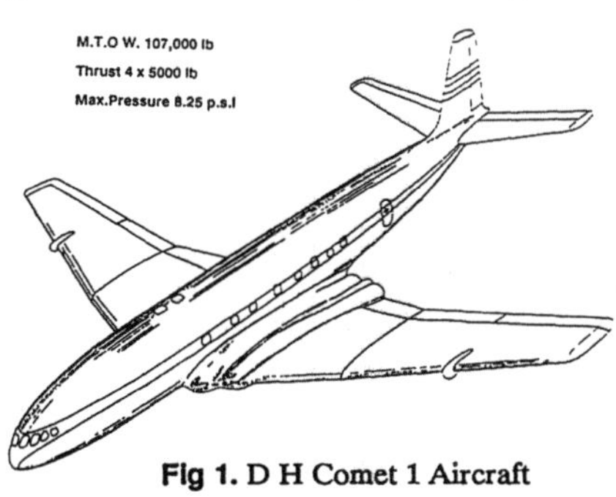

Fig 1. D H Comet 1 Aircraft

Springer Series in Computational Mechanics
Atluri, Sampath, Tong (Eds.)
Structural Integrity of Aging Airplanes

This aircraft suffered two in flight fuselage structural failures in 1954. Table 1 shows the chronological history of fatigue testing and of three of the aircraft fleet, the data is taken from [6]. A number of points are clear from the table. Firstly, the fatigue test specimen result was greatly influenced by the application of 2 proof tests of twice the working pressure p. It survived 18000 pressure cycles whereas the second test specimen without aptitude of 2 x Δp failed after 3000 cycles. Secondly the in service failures were at 900 and 1290 cycles even lower than the second test result. Thirdly, and perhaps the most significant, both the aircraft that crashed had been proof pressure tested on the ground approximately 160 and 120 flights before failure in the air. (These figures being estimated from the flying hours and the average utilisation of the aircraft).

	TEST	G-ALYP	G-ALYU	G-ALYY
	2000 +2 @ 2Δp	$1^1/_3$Δp		$1^1/_3$Δp
02 Mar 52		C of A		
30 Sep 52				C of A
30 Sep 53	18000 (failed)			
11 Nov 53		$1^1/_3$Δp		
10 Jan 54		1290 (crashed)		
15 Feb 54				$1^1/_3$Δp
08 Apr 54			1230 (Grounded)	900 (Crashed)
20 Dec 54			3060 (failed test)	
Cycles after proof		~160		~120

Table 1. Pressure cycles on 3 Comet 1 aircraft

In the analysis of crack growth during and following proof testing consideration to the effects of corrosion within lap joints has not been properly addressed. If the cracks extended into a corroded patch during overpressure the subsequent retardation due to tip plasticity affects would be reduced and furthermore the subsequent growth rate might be more rapid see ref [8]. This would render the projected repeat proof test interval analysis suspect for the real situation of the high time aircraft fuselage.

A third aspect of proof pressure testing that must be appreciated is that a pressure fuselage is not just a symmetric thin shell it contains many other features. In particular over the wheel bays and wing box there are longitudinal floor beams, flat pressure decks and flat pressure bulkheads, see fig 2 below.

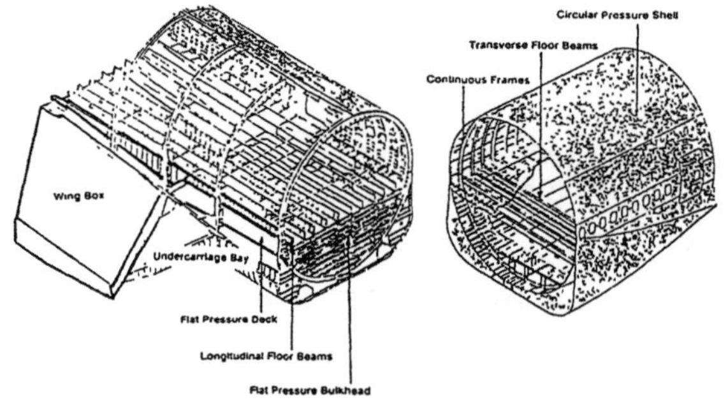

Fig 2. Section of typical fuselage

An incident occurred in 1980 [7] where service bulletin modification action was being carried out to both the web and booms of the longitudinal floor beams, see figure 3. New booms were installed but the web replacement was delayed to allow the aircraft to be ferried to the U.S.A. for door and window modifications. These latter modifications were followed by a pressure test to 1.33 Δp. The aircraft was returned to U.K. for completion of the maintenance check, upon inspection it was found that one of the webs of the longitudinal floor beams was cracked through as shown in fig 3. The aircraft had only had 2 flights and the proof pressure applied. Had not re-inspection been necessary for the other maintenance reasons this significant crack caused by the proof pressure would have gone undetected for some considerable time.

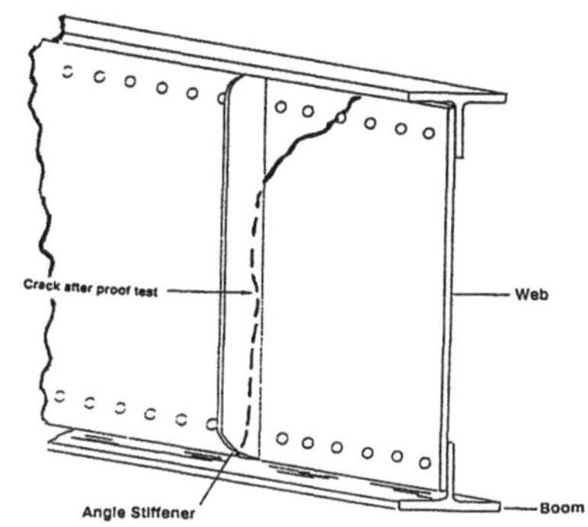

Fig 3. Sketch of longitudinal floor beam

For any proof pressure testing to be a viable tool to assess structural integrity it would have to be accompanied by detailed inspection both before and after the pressure test. Also adequate allowance would have to be made for the presence of corrosion which as yet has not been done. Furthermore the projected repeat intervals for the pressure test of the order of 500 cycles render the concept impractical for the airworthiness assurance of civil transport aircraft.

Inspection reliability in relation to the task.

In the early 1980's the CAA instigated a study into inspection reliability in relation to transport aircraft structures. The study was carried out by a small team of non-aviation engineers/scientists from the Cranfield Institute of Technology. Two U.K. airlines were studied, one an international Scheduled Carrier and the other a Charter Operator. Four techniques were used; direct discussion and interviews, a questionnaire for inspectors, a survey of structural occurrence reports and a task description/analysis of inspectors at work.

A detailed summary of this work [9] is included as appendix 1. The major observation was that there was no universal panacea for inspection reliability but that small improvements could be made in many areas such as :- hangar lighting and noise level, better access by staging and hoists, a standard for torches and mirrors, an eyesight test for inspectors, refresher training for inspectors and better cleaning during inspection.

The analysis of crack sizes found by visual inspection on the CAA mandatory occurrence report data base over a ten year period showed that U.K. findings were similar to U.S.A. ones [10], see fig 4 below.

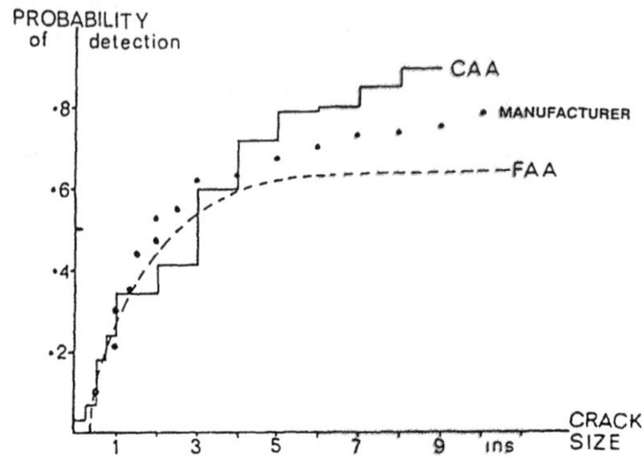

Fig 4. Cracks reported found by visual inspection

Potential effects of corrosion on crack growth and toughness.

A realistic model for the ageing aircraft structure is that cracks can initiate independently from corrosion, but may encounter it in their subsequent growth. CAA instigated studies propagating both monotonic and cyclic cracks through exfoliation corrosion in 7178-T6 alloy taken from an aircraft that had seen 15 years in service. This work is fully described in another paper [8] in this publication. No effect on toughness was observed, but a threefold increase in cyclic crack growth rate was noted, see Fig 5 below.

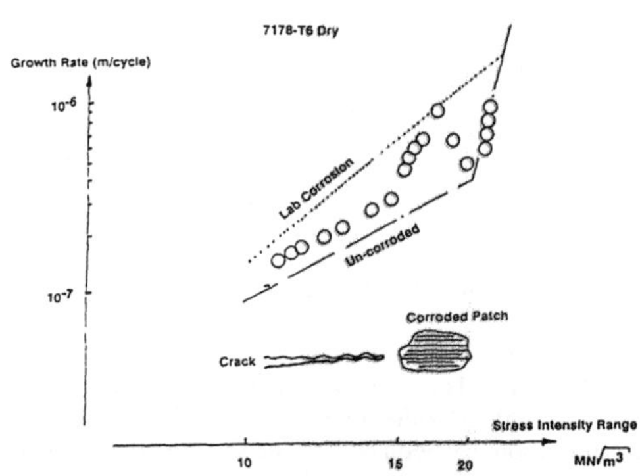

Fig 5. Crack growth across real-time corrosion

Definition of corrosion inhibiting fluids

In the context of safety related issues in the current "Ageing Aircraft" programmes for commercial transport aircraft the CAA believes that a clearer understanding is needed of the technical background to the nature, usage and efficacy of corrosion inhibiting fluids used in corrosion control programmes. Particular areas of interest for any product in a range are; the composition and specification; the quality control in manufacture; the release and acceptance standards; the capability of the product and how it is substantiated; the life of the product in storage, service and/repeat application interval; the method of application and removal; and the fire; smoke and toxicity in service and other environmental considerations.

The ideal performance of a corrosion inhibiting fluid is multi factored it needs to provide corrosion protection; be acid and alkali resistant; provide water displacement; be stable at high and low temperature; be inert on wiring, seals, glazing, furnishings, sealants, paints etc; have good coverage and adhesion, be transparent, easy to clean, and durable but easy to remove.

Such fluids are primarily made up of petrochemical derivatives and may have the following typical constituents:- a solvent base which evaporates off after application which is about 75% of the total initial volume, a wax or grease to form the film which makes up 20% of the bulk, plus small quantities of corrosion inhibitors (possibly petrochemical sulphonates), de-watering agent (a volatile alcohol), anti-oxidants to prevent long term cracking of film and a surface active agent to promote good contract with the substrate.

These complex materials are supplied to a wide range of specifications in the industry. There are several different national codes e.g. MIL-C-16173D, Def.Stan 80-83/1 and each aircraft constructor has its own e.g. BMS 3-26, CML 05-27D. There is a strong need for uniform standards within the industry so that these components may be used not necessarily interchangeably but at least universally across a fleet of different aircraft types. It is suggested that either the operators under ATA or IATA should assemble all interested parties to agree a common standard or perhaps an FAA research contract should be let to achieve the same goal.

Implementation management of ageing aircraft programmes

As with all the parties directly involved in the ageing aircraft activity in the last two years the CAA expects a significant increase in workload in aircraft maintenance and its supervision as the various programmes are implemented. In terms of fleet size the task in the U.K. is approximately one tenth that of the USA. see Table 2 below.

	UK	USA
10-15 years	55	578
15-20 years	64	556
Over 20 years	109	1241

Table 2. Number of aircraft in ageing fleet 1989.

This ageing aircraft fleet falls into five type groups which will be handled separately but along similar lines. "ATA Airliners" are the first group to be handled these include B747, B737, DC9 and BAe 1-11 in the UK. Commuter aircraft such as the Jetstream and Shorts 3-30 will be considered next. Other UK built transports such as the Viscount and Concorde will be appraised along with other foreign transports such as F27 and Electra. Finally the "Golden Oldies" such as DC-3,DC-6 etc will need special attention.

To expedite the consultative process over compliance with the ageing aircraft programmes currently being evolved for the "ATA Airliners" the CAA has set up a special team known as ACT (Ageing Compliance Team) to handle the problem in a systematic way. Figure 6 shows a flow chart of the normal interaction between the aircraft, the maintenance organisation, the aircraft constructor and the CAA Maintenance Surveyor. For ageing aircraft programmes reference through the ACT will be available to provide a consistent interpretation of requirements including inspection of selected aircraft. The ACT is a 4 man team made up of both maintenance and structural design staff each with many years experience in the industry.

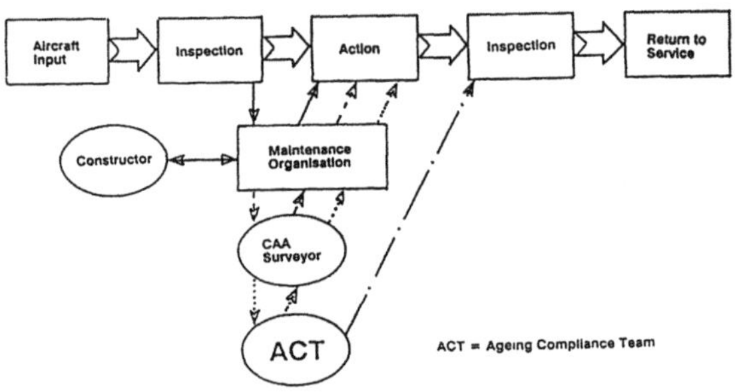

Fig.6. Flow chart of ageing aircraft compliance management

Comment in Conclusion

The author would like to thank the Civil Aviation Authority Safety Regulation Group for permission to present this paper. The views expressed are personally held ones and may not necessarily reflect the position of the CAA.

From the work presented a number of conclusions may be drawn.

1, For compliance with ageing aircraft programmes strengthened organisation is necessary to cope with the increased workload. The A.C.T. set up by CAA may be particularly appropriately named as the only way to solve the ageing problem of aircraft is to Act now.

2. Proof pressure testing of ageing fuselage structures is not a viable airworthiness procedure for aircraft in commercial transport service today.

3. Further work needs to be done on the interaction of corrosion with fatigue crack propagation.

4. The recommendations for improvements in inspection reliability are to be considered for incorporation in forthcoming JAA regulations.

5. There should be an industry-wide effort to rationalise corrosion inhibiting fluid specifications.

References

[1] Williams, R., Some key aspects to improve aircraft maintenance. A.A.A.F. Conference on Maintenance and Ageing of Aircraft Structures, Bordeaux-Lac, March 1990.

[2] James, D.O.N., Airworthiness of ageing aircraft and structural repairs. A.A.A.F. Conference on Maintenance and Ageing of Aircraft Structures, Bordeaux-Lac, March 1990.

[3] Broek, D., The feasibility of proof testing, TR 89126. Fracteurearch, Ohio 1989.

[4] Newman, J.C., Poe, C.C., Dawicke, D.S., Evaluation of Pressure Proof Test Concept for This Sheet 2024-T3. International Symposium on Structural Integrity of Ageing Airplanes, Atlanta, March 1990.

[5] Broek, D., Impact of Damage Tolerance Requirements in Theory an in Practice, International Symposium on Structural Integrity of Ageing Airplanes, Atlanta, March 1990.

[6] C.A.P. 127, Civil Aircraft Accident Report of the Court of Inquiry into the Accident to Comet G-ALYP and Comet G-ALYY. Her Majesty's Stationery Office London 1955.

[7] C.A.A. Mandatory Occurrence Report No.80/00568C, 1980.

[8] Chubb, J.P., Hockenhull, B., Morad, T., Bristow, J.W., The Effect of Exfoliation Corrosion on the Fatigue Behaviour of Structural Aluminium Alloys. International Symposium on Structural Integrity of Ageing Airplanes, Atlanta, March 1990.

[9] Lock, M.W.B., Strutt, J.E., Reliability of In-Service Inspection of Transport Aircraft Structures. CAA Paper 85013, Civil Aviation Authority London 1985.

[10] Bristow, J.W., Lavers, B.F., Aircraft ageing as seen by an authority. Airmec 85 Conference, Dusseldorf 1985.

Appendix 1

The following 12 pages reproduce a summary of the independent investigation on inspection reliability carried out by Lock & Strutt [9]. The views expressed are those of the researchers and do not necessarily reflect the CAA position.

A THREE PART STUDY OF INSPECTION RELIABILITY FOR TRANSPORT AIRCRAFT STRUCTURES

To be published as CAA Paper 90003

Dr. M. W. B. Lock
Dr. J. E. Strutt

March 1990

CONTENTS

2

SECTION 1 - ABSTRACT

A study was made of the reliability of structural inspection of transport aircraft with particular emphasis being laid on the influence of human factors on the performance of inspection. Data for the study was obtained from various sources: discussions with and visits to operators; mandatory occurrence reports; a questionnaire circulated amongst selected aircraft operating companies and by task description and analysis at shop floor level. The importance of various performance-shaping factors to the inspection activity has been established. The results suggest that the most critical steps are defect searching and recognition. To this end, recommendations are made regarding primary and secondary access, lighting and eyesight.

The role of the inspector, his training and the practical feedback of defect criticality is discussed.

This paper is a shortened version of CAA paper 85013 'Reliability on In-Service Inspection of Transport Aircraft Structures', August 1985.

The study was carried out within the UK under an ARB fellowship by 2 lecturers from the Cranfield Institute of Technology in the early 1980s.

Thus, a view of aircraft structural inspection was taken by engineers/scientists from outside the industry.

The condensed version has been prepared to make it more accessable to a wider readership and to provide a starting point for new work in the field in 1990.

There are two obvious areas which will have changed since the 1980s. First, the advent of the modern small computer into planning and data storage for records etc.. Secondly, a far greater use of much improved NDT techniques.

The authors would like to acknowledge the support and co-operation received from the UK aircraft operating industry.

3

SECTION 2 - INTRODUCTION

The high level of safety required of aircraft structures has always imposed stringent requirements on reliability and fail-safe designs. However, after manufacture, the continued integrity of the aircraft structure is normally ensured by periodic inspection and maintenance operations, carried out during the working life of the aircraft. Whether the aircraft structure design philosophy tends towards fail-safe or safe life, much reliance must be placed on inspection techniques to detect structural damage before it reaches a critical condition.

Many structural failures can be traced to a defect or defective area which often arise from degradation processes associated with the normal wear and tear of service life. With increasing time the likelihood of occurrence of such defects will increase, leading to a higher probability of failure. At present, it is not possible to predict where or when all defects will occur on structures with any precision, although relevant areas or times can be predicted, using fracture mechanics/crack growth curves etc. Inspection is, therefore the only method of ensuring that the structure is sound and safe for continued operation.

On aircraft structures, the main degradation modes of concern are the development of cracks and, on older aircraft, corrosion. The inspection function is, therefore, mainly one of finding and assessing cracks and corrosion. In practice, however, it is difficult to completely achieve the objectives of inspection. the main difficulty is that in order to assess defects and estimate consequences, defects must first be detected and located. In the aircraft industry, as in other industries, defect location is often only through the eye of the inspector and is, therefore, of limited reliability.

Many factors are relevant to the reliability of inspection:

1. The large size of structures to be inspected.
2. The time available to inspect the required area.
3. Human factors, such as conditions under which inspections take place, attitude of the inspector, background knowledge of the inspector, organisational structure, etc.
4. The non-availability of the defect location aids as one goes down the line.
5. Geographical location which may involve extreme heat/cold/wet/dusty conditions.
6. Local conditions such as poor access, bad lighting conditions, etc.
7. Commercial pressures requiring aircraft to be operational, i.e. financial and/or passenger consideration.

The increasing size and height of aircraft have tended to make visual inspection more difficult and time-consuming. The time available for inspection is in many cases very short, particularly for fast turn-around line inspections, which are

4

often carried out under difficult environmental conditions and are, in any case, unlikely to contain checks for cracks and corrosion.

There are many aspects to consider in assessing inspection reliability. For example, what is the chance of missing defects in hitherto unsuspected areas by concentrating on those areas which experience shows to be critical? How long should an inspector stay on one part of the aircraft? What affects the inspector's concentration?

Another consideration is the impact of instrumentation in helping to locate and size defects more quickly and effectively. At present there are a wide range of techniques available and/or under development for sizing and assessing cracks and corrosion. Experience has shown us that there are, however, few techniques for rapid location of defects in large areas of structures. It seems unlikely, in the short term, that significant advances on the eyeball method of defect location over large areas will be found. The situation may change in this rapidly developing field and should be regularly reviewed.

It is considered that at the present time, any significant improvement in the reliability of inspection will be the direct result of such improvements in the reliability and working methods of the inspectors.

SECTION 3 - SCOPE OF PROJECT

In view of the comments in Section 2, a systematic study was made of the reliability of inspection along several lines.

3.1 The Maintenance Schedule (MS) and Its Operation

A detailed survey and comparison was made of the production and operation of MSs by two companies with contrasting needs. The first, a large carrier with international scheduled flights necessitating reasonably equal inspections throughout the year. The second, a company concentrating on the holiday trade in summer and to a less extend for winter skiing. The most convenient period for major maintenance was late autumn and early spring with much shorter checks done in high season.

Many visits were made to Airline Operators in different locations over a period of one year.

Many hours were spent talking with, and watching the duties of, maintenance engineers, planners, inspectors, supervisors, mechanics and quality control personnel. The overall impression was of a conscientious workforce with a real pride in its work.

3.1.1 Comments

Properly qualified personnel were seen to be used in all key positions such as in planning, inspecting and quality control.

Providing that it is adhered to, the system for determining and allocating tasks is adequate. Errors which could occur in the card system are considered in Section 4.

A random inspection of up to 10% of the scheduled work can be, and is, made by Quality Control and a further investigation can be made by the CAA during or after inspection period.

The possibility of error in work at the basic level of fitter, etc., is covered by subsequent inspection at many stages. thus it would be difficult for even deliberate errors to occur. Many tasks, e.g. lubrication, cannot be inspected adequately except perhaps whilst they are in progress. Crucial functions such as moving surfaces and their controls are inspected by two separate inspectors.

The overall impression of the operation was that of a highly responsible attitude towards aircraft maintenance/inspection.

3.2 Analysis of Mandatory Occurrence Reports (MOR)

To establish the nature and extent of the defects in aircraft structure and the proportion that could be assigned as an error of maintenance or other source, a survey was made of the CAA MOR reports 1976-1981. The aim was to compare occurrences found during maintenance operations and those arising at other times.

An analysis was undertaken on the data mentioned for the following points:

(a) Location of defect
(b) Type of defect (e.g. flaw, corrosion, broken fastener)
(c) Whether flaws were sized or associated with corrosion, fasteners, etc.
(d) Hours since overhaul, hours since new
(e) Other circumstances (e.g. prior heavy landing, lightning strikes)
(f) Allocation of cause (e.g. error by manufacturer, overhaul, maintenance)

3.2.1 Comments

Several interesting points were noted during compilation and analysis of the data which led us to make the following observations.

In many cases, although a defect and its location were described well, the size was not given, especially where complete fracture occurred.

We feel that one of the more important uses to which the computer files could be put, would be to provide data for simple fracture mechanics analysis of flaws. This would require at least the size of flaw etc., and the structural geometry and the stress level to be expected there, together with some details of the aircraft's life since the relevant part of the structure was last inspected.

Area and depth of corrosion was not generally given and only very rarely was mention made of corrosion in connection with a flaw. This is in contradiction to subsequent information received concerning some areas.

The age of the aircraft was available in MORs in less than half of the cases found during maintenance. For occurrences found outside maintenance, less than 20% were in the MORs. Some companies were noticeably worse than others at giving this information. All companies were poor in reporting 'time since overhaul'.

The number of reports containing the time (or number of landings) since the last relevant maintenance check was even lower, only 14% and 2% being in the MORs arising during and after maintenance respectively, of which less than half reached the computer files.

One aim of this part of the programme was, if possible, to quantify the number of serious defects arising between inspections. These defects can be due to maintenance, overhaul/repair and manufacture errors or intrinsic to the aircraft and designated 'unassigned'. Of about 360 MORs relevant to this data collection, about 90 were assigned to inspection/maintenance error. A simple calculation gives a

7

figure of about one mistake in 20 years for each inspector.

A disturbing statistic is that 67% of all MORs relevant to inspection are 'unassigned', i.e. 90% of all MORs arising during maintenance and 44% afterwards. It would seem reasonable that any occurrence that endangers an aircraft must have a cause and that the cause is due to an error, be it human, metallurgical or otherwise.

Of these unassigned MORs many could be due to material failures, either through choice or use, which should be the concern and responsibility of the manufacturer. The rate of assignment or error to the manufacturer, 2% overall, seems very low.

While one considers the above section in the whole context of the reliability of structures it has to be admitted that one is not trying to reduce a failure rate from 50% to 40% but more in the region of 10^{-7} to zero.

As a final comment, it must be said that the requirements under which operators etc. initiate MORs are not such as to provide a total survey of aircraft structural failure. It appears that a subsequent similar occurrence of a defect previously reported as an MOR is most often not reported again, especially if the CAA conclusion was that no further action should be taken after repair etc.

3.3 <u>Questionnaire on Performance Shaping Factors</u>

The purpose of this was to complement the opinions on performance-shaping factors (PSFs) we had obtained from talking in and around the aircraft maintenance organisations. The result would be to provide more quantitative data on the respective PSFs so that their relative importance could be determined.

The questions were framed so that in all relevant cases a number was required as well as room for a comment. The number might be, for example, the temperature at which inspection begins to be affected, or the number of hours a man could work continuously without significantly affecting his error rate.

It was realised at the outset that an accurate answer was not available for most of the questions but it was expected that the range of the answers would give an indication of whether the question highlighted a point of common agreement or dissension.

Most importantly the comments made would be valuable in assessing the relevance of the PSFs to inspection reliability. We would like to take the opportunity here or thanking those inspectors, managers etc. who contributed so much time in filling in these 60-odd questions.

8

A detailed analysis of the questionnaire is clearly not possible in this brief report but the conclusions have been incorporated into Section 4. Further details are available in the full report, CAA paper 85013.

3.4 <u>Human Factors and the Inspection Process</u>

To this end the authors looked, in detail, at the human factors involved in inspection, such as the organisation of inspections and who carried out specific activities, the reporting, recording and monitoring of inspection results, the type of information recorded, conditions under which they are carried out, communications between inspectors and other workers, etc..

It would appear that there are three objectives implied in structural inspection. The first and foremost aim is to find any defects or damaged areas, over the structure as a whole, which are likely to impair structural integrity or serviceability. Secondly, any defects found should be assessed so that, thirdly, an estimate can be made of the possible consequences of not repairing the damage, i.e. is there sufficient life to the next inspection?

The above definition and meaning of inspection highlights an important point, that inspection is not simply concerned with finding defects, it also involved interpretation and decision-making processes; an important factor in the reliability of inspection.

The instructions in the maintenance schedule are important in prompting the inspector where to look and what to look for. However, because an aircraft is a very large and complex structure, it is not possible to give specific instructions to cover all eventualities. Typically, therefore, the MS is composed of both specific and general instructions.

The specific instructions are directed to areas which are known to be problem areas and/or which are critical to airworthiness. The more general instructions cover all other eventualities. From a safety point of view, the general area inspections are very important as it is impossible to ensure that all significant defects will arise at areas or on items which are specifically mentioned in the MS.

Because of the large number of items and areas to be inspected on an aircraft, a complete task analysis of the whole aircraft was not practicable. It was decided therefore to select a small area of a representative transport aircraft for study. The tailplane area appeared to be ideal representing a typical, yet critical, area on which to model the inspection process.

9

A task description was obtained by working with the inspectors in the execution of their duties on the shop-floor, observing the way in which the tasks were carried out. An analysis of parts of this task description was made and comparisons drawn with an idealised 'inspection model'.

3.4.1 Comments

From a human point of view the following comments seemed most appropriate; most are expanded in the conclusions in Section 4.

A heavy reliance on visual detection was evident throughout, particularly for area checks.

In some instances the lighting was less than ideal for visual detection of condition, particularly where access was difficult. Inspection of areas in shadow or areas out of the line of sight appear to be particularly troublesome.

A surprising (to us) amount of information on the condition of items arises through other senses, particularly the sense of touch. Roughness, strange changes in curvature, freedom from movement, looseness, flakiness, etc.; such information often comes through the fingers; frequently this is triggered by visual signals in the first instance.

Individual skills and working methods appear to be important. For example, one inspector discussing the work method of one of his colleagues commented 'He carries an inhibitor-soaked rag with him and he finds lots of cracks'.

It is not known how widely special individual skills are known or used but it is clearly to the advantage of all operators for proven successful skills to be more widely spread amongst other inspectors, possibly through a shop-floor intercompany news sheet.

Access is critical to the reliability and efficiency of the inspection. Those areas which are awkward to inspect also tend to be uncomfortable. The use of the man hoist in the tail inspection was such a help in the inspection it is difficult to imagine how an effective inspection can be carried out without one, unless custom-built staging is available.

General observations indicate that comfort appears beneficial to concentration and therefore should improve the reliability of the inspection process.

A feeling of physical security is quite important. One of the main differences between the inspection of the two horizontal stabilisers and the fin was that access was provided by scaffolding for the latter. The inspector commented "I am not very happy with the scaffolding round here to carry out a thorough check. I like to feel 'comfy' so that I can push and pull as required, so I will have a quick look for anything obvious and wait until I can get up with the hoist".

10

Having to return to re-inspect an area which has had a cursory inspection could result in the re-inspection not being carried out, particularly if the cursory inspection indicated that the part was in good condition.

Deterioration in condition of items is a continuous process which gradually changes from acceptable to unacceptable. The inspector appears to possess a 'built-in' mental model of what is acceptable and what is not. Training and experience is obviously crucial in this respect and it was evident in almost every task and subtask carried out. A sound engineering knowledge of the aircraft is vital.

3.5 Final Comments on the Scope

The areas with which we have chosen to concern ourselves are those which, as scientists and engineers outside the aircraft industry, we see as important. The way the MS is produced, the way it is used and the very individual interpretation each inspector makes of it are the source of reliability.

In such a complex structure as an airframe there are obviously as many ways to organise the inspection as there are inspectors. What we have seen shows us that the will is there to do a first rate job and also, and more important, to consider any way to improve it.

11

SECTION 4 - CONCLUSIONS AND RECOMMENDATIONS

This section aims to condense what we have seen, heard and been told of into a set of conclusions and recommendations. In some cases the conclusion will be that a certain aspect of inspection is well organised and offers little opportunity for any modification which could lead to an increase in inspection reliability; in others, where an area permits of improvement, it is the authors' aim to make constructive recommendations, where possible.

The problem of increasing reliability is made extremely difficult because of the very good reliability which already exists. there is no universal panacea, improvement will only be made by small improvements in many areas. It may be that a new inspection philosophy will provide this by analysing each task carefully.

New aircraft types may be better designed for automatic NDT techniques etc., but visual inspection is still 99% of the job at present.

Facilitating the visual inspection task with better eyes, tools and knowledge are the areas where the small gaps in reliability will be closed.

In the following pages factors affecting inspection are split into several groups; human, workplace, scheduling, aircraft and reliability, for convenience although many factors occur in more than one group.

4.1 People

(a) **Eyesight:** Obviously, as 99% of inspection is visual, the vision of the inspector must be considered of prime importance.

The Questionnaire revealed that 50% of the sample of inspectors could not read typescript more than 4 feet away. the likely effects of this are easily imagined in restricted areas, where inspection is only possible obliquely or at a distance and in cases where unsatisfactory access may lead to distant viewing. Similar comments may be made for short distance vision defects. No correlation was seen between this simple vision test and whether an inspector wore glasses or had visited an optician!

Consultation with professional optical advice suggested that a set of vision requirements could be tabulated for inspectors who could then be made responsible for obtaining a yearly certificate of vision and this would be covered under the NHS*.

Recommendation: Produce vision requirements for inspectors.

* The NHS no longer provides free eye testing.

12

(b) **Motivation and Attitude:** On the whole the motivation of the inspector is rarely seen to be a factor which adversely affects reliability. He is, in the overwhelming majority of cases, dedicated to aircraft and enjoys his job.

The inspector's attitude can be affected by a multitude of factors such as his bodily comfort on the job, home matters, operator attitudes, wages, health, etc..

The authors' experiences and the questionnaire answers indicate that, as with motivation, attitude has little or no effect on reliability. The reason for this seems to be that the major effect that any factors adversely affecting the inspector's attitude towards his job increase his inspection time and/or his snagging rate with the consequent extra cost to the operator.

A major cause for dissatisfaction is the tremendous gulf between aircrew and ground staff conditions and remuneration. This seemed out of proportion to their respective responsibilities. The operator could alleviate this to some extent by providing comparable facilities in the crewroom, workshop, etc.. Minor perks, such as perhaps the aircrew's duty-free allowances, could be offset by the stores buying an inspector's DIY requirements at trade prices.

(c) **Training and Experience:** On the whole, the approach to engineering training seems first class in the aircraft industry with a good balance between knowledge of engineering as a whole and particular aircraft.

Whilst manufacturers are making the new generation of aircraft easier to maintain, there remains the need for a thorough knowledge of particular types by an individual inspector and the authors feel that the abandonment of 'with type rating' could lead to a lessening of inspection standards and, without doubt, a lengthening of the period during which the inspector gains his experience because of the lack of formal training.

Examinations on types have the effect that the inspector has to 'get it all together' for one occasion. In this way he sees the total picture in a way that he will not if experience is gained piecemeal.

Many comments have been made throughout the work that the growth of experience is to some extent limited unless the man at the 'sharp end' is kept in the picture with regards to inspection and maintenance happenings elsewhere in the industry.

To some extent this involves feedback of the effects of a man's own major defect reports, i.e. what was the eventual effect of informing the manufacturer, what did QA do, did the schedule get amended, was it really a dangerous occurrence? This is dealt with in the next sub-section.

13

It seems that there may also be a need for some form of on-going training. Inspectors would welcome a few hours a month spent refreshing their memories on specific areas or informing them of changes in the schedules and the reasons.

Demonstration of new equipment for both inspection and engineering could be arranged. this might help rid the industry of its cache of poor torches, mirrors and other tools, especially excessively noisy ones.

(d) **Feedback:** The efficiency of an inspector depends, amongst other parameters, on his basic training and his subsequent experience. the latter increases especially with regard to the number of defect occurrences he sees.

The quality of his judgement in any one situation depends not only on the times he has seen it before, but on the reinforcement he has received through feedback from those concerned with the subsequent analysis.

There seems little effort to inform the inspector of the results of this analysis. Naturally he discusses the defect etc. with his foreman/quality control man etc. at the time, but there seems to be little effort to reinforce his experience in a quantitative way in the light of, say, the manufacturer's report.

An inspector, if he finds a 1" crack in a part, may have a 'feeling' that this is nasty and should be eradicated. After the manufacturer has looked at this he may say for instance that even a 12" crack in this place would propagate too slowly for replacement to be necessary. Conversely, he might be of the opinion that 0.1" was too long and only a miracle......etc.. If the former is true, the operator could be saved expense; if the latter, the inspector will know to look critically next time.

One useful idea might be that video equipment could be used to circulate a 'visual' of inspection procedures. This would be especially useful where a procedure was difficult or lengthy to describe. In this way, an experienced inspector will be able to demonstrate and spread his personal skills more widely amongst other inspectors.

These video-tapes could be circulated within a company or even by a manufacturer to their customers and could form a solid and useful basis for 'refresher' training.

A major fault usually percolates through the systems rapidly but it is also the small defects which may one day, if continually missed, prove significant.

In connection with this, and properly classed as feed-forward rather than feedback, a case arose during the task description work where an inspector saw a small defect and decided that this could wait until the next check before action need be taken.

14

Whilst we are assured that all flaws/cracks in primary structure and all corrosion anywhere are rectified as and when found, it remains a fact that more defect cards arise during minor checks taken with block hangar checks than when only the minor check is done. The conclusion is that rectification of minor defects is postponed.

Although any increase in paperwork is not popular, it is our feeling that something similar to that arising for a defect should be raised for these minor postponed defects. A simple card system would suffice. On the next relevant inspection these 'watch-out' cards could be added to the other jobs carried forward on flight and tech logs etc..

Recommendations:

(i) Final CAA decision etc. on MORs should percolate to inspector level within operators.

(ii) Formation of a video unit to record major structural defects for circulation to other units and for record purposes, and for refresher training.

(e) **Inspector Quality:** It is thought that the quality of inspection is not affected at present by wage rates, promotion prospects or bonus payments. However, there is evidence of a gradual decrease in relative status, rates etc. which may, if prolonged, lead to a lowering in inspector quality by encouraging departure from the industry and, worse still, discouraging the brighter school leaver from coming into the industry.

The road to qualification is not easy and is relatively low paid. Training costs the man in both leisure time and money. If he sees the cleaners in the local motor factory earning similar wages he may not feel like making the effort.

The inspector needs more respect and better working conditions than are currently visible to the authors. In some operating companies there is insufficient room for him to do his paperwork, poor washing facilities and because he does not actually 'do' something or produce a result, he is taken for granted.

The inspector is as much a tool as his torch and needs looking after for best performance.

On-going training should be the rule rather than the exception. More formal data concerning his own operator's maintenance problems and those of others would be of benefit throughout the industry.

Above all, the inspector feels on the side of the aircraft and not the operator. This means he is 'with the CAA' and this independence should be fostered by contact. The average inspector does not like the possibility of gradual transference of power from the CAA to the operator upon which he is dependent.

There is a dichotomy of views in the industry over the inspection and supervision roles. It almost seems that operators with inspector/supervisors are splitting the roles and those with separate parts are amalgamating them.

Most inspectors would prefer separate roles for two main reasons. First, that having the joint role interrupts inspection continuity and secondly, that one role tends to dominate the other.

4.2 <u>Workplace Factors</u>

These factors concern the condition of the hangar, the environment and inspection aids.

(a) **Access:** Repeatedly the subject of access arises. Poor access can either entail difficulty in getting into the position for inspection (primary access) or difficulty seeing or feeling an area when in the best position possible (secondary access).

Ideally, custom-built staging should be provided for inspectors, giving primary access to all areas. In practice a melange of scaffolding, planks and hop-up towers is frequently all that exists.

The use of man-hoists or 'cherry-pickers' is the best way to get a man where he should be if custom staging is not available. A distance on one foot up, down or to the side can make the world of difference to operator comfort and efficiency and is not easily achieved with scaffolding.

Easy primary access must reduce error, for the inspector is left free with both hands to inspect. The authors have been on some so-called access platforms, etc., where they were concerned mostly with holding on or balancing. Under these conditions it is just not possible to push or pull fasteners and structure to the extent one needs for proper examination.

Secondary access problems can be caused by insufficient panel removal, lack of headroom or width, poor component design or location, poor inspection tools etc.. Little can be done about the first three on current aircraft but the tools can be improved and further recommendations will be made in the next paragraph (b).

Bad access costs time and shortening of time-of-inspection with consequent lesser coverage, especially in area checks.

A recommendation to improve access is easy to say but difficult to carry out without major costs or redesign and immediate cost may not be acceptable. However, the improvement in inspection and turnaround time of many subsequent aircraft being serviced should be offset against that cost.

<u>Recommendation</u>: Adherence to HSE regulations for scaffolding etc..
Improvement in general standard of staging and hoists

(b) **Lighting:** Adverse comments on lighting were found throughout our study. Sharpness of vision depends on the limit of the aperture, the size of the iris of the human eye and this is directly dependent on the intensity of light.

Good general lighting in the hangar is difficult where custom staging is not available because of the ingress etc. of aircraft into the hangar. It could therefore be possible to raise and lower strip lighting. Spotlight arrays are frequently seen but they are unwieldy and serve more as a substitute for torches; they are an aid mostly to primary access. Poor lighting reduces the inspector's concentration and may induce minor headaches also.

<u>Recommendation</u>: Improvement in hangar lighting near aircraft if required.

(c) **Noise:** For the majority of inspectors, noise has a major effect on their efficiency, reducing concentration, creating tension and shortening the effective inspection time.

The prime sources of noise are auxiliary power units (APU) and riveting guns. Better co-ordination with inspection periods and choice of testing areas for APUs and other engines would reduce this hazard, as would better silencing and co-ordination of operating times of noisy equipment; a noisy hour might be instituted once a day.

There is no doubt that an inspector inside a wing tank on the outside of which there is a riveting gun operating is receiving far more noise than that recommended in the HSE Code of Practice. Even levels that are permitted may not be low enough to enable sensible inspection decisions.

<u>Recommendations</u>:
(i) test engines away from hangars
(ii) install silencers on equipment and hydraulic pumps near inspection areas

17

periods to avoid the simultaneous operation of noisy tools

(iv) adherence to, at the very least, the HSE Code of Practice concerning noise

(d) **Equipment:** Inspection equipment can be divided conveniently into two parts, electronic and manual.

In the electronic section, e.g. Ultrasonic, eddy current, X-ray etc., the systems are well chosen because of cost, regularly serviced, and their sensitivities checked daily as a matter of course. Discussion of this aspect must be left to another time and place.

The manual systems are not serviced or calibrated in the same way. The equipment used by the inspector to aid the actual inspection is mostly simple; a torch, a swivel mirror, a hand lens, a penknife and a coin (to check for delamination). With his eyes and fingers these 'tools' make up 99% of the inspection kit.

In general, the lens, knife and coin do their jobs well; the torch and mirror less so.

The light pattern of the average inspection torch we saw was mottled in intensity, dim and frequently had a dark centre. Coupled with torch cases that get sticky and soften in 'skydrol' or break if dropped, it is a poor display. It is no defence for the company to complain about good torches 'walking away'.

The mirror was often too small (except for very awkward access) of poor reflecting ability (due to dirt, cracks, worn coating etc.) and frequently had a loose swivel. Inspection 'backwards' through a mirror is difficult mentally at the best of times; through a glass darkly, impossible.

It cannot be stressed too highly that eye-mirror-torch is the spearhead of the inspection process. Critical structure is, by the nature of aircraft design, not often on the surface but buried in skin, cables, pipes, control rods, sensors, pumps, actuators, etc., etc....

Good secondary access by reflection is essential. A quality standard needs setting for the torch/mirror (or useful combination).

<u>Recommendations</u>: (i) Equipment evaluation exercise resulting in a code of practice for torches incorporating case material, beam intensity, uniformity of intensity etc.

(ii) Similar for mirrors

18

(iii) Provision of these by engineering company, not the inspector

(e) **Safety:** This is a contributory factor to inspection reliability in so far as a comfortable, 'safe' inspector is likely to do a more efficient job than if he feels in danger of falling or being fallen on.

Insufficient attention is paid to primary access safety; one support at each end of a plank 20 feet from the ground is insufficient and the absence of toe-boards seems to be 'de rigueur' on scaffolding.

In passing, it is worth mentioning one point which could affect continuing structure reliability. the absence of any apparent swarf control procedures during rivet removal causes surface scratching on structure and the possibility of mechanical control blockage; the finals and functions procedures do not necessarily reveal the presence of old rivet heads, stems, etc., which may vibrate into the wrong position during subsequent flight.

Although wing and tail plane surfaces may be covered during work above them, this covering can be subsequently moved before that work is finished, e.g. for a final inspection or by wind if the hangar door is opened, etc..

<u>Recommendations</u>: (i) Better safety regulations, especially with regard to access platforms, scaffolding, toe-boarding, plant support.

(ii) Better general tidiness of fitters.

4.3 <u>Scheduling and Paperwork</u>

This paragraph contains factors affected by maintenance scheduling and the associated paperwork.

(a) **The Maintenance Schedule:** This is seen to be a document of great complexity having been built up from so many sources, e.g. manufacturer's initial data, operators' experience, various aviation authorities, planners, QA, SBs and so on.

The possibility of error in this complexity is self-evident and will exist in spite of the great and conscientious efforts directed towards it.

Simplicity of instructions should be aimed at in MS documentation; it is no good saying or thinking 'the inspector will know what I mean'. Some instructions and diagrams are

19

difficult or at least time-consuming to interpret and should be simplified. The topic concerning area checks versus specific highlighting is discussed in paragraph 4.3 (c).

One error which was missed on at least three levels was described to us. Here, the MS omitted the instruction to remove an access panel and a grease nipple was missed resulting in a seized tailplane bearing.

A greaser and inspector subsequently failed to notice on twelve separate occasions that one nipple out of three to be greased was inaccessible through the access panel. Their main error was that they considered the MS to be infallible!

Other MS errors likely to occur, besides typographical errors, are due to maintenance planning departments failing to call up the procedure relevant to the check being performed.

Although the authors are aware of all the drawbacks of the computer, it is their opinion that scheduling should be computerised whenever economically feasible. The main drawback can be quick access to the information unless presentation and recovery of the data is made of prime consideration, as should be ease of modification of the schedule.

It is essential that the programmes used should be written by someone with maintenance planning experience even at the expense of a less 'elegant' final format.

Comments on structure sampling are made in paragraph 3 (d).

Recommendations: (i) Aircraft checks should be compiled with the aid of computer data banks.

(ii) The MS should be checked against the aircraft periodically, e.g. after a major type alteration or a set number of modifications.

(b) The 'Card' System: This is a method by which the inspector received his instructions. It generally consists of a series of cards carrying explicit or general area checks to be made.

Several minor problems could arise by misuse of the card system such as loss of a card or a failure to check against master listing. Existing checks, if effectively applied, will mean that these will only result in inconvenience and cost to the operator.

However, the loss of a page of the usual rough note book or an interruption or series of interruptions in work pattern, e.g. a fitter requiring advice on a separate area of the aircraft, QA/CAA personnel wanting information etc., could cause an inspector to 'remember' wrongly when filling in a card some time after the inspection.

20

Separation of the inspection and supervision roles either by having separate personnel or by allocating 'supervision-free' periods would go a long way to preventing inaccurate reporting as well as helping preserve the concentration of the inspector. Only the difficulty of instituting such a step prevents the authors from making it an itemised recommendation.

Recommendation: Issue of 'official' bound rough note books for initial note of defect before recording. Note that 'perfect' binding by edge gluing sheets is not sufficient.

(c) **Area Check or Explicit Instruction?:** Throughout the investigation it was clear that there was some debate whether area checks were wholly valid. All major structural parts and/or joints are explicitly defined by schedule references but an area check implies look at everything.

It is obviously not desirable or necessary to explicitly define every pipe-clip, row of rivets, cable, pulley, etc., to be inspected but inspectors would like more highlighting of all components/structure that have caused trouble in the past and not just the major problems; these get an SB or similar in any case.

Component history analysis and MOR searches will not highlight a cable which is always found to be frayed or a clip always broken. The individual inspector has not the time to report this sort of occurrence nor perhaps the knowledge that this minor fault is repetitive. A system exists which highlights this sort of defect; it is by waiting until the stores runs out of the item/repair kit etc.. A better system of written 'one liners' would help prevent this time-consuming type of error. The scheduling computer recommended in paragraph 3 (a) could be used for this. It could 'automatically' collate areas with the relevant defect for easy display to future inspectors in order of number of occurrences.

Recommendation: A simple 'imminent defect' file collated by area is advocated. Possibly colour coded area sketches could highlight problem areas.

4.4 Aircraft

In this paragraph the factors affecting inspection, etc., of the aircraft itself are considered.

(a) **Labelling:** During the inspection task the inspector is frequently presented with an instruction to check, say, the area bounded by two bulkheads. These are numbered in the instructions and in the diagrams which, with some loss of time, the inspector can/obtain.

21

When he gets to the aircraft, entry may be into a gloomy hole by small access panel and the room to manoeuvre small. The inspector may have not inspected this area for some time and an aide mémoire, such as numbering on the bulkhead, would reinforce the location identity.

Similarly, critical structural areas might be labelled or even patch-coloured to ensure their being correctly located and identified at inspection time.

The authors realise that most inspectors are very conversant with the type they are inspecting but any relaxation of 'type rating' on the licence would increase inter-type mobility and consequently less familiarity.

Recommendation: Labelling of aircraft locations, datum points and critical structure.

(b) **Crack Sizes and Fracture Mechanics:** During the analysis of MOR and then during the co-ordination of the Questionnaire, much data was amassed concerning detectability of cracks.

The Questionnaire answers suggested that a flaw size of ¼" - ½" depending on area condition etc., was on the limits of visual detectability. A histogram of flaws found during the MOR exercise shows a very similar result.

The scope of the report did not allow the authors to analyse the data more fully from each occurrence and therefore the relationship between the crack size and the material geometry was not obtained. had it been possible to extract this data easily from the files, i.e. total cross-sectional area and shape together with part geometries etc., then some guide as to the seriousness of the crack could have been made by relating these, through simple fracture mechanics calculations, to the forces experienced by that part.

In general, the reply to a query sent to the manufacturer concerning a defective part, area etc., is solely how to repair, replace, etc., and contains no data as to the criticality of this defect.

Not only should the large dramatic failures be assessed but the lesser everyday defects, for it is these that may point to overstressing in the area and possibly more serious failure later.

Recommendation: MOR computer files should contain such defect data as would allow fracture mechanics assessment where relevant.

22

(c) **Physical Obstructions:** This is a general point that, at the design stage, pipework, cabling, trim etc., should be located/bent/routed so that primary structure is left accessible to the inspector wherever practicable.

A frequent complaint arising from the Questionnaire etc., was of poorly designed oil-sodden soundproofing etc., preventing easy access to structure especially within the passenger compartments.

Over-enthusiastic use of polysulphide sealant and anticorrosive agent was also a problem which may disguise flaw indications. There is also still controversy on whether such agents act as crack-growth accelerators through a stress-corrosion mechanism.

(d) **Sealing of Public Areas:** A frequent sight during maintenance is exfoliated material being replaced especially around the toilet/kitchen areas. Although some effort is made to seal off these areas from the structure with plastic sheet interfaces, it is clearly ineffective.

It does not seem improbable that, one day, instead of a floor support being exfoliated, it will be the pressure hull. A small exploratory study of the diffusion through various polymers of the relevant effluents etc. from these areas could be made.

Recommendation: Check diffusion of toilet/kitchen area effluents through various polymer sheetings to ensure applicability as area sealants.

(e) **Cleaning:** Much is said concerning cleaning of aircraft structures/areas throughout this report. Poor inspector/cleaner liaison can result in defects obliterated or more difficult to find. Initial corrosion products could be removed, leaking cracks disguised.

Shortage of time allotted in the programme for proper cleaning is a frequently-heard complaint and it is the opinion of the authors that more detailed cleaning procedures should be laid down with final surface appearance graded such as visually dirt-free, lubricant-free, all corners penetrated etc.. Cleaning should be considered as a proper contributory factor to reliable inspection.

Recommendation: Recognition of cleaning as an essential adjunct to easy, reliable inspection by more detailed procedures.

23

4.5 Quantification of Reliability

The most critical part of the inspection process is the visual scanning and defect recognition elements of the inspection task. any efforts to improve the reliability of inspection should therefore concentrate on this.

A quantified estimate of inspection reliability is impossible at present. Not only are there inherent difficulties but also the type of data available is inadequate.

A 100% reliability inspection is, by definition, one in which all existing defects affecting airworthiness are found and reported. Unfortunately, there is no way of knowing the total population, only those found.

An alternative approach to quantification of the reliability of inspection is through the fault tree and task analysis approach. this could be done if the reliability of each task could be estimated. The total reliability of the inspector of an aircraft could then be obtained by summing the reliability of the individual tasks. However, the important variables affecting the reliability of each task, i.e. the probability of looking in the correct location, the possibility that the defect is visible and the probability that it will be recognised, are also very difficult parameters to estimate and much research work would be required if progress is to be made by this line of approach.

Recommendation: Research studies on:

(i) Defect visibility with particular emphasis on the influence of access, range and lighting.

(ii) Defect recognition with particular emphasis on psychological factors such as expectation, anticipation and perception.

4.6 Reliability Data

In common with many other reliability exercises, the dearth of data of the right type was a severe hindrance to the quantification of the reliability of inspection. As mentioned in the section on analysis of MOR data, there is very little data collected which contains information on the size/extent of a defect, how it was found, what access problems there were and in some cases even, when it was found. A different type of data bank from those currently existing to store MOR data is required for inspection reliability purposes.

24

Recommendation: A small pilot project to devise a defect data collection system and a trial period at one of the operators.

4.7 Final Comments

The foregoing conclusions and recommendations take, on the whole, a broad-brush approach. We saw many instances where we would be tempted to suggest small changes in division of tasks or layout of instructions. However, it is not our belief that minor tinkering will rectify the very small 'unreliability' left in the system. The answer may lie in the 'new' computer technologies but we would stress the importance of having the programmes written by experts in aircraft and not in computing - there is plenty of room in modern computers for the 'less than elegant' programme but no room for less than total knowledge of the aircraft.

5. The Civil Damage Tolerance Requirements in Theory and Practice

David Broek

FractuREsearch Inc.
9049 Cupstone Drive
Galena, Ohio 43021

Summary

This paper presents a brief review of the Damage Tolerance Requirements (DTR) for civil aircraft, and of their impact and significance for both newly developed and ageing aircraft. Some improvements to the DTR are suggested. In principle, fracture control of aircraft is ensured by non-destructive inspection, although destructive inspection has been suggested for certain ageing aircraft, as will be discussed briefly as well.

Introduction

Fatigue cracks do develop in airframes despite careful detail-design. As crack initiation takes time, cracks tend to occur more frequently in older airplanes. Cracks impair the strength, but safety is ensured if they are detected and repaired before reaching the largest permissible size. The size of the detected crack is of no relevance if much larger cracks can be tolerated: when the structure can sustain a 20-inch crack, there is no reason for alarm if a 10-inch crack is detected.

One manufacturer has reports of 35,000 detected cracks [1]. Obviously, all of these were detected timely. On the other hand, since WW II well over 200 airplanes had fatal accidents due to cracks [2]. This shows that cracking is not limited to old airplanes: the Damage Tolerance Requirements (FAR-25), implemented a decade ago, do serve a purpose. If inspection procedures and intervals are set in a rational manner, it matters not when cracks occur, because they will be timely detected. Then there is no need to restrict the service life. Provided they are found, it does not matter by which means cracks are detected. If small cracks already present a problem, more sophisticated techniques than visual inspection must be used, such as ultrasonic or eddy current inspection, all ***non-destructive***. Another alternative is proof testing, a ***destructive*** inspection [3,4].

The Damage Tolerance Requirements and Their Use

The DTR, covering less than 1 page in the Airworthiness Requirements of FAR 25, can be summarized in two sentences. In essence, they stipulate (1) that the residual strength under the presence of cracks may not drop below a specified level (generally limit load, but for fuselages 1.1

Springer Series in Computational Mechanics
Atluri, Sampath, Tong (Eds.)
Structural Integrity of Aging Airplanes

times operating pressure plus aerodynamic suction, and/or operation pressure plus suction plus flight loads), and (2) that inspections must allow crack detection before the size associated with the above stress level is reached. ***MSD is mentioned in the DTR.*** The crack size associated with the above permissible residual strength is called the ***permissible*** crack size. The following paragraphs show how the DTR are applied.

Airframes are designed to have sufficient strength to sustain the highest service loads with a safety factor of 1.5. When cracked, the structure has lower strength: the longer the crack, the less the residual strength, as shown in Figure 1a. A cracks grows slowly by fatigue. If left unchecked it will eventually reach critical size. It must be detected and repaired before it reaches the ***permissible*** size, which is ***smaller than the critical size.*** The DTR specify the minimum permissible residual strength. The manufacturer calculates the residual strength diagram as in Figure 1a, and finds the permissible crack size on the basis of the DTR. The (smallest) detectable crack is defined here as the 'detectable size', but a more precise definition of detectability appears later.

Figure 1b shows how a fatigue crack grows from 'detectable' to 'permissible' size. At an inspection at time ***t1*** (Figure 1b), the crack - being below 'detectable'- will not be found. It reaches permissible size at time ***t3***, ***H*** hours (or ***H*** flights for a fuselage) after this unsuccessful inspection. Detection before that time requires another inspection at e.g. ***t2***, where the crack is detectable, so that it can be found and repaired. Hence, the inspection interval, ***I***, must be less than ***H***, e.g. ***H/2***. (i.e. If the crack grows from 'detectable' to 'permissible' in 10,000 hours, inspections would be repeated at 5,000 hours intervals.) Given the structural configuration and the stresses, crack growth curve and residual strength diagram can be calculated. From these and the DTR, the time ***H*** is found, and the inspection interval prescribed to ensure timely detection.

Brief as they are in comparison with the Military DTR, the civil DTR say practically everything needed. With the calculated interval of e.g. ***H/2***, a crack always will be inspected twice before reaching the permissible size, regardless of when it starts. It means that the aircraft can be used indefinitely: if cracks occur they ***will*** be detected. Should the number of cracks and repairs become too large, the operator may decide to retire the airplane for ***economic*** reasons (not for ***safety*** reasons). It is unimportant for safety whether the permissible crack size is small or large, or the inspection interval short or long. In either case there are two opportunities for detection and repair or replacement.

For civil aircraft it is in the manufacturer's best interest to make the inspection interval long, because low maintenance promotes sales. Should the interval be short, the manufacturer has various options to increase ***H*** and ***I*** (Figure 2):

- more sophisticated inspection (e.g. eddy current instead of visual) with smaller a_p .
- reduce stresses or stress concentrations to get slower crack growth.
- use material with better crack growth properties (slower growth).
- build fail-safe structures.

Fracture arrest capability (fail-safety) is provided in practically every modern jet in wing as well as fuselage. In these stiffened-skin structures arrest at the permissible residual strength level, is ensured by proper selection of stringers, frames or straps, their sizes, spacings, materials, and fastening [3]. The fuselage structure can sustain a broken frame (or strap) plus a skin crack through the two adjacent bays. Thus, the permissible crack size is on the order of 20-40 inches, making inspection relatively easy, because one should be looking for cracks on the order of a few inches in size. It is desirable that these are found and repaired; but if they are missed, an eventual fracture will be arrested at adjacent frames or straps, causing a so-called 'flap' and decompression. Though inconvenient for the passengers, it is ***not an accident***, despite the sensation it stirs. Several such incidents have occurred. Obviously the concept works, as demonstrated by tests [4-7], and flapping incidents. It did not work in Aloha's case, the reason for which will be discussed later.

Unsupportable Objections Against the DTR and Analysis

Most objections against the DTR are without ground, the one or two recommendable improvements overlooked. Objections concerning analysis concepts can be dismissed. They come from the academe claiming Linear Elastic Fracture Mechanics (LEFM) is not applicable, and are parroted by the uninitiated. Anybody calculating J_{el} (G) and J_{pl} for specific aircraft structural cases will find that J_{pl} is usually less than 0.1 J_{el}, often less than 0.01 J_{el}, and therefore negligeable, particularly during fatigue crack growth, where plasticity is extremely limited. Hence, the use of J_{el} (i.e. K) is justified, especially because the inspection interval is decided primarily by crack growth.

Fracture arrest analysis is also satisfactory. Hundreds of tests [4-7] and flapping incidents are witness that there is no need for dynamic analysis: plastic deformation of frames or straps can easily absorb the kinetic energy [3], the fracture speed in aluminum alloys (200-300 m/s) being low [8] (high R-curve).

If DT analysis is sometimes questionable, it is ***not because of alleged shortcomings of the concepts***, but because of necessary engineering judgements in the analysis. In the crack growth analysis large errors may be introduced by assumptions and idealizations; assumption of a circular versus an elliptical crack may account for a factor of 2 or 3 on life; the assumption of upper-bound rate data for a factor of 2; the neglect of varying environments at different altitudes for a factor of 2; the assumption about clipping for a factor of 2-3. Loads (stresses) and sequencing are a projection in the future (assumption), and inaccuracies here may account for another factor of 2. All these are extensively discussed in my latest book [3]. Not all the above factors will work in the same direction, but it is obvious that a factor of 2 - 3 in life (***H and I***) can be easily introduced. Academic refinements of analysis concepts, that might account for 5 or 10 per cent, can be dismissed as secondary; they would not make things better, only more complicated.

Recommendable Improvements to DTR

The analysis may be uncertain by a factor of 2, due to the necessary idealizations and judgements. This will not be changed by more sophisticated concepts and computers: accuracy will always be decided by the necessary idealizations. This is nothing new in engineering: in the static analysis of structures many assumptions and idealizations must be made as well, despite the overrated finite element analysis. One hardly hears objections here; they are worries of the past, simply dealt with by means of a safety factor. In the same vein, the uncertainties in the damage tolerance analysis must be dealt with through a safety factor. This is essentially done by taking $I = H/2$ (safety factor of 2). A ***real improvement*** of the DTR then would be a ***prescribed safety factor***, adequate to cover uncertainties in analysis.

A practical objection has been: the DTR do not work, as demonstrated by the Aloha case. This is a crude generalization: the Aloha case is a special one as discussed later. But supposing it were true in general, it would be because cracks are not timely detected. This can mean either of three things:

a. The inspection interval taken as $H/2$ was too long (e.g. the safety factor of 2 is too small).
b. The inspections were not prescribed properly
c. The inspections were not performed properly.

The reason for ***a*** is that either wrong assumptions were made in the analysis, thus overestimating H and I, or that the DTR do not prescribe an adequate safety factor. The reason for ***b*** may be due to the inspection documents or the DTR. Inspection documents are hard to read, and consist of more legalese than technical information. It is insufficient to prescribe general inspection procedure and interval. A pictorial representation should be provided, showing the inspector where to look and how, the direction of the potential crack, the approximate size of crack to look for (whether .1 inch or 1 inch), and all critical locations be prescribed instead of some being made optional. The DTR or supplementary '***Means of Compliance***' could specify the contents of inspection documents. Concerning ***c***, the operator must realize that (costly) inspections are the crucial basis of the DTR. He may reject the DTR and associated inspections (or SID), but then should adhere to the old requirements of mandatory retirement after 60,000 hours, which is costly too. Taking advantage of the DTR also has a price: higher inspection costs, better inspections, and more repairs.

Suggested Improvements of DTR

As shown, an inspection interval of $I = H/2$ is ***inadequate for practical reasons.*** It is also ***unsatisfactory for more fundamental reasons.*** Consider the two cracks in Figure 3. They have equal inspection intervals $H/2$. At any inspection the case 1 crack is always longer than the case 2 one, making the case 2 crack more likely to be missed. The same would hold if $I = H/3$ or $H/4$. Thus, larger safety factors would ***still*** lead to 'safe' and 'unsafe' cracks. A more rational approach to determine inspection intervals is available [9,10,3].

As shown in Figure 4, the probability of detection rises from zero for small cracks, to approach 1 for large cracks. More sophisticated techniques provide higher probability curves. The probability also depends on ***accessibility.*** When the location is difficult to reach, the probability of detection is lower (Figure 4). Equally important is the ***specificity*** of the inspection. If one location, e.g. a door corner, must be inspected the probability of detection is higher than if 1000 fasteners of a splice must be examined (Figure 4). The curves in Figure 4 are decided by 'human factors', but there is no need for esoteric terms: curves as in Figure 4 can be obtained from analysis [9,10,3] of ***available data,*** be it that additional data are desirable.

The probability of detection being less than 1, any crack may be missed at any inspection. At the next inspection the crack is longer, the probability that it is missed lower. The ***cumulative*** probability that a crack will have been detected after n inspections is [9,10,3]: $P = 1 - \prod q_i$, where q_i is the probability to miss in inspection i. With probability curves for specificity and accessibility, and the calculated crack growth curve, the cumulative probability of detection can be calculated, and the inspection interval providing e.g. a 95 per cent cumulative probability of detection - or any other probability desired - can be determined. As shown in Figure 5, this leads to a shorter interval of the case 2 crack in Figure 3.

The DTR would be greatly improved if they prescribed that the inspection interval be determined to give a certain cumulative probability of detection. It would ascertain consistent safety levels. This procedure has been applied already for the certification of aircraft. As the damage tolerance analysis of an airframe may require 50,000 man hours or more, the primitive use of the results by using $I = H/j$ is unsatisfactory.

Multi-Site-Damage (MSD)

The DTR (and SID programs based on the DTR) provide for indefinite use of (ageing) aircraft. This is ***at the price of higher maintenance, without affecting safety***; the alternative is the higher cost of retirement.

Things went wrong in the Aloha case for very specific reasons. The fuselage lap-splices in older 737 models were joined with a cold adhesive, which tends to disbond over time. Then the fasteners must transfer the load across the splice (Figure 6a), ***a task for which they were not designed from the point of view of fatigue***. This causes simultaneous fatigue cracks at many fasteners, the so called ***Multi-Site-Damage (MSD)***, which tend to link-up into longer cracks. There is no question that the fuselage can arrest a two-bay skin fracture with broken strap. However, when there is MSD ahead of such a fracture, it will ***continue*** running by 'playing Packman' with adjacent MSD, opening up the whole fuselage (Figures 6 b,c).

Whereas under normal circumstances inspection could be based upon cracks of 1 to several inches (previous discussion), it must now be based upon cracks of 0.1 inch which are much harder to detect. Moreover, instead of looking (for a ***large*** crack) at ***one*** fastener, one must search (for ***tiny*** cracks) at ***thousands*** of fasteners; such a low specificity inspection provides low probability of detection (Figure 4). Nevertheless, Aloha ***did detect*** such cracks at ***stringer 4*** and repaired them (Figure 7a). The AD ***did not specify inspections at stringer 10 where the failure occurred*** (Figure 7).

Because of the low probability of detection of small cracks at many fasteners by NDI it has been suggested to do proof testing. While with ***NDI*** cracks are betrayed by direct observation or electronic signals, in proof testing they are betrayed by structural failure (***on the ground; not in the air***). The potential of proof tests for crack detection was recognized decades ago, but its only application to aircraft has been to the wing pivot fitting of the military F-111 aircraft.

The principle of proof testing is shown in Figure 8. At a certain time a proof load is applied, inducing a very high stress. If a crack equal or larger than the 'proof size' is present, the structure fails (Figure 8a): the proof test detects all cracks larger than the 'proof size'. Cracks present after a successful proof test may well be much smaller than the 'proof size' or there may be no cracks at all. But the only certainty is that no crack is larger than the 'proof size'. As shown in Figure 8b such a crack would grow to the 'permissible size' in H hours (flights for a fuselage). Therefore, the proof test must be repeated before H hours have elapsed. The interval is established by calculating H. If there was indeed a crack equal to the 'proof size', it will grow to permissible size during the next interval, and be detected by a failure during the next proof test. If in a certain test the crack is much smaller than the proof size, it will not have reached proof size yet at the subsequent test. Then - that proof test being passed successfully - ***still*** no cracks larger than the proof size exist, and operation can be safely continued over another interval. This can be repeated until the proof test 'detects' a crack. Then a repair must be made, the damage (and crack) thus being eliminated. The higher the proof load, the smaller the crack it discovers, and the longer the interval can be. Lower proof loads must be repeated more often.

A fuselage is eminently suitable for proof testing, because it can be loaded easily on the ground by pressurization. The proof level must be as high as practical, but may be limited for the following reasons:

a. A proof test failure is justified because it prevents one in service, but preferably the failure should be reparable.
b. The test should not damage other parts, making the cure worse than the desease.
c. It should be more economical than alternatives, that provide ***equal safety***.

A rational analysis of all problems involved has been performed [4], where due account was taken of retardation and stable fracture. Results are summarized in Figures 9 and 10. These show that the proof test interval must be about 250 flights if the proof pressure is 1.33 times the normal pressure, or 600 flights if the proof pressure is 1.5 times normal. It is important that the 95 per cent survival rate in Figure 9, is for the case of a critical crack (proof size) at the time of proof testing. In the general case, survivability is practically unlimited due to retardation. However, life extension is not the aim of proof testing; crack detection is, and it dictates the short interval.

The analysis in Ref [4] considers all objections against proof testing in a rational manner. All of these appear non-sustainable. If there is one objection against proof testing it is an economical one: the interval is short, so that proof testing is costly. The alternatives are retirement or crack detection by NDI, ***which are costly too***, because the inspection interval will have to be ***short*** to be ***equally as safe as proof testing***. This matter is discussed in another paper in this symposium (Sampath and Broek), which shows that the present interval of 3500 flights should be reduced to about 1500 flights to provide adequate probability of detection (Figures 11, 12).

Aging Aircraft

The following considerations deal ***only*** with the effects of ***cracks, not*** with ***corrosion and other*** non-structural problems. As demonstrated, when the DTR are ***properly applied***, old airplanes can be equally as safe as new airplanes and can be used indefinitely. Older aircraft will develop more cracks and require intensive inspections and repairs, but that is the price for the life extension. When this burden becomes too heavy, the airplane should be retired. The decision can be left to the operator, because it is an ***economical*** one. Even in the case of MSD safe operation is guaranteed with ***more frequent inspection or proof testing***, whichever is more economical.

The occurrence of MSD is limited to early series of a few airplane types, because of specific design and production procedures. There is no reason to expect MSD on a large scale in other airplanes. A more or less uniform stress field is required for MSD. This makes it almost impossible in wings. In fuselages, with properly bonded or scalloped lap-joints, MSD is extremely unlikely. In general the damage tolerance approach is adequate, be it that some improvements in the DTR are desirable, and that the requirements should be strictly enforced. In cases where MSD is possible, shorter inspection intervals, accounting for the low specifity inspection for very small cracks, are needed.

Conclusions

1. Most objections against Damage Tolerance Requirements and Damage Tolerance Analysis are unfounded. A supportable objection is that safety factors presently used are inadequate. This can be cured by a few improvements in the DTR, namely:

a. Implementing a safety factor of $j > 2$ for the determination of the inspection interval, I, where $I = H/j$, and where j should be ***dependent upon access and specificity.***

b. Instead of a it would be ***better to*** specify a ***cumulative*** probability of detection of e.g. 95 or 98 per cent. This permits a rational prescription of the inspection interval and ensures consistent safety levels, provided due account is taken of accessibility and specificity. Manufacturers should be ***urged*** to pool their crack reports, because these provide an enormous data base for the establishment of reliable base-line probability curves for crack detection (Figure 4).

c. Making it mandatory (through a 'Means of Compliance' document) that inspection manuals show pictures of the location, direction of anticipated cracks, specify the approximate size of crack inspectors should look for, and that extraneous legalese be put in an appendix so that the inspection document becomes the technical document it is supposed to be.

d. Operators should be made aware that indefinite use under the DTR is not for free; the cost is more maintenance and inspection; the alternative is to buy new.

2. The aging aircraft "problem" is but a specific case of general damage tolerance. It can be dealt with by established methods without special research and large public expenditures. The problem is ***one of economics not of structural safety.*** It is a manageable "problem", ***made*** to seem overly important because of a specific incident, due to MSD in a few aircraft types and serial numbers, which can be corrected rationally.

References

1. Boeing testimony, NTSB hearings, Seattle (1988)

2. Campbell, S., A survey of fatal aircraft accidents involving metal fatigue, National Aeronautical Establishment, Ottawa, LTR-ST-1219 (1980)

3. Broek, D., The Practical Use of Fracture Mechanics, Kluwer (1988)

4. Broek, D., The feasibility of proof testing, FractuREsaerch Inc., Galena, Ohio, TR 8916; Report to USDOT/TSC (1989)

5. Swift, T., Applications of fracture mechanics to the DC-10 fuselage, AGARDOgraph 176, pp 226-287 (1974)

6. Vlieger, H., Residual strength of cracked stiffened panels, Eng. Fract.Mech.(1973) pp 447-478

7. Broek, D., Elementary Engineering Fracture Mechanics, Kluwer, 4th ed.(1986)

8. Broek, D., and Nederveen, A., The influence of the loading rate on the residual strength of alluminum alloy sheet, Natl. Aerospace Lab (Amsterdam) TR-M-2154 (1964)

9. Broek, D., Fracture control by periodic inspections, in Structural Failure, Product Liability and Technical Insurance, Rosmanith Ed., pp238-268, Inter Science (1987)

10 Broek, D., USAF conference on damage tolerance, Dayton (1986)

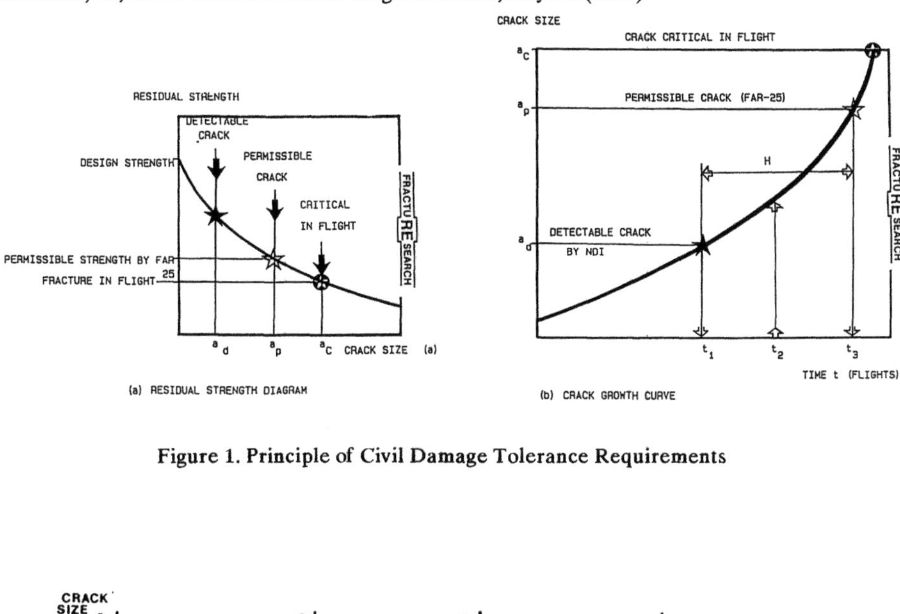

Figure 1. Principle of Civil Damage Tolerance Requirements

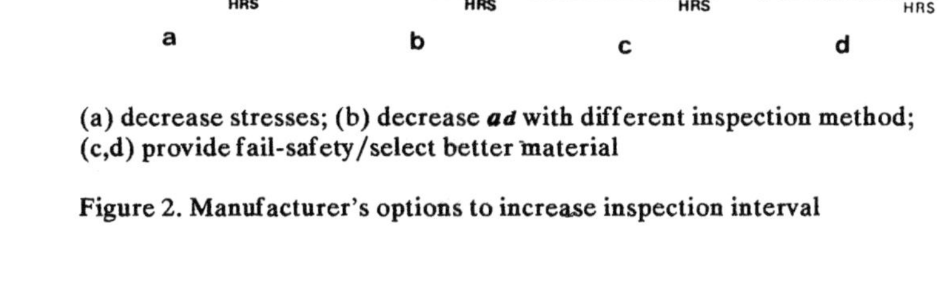

(a) decrease stresses; (b) decrease *ad* with different inspection method; (c,d) provide fail-safety/select better material

Figure 2. Manufacturer's options to increase inspection interval

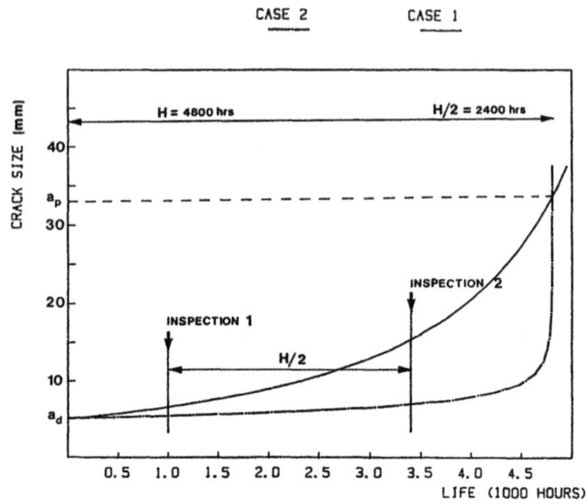

Figure 3. Two cracks with different growth curves (due to geometry)

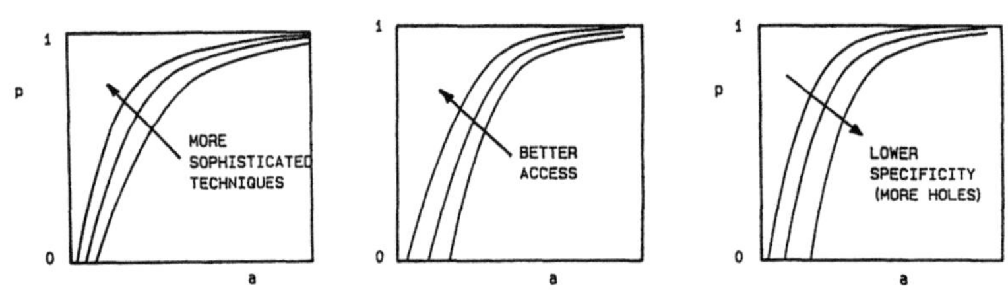

Figure 4. Effect of various parameters on probability of crack detection in one inspection

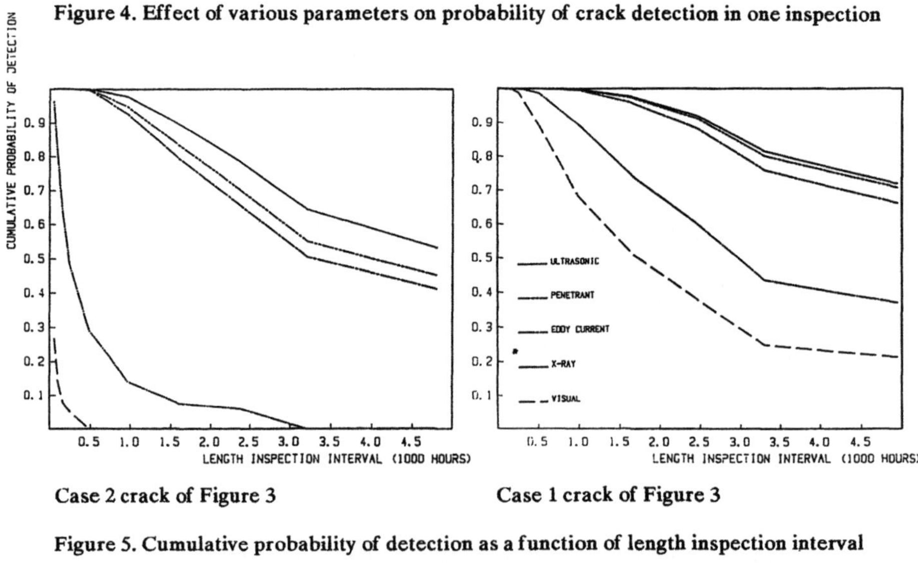

Case 2 crack of Figure 3

Case 1 crack of Figure 3

Figure 5. Cumulative probability of detection as a function of length inspection interval

a. MSD

b. Structure

c. Sequence of events

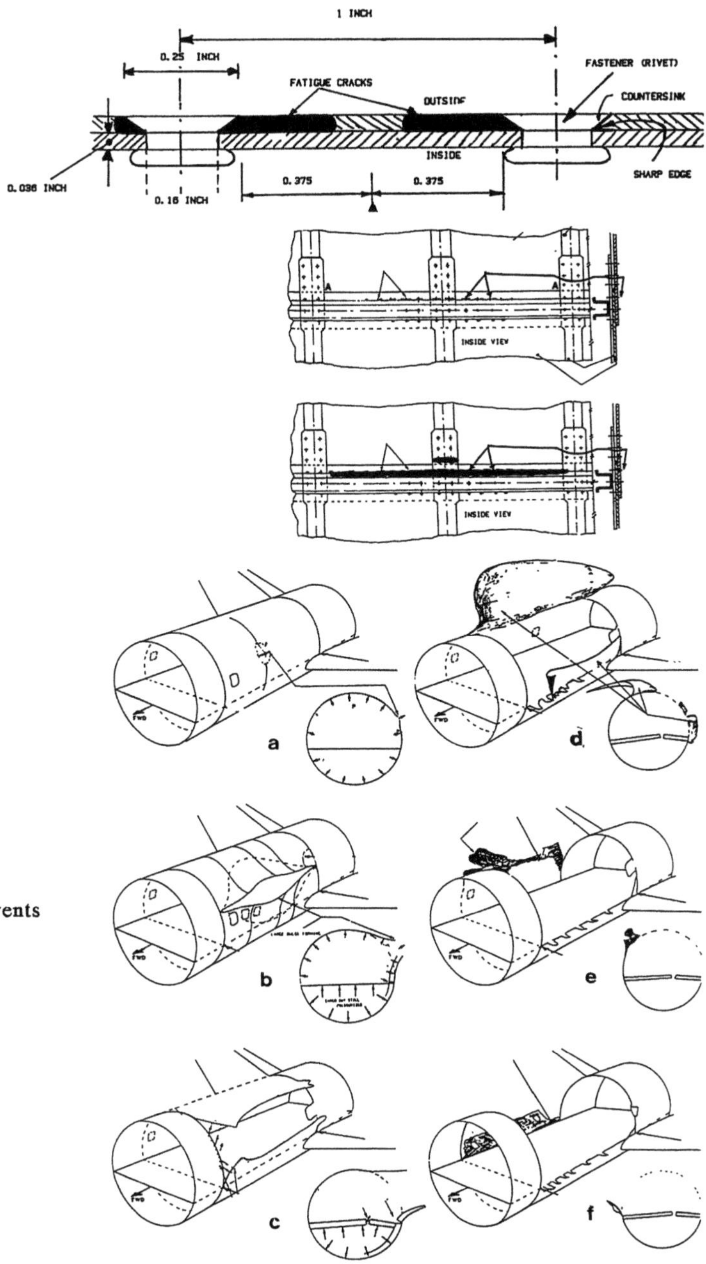

Figure 6. The Aloha case

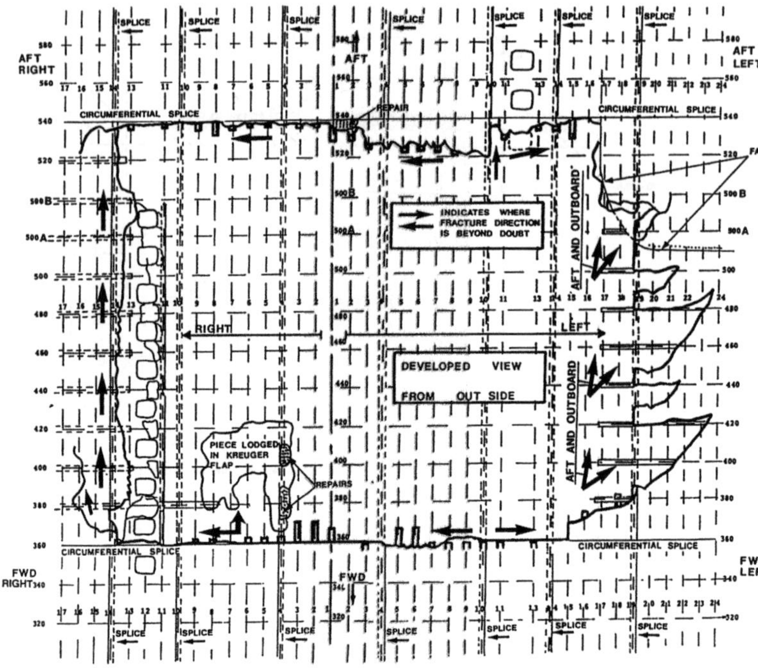

a. Fractures. Arrows show fracture directions, all emanating from stringer 10, showing that fracture started at stringer 10.

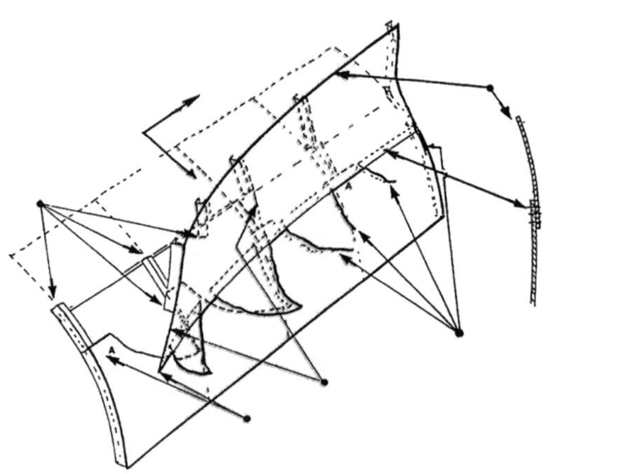

b. View of left side fracture (origin at stringer 10 at top of figure)

Figure 7. Aloha fracture originating at stringer 10

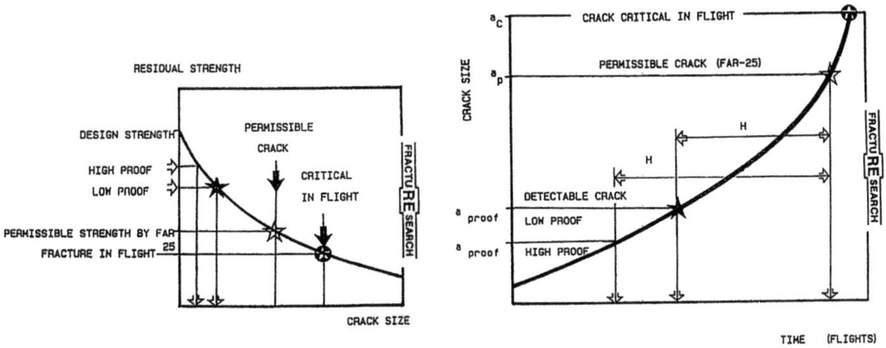

Figure 8. Destructive inspection by Proof Test; Left: residual strength; Right: Crack growth curve

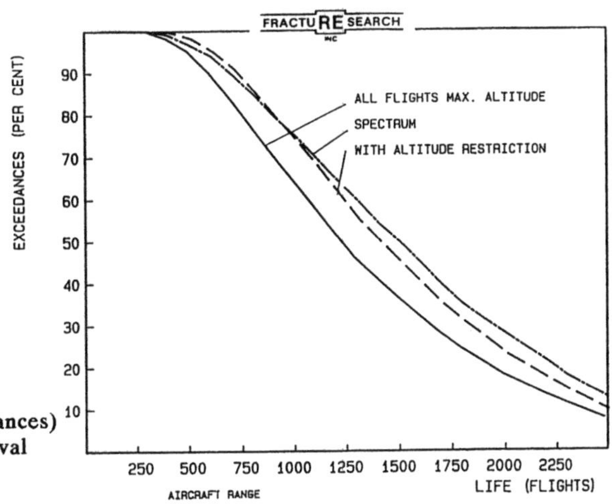

Figure 9. Survivability (exceedances) as a function of proof test interval

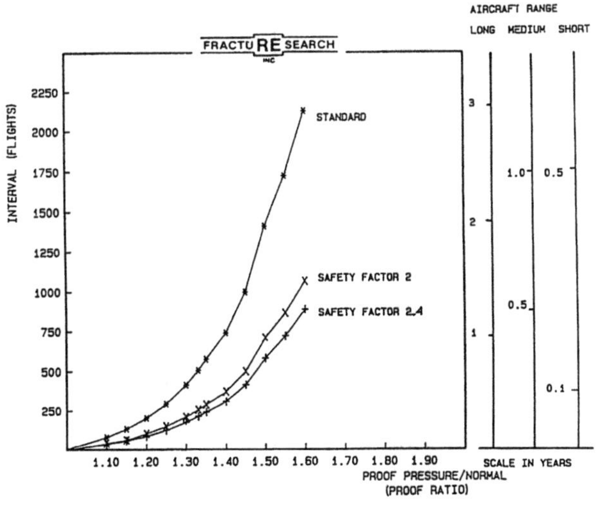

Figure 10. Required proof test interval for various conditions

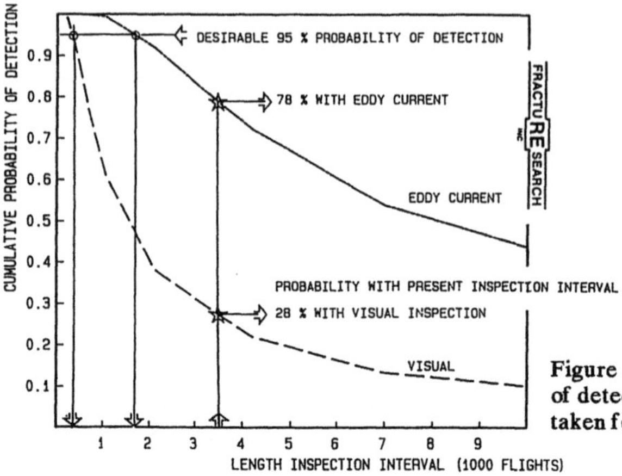

Figure 11. Cumulative probability of detection of MSD [4], if no credit taken for previously detected cracks

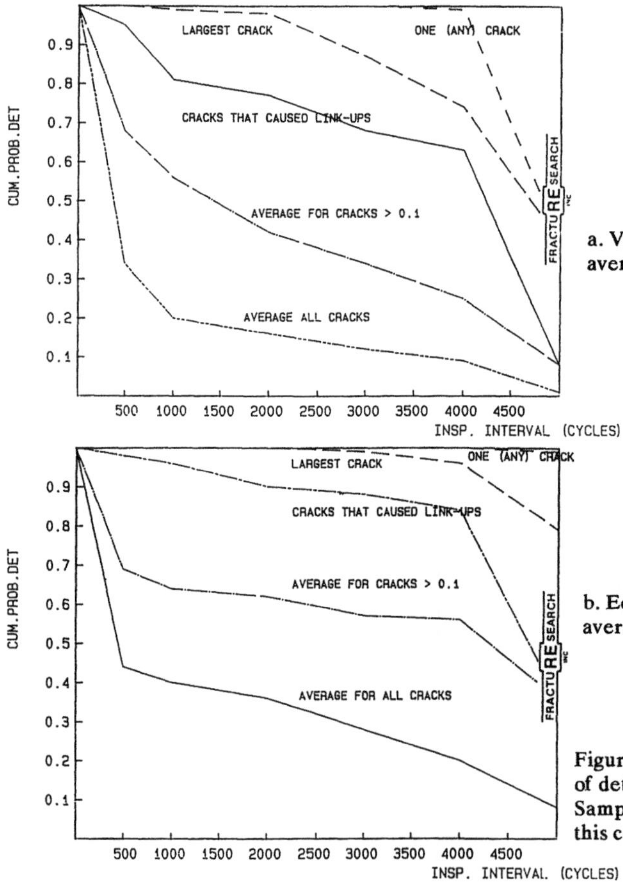

a. Visual inspection averages for 6 MSD cases

b. Eddy current inspection averages for 6 MSD cases

Figure 12. Cumulative probability of detection of MSD (see paper Sampath and Broek, this conference)

6. The Effect of Exfoliation Corrosion on the Fatigue Behavior of Structural Aluminium Alloys

J. P. CHUBB, T. A. MORAD, B. S. HOCKENHULL AND J. W. BRISTOW*

School of Industrial Science
Cranfield Institute of Technology
Cranfield, Bedford, UK

* Civil Aviation Authority
Gatwick Airport
London. U.K.

Summary

The effect of prior exfoliation corrosion on the fatigue behaviour of 7178-T6 and 2024-T351 aluminium alloys is investigated. The fatigue crack growth rate in both as received and prior corroded material is measured in both wet and dry environments. The exfoliation corrosion is shown to enhance the fatigue crack growth rate in the 7178-T6 material whereas the same effect was not observed in 2024-T351.

Introduction

In recent years the demand for aircraft has outstripped supply necessitating longer operational lives for existing aircraft. This has lead to a large fleet of ageing aircraft and therefore the need to re-evaluate the fail safe characteristics of this fleet.

In the fail safe approach it is assumed that a crack will be detected before it is likely to cause failure. The approach requires regular inspection of different parts of the airframe and the inspection intervals are based on fatigue crack growth data for the materials of interest measured in air in the laboratory. In new aircraft this laboratory data is reasonably accurate, however as the aircraft gets older corrosion of the airframe becomes a greater problem and doubt is cast on the validity of using fatigue data measured in air when analysing a structure which sees a corrosive environment.

The common way of dealing with fatigue in a corrosive environment is to use corrosion fatigue data. It is known that the presence of a corrosive medium at the crack tip can have a significant effect on the fatigue crack growth rate [1-4]. If corrosion fatigue data was to be used to determine inspection intervals for aircraft then these intervals would be reduced, decreasing the operational efficiency of the aircraft. Data from in service fatigue suggests that the rate of crack growth is below that predicted by corrosion fatigue experiments and closer to that seen in air. On reflection this may be expected because in an operational airframe the processes of corrosion and fatigue tend to occur sequentially rather than concurrently.

Springer Series in Computational Mechanics
Atluri, Sampath, Tong (Eds.)
Structural Integrity of Aging Airplanes

With this in mind the present work is aimed at determining what happens if a growing fatigue crack encounters a patch of corrosion. The type of corrosion chosen was exfoliation, because it is commonly seen in airframes and is of concern due to its insidious nature [5,6]. Two structural aluminium alloys were chosen, 7178-T6 an aluminium zinc alloy commonly used in the early jet airliners, now the ageing fleet, and 2024-T351 an aluminium copper alloy widely used in commercial airframes.

Experimental

The 7178 material was cut from the upper wing skin of an aircraft which had seen 15 years service. The plate was 2.5mm thick and contained some regions of in-service exfoliation corrosion around rivet holes. The 2024 material was supplied as a plate 12.5mm thick. In order to enable comparison between the two materials the plate was machined into three sheets each 3mm thick. The sheet taken from the centre portion of the original plate was used for the experiments to ensure that any recrystallised grains at the original plate surface were removed making the specimens more susceptible to exfoliation corrosion.

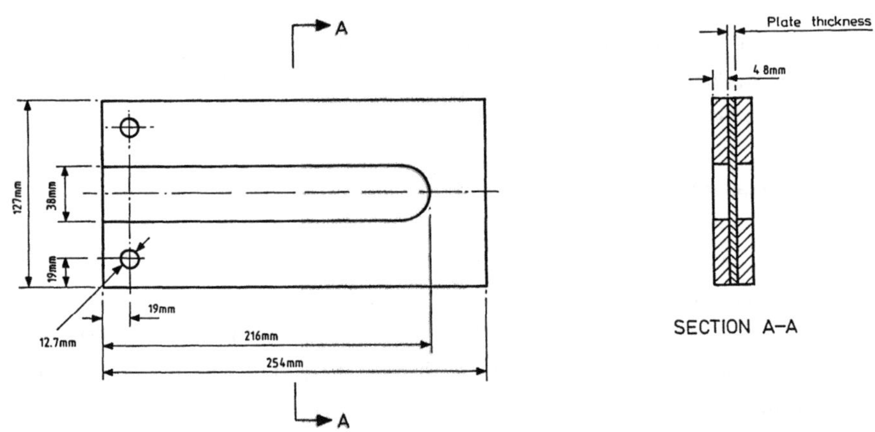

Fig. 1 The DCB specimen configuration.

The fatigue experiments were carried out using a DCB specimen configuration. Anti-buckling restraints made from 4.8mm thick 7075-T6 aluminium were bonded to the specimens using an epoxy resin. The specimen configuration is shown in Figure 1 where it can be seen that the restraints left a window of approximately 38mm by 216mm in which the crack was grown. An experimental compliance calibration was produced [7] for the specimen giving the relationship between K and crack length as:

$$K^2 = \frac{P^2}{2B^2W}[445.8(a/W)-768.3(a/W)^2+975.7(a/W)^3]$$

where K is the stress intensity, P the load, B the sheet thickness, a the crack length and W the specimen width from the loading point to the anti-buckling restraint.

Specimens were fatigued using a servo-hydraulic machine at a frequency of 5 Hz with a sinusoidal waveform and a stress ratio of 0.1. The tests were carried out in a perspex chamber to control the humidity. The chamber either contained silica gel with air being pumped into the chamber through a drying bottle to keep the relative humidity close to 0% or no silica gel was used and air was pumped through water to keep the relative humidity near to 100%, referred to as dry and wet conditions respectively. During the test the crack length was measured using a travelling microscope and the cycles recorded by an electronic counter.

Selected specimens were corroded on one side prior to testing. The corroded area ranged in size from 20 to 180mm. The corrosive medium was Exco (4.0M NaCl, 0.5M KNO_3 0.1M HNO_3) a solution recommended to assess susceptibility of materials to exfoliation corrosion [8]. The time of exposure used for the fatigue specimens, 96 hours, was chosen following corrosion immersion tests [7]. A comparison was made between a piece of material from around a rivet hole in the original 7178 sheet, which showed in-service corrosion, and a piece of the same sheet corroded in the laboratory using Exco. The resulting micrographs are shown in Figure 2.

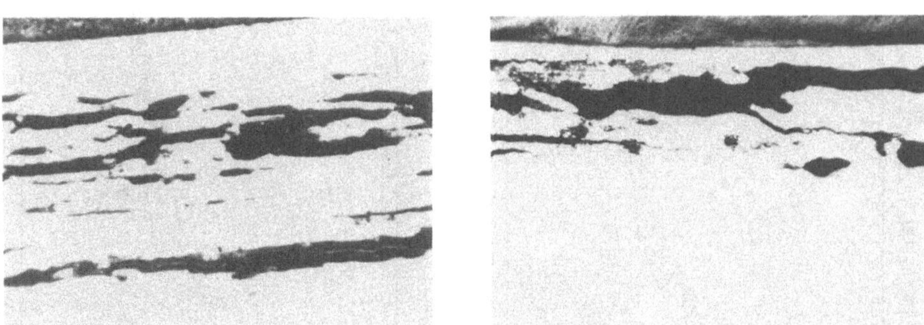

a b

Fig. 2 A section of exfoliated material: a. In-service corrosion, b. Laboratory corrosion.

Results

The fatigue data were plotted as a versus N and the best fit curve, judged by eye, was drawn through the data points. The slope of the curve at any point was found by drawing tangents. The corresponding stress intensity factor range was calculated from the load range, the appropriate crack length and the compliance calibration. The results are presented as plots of log da/dN versus log ΔK.

Fatigue results from as received 7178-T6 tested in both dry and wet air showed, Figure 3, the increased crack growth rate in the wet environment particularly at the lower ΔK values. Results from corroded material showed an increased growth rate over as received material at the lower ΔK values, both tested in dry air, Figure 4. The increase in growth rate was similar to the difference observed between as received material tested dry and wet and gave three times the growth rate at a ΔK of 15MPam$^{1/2}$. It should be noted that at ΔK values above 28MPam$^{1/2}$ the results from specimens tested wet, dry or corroded were similar.

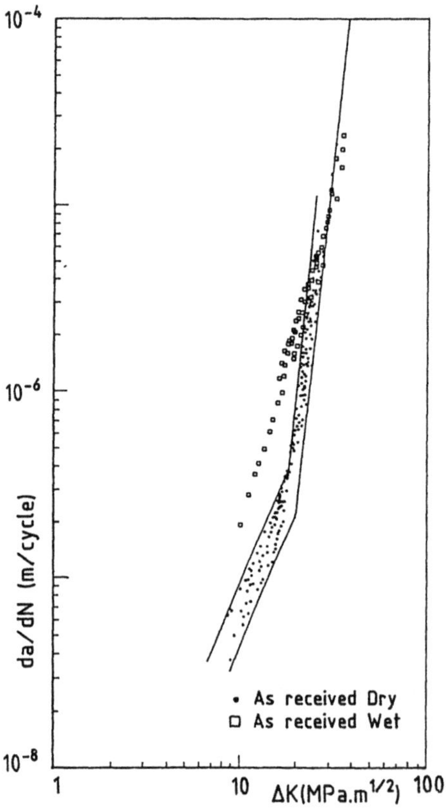

Fig.3 Fatigue crack growth data for as received 7178-T6.

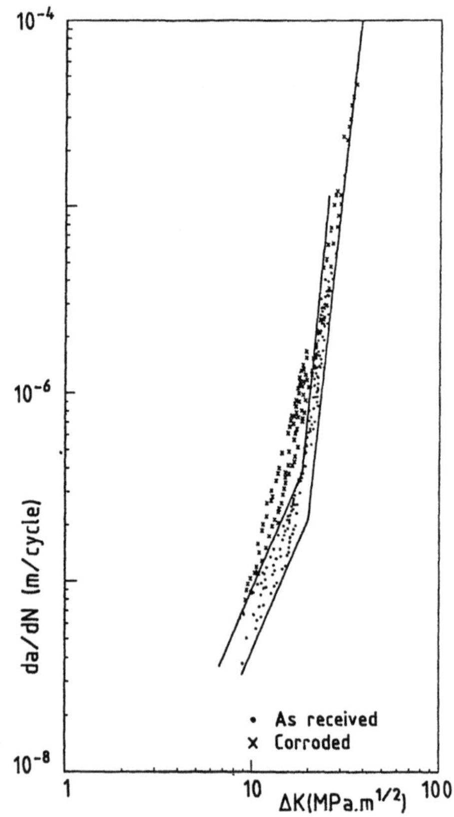

Fig. 4 Fatigue crack growth data from corroded 7178-T6 tested in dry air.

In order to complete the information, a corroded sample was tested in wet air. The results of all the tests on 7178 both as received and corroded in wet and dry environments are shown in Figure 5. The growth rates at the lower ΔK end increased in the order as received dry, corroded dry, as received wet and corroded wet, while all data were similar at the higher ΔK values.

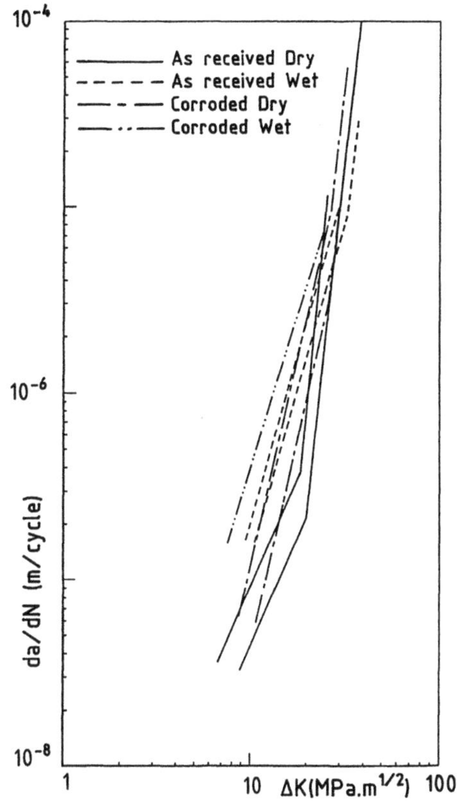

Fig. 5 Fatigue crack growth data for as-received and corroded 7178-T6.

A final experiment on 7178 involved the testing of a specimen containing a patch of in-service corrosion in dry air. The results are shown in Figure 6 along with the scatter bands for as received and corroded material tested at the same relative humidity. The results from the in-service specimen, which contained a small patch of corrosion showed a discontinuity of growth as the crack entered and left the corroded patch. This was a similar but more noticeable effect to that observed in laboratory corroded material with small patches. Subsequent sectioning of the in-service sample showed the depth of attack to be greater than that obtained in the laboratory particularly at the point furthest from the loading points and it was felt that this accounted for the more pronounced behaviour in this sample.

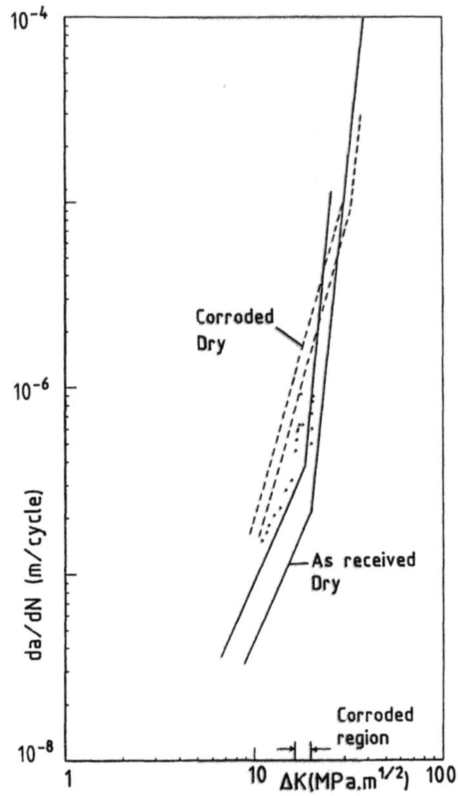

Fig.6 Fatigue results from a specimen of 7178-T6 containing in-service corrosion. The extent of the corroded patch is indicated. Scatter bands are for as-received and corroded material.

Fatigue results for as received 2024-T351 tested in both wet and dry environments showed, Figure 7, no significant effect of moisture on the crack growth rate in this alloy. Results from corroded material tested in dry air were not significantly different from as received 2024, Figure 8. However, a decrease in the crack growth rate was observed once the crack had re-entered as received material after the corroded patch. The effect is shown in Figure 9 where the results from as received specimens are compared to results from as received material where the crack had passed through a corroded patch.

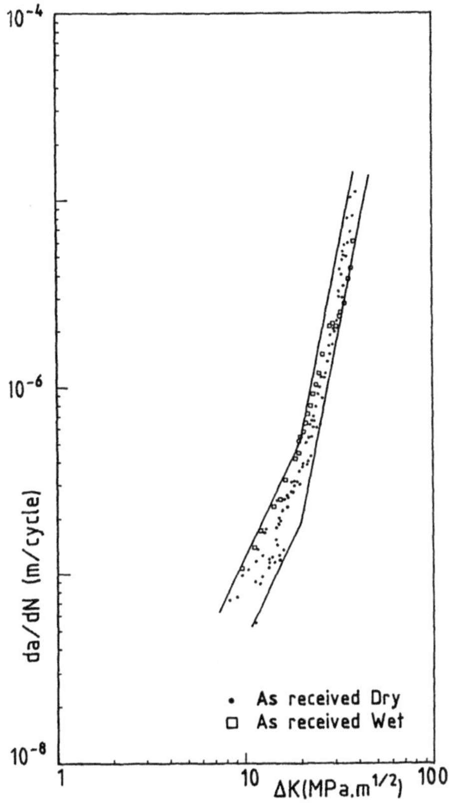

Fig. 7 Fatigue crack growth data for as received 2024-T351

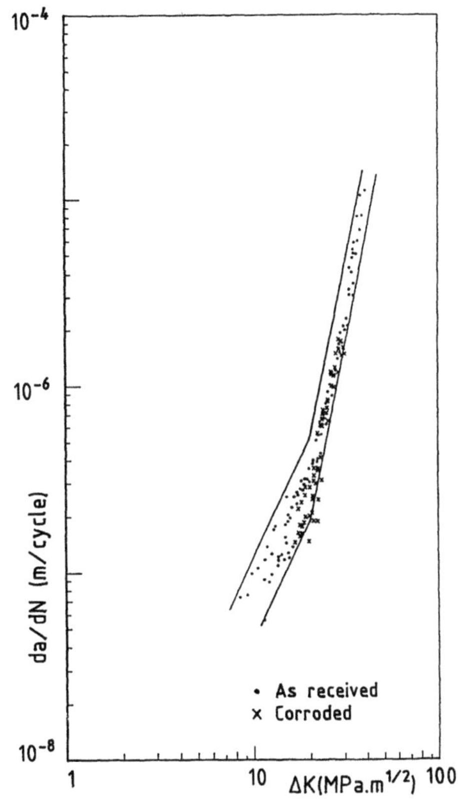

Fig. 8 Fatigue crack growth data for corroded 2024-T351 compared with as-received.

Examination of fatigue fracture surfaces in both alloys showed that the basic mechanism of crack growth at lower ΔK values was striations while at higher ΔK values ductile tearing was the dominant mechanism. The transition between the two mechanisms depended upon the testing environment and material and occurred at higher ΔK values as the humidity increased in 7178 but not in 2024.

Discussion

Examination of in-service and laboratory exfoliation corrosion has shown them to be similar,

Figure 2. Hence the results from laboratory experiments should relate to fatigue crack growth in an ageing airframe.

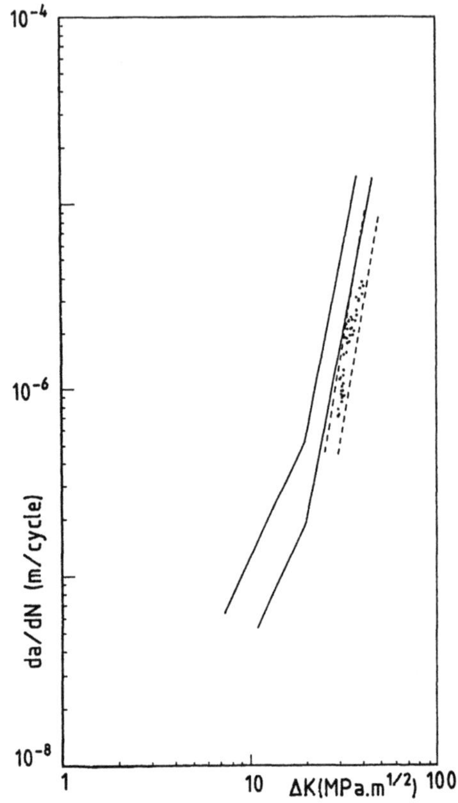

Fig. 9 Fatigue crack growth data for 2024-T351 showing a comparison between the growth rate in as received material and in as received material after the crack has passed through a corroded patch.

Fatigue results from 7178-T6 aluminium showed that in dry air the presence of exfoliation corrosion enhanced the fatigue crack growth rate by a factor of about 3 at low ΔK values, Figure 4. This was shown to be similar to the effect of humidity, Figure 5. The results suggest that the role of exfoliation was to trap moisture in the blistered material hence making the local environment at the fatigue crack tip wet. The result from a corroded sample tested in wet air showed that the role of exfoliation was not that simple because with both high humidity and corrosion the growth rate was higher than with either alone. It was also noticeable that exfoliation only had a significant effect at lower ΔK values where the fatigue crack growth was predominantly by striation formation. Once ductile tearing became dominant, the effect of both exfoliation and humidity were not significant. These observations inferred that exfoliation in 7178-T6 aluminium lead to chemical entrapment which caused a reaction at the fatigue crack tip increasing the growth rate as it passed through the corroded region. In wet environments the effect was enhanced by the additional moisture that was available.

Results from the specimen showing in-service corrosion were similar to those from laboratory corrosion with a small patch, where the crack tended to accelerate into the patch as the tip encountered the entrapped chemicals and decelerate out, Figure 6. The deceleration was modelled by mechanically thinning the material instead of corroding it and was thus thought to be due to the thinning effect of exfoliation. When the crack front plastic zone left the thinner region in the corroded patch it needed to adjust to the thicker material it encountered and hence the crack was slowed. The results from the in-service corrosion related reasonably well to those from laboratory corrosion. The slight shift of the results from the in-service sample out of the scatterbands for as received material was thought to be due to the role of rivet holes close to the crack tip. These holes could not be avoided because of the position of the in-service corrosion in the sheet supplied.

The fatigue behaviour of 2024-T351 was similar in both dry and wet air, Figure 7, in contrast to the 7178, indicating a lower susceptibility to corrosion fatigue. This is in agreement with previous work [9] that showed the effect of moisture on fatigue in aluminium copper alloys is less than in aluminium zinc alloys. The result was further emphasised by fractographic observations which showed that the change in growth mechanism from striations to ductile tearing in the 2024 occurred at the same ΔK in both dry and wet air. In common with the lack of effect of humidity on fatigue in the aluminium copper alloy there was no effect of exfoliation, Figure 8. However, results from corroded samples did show that on re-entering as received material after the corroded patch the crack growth rate was lower than if the corrosion was not present, Figure 9. It was thought that this behaviour was due to the thinning effect of the exfoliation causing extra plastic deformation at the crack tip in the exfoliated patch leading to enhanced crack closure effects [10] once the crack emerged from the corrosion. In order to test this idea two specimens of 2024-T351 were tested with regions, similar in size to the corroded patches, mechanically thinned to try to model one effect of exfoliation. The results showed that as the crack emerged from the thinned region the growth rate was retarded in the same way as following a corroded region, Figure 9. These results are based on a limited number of tests and work is continuing to further elucidate the role of exfoliation corrosion in 2024 material.

Conclusions

1. Exfoliation corrosion enhanced the fatigue crack growth rate in 7178-T6 aluminium at the lower ΔK values where the dominant growth mechanism was striation formation. The increase in growth rate was a factor of 3 at a ΔK of $15MPam^{1/2}$ in dry air.

2. Exfoliation corrosion has not been seen to significantly affect the fatigue crack growth rate in 2024-T351 aluminium.

3. The difference in behaviour between the aluminium copper and aluminium zinc alloys was due to their different susceptibility to humidity. The aluminium zinc alloy showed an enhanced fatigue crack growth rate in wet atmospheres when compared to dry showing a susceptibility to corrosion fatigue that was not as apparent in the aluminium copper alloy.

4. The behaviour of the aluminium copper alloy when the crack growth was lower than expected following corrosion was due to the thinning effect of exfoliation corrosion leading to enhanced crack closure following the corroded region.

Acknowledgement

The authors would like to thank the Civil Aviation Authority for financial support.

References

1. Williams, D. N. : Environmental Corrosion Fatigue Behaviour of Aluminium Alloys. DMIC Memorandum No.249 (1970) 14-26.

2. Bradshaw, F. J.; Wheeler, C. : The Influence of Gaseous Environment and Fatigue Frequency on the Growth on Fatigue Cracks in Some Aluminium Alloys. Int. J. of Fract. Mech. 5 (1969) 255-268.

3. Wei, R. P. : Fatigue Crack Propagation in a High Strength Aluminium Alloy. Int. J. Fract. Mech. 4 (1968) 159-170.

4. Wei, R. P. ; Simmons, G. W. : Recent Progress in Understanding Environment Assisted Fatigue Crack Growth. Int. J. Fract. 17 (1981) 235-247.

5. Mitchell, R. G. : An Airline View of the Corrosion Problem. Ind. Corr. 1 (1983) 11-16.

6. Doruk, M. ; Tufekcioglu, A. : Maintenance and the Military Environment. AGARD (NATO) May 1985 (LS 141).

7. Morad, T. A. : The Effect of Exfoliation Corrosion on the Fatigue and Fracture Behaviour of Thin Aluminium Alloys. Ph.D. Thesis, Cranfield Institute of Technology, 1982.

8. ASTM Standard G34: Exfoliation Corrosion Susceptibility in 2xxx and 7xxx Series Aluminium Alloys. ASTM Standards Part 10 1980.

9. Feeney, J. A. : McMillan, J. C. : West, R. P. : Environmental Fatigue Crack Propagation of Aluminium Alloys at Low Stress Intensity Levels. Met. Trans. 1 (1970) 1741-1757.

10. Elber, W. : Significance of Fatigue Crack Closure, Damage Tolerance in Aircraft Structures. ASTM STP 486 (1971) 230-242.

7. A Damage Tolerance Approach for Management of Aging Gas Turbine Engines

W.D. Cowie

Chief, Structures and Durability Group
Propulsion SPO
Aeronautical Systems Division
Wright-Patterson Air Force Base, Ohio

Abstract
Damage Tolerance Philosophies are being used today in the Design, Development, and Life Management of USAF turbine engines. The establishment of a Damage Tolerance Philosophy and implementation of a Fracture Control Program has been an integral part of the USAF turbine engine development process since 1978. It has been applied to new engine programs as part of the USAF's Engine Structural Integrity Program (ENSIP) described in military standard 1783 issued formally in 1984. This military standard was reviewed and approved by the Aerospace Industries Association of America, Inc (AIA) in 1982. Damage Tolerance philosophy has also been applied to existing inventory USAF engines through structural durability and damage tolerance assessments. In all, fracture control programs have been applied to the F-100, TF-34, F100-PW-220, F100-PW-229, F110-GE-100, F110-GE-129, F101-GE-102, F109-GA-100, F-119, F120 and T406 engines and have resulted in the generation of NDE flaw size quantification probability of detection data and in implementation of enhanced nondestructive evaluation methods (i.e., eddy current) at manufacturing and at field/depot. Experience has shown that these inspection have been successful in detecting early cracking and in accelerating corrective actions. It has been highlighted that several development efforts in the last five years have been identified FPI process improvements that must be implemented within industry and Air Force depots to improve flaw detection reliability. The need to quantify detection reliability for imbedded defects is also identified. The engine development process has been evolutionary in terms of application of upgraded requirements. The new process of fracture control, sometimes referred to as damage tolerance, is contained in ENSIP and selects material as well as governs design and life management. Recent experience clearly demonstrates that the damage tolerance requirement is cost effective when assessed on a life cycle basis.

Overview of ENSIP

In the early 70's, a large number of structural problems occurred in Air Force turbine engines. Many of these were safety problems causing loss of aircraft and an even greater number affected durability, causing a high level of maintenance and modification costs. All of these problems adversely affected fleet readiness.

Springer Series in Computational Mechanics
Atluri, Sampath, Tong (Eds.)
Structural Integrity of Aging Airplanes

The Engine Structural Integrity Program (ENSIP) was intended to substantially reduce these problems. As recommended by the Scientific Advisory Board, in 1976 ENSIP was established by the Air Force to provide an organized and disciplined approach to the structural design, analysis, development, production and life management of gas turbine engines with the goal of assuring engine structural safety, increased service readiness and reduced life cycle costs.

As shown in Figure 1, there are five major tasks associated with ENSIP; Each major task is subdivided into a number of subtasks which guide the development process.

Engine Structural Maintenance Plan

The engine structural maintenance plan represents the output of the engine ENSIP program. This plan will identify and define the individual part life limits, the necessary inspection periods for each fracture critical part as well as the inspection procedure.

Conventional Life Management

The conventional fatigue design approach is shown in Figure 2. Components are designed so that the Low Cycle Fatigue (LCF) limit exceeds the required usage interval in terms of engine flight hours or cycles. The LFC limit is based on the lower bound (-3 sigma or 1/1000) of the distribution of crack initiation times. The lower bound limit has been chosen to prevent occurrence of cracking and resultant failure and the need for repair (economics). The main concern with the conventional approach is that no recognition or provision exists regarding the impact initial defects can have on total component life (i.e., component failure can and has occurred prior to reaching the LCF limit). Implementation of the conventional life philosophy requires that all like components be removed from service upon expiration of the lower bound life. Since there is wide scatter in materials data for fatigue, the analysis is usually based on -3 sigma material properties which implies that all components are treated like the worst case component in the population.

It can further be noted that the fatigue design philosophy is strongly based on the premise that the material is free of initial defects. This philosophy makes no allowance for material, manufacturing, producibility, and handling anomalies or rogue flaw defects. In MIL-STD-1783, the damage tolerance philosophy and policy requirement has been incorporated to prevent safety or flight structural failures caused by material defects, manufacturing defects or fatigue induced cracks. There have been numerous engine failures in the past where initial defects grew in size due to the use of high strength, low fracture toughness material and improper detail design resulting in high stress levels and structural discontinuities, and lack of adequate quality control.

The primary need for damage tolerance requirements has occurred due to the ever present drive to minimize engine weight through development of materials with increased strength and resistance to crack initiation. An undesirable but attendant feature of these new materials has been a decrease resistance to crack propagation (Figures 3 and 4). As a result of these trends, the F100 and TF#$ engine durability and damage tolerance assessments were preformed.

ENSIP and Fracture Control Philosophy/Policy

GENERAL: Damage tolerance is defined as the ability of the engine to resist failure due to the presence of flaws, cracks or other damage for a specified period of usage. The damage tolerance or fracture control philosophy used in ENSIP is shown in Figures 5 & 6. Components are designed for crack growth so that the safety limit exceed 2X the required inspection interval. The safety limit or residual life is the time for assumed initial flaws to grow and cause failure. Since the requirement is to inspect at one-half the safety limit, the design goal for the safety limit is 2X the required design life (i.e., no inspections). The minimum design requirement for the safety limit is 2X the planned depot visit interval. An important aspect of the damage tolerance requirement is that it applies only to fracture critical parts. Also, experience has shown (Fig. 7) some of the secondary benefits are quite meaningful.

FRACTURE CRITICAL COMPONENTS: Fracture critical components are defined as those components whose failure will result in probable loss of the aircraft as a result of noncontainment or, for single engine aircraft, power loss preventing sustained flight due to direct part failure or by causing other progressive part failures. Damage tolerance requirements are applied only to fracture critical components and not, in general, to durability critical components. As can be expected, component classification is affected by aircraft engine configuration, i.e., single engine or multi-engine. Component classification is established early and is identified in the contract.

INITIAL FLAW SIZE: Initial flaws are assumed to exist in fracture critical components. Experience has shown that premature cracking (i.e., crack initiation prior to the LCF limit) occurs at high stressed areas and initial conditions have included both material and manufacturing related quality variations (i.e., voids, inclusions, machining marks, scratches, sharp cracks, etc.). The fracture control or damage tolerance requirement uses a sharp crack initial flaw assumption to characterize these abnormal initial conditions. Assumed initial imbedded flaw sizes are based on the intrinsic material defect distribution or the Nondestructive Evaluation (NDE) methods to be used during manufacture. An inspection reliability of 90 percent Probability of Detection (POD) at the lower bound 95 percent Confidence Level (CL) is required for the assumed initial flaw sizes.

The assumed initial flaw size to account for intrinsic material defect distribution should encompass 99.99 percent of the defect population if a scatter factor of two is used to establish the inspection interval or 99.9 percent of a scatter factor of one is used.

Initial flaw sizes for surface locations (holes, fillets, scallops, etc.) will be consistent with the demonstrated capability (90% POD/95% CL) of the inspection systems proposed for use. In no case will the initial flaw be allowed to be smaller that .005 in. X .010 in. or .005 in X .005 in. corner crack.

INSPECTION INTERVALS: It is highly desirable to have no damage tolerance inspections required during the design lifetime of the engine. This inservice non-inspectable classification requires that components be designed such that the residual life or safety limit be twice the design life. Designing components as inservice non-inspectable is a requirement for those components or locations which cannot be inspected during the depot maintenance cycle.

It is recognized though that the weight penalty incurred to achieve a safety limit/residual life/damage growth interval twice the design life may be prohibitive on some components/locations. Therefore, inservice inspections will be allowed on some components subject to justification. The basis for the justification is characterization of the costs as a function of the requirements as established by trade studies. Cost is usually in terms of weight or Life Cycle Cost (LCC) and the requirement in terms of safety limit/residual life/damage growth interval.

The depot or base level inspection interval for damage tolerance considerations should be compatible with the overall engine maintenance plan. Once again, it is highly desirable that the inspection interval be equal to the hot part design service life as this is the expected minimum depot or maintenance interval for the engine or module. It is required that the damage tolerance inspection interval be contained in the contract specification.

Fracture Control Verification

Verification that the fracture control policy is met is accomplished through development and implementation of a damage tolerance control plan, by analysis and test, and by implementation of reliable inspection methods during manufacture and field/depot maintenance.

ANALYSIS AND TEST CORRELATION REQUIREMENT: Schematics for damage tolerance are shown in Figure 8. Particular emphasis is placed on establishing correlation between analytical predictions and test measurements for growth of cracks in critical areas. Refined analysis models that predict the stress state at and away from the surface as well as multiple cyclic test of coupons, subcomponents and full scale components in the presence of initial damage are required. As with the overall ENSIP development philosophy, damage tolerance analysis and tests are conducted early with minimum impact. Test requirements include specimen tests to define basic materials fracture data (i.e., K_{IC}, Kc, K_{ISCCA}, da/dn), subcomponent tests to evaluate crack growth at typical critical features (i.e., boltholes, fillet radii, snaps, etc.), spin pit test of full scale components (e.g., disks, spacers, etc.) and even engine test of pre-flawed components. Verification requirements also include engine test with components that are pre-flawed or cracked in critical locations to determine the effects of the "real" environment (temperature and gradient, vibration, etc.). Such tests as shown in Figure 9 must be closely controlled and monitored using the inspection requirements planned for service engines to assure safety of the test engine.

Non-Destructive Evaluation (NDE) requirements are implemented on fracture critical parts during manufacture and during field/depot inspection to protect safety. Figure 10 shows the status of NDE implementation for various USAF engines.

RETIREMENT-FOR-CAUSE: Traditionally, components whose dominant failure mode is Low Cycle Fatigue (LCF) have been designed to a "crack initiation" criterion. Using this approach, only one component in a population of 1000 would have actually initiated a crack and the remaining 999 components would be discarded with substantial undefined useful life to crack initiation remaining. Under the Retirement-for-Cause (RFC), this additional useful life can be utilized by adopting a refection criterion that uses each component in a population until it specifically initiates a crack rather than rejecting the entire population on the behavior of the statistical minimum. The development of fracture mechanics concepts and automated NDE techniques over the last several years has permitted the implementation of such an approach on a safe basis. The RFC concept would use the damage tolerance philosophy (shown in Figure 6) to life management.

It is clear that not all fatigue-limited components may be handled in this way, and that each component must be evaluated individually to determine the economic feasibility of RFC. The inspection interval must be such that it does not place undue constraints on the operation of the component or that the cost of the necessary "tear-down" and inspection does not negate the advantage the life extension gained. It seems unlikely that RFC can be applied to components limited by high cycle fatigue considerations, but for many high cost components limited by LFC, such as engine disks, this approach does offer significant economic advantages. It is clear that in applying RFC, Non-

Destructive Evaluation (NDE) becomes a critical factor. Figure 11 shows how damage tolerance is applied in ENSIP to achieve safety from rogue flaw induced failures, inspecting at production and at half life when the hot section components are replaced, and how life extension can occur using the retirement for cause option.

ENSIP benefits on hot section life for fighter engines are shown in Figure 12. Cold section life is in excess of twice that shown for hot section. Ground run accelerated mission tests (AMT) have shown this advance as well as lead the fleet operational engines.

Advances in State of Art

Figure 13 shows about a 5X improvement in eddy current flaw detection capability for 90% POD, 95% confidence level in nickle base, titanium and powder metals in the past three years for fully automated inspection systems. Flaw sizes for various contours and shapes, i.e. bolt holes, rabbits, bores, scallops, etc. on the order of 5 to 7 mil deep can reliably be found. These flaw sizes are used as the initial flaw sizes in the damage tolerance analysis to establish inspection intervals. Figure 14 is an example of the fully automated "retirement for cause" eddy current inspection system POD capability now installed in two depots.

Figure 15 shows the flaw size detection capability of a fully automated penetrant inspection system. Currently a .040 in. flaw length can be detected. It is felt by 1992, this flaw size detection capability will be .030 in. length. These systems will be incorporated in depot and production beginning in 1992.

A Military Standard has been written to standardize NDE reliability demonstrations covering: Demonstration design; Specimen fabrication and maintenance; Testing procedures; Data analysis; Presentation of results; Retesting (demonstration pass/fail criteria); Limited to EC, FPI, MPI and UT.

This standard was written by a government/industry NDE reliability working group made up from representatives from P&W, GE, Systems Research Laboratory (SRL), University of Dayton Research Institute and Garrett Turbine Engine Co.

Figure 16 shows an example of advances made in the material area that enables a modification to be made to an existing material process to enhance crack growth characteristics to better meet ENSIP damage tolerance requirements while virtually maintaining the other desirable material characteristics.

Conclusion

ENSIP with damage tolerance has been applied since 1978 - to at least 15 engine programs.

Increases in safety, reliability, quality assurance and cost savings have been realized.

Advances in automated NDI, materials, POD standards allow damage tolerance concepts for design, development and life management: A VIABLE APPROACH FOR COMMERCIAL GAS TURBINE ENGINES.

References

1. Tiffany, C.F., Cowie, W.D.: Progress on the ENSIP approach to improved structural integrity in gas turbine engine/an overview. (ASME), 1978, 78-WA/GT-13; presented at the Winter Annual Meeting, San Francisco, CA.

2. Cowie, W.D., Stein, T.A.: Damage tolerant design and test considerations in the engine structural integrity program. (AIAA), 1980; presented at the propulsion session of the 21st Structures, Structural Dynamics, and Materials Conference in Seattle, WA.

3. King, T.T., Cowie, W.D., and Reimann, W.H.: Damage tolerance concepts for critical engine components. 1984; presented at the Agard Conference, San Antonio, TX.

4. Military Standard, Engine Structural Integrity Program (ENSIP), MIL-STD-1783 (USAF) dated 30 November 1984.

INTRODUCTION TO ENSIP

THE ENSIP TASKS

TASK I	TASK II	TASK III	TASK IV	TASK V
DESIGN INFORMATION	DESIGN ANAL. COMPNT. & MAT. CHARAC.	COMPONENT & CORE ENG. TESTING	GROUND & FLIGHT ENG. TESTS	PROD. QUAL. CONTROL & ENG. LIFE MGT.
• ENSIP MASTER PLAN • DESIGN SERV. LIFE & USAGE REQUIREMENTS • DESIGN CRITERIA	• DESIGN DUTY CYCLE • MAT'LS AND PROCESSES DESIGN DATA CHARACTERIZED • STRUCTURAL/ THERMAL ANALYSIS • MFG. AND QUALITY CONTROL	• STRENGTH TESTING • DAMAGE TOLERANCE TESTS • DURABILITY TESTS • THERMAL SURVEY • VIBRATORY STRAIN & FLUTTER BOUNDARY SURVEY	• ENVIR. VERIF. TESTING • (AMT) TEST SPEC. DERIV. • DURABILITY TEST (AMT) • DAMAGE TOL. TESTS • FLIGHT TEST STRAIN SURVEY • UPDATED DURA. & DAM. TOL. CONTROL PLAN • PERFRM. DETERIOR. STRUC. IMPACT ASSESSMENT • CRITCL. PART UPDATE	• PROD. ENG. ANALYSIS • STRUC. SAFETY & DURAB. SUM. • ENG. STRUC. MAINT. PLAN • INDIV. ENG. TRACKING • LEAD THE FORCE PROG. (USAGE) • DURA. & DAM. TOL. CONTROL PLAN IMPL. • TECHNICAL ORDER UPDATE

FIGURE 1

CONVENTIONAL APPROACH TO LIFE MANAGEMENT OF CYCLIC LIMITED ENGINE COMPONENTS

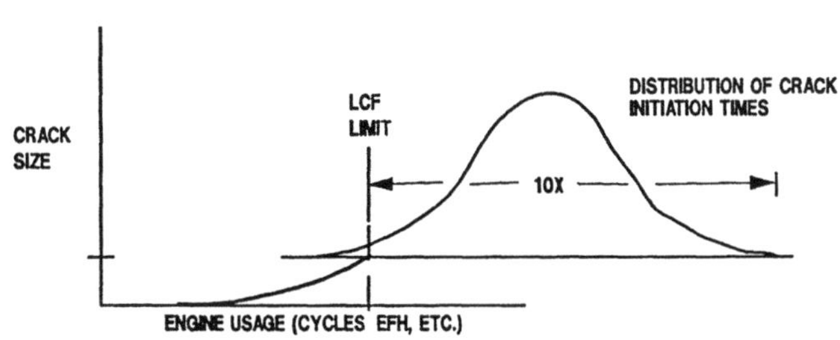

CRITERIA: LOW CYCLE FATIGUE (LCF) LIMIT BASED ON LOWER BOUND (-3σ OR 1/1000) DISTRIBUTION OF CRACK INITIATION TIME

ACTION: 100% PART REPLACEMENT AT LCF LIMIT

CONCERNS: NO RECOGNITION OF IMPACT INITIAL DEFECTS CAN HAVE ON TOTAL PART LIFE (PART FAILURE CAN OCCUR PRIOR TO LCF LIMIT)

MAJORITY OF PARTS MAY BE DISCARDED PRIOR TO REACHING THEIR INDIVIDUAL CRACK INITIATION TIME (I.E., IF LCF LIMIT IS LESS THAN FULL LIFE)

FIGURE 2

SOME MATERIAL TRENDS

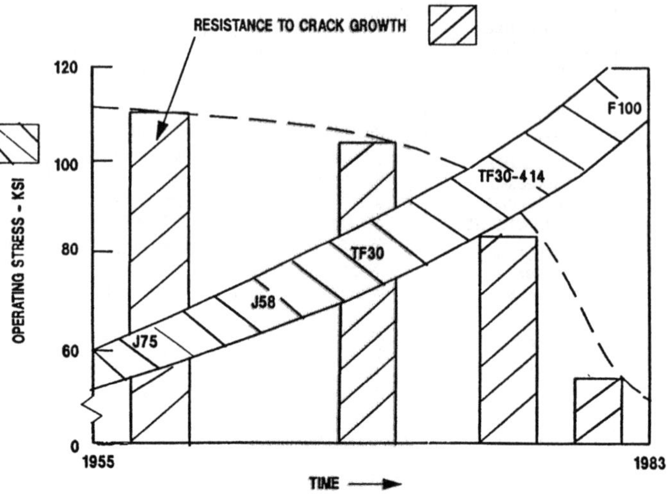

FIGURE 3

SOME MATERIAL TRENDS

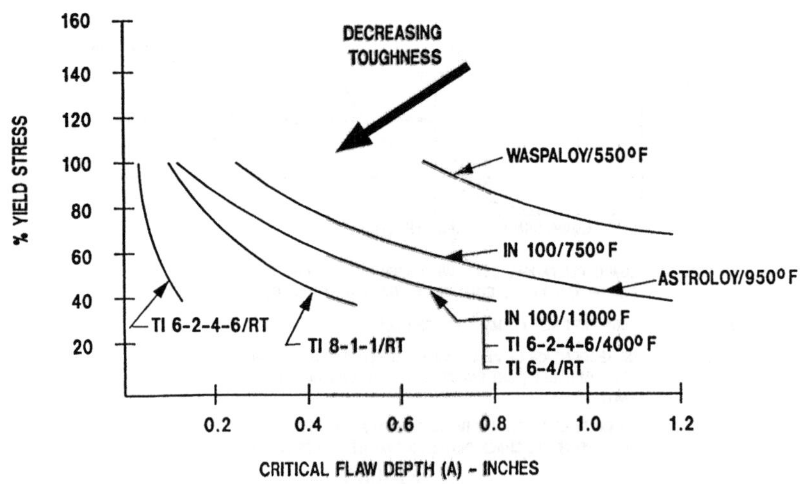

FIGURE 4

ENSIP DAMAGE TOLERANCE PHILOSOPHY

DAMAGE TOLERANCE CONCEPTS ARE USED IN ENSIP TO DEFINE CRITICAL PART INSPECTION REQUIREMENTS!

- •WHERE TO INSPECT
- •WHEN TO INSPECT
- •HOW TO INSPECT

THIS ENHANCES SAFETY, RELIABILITY AND QUALITY ASSURANCE THROUGH DEFINING INTELLIGENT INSPECTION REQUIREMENTS

FIGURE 5

DAMAGE TOLERANCE APPROACH TO LIFE MANAGEMENT OF CYCLIC LIMITED ENGINE COMPONENTS

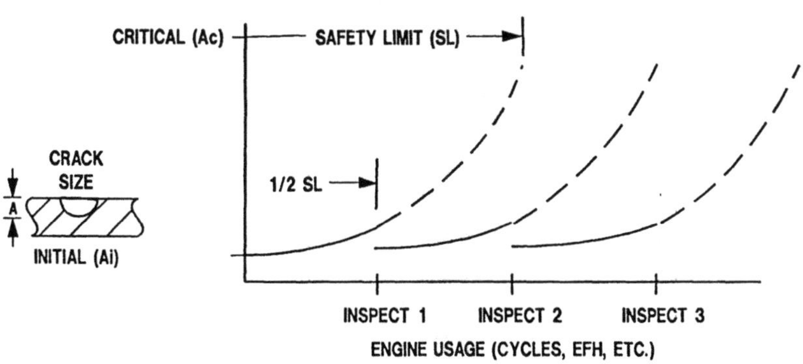

CRITERIA:

SAFETY LIMIT OR RESIDUAL LIFE IS TIME FOR INITIAL FLAW (Ai) TO GROW AND CAUSE PART FAILURE

INSPECT AT 1/2 SL. INSPECTION INTERVAL≥1 LT (DESIGN GOAL) OR 1 DEPOT INTERVAL (MIN DESIGN REQMT)

APPLIES ONLY TO FRACTURE CRITICAL PARTS

Ai BASED ON NDI METHOD OR MATERIAL DEFECT DISTRIBUTION (FOR IMBEDDED DEFECTS)

DTATL-14 De107

FIGURE 6

DAMAGE TOLERANCE BENEFITS

PRIMARY

- SAFETY
- LOWER OWNERSHIP COSTS
- OPTIONAL LIFE EXTENSION
- INCREASED READINESS

SECONDARY

- SELECTS MATERIAL DURING DESIGN PHASE
- REQUIRES REFINED THERMAL / STRESS ANALYSIS
- DRIVES NDI STATE OF ART/NDI IS PART OF DESIGN
- FORCES DESIGNERS TO COMMUNICATE WITH MFG PERSONNEL
- PART QUALITY IS ENHANCED
- LOWERS STRESS LEVEL ON CRITICAL PARTS
- MATERIALS ARE CHARACTERIZED BETTER
- DRIVES RESEARCH FOR IMPROVED DAMAGE TOLERANCE IN MATERIAL
- CORRELATION OF CRACK GROWTH ANALYSIS WITH TEST INSTILL CONFIDENCE
- DEFINES INTELLIGENT AND COST EFFECTIVE WAY TO INSPECT

FIGURE 7

RESIDUAL LIFE ANALYSIS AND TEST PROCEDURE

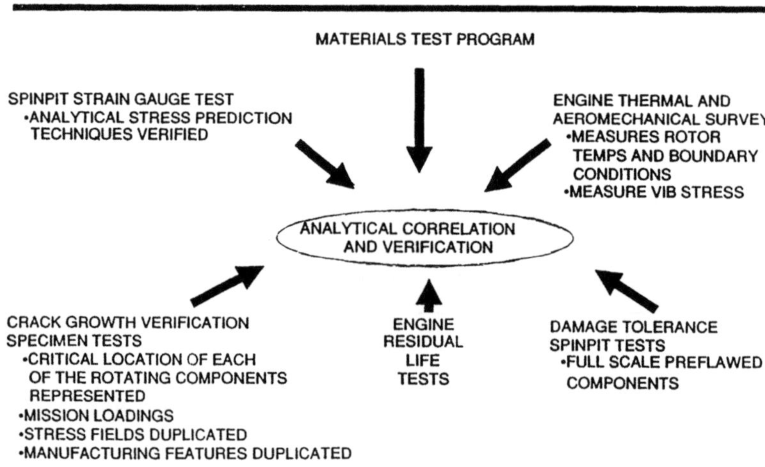

FIGURE 8

ACTUAL AND PREDICTED FLAW GROWTH (ENGINE TEST)

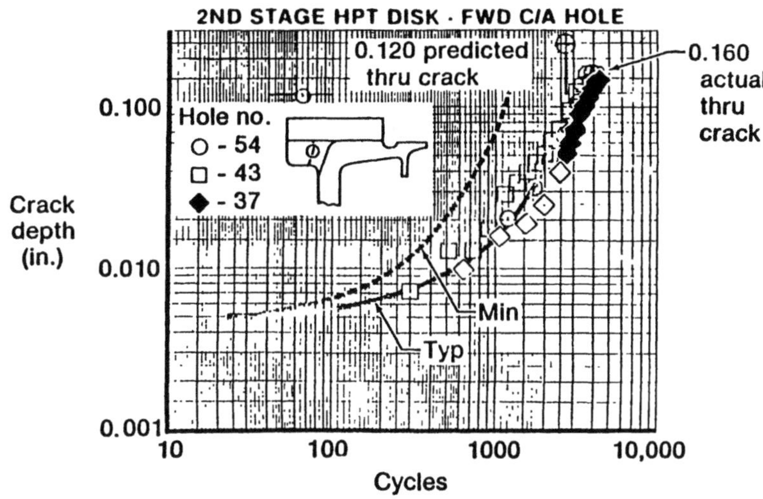

FIGURE 9

STATUS

IMPLEMENTATION OF UPGRADED NDI REQUIREMENTS

F100 ENGINE

03/80-SEMI AUTO EDDY CURRENT @ DEPOT
11/82-AUTOMATED EDDY CURRENT @ PROD
01/84-FAN DISK CRYOGENIC PROOF TEST

F100-PW-220

01/85-AUTO EDDY CURRENT @ PROD
01/85-AUTO EDDY CURRENT @ DEPOT

TF34 ENGINE

01/84-AUTO EDDY CURRENT @ DEPOT

F101 ENGINE

01/84-AUTO EDDY CURRENT @ PROD
01/88-AUTO EDDY CURRENT @ DEPOT

F110-GE-100

01/85-AUTO EDDY CURRENT @ PROD
01/88-AUTO EDDY CURRENT @ DEPOT

F109

01/86-AUTO EDDY CURRENT @ PRODUCTION

F110-GE-129

01/89-AUTO EDDY CURRENT @ PRODUCTION
01/88-AUTO EDDY CURRENT @ DEPOT
01/92-AUTO FPI @ DEPOT AND PRODUCTION

F100-PW-229

01/89-AUTO EDDY CURRENT @ PRODUCTION
01/88-AUTO EDDY CURRENT @ DEPOT
01/92-AUTOMATED FPI

F120/F119

DEPOT/PRODUCTION BEING DEFINED

FIGURE 10

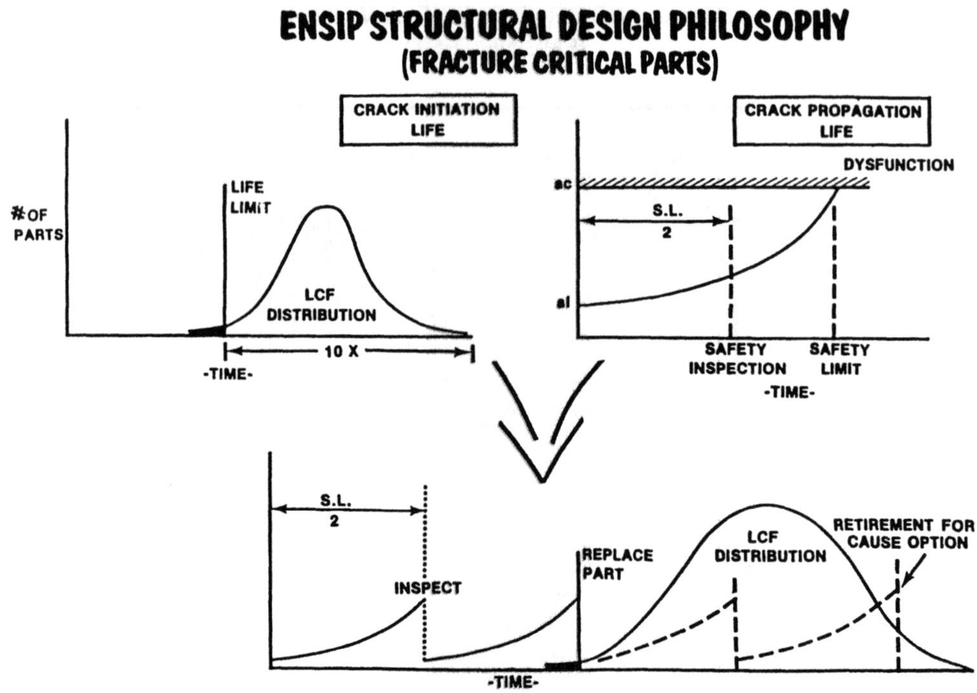

FIGURE 11

ENSIP BENEFITS

USEFUL LIFE OF CRITICAL COMPONENTS SIGNIFICANTLY IMPROVED RESULTING IN REDUCED COST OF OWNERSHIP

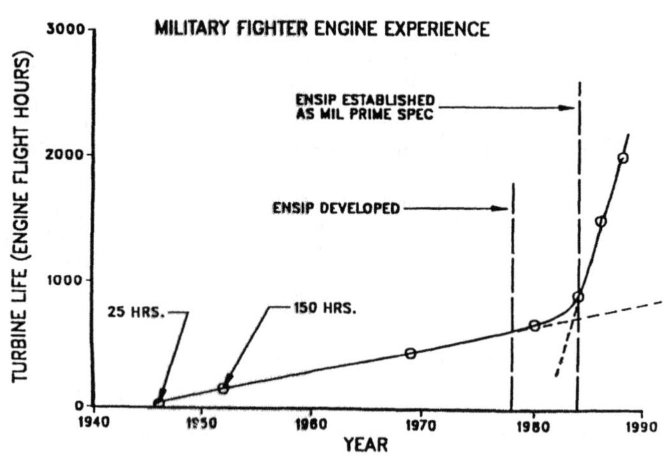

FIGURE 12

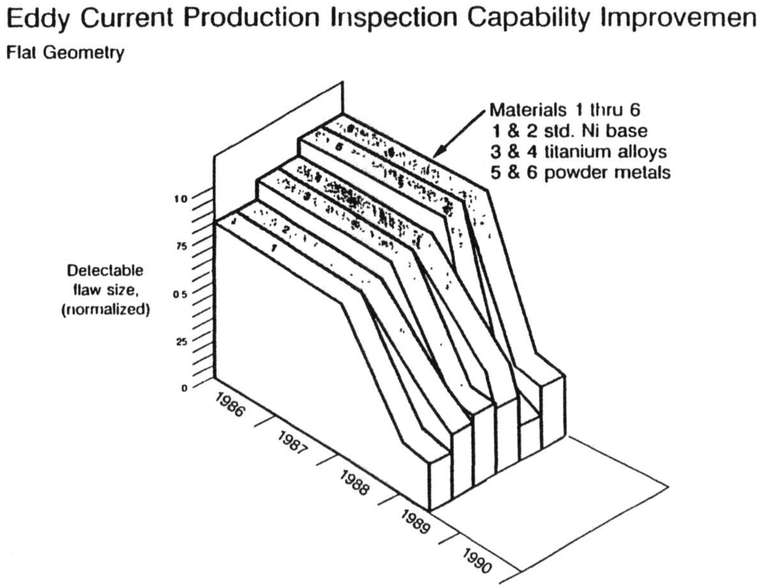

FIGURE 13

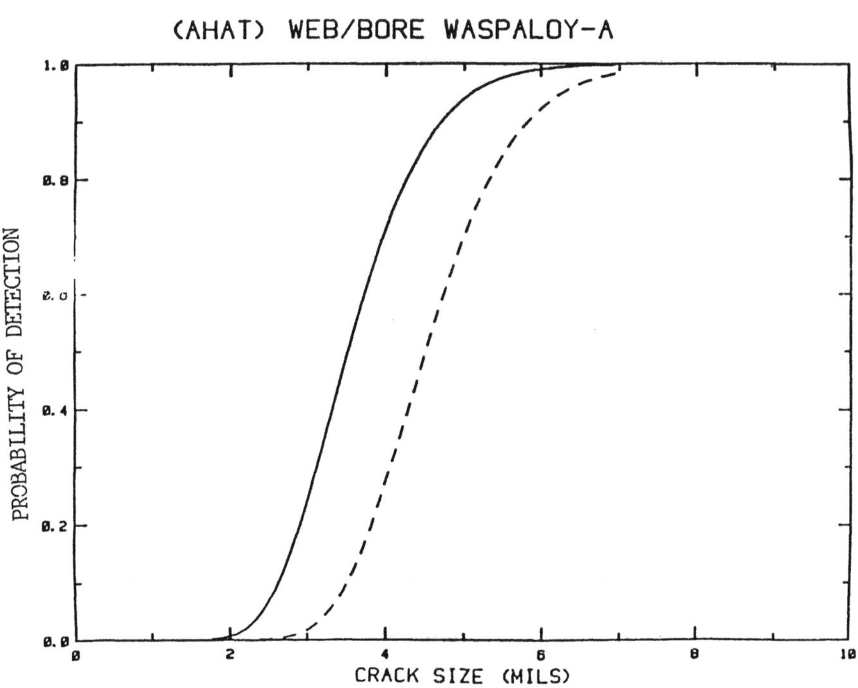

FIGURE 14

ADVANCES

AUTOMATED PENETRANT NDI SYSTEM 90% POD, 50% CONFIDENCE LEVEL

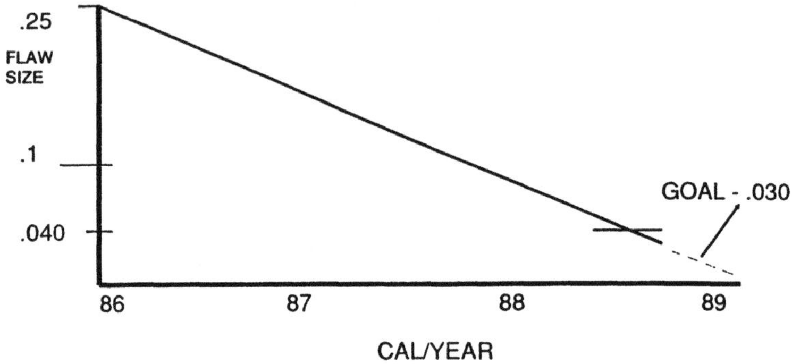

FIGURE 15

CRACK GROWTH LIFE COMPARISON

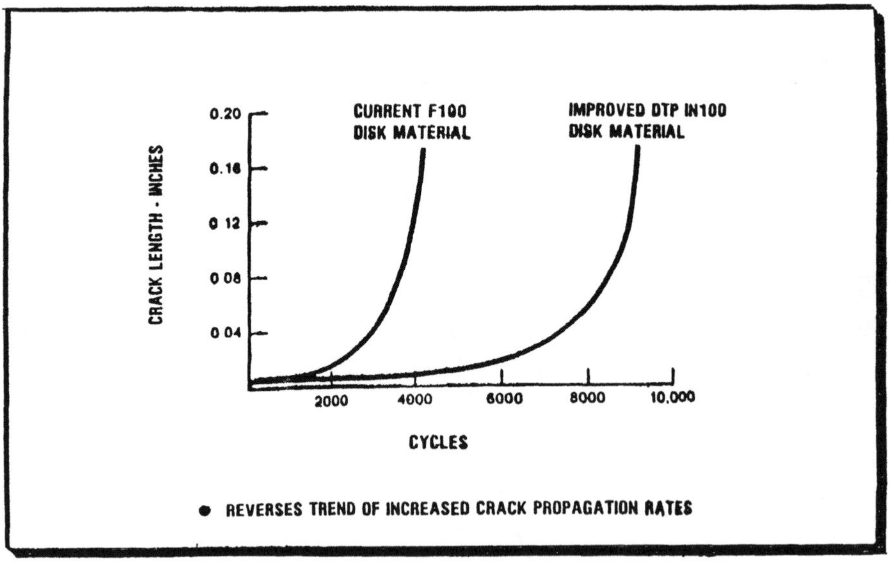

FIGURE 16

8. An Evaluation of the Pressure Proof Test Concept for 2024-T3 Aluminium Alloy Sheet

D.S. Dawicke*, C.C. Poe, Jr.**, J.C. Newman, Jr.**, and C.E. Harris**

* Analytical Services and Materials, Inc., Hampton, VA 23666
** NASA Langley Research Center, Hampton, VA 23665

ABSTRACT

The concept of pressure proof testing of fuselage structures with fatigue cracks to insure structural integrity was evaluated from a fracture mechanics viewpoint. A generic analytical and experimental investigation was conducted on uniaxially loaded flat panels with crack configurations and stress levels typical of longitudinal lap splice joints in commercial transport aircraft fuselages. The results revealed that the remaining fatigue life after a proof cycle was longer than that without the proof cycle because of crack growth retardation due to increased crack closure. However, based on a crack length that is slightly less than the critical value at the maximum proof stress, the minimum assured life or proof test interval must be no more than 550 pressure cycles for a 1.33 proof factor and 1530 pressure cycles for a 1.5 proof factor to prevent in-flight failures.

INTRODUCTION

Several recent incidents involving fatigue cracking have raised concerns about the structural integrity of the aging commercial transport aircraft fleet. The development of multiple site damage (MSD) in lap-splice joints and other structural components is of great concern to the aerospace community. MSD is the formation of a row of cracks, such as along the top line of rivets in a lap-splice joint. The critical size of the individual cracks may be relatively small, making their detection with current nondestructive examination (NDE) methods difficult.

The procedure of overpressurizing the fuselage of commercial transport aircraft has been postulated as a proof test that will insure the continued safe operation of aircraft with MSD fatigue cracks. A precedent for conducting the proof test at a pressure above the normal in-flight pressure exists because all new commercial transport aircraft are subjected to the design limit pressure certification of 1.33P, where P is the normal in-flight pressure. However, after certification the fuselage is only required to be fail-safe or damage tolerant at 1.10P. At no other time in the life of the airplane would the fuselage be subjected to the design limit pressure unless a major structural repair or alteration requires a new certification.

Springer Series in Computational Mechanics
Atluri, Sampath, Tong (Eds.)
Structural Integrity of Aging Airplanes

The strategy for evaluating the concept of pressure proof testing is based on the assumption that the proof test is an alternative to NDE to insure that critical fatigue cracks are not present in the fuselage. The strategy assumes that, first, considering the principles of fracture mechanics and the ductile fracture behavior of 2024-T3 aluminum alloy, a fatigue crack will extend by stable tearing on each load cycle prior to reaching a critical crack length. When the critical crack length is reached, unstable or catastrophic crack extension will occur. Second, for the proof test to function as an alternative to NDE, the assumption must be made that after a successful test, defined as no catastrophic structural failure, cracks exist in the fuselage which are just smaller than the critical crack at the proof test stress. Given these two basic tenets, a proof test logic was established, as illustrated in Figure 1. In this illustration, the proof test pressure was taken to be 1.33P (a proof factor of 1.33). At this pressure, the critical crack length, c_r, can be calculated by using the crack growth resistance curve (R-curve) for 2024-T3 aluminum alloy that specifies the amount of stable tearing which will take place during the proof test. The minimum assured life of the fuselage is then determined by predicting the fatigue crack growth until the residual strength is degraded below the required value for the normal in-flight pressure. It should be noted that the actual remaining life must be defined as the life at which the residual strength degrades below the 1.10P fail-safe pressure. This requirement was not used because the actual residual strength could not be checked while cycling at the 1.0P stress in the experimental verification tests. Therefore, the reported lives are slightly higher than the actual minimum assured fatigue lives.

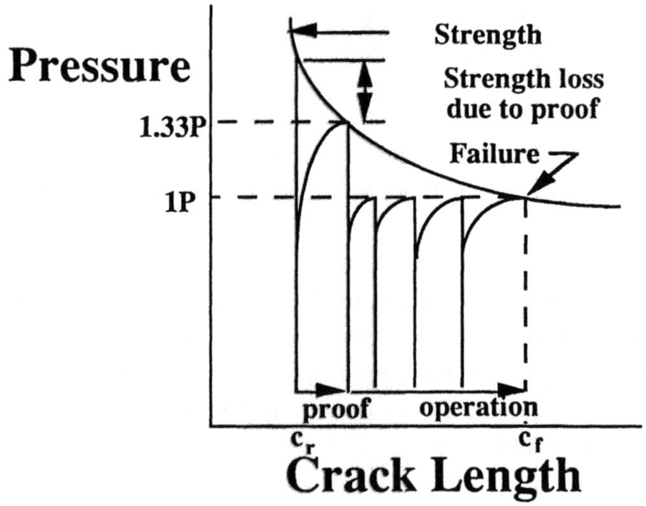

Figure 1. Proof test logic for pressurized fuselage structures.

The residual strength of the fuselage will be decreased by a single proof test cycle, as shown in Figure 1. However, the net effect of the proof test is to increase fatigue life as a result of the crack growth retardation resulting from elevated crack closure due to the proof cycle, as shown in Figure 2. The amount of increase in fatigue life is a function of proof factor, operational pressure level, and initial crack length.

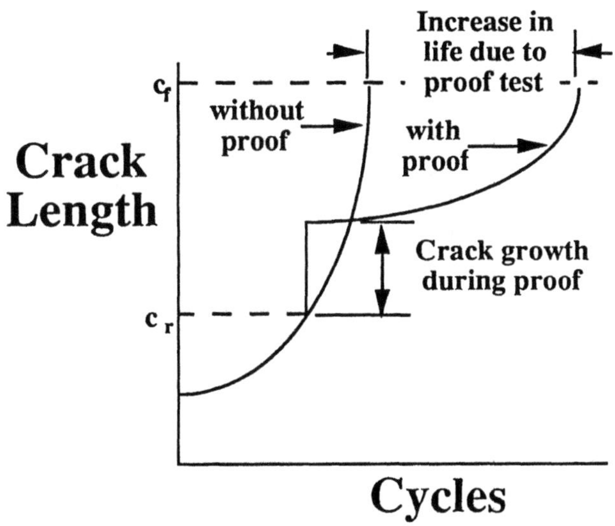

Figure 2. Effect of proof test on life.

The technical evaluation of the concept of pressure proof testing components with MSD is summarized in this paper. Proof test simulations were conducted on flat center crack tension (CCT) specimens, multiple-open hole specimens, and riveted lap-splice joint specimens, which are illustrated in Figure 3. Fatigue life predictions were made using a fatigue crack closure model with verification derived from comparisons with experiments [1]. The generic experimental and analytical investigation was conducted for several issues considered to be critical for making the decision to employ proof testing. However, all of the issues that must be considered in proof testing fuselage structures could not be technically evaluated because NASA has neither the resources nor access to fuselage design details and stress analyses of the jet transports in the commercial fleet. The conclusions stated herein are based solely on the technical results of the subject evaluation as reported in this document.

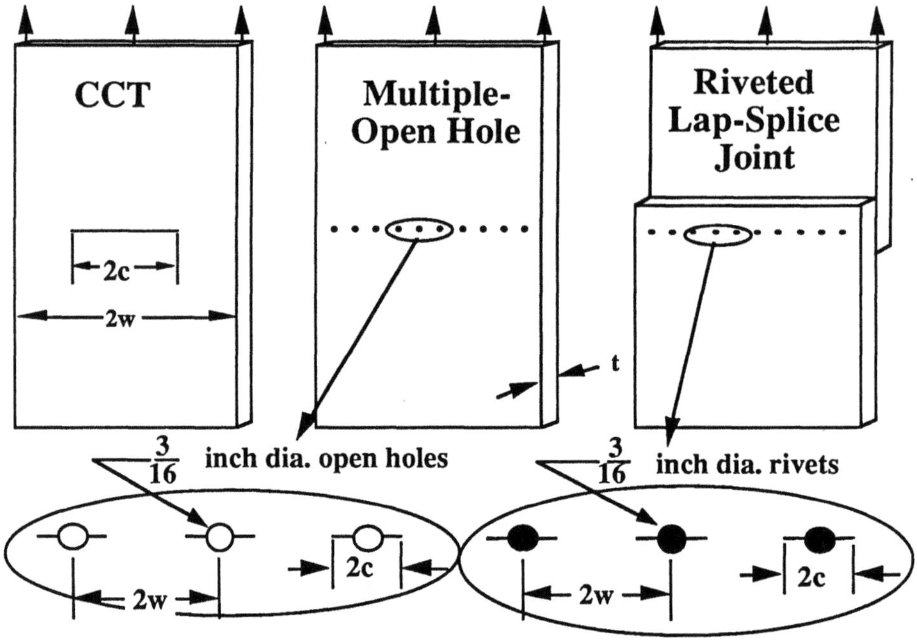

Figure 3. Specimen configurations used in the proof test evaluation.

EXPERIMENTS

The experiments conducted in the proof test evaluation were intended to simulate the effect of a proof test on a typical commercial transport aircraft fuselage material and structural configuration. The structural configuration examined was a riveted lap-splice joint. Tests were conducted on flat uniaxially-loaded CCT specimens, multiple-open hole specimens, and lap-splice specimens with a single row of rivets. The specimen configurations were chosen to provide a progression of increasingly more complex configurations and bring into account the important structural details of MSD and rivet-loading, while at the same time maintaining the simplicity of coupon specimens. The operational stress levels were in the range of 10-14 ksi, which are typical of the levels present in fuselage skins. The actual fuselage structures are curved, supported by other load carrying structures (such as tear straps and stringers), and subjected to pressure loading which results in biaxial stresses in the fuselage skin. These details were not examined in this evaluation.

The proof tests were conducted by first precracking, from an initial saw cut(s) of half length c_n, to obtain natural cracks. The specimens were then loaded under stroke control until the onset of maximum load (unstable crack extension under load control). If the specimen survived the proof cycle, the crack would have been slightly less than critical. The fatigue (or operational) stress was obtained by dividing the proof stress by the proof factor. The specimens were then cycled at the operational stress until failure (minimum stress equal to zero). The half crack length at failure (c_f) was recorded.

The desired operational stress range was obtained by estimating the crack length (c_r) which would result in a near critical crack for the given proof conditions. The 2024-T3 aluminum alloy fractures at about net-section stress equal to the yield stress, making the critical crack length estimation very simple. Some of the proof tests were followed by fatigue tests which had the same initial crack length and operational stress, but had no proof cycle applied. These tests served to define the increase in life due to beneficial crack-growth retardation from the proof cycle. Additional details of the proof and fatigue tests for each of the three specimen configurations are provided in reference 1.

ANALYSIS

The life predictions for the fatigue tests, with and without the proof cycles, conducted on the CCT, multiple-open hole, and riveted lap-splice joint specimens were made using a plasticity-induced crack closure model [2]. The model was based on Elber's crack closure phenomenon [3-4] and the Dugdale plastic-zone model [5]. Elber observed that fatigue cracks close at loads above the minimum load, reducing the stress intensity factor range over which damage occurs. The reduced stress intensity factor range was called the "effective stress intensity factor range" (ΔK_{eff}). The closure behavior was attributed to the development and contact of a region of plastically deformed material behind the crack tip (crack wake) resulting from the crack propagating through the plastic zone.

At the high stress levels examined in this study, the plastic zone is no longer small compared to the crack length, making linear-elastic analyses inadequate. The Dugdale plastic zone length was added to the crack length to correct for plasticity effects, in the same manner as the Irwin plastic zone correction [6]. The effective stress intensity factor was defined in terms of the current crack length (including the current crack growth increment) to model stable crack growth during proof. An infinite periodic array of Dugdale models [7] as modified in references 6 and 8 was used to determine the plastic zone length for the MSD configurations. Additional details of the crack closure model are presented in references 1, 2, and 9.

The crack growth rate description was developed from baseline data covering nine orders of magnitude, including the threshold regime (10^{-7} to 10^{-9} inch/cycle) [11], the mid-rate regime (10^{-3} to 10^{-7} inch/cycle) [12], the high-rate regime (>10^{-3} inch/cycle) [13], and additional high growth rate data from R-curve tests. The use of the plasticity corrected effective stress intensity factor correlated all of the crack growth rate data, including that from the R-curve tests, as shown in Figure 4.

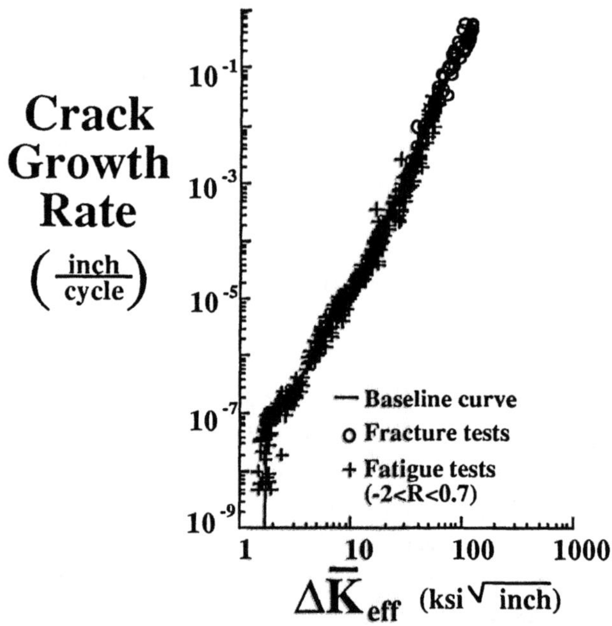

Figure 4. The effective stress intensity factor range against crack growth rate behavior of 2024-T3 aluminum alloy for thicknesses of 0.071-0.090 inches.

A life prediction based on the crack closure analysis was made for each test conducted in the experimental portion of the study. The MSD cracking of the multiple-open hole and riveted single lap-splice specimens was analyzed by assuming an average crack length (equal cracking at each hole) and was based on a solution for an infinite linear array of cracks [7,10]. The riveted lap-splice joint tests had local bending present elevating the crack tip stresses. The stress intensity factor range was modified by multiplying the applied stress in the stress intensity factor solution by $(1 + \gamma/2)$, where γ is the ratio of the outer fiber bending stress to the axial stress [14].

RESULTS AND DISCUSSION

The closure based life predictions were compared with the experimental results for each of the three configurations. Minimum assured lives were also obtained analytically for other conditions not addressed experimentally. The minimum assured lives were used to examine the effect of the proof cycle interval on the life of non-critical cracks.

Center Crack Tension Specimens

A comparison of the closure based life predictions and the experimental results from the CCT specimen proof and fatigue tests indicated that the average ratio of predicted-to-test life was 1.06 with a standard deviation of 0.33. The average ratio of predicted-to-test crack extension in the proof cycle was 0.96 with a standard deviation of 0.3. Crack lengths at failure were predicted within 5% of the test values.

The effects of the proof cycle and proof factor on fatigue life of CCT specimens are shown in Figure 5 for tests conducted at operational stress levels around 11.5 ksi. The tests conducted with a 1.33 proof factor resulted in an average minimum assured life of 146 cycles, which was almost 3 times the life of identical tests conducted without the proof cycle. The large difference in lives was a result of beneficial crack growth retardation due to the single proof cycle. The predicted fatigue lives were higher than that obtained from the actual tests (but within a factor of 1.3). A single test was conducted at a proof factor of 1.5 and resulted in a fatigue life of 604 cycles, over 4 times the life of the tests conducted at the 1.33 proof factor. The predicted results gave 508 cycles to failure. (The predicted life without the proof was larger than that given for the 1.33 proof factor because the initial crack length was smaller).

Multiple-Open Hole Specimens

The length of the cracks at each hole was measured before and after the application of the proof cycle and at several intervals prior to failure. Four of the tests had unequal cracking resulting in crack link-up prior to the application of the proof cycle and experienced additional crack link-ups prior to failure. The other five tests had nearly equal cracking before and after the proof cycle, with failure resulting shortly (1 to 20 cycles) after the first occurrence of crack link-up.

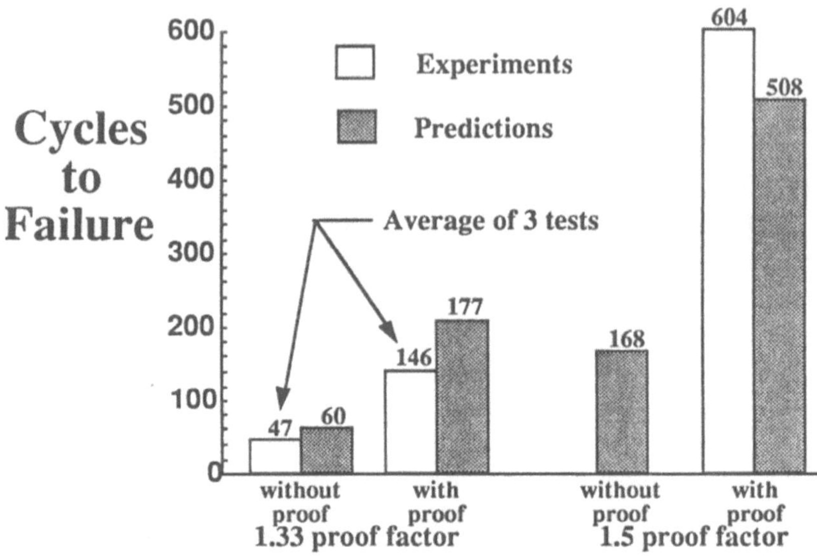

Figure 5. Comparison of fatigue life with and without proof for various proof factors for tests conducted at a stress level of about 11.5 ksi.

The life predictions based on equal cracking tended to predict longer lives than observed experimentally for tests, especially for specimens with unequal cracking and crack link-up. The most significant shortcoming of the current analysis is the inability to describe unequal cracking and crack link-up. The assumption of equal cracking may not be the most critical case and as a result, the predicted lives tended to be longer than observed experimentally. This is best illustrated by examining the crack length against cycle behavior of two tests , as shown by the symbols in Figure 6. These two tests had nearly the same initial total crack length and loading conditions, but one lasted 372 cycles while the other lasted 501 cycles. The closure based life prediction, shown as the solid line in Figure 6, follows the behavior of the experimentally measured crack lengths up to the point of link-up of the longest cracks. The sudden increase in crack length caused rapid crack extension and failure within the next 2 cycles. The analysis currently cannot predict this type of failure mode due to the assumption of equal cracking at each hole. Crack link-up prior to the proof presents a different type of problem for the analysis. The link-up of two or more cracks greatly increases the stress intensity factor of the linked-up crack, which results in large crack extensions during proof and high crack growth rates in fatigue.

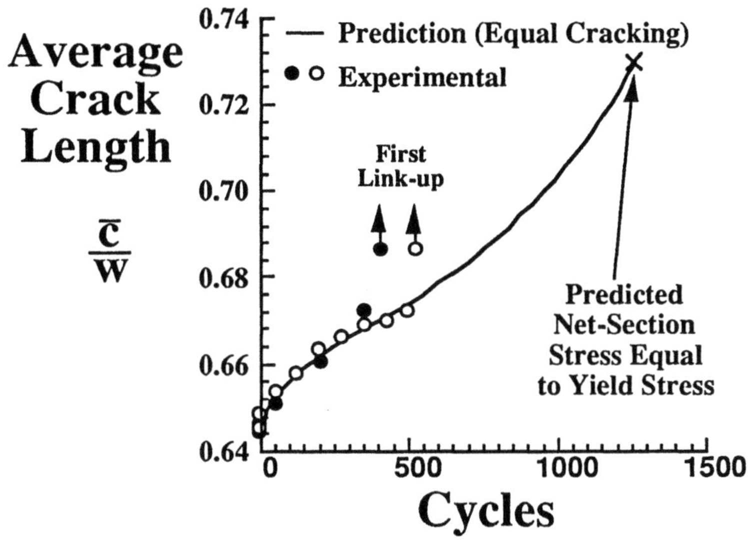

Figure 6. Experimental and predicted crack-length-against-cycle behavior for two identical single proof cycle (1.33 proof factor) fatigue tests of multiple-open hole specimens with nearly equal cracking at each hole.

The analysis indicated that accurate description of the specific crack configurations are required for accurate MSD life predictions. It is believed that a fracture mechanics approach, tracking individual cracks and accounting for crack interaction and link-up could predict crack growth behavior of MSD cracking.

Riveted Lap-Splice Joint Specimens

Again, the crack length for each hole was measured before the application of the proof stress, after the proof stress and at several intervals during the test. Two of the tests had unequal cracking, resulting in crack link-up prior to the application of the proof cycle and experienced additional crack link-ups prior to failure. The other four tests had nearly equal cracking, with failure resulting shortly (1 to 20 cycles) after the first occurrence of crack link-up.

The riveted lap-splice joints were tested with guide plates to prevent buckling and out-of-plane bending. However, localized bending was observed along the row of rivets and life predictions which neglected bending produced longer lives than were observed experimentally. A simple strength of materials estimation of bending, neglecting the contribution of the guide plates, indicated that the parameter describing the elevation of stresses (γ) should be 3. Two tests had severe bending present along the row of rivets, while the other tests had noticeably less bending.

The life predictions for two riveted lap-splice joint specimens with severe and minor rivet bending are shown in Figures 7 and 8, respectively. The different levels of bending was obtained through changing the amount of rivet compaction and increasing the effectiveness of the guide plates. The severity of bending was based on observations of the local out-of-plane rivet rotations and deformations during the tests. Both tests had nearly the same initial total crack length, proof factor, and operational fatigue stress, yet the lives ranged from 160 to 2460 cycles. Predictions of the crack growth behavior were made assuming both no bending ($\gamma = 0$) and a strength of materials estimate of the amount of bending present ($\gamma = 3$). The prediction with $\gamma = 3$ agreed with the experimental measurements for the test with severe bending until the rivets failed, as shown in Figure 7. The prediction with no bending ($\gamma = 0$) agreed with the experimental measurements for the test which exhibited very little bending until first crack link-up occurred, as shown in Figure 8.

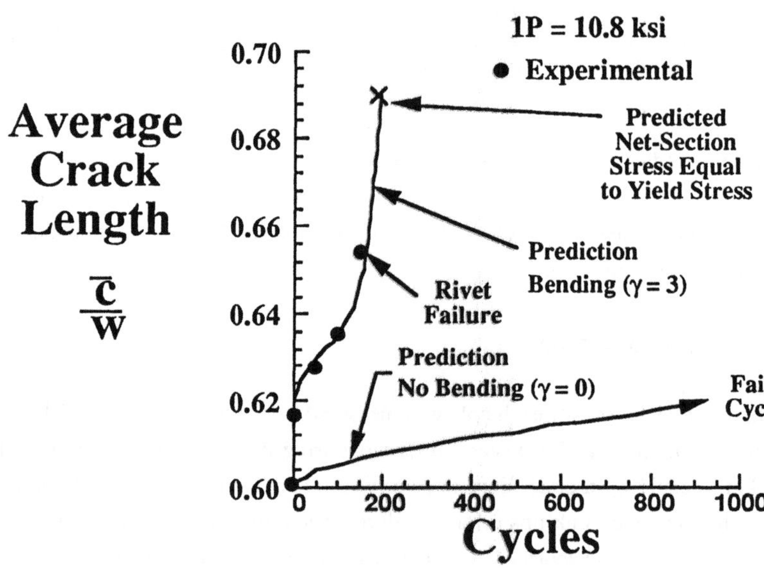

Figure 7. Experimental and predicted (with and without bending) crack-length-against-cycle behavior for a single proof cycle (1.33 proof factor) fatigue test of a riveted lap-splice joint specimen.

Analysis of MSD with Equal Spaced Cracks

Critical Cracks. The extent of crack growth retardation observed in the multiple-open hole and riveted lap-splice joint MSD experiments and predictions was less than that observed in the CCT specimens. Direct experimental comparison of life with and without the proof cycle was not possible in the multiple-open hole or riveted lap-splice specimens because exact crack distributions could not be duplicated. However, the direct comparison could be made with the analysis, as shown in Figure 9. The analysis assumed equal crack lengths at each hole for the multiple-open hole and riveted configurations and considered the crack length just critical at the application of the proof stress. At an operational stress of 1P=12 ksi, the crack growth retardation due to the single 1.33 proof cycle resulted in a life twice as long as the test without the proof cycle for the CCT configuration, while the increase in life was only 1.5 times for the multiple-open hole and riveted configurations.

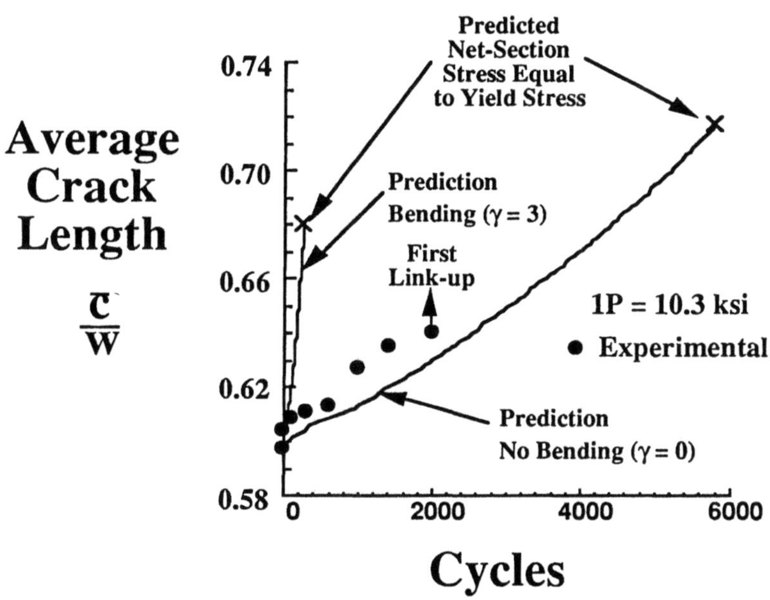

Figure 8. Experimental and predicted (with and without bending) crack-length-against-cycle behavior for a single proof cycle (1.33 proof factor) fatigue test of a riveted lap-splice joint specimen.

The fatigue crack closure analysis was used to examine other operational conditions not addressed experimentally. The analysis considered the riveted lap-splice configuration and varied the proof factor and operational stress level to observe their effect on the cycles to failure, as shown in Figure 10. No bending was considered in the analysis and equal cracking at each rivet hole was assumed. In each case, the initial crack length was chosen to be just critical during the application of the proof cycle. The predicted cycles to failure increased as the proof factor increased and/or the operational stress decreased. The values given in Figure 10 represent the minimum assured life for the various operational conditions. To determine a safe proof test interval, the indicated minimum assured lives would have to be divided by an appropriate safety factor.

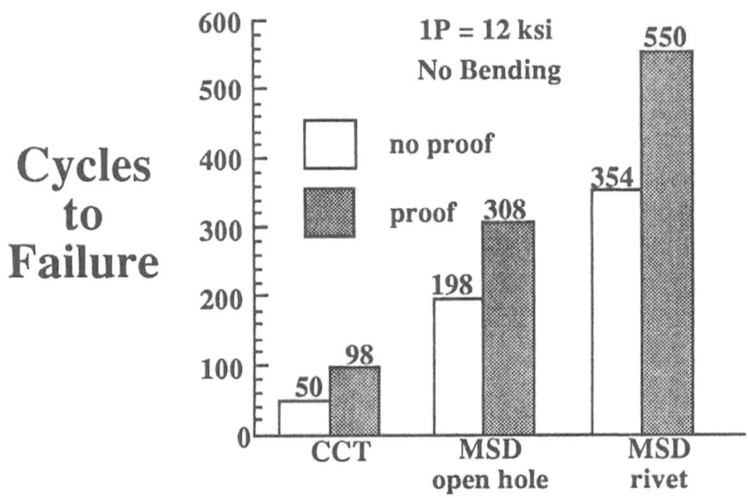

Figure 9. Predicted effect of proof stress on life for CCT, open hole, and riveted lap-splice specimens without bending (1P = 12 ksi, 1.33 proof factor).

Non-Critical Cracks. The effect of multiple proof cycles on the life of non-critical cracks was examined for the riveted lap-splice joint configuration. The fatigue crack closure model was used to predict the cycles to failure of non-critical cracks based on the assumption of equal crack lengths at every rivet. The initial crack lengths examined ranged from 80 to 100% of the crack length just critical at the proof stress (c_T). The critical crack length at proof was calculated using the Two Parameter Fracture Criterion [15] and found to be 0.376 inches for the 1.33 proof factor. The minimum assured life for a fatigue test with a single 1.33 proof cycle was 550 cycles. Thus, the proof cycles were applied at the first cycle and repeated every 550 cycles. A proof cycle interval of 1000 was also analyzed to examine the effect of choosing too large of a proof cycle interval. The 1P operational stress was 12 ksi, no bending was considered, and equal cracking at each hole was assumed in the analysis. The life without any proof cycles was also predicted. The predicted lives, in terms of cycles to failure as a function of the initial crack length normalized to the crack length just critical under the proof cycle (c_i/c_r), are shown in Figure 11.

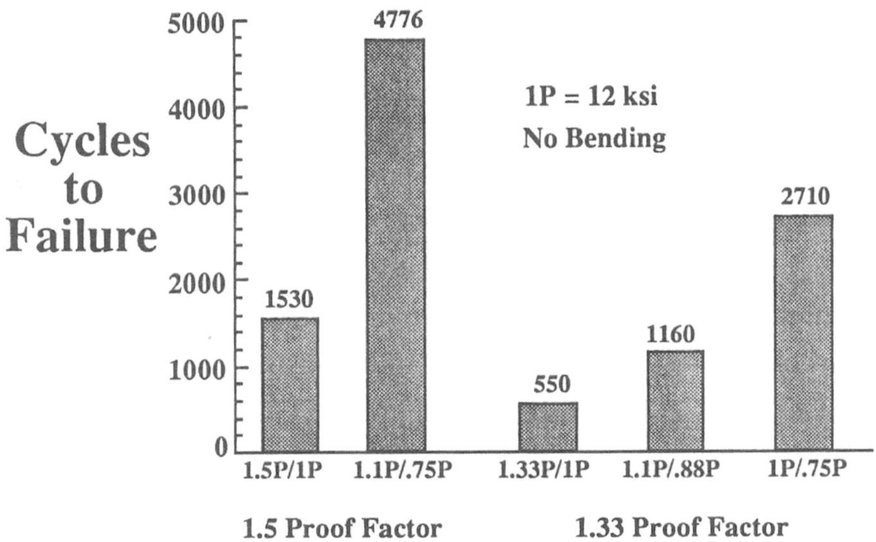

Figure 10. Predicted cycles to failure for the riveted lap-splice joint configuration, without bending, with cracks just critical for 1.5 and 1.33 proof factors and subjected to various operating conditions (proof stress/operational stress).

The crack length against life behavior for a successful proof factor and proof cycle interval combination appears as a series of step functions, indicating that failure occurs only during the application of a proof cycle. This type of behavior is seen in Figure 11 for the 1.33 proof factor applied every 550 cycles. If the interval between applications of the proof stresses is too large, then it would be possible to have an in-flight failure between applications of the proof cycles. This type of behavior is indicated as the small dashed lines in Figure 11 for the 1.33 proof factor applied every 1000 cycles . The predicted cycles to failure for the proof test interval based on the minimum assured life follows the trend of the predicted cycles to failure for the fatigue test without any proof cycles (the large dashed curves of Figure 11). This indicates that for crack lengths near critical, the proof test does not provide substantial increases in life. However, as the initial crack length decreases, the retardation of the proof cycles does provide an increase in life.

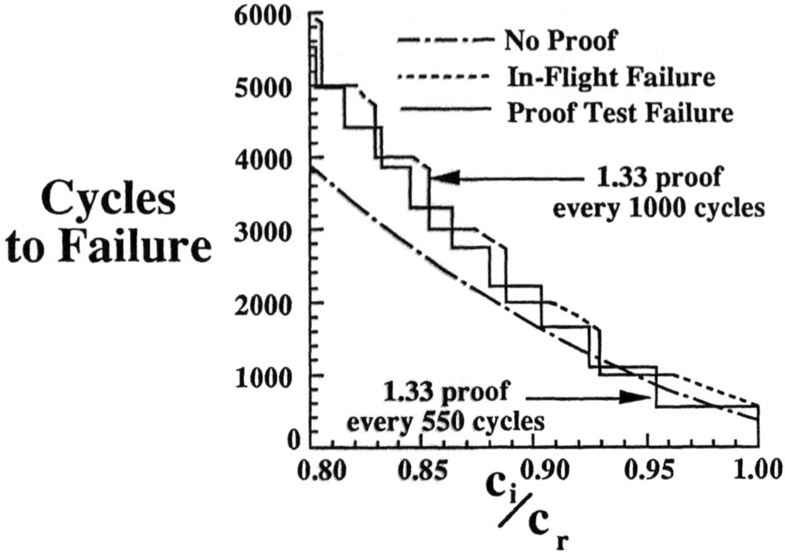

Figure 11. Predicted cycles to failure as a function of initial crack length for a riveted lap-splice joint, assuming no bending, subjected to an operational stress of 1P=12.0 ksi and a 1.33 proof factor applied every 550 or 1000 cycles.

CONCLUDING REMARKS

The concept of pressure proof testing of fuselage structures with fatigue cracks to insure structural integrity was evaluated from a fracture mechanics viewpoint. A generic investigation was conducted on the behavior of uniaxially-loaded, unstiffened flat panels with crack configurations and stress levels typical of the longitudinal lap splice joints in the fuselage of commercial transport aircraft. The specific conclusions from this investigation are as follows:

1. The remaining fatigue life in tests with a proof cycle is longer than that without the proof cycle because of the effect of crack growth rate retardation.
2. The remaining life after the proof cycle increases with increasing proof factor and also increases with decreasing normal operating pressure for a constant proof factor.
3. Fatigue life calculations from the crack-closure model agreed well with tests on center crack tension specimens (mean of ratio of predicted-to-test life on center crack tension specimens was 1.06 with a standard deviation of 0.33).

4. The prediction of MSD fatigue crack growth behavior is strongly dependent upon accurate stress intensity factor descriptions of specific cracking patterns. The current analysis can describe the behavior for equal cracking, but cannot describe unequal cracking and crack link-up.

5. The proof test interval can be no more than 550 pressure cycles for a proof factor of 1.33 and 1530 pressure cycles for a proof factor of 1.5 to prevent in-flight failures. These values may be unconservative due to the possibility of unequal cracking and crack link-up, and also because nominal material property values were used in the analysis.

6. In-flight failures are possible if the proof test interval is too large.

The decision to conduct a pressure proof test of the fuselage cannot be made solely on the basis of this investigation because all aspects of the problems were not considered. For example, a proof test produced by cabin pressure will not result in the critical stress state in several regions of the fuselage structure. Furthermore, the possibility of developing critical cracks in the components of the fuselage stiffener system such as window frames, circumferential frames, longitudinal stiffeners, and the aft pressure bulkhead must be evaluated. Finally, the results of this investigation may not be conservative because of the assumption of equal MSD cracking and also because the effects of corrosion on the measured and predicted fatigue lives were not addressed.

REFERENCES

[1] Dawicke, D.S.; Poe, C.C, Jr.; Newman, J.C., Jr. and Harris, C.E., NASA TM 101675,1990.
[2] Newman, J.C., Jr., ASTM STP 748, 1981, pp. 53-81.
[3] Elber, W., Engng Fracture Mech. J., Vol. 2, No. 1, 1970, pp. 37-45.
[4] Elber, W., , ASTM STP 486, 1971, pp. 230-242.
[5] Dugdale, D.S., J. of the Mech. and Phys. of Solids, Vol. 8, No. 2, 1960, pp. 100-104.
[6] Newman, J.C., Jr., AGARD CP-328, 1982, pp. 6.1-6.26.
[7] Smith, E., First Int. Conf. on Fracture, 1965, pp. 139-151.
[8] Newman, J.C., Jr., ASTM STP 896, 1985, pp. 139-166.
[9] Newman, J.C., Jr.; Poe, C.C., Jr. and Dawicke, D.S., Submitted for Publication at the Fourth Int. Conf. on Fatigue, 1990.
[10] Newman, J.C., Jr., NASA TN D-8244, 1976.
[11] Phillips, E.P., ASTM STP 982,1988, pp. 505-515.
[12] Hudson, C.M., NASA TN D-5390, 1969.
[13] Dubensky, R.G., NASA CR-1732, 1971.
[14] Roberts, R. and Erdogan, F., J. of Basic Engng., 1967, pp. 885-892.
[15] Newman, J.C., Jr., ASTM STP 605, 1976, pp. 104-123.

9. Aging Jet Transport Structural Evaluation Programs

Ulf G. Goranson and Matthew Miller
Boeing Commercial Airplane Group
Seattle, Washington, U.S.A.

ABSTRACT

Economic and market conditions have resulted in the use of commercial jet airplanes beyond their original economic design life objectives. The average age of the world airline jet transport fleet has increased from 8 to 12 years since 1980. Standard Boeing practices to ensure continuing airplane structural integrity include inspection and overhaul recommendations contained in maintenance manuals and service bulletins. As airplanes exceed their economic design life objectives, the incidence of fatigue increases and corrosion may become more widespread. This presentation is focused on principles of durability and damage tolerance technology standards suitable for large teams of structural engineers; major tests and teardowns of airplanes retired from service; results from worldwide surveys of over 80 aging fleet airplanes and joint Boeing, airline and airworthiness authority reviews of service bulletins, corrosion control programs, basic maintenance and supplemental structural inspection programs, and structural repair quality. These initiatives have provided timely preventive maintenance recommendations that will permit continued safe operation of aging jet transports until their retirement from service for economic reasons.

INTRODUCTION

Boeing continually reviews reported service data and other firsthand information from customer airlines in order to promote safe and economic operation of the worldwide fleet. The active service life of commercial airplanes has increased in recent years as a result of low fuel costs, and increasing costs and delivery times for fleet replacements. Air transport industry consensus is that older jet transports will continue in service despite anticipated substantial increases in required maintenance. This paper focuses on updating recent initiatives to enhance aging airplane safety in addition to standard Boeing practices to support continuing structural integrity by inspection and overhaul recommendations contained in maintenance manuals and service bulletins. These actions include fatigue testing and teardown inspections of high-time airframes retired from service, timely development of Supplemental Structural Inspection Documents applicable to selected older airplanes, and surveillance of airplanes undergoing maintenance to gain an engineering assessment of their condition. Recent aging fleet concerns have also resulted in specific Boeing initiatives and joint manufacturer, airline, and airworthiness authority actions. These actions are focused on mandatory modification rather than continued inspection defined in service bulletins for airplanes exceeding design life objectives, development of improved mandatory corrosion control programs, reviews of basic maintenance programs and supplemental structural inspection programs, and development of guidelines to determine the adequacy of structural repairs in relation to damage tolerance and long-term operation.

All Boeing jet transports are designed to be damage tolerant, which has evolved from the earlier fail-safe principle. On the whole, service experience with fail-safe designs has worked very well; in thousands of cases, fatigue and other types of damage have been detected and

Springer Series in Computational Mechanics
Atluri, Sampath, Tong (Eds.)
Structural Integrity of Aging Airplanes

repaired. Boeing jet transports are designed for a minimum 20 years of economic operational service. It should be noted that there is no limit to the service life of damage-tolerant designed airplane structure, provided the necessary inspections are carried out along with timely repair and/or replacement of damaged structure or preventive modifications for airplanes exceeding economic design life objectives. Operational efficiency is impacted by the cost and frequency of repair; durability may, therefore, limit the productive life of the structure.

TEST VERIFICATION

It is impractical to conduct verification testing of all critical conditions and portions of the airplane structure. Analysis methods and allowables are verified by test and the airworthiness certification substantiated by analyses. Development and verification tests comprise small laboratory test specimens, large panels, and major components representing wing, empennage, and fuselage structure, and full-scale airframes.

Full-scale static testing of new models is conducted to verify limit load carrying capability and to satisfy certification requirements. In addition, full-scale fatigue tests are conducted. The primary objective of these major cyclic tests is to locate areas that may exhibit early fatigue problems. The testing is accelerated relative to fleet usage with more than 10 years of repeated load experience completed before the first revenue flight. This provides timely opportunities to develop necessary modification of details that might exhibit early cracking and to demonstrate compliance with economic fatigue design life goals. Additional objectives of the full-scale fatigue test are to help develop and verify inspection and maintenance procedures. Full-scale fatigue tests are not intended to demonstrate "safe- life" limits of structures certified as damage tolerant, nor are they an alternative to the inspections required for continued safe operation of aging airplanes. Testing of new airplane structures does not incorporate corrosion and/or accidental damage that can accelerate fatigue cracking. In addition to new model fatigue tests conducted prior to service introduction, similar tests are conducted on older airframes to gain insight into the problems that might be experienced on high-time airplanes with repairs and service caused defects. These tests do provide valuable information but do not account for typical airplane-to-airplane variability.

Details of recent test activities are presented in reference 1.

GENERIC PRESSURE VESSEL

To supplement airframe testing, two generic pressure vessels are being fabricated. The first specimen has a radius of curvature of 74 in (1.9m) to match narrow-body airplanes such as the 727, 737, and 757, Figure 1. The second has a radius of 127 in (3.2m) to match the widebody 747 and 767 airplanes. Removable test sections will be inserted in a cutout in the cylinder. Tests will be conducted under pressure loading only, using air as the pressurizing medium. A typical test section will be reused for several fatigue, fatigue crack growth, or residual strength tests. Where appropriate, one crack will be inserted in the test article at a time, and the specimen will be repaired before a new crack is inserted, Figure 2.

In parallel with the test program, each test section will be modeled using finite element techniques. The analyses involve the use of solution procedures for geometrically nonlinear displacements. The analysis results will be compared with a comprehensive set of strain gauge and crack opening displacement measurements obtained from the test specimen. Crack

Fig. 1. 74-in (1.9m) Radius Pressure Test Article

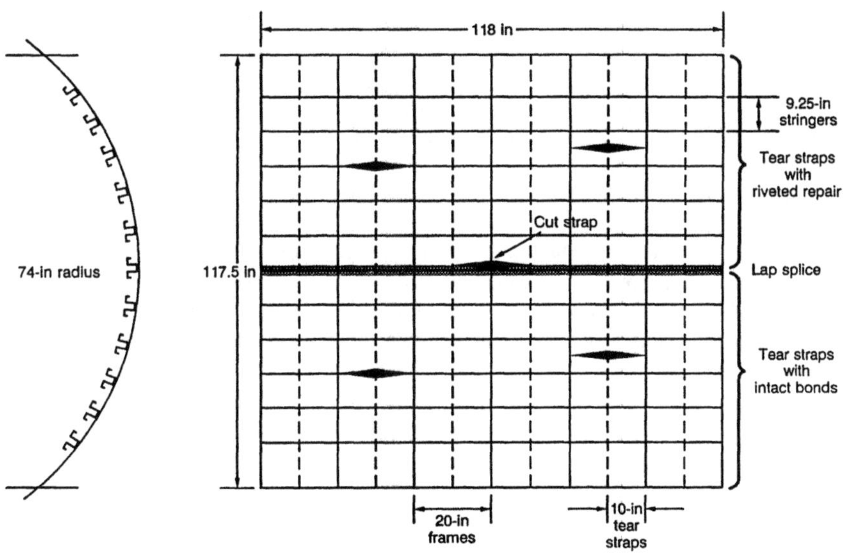

Fig. 2. Typical Test Panel

behavior will be predicted and compared with test experience. The structural modeling will be refined as necessary to provide a validated tool for future commercial airplane design.

TEARDOWN INSPECTIONS

Since the introduction of the 707, Boeing has conducted several teardown inspections and evaluations of high-time airplanes as a part of a continuing assessment of airplane structure.

These inspections permit a detailed examination of structural performance and provide much useful information for forecasting future structural maintenance requirements. Sophisticated inspection techniques, capable of finding smaller cracks than those typically found during routine airline inspections, are used on the disassembled structure. Teardowns also provide an excellent database for calibrating analysis tools and, if required, to develop structural modifications on future production airplanes. Major teardown inspections supplementing normal fleet surveillance activities have been conducted on several Boeing models:

- 707 wings, including center sections, 1965 and 1968.
- 707 wing, center section, and fuselage, 1973.
- 707 empennage, 1978.
- 727 forward fuselage (after fatigue testing), 1978.
- 737 forward fuselage, wing, and empennage, 1987.
- 737 rear fuselage (after fatigue testing), 1988.
- 747 wing and empennage, and fuselage (after fatigue testing), 1988 through 1990.

Details of findings on recent test articles are given in Reference 1.

SUPPLEMENTAL STRUCTURAL INSPECTION PROGRAMS

CAA Airworthiness Notice 89 and FAA Advisory Circular 91-56 allow structural reassessments, including recommendations for any necessary supplemental structural inspections, as an alternative to imposing service life, operational, or inspection limitations. The resulting structural inspection program supplements existing operator maintenance programs that generally have been very effective in maintaining the inherent damage tolerance of Boeing jet transport structures. Model-specific documents provide inspection options for selected structurally significant items (SSI) that will ensure timely detection of fatigue damage in older airplanes. This is achieved, where required, by use of more refined inspection techniques within the operator's existing scheduled maintenance program or by supplemental inspections. These inspection requirements are applicable to those high flight cycle airplanes termed candidates, since these are the most likely to experience the earliest fatigue cracking in the fleet, Figure 3. When cracking is detected and reported, necessary action is taken to safeguard the total fleet.

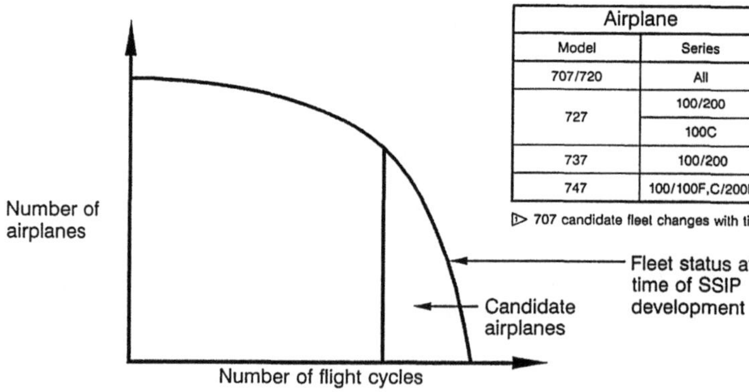

Airplane: Model	Airplane: Series	Number of candidate airplanes
707/720	All	141 ▷
727	100/200	389
	100C	72
737	100/200	123
747	100/100F,C/200B	117

▷ 707 candidate fleet changes with time.

Fig. 3. Supplemental Inspection Candidate Airplanes

The supplemental structural inspection documents (SSID) for models 727, 737, and 747 were released in 1983 and contain usage procedures, lists of candidate airplanes, SSI information with example maintenance program requirements, and a discrepancy reporting procedure. The SSID for model 707 was first released in 1979 but is less sophisticated in terms of the inspection options that are provided. The basic information and approach is, however, similar and several updates have been issued to reflect fleet composition changes. SSID inspections have been mandated on selected 727, 737, and 747 airplanes since 1984 and on selected 707 airplanes since 1985.

When a discrepancy is detected in an SSI, it is essential that timely action be taken to safeguard the total fleet. Reported information is used to establish a threshold and a method and repeat interval for inspections to find all discrepancies in the fleet. Service bulletins are issued to ensure proper attention in the total fleet based on discrepancies that have been found.

FLEET SURVEY PROGRAMS

In order to further expand existing Boeing knowledge of aging airplanes and those factors influencing maintainability of structures and systems, a 2-year Aging Fleet Evaluation Program was implemented in June 1986. The specific objectives of this program were to—

- Gain an engineering assessment of the condition of older airplanes with emphasis on structures and systems.
- Observe the effectiveness of Boeing corrosion prevention features and other corrosion control actions taken by the operators.
- Acquire additional fleet data that will be used to improve maintenance recommendations and promote improved design of new airplanes.

Because of its success, the program has continued and has been expanded to include all current production models, Figure 4. Details of the findings from this program are provided in reference 1.

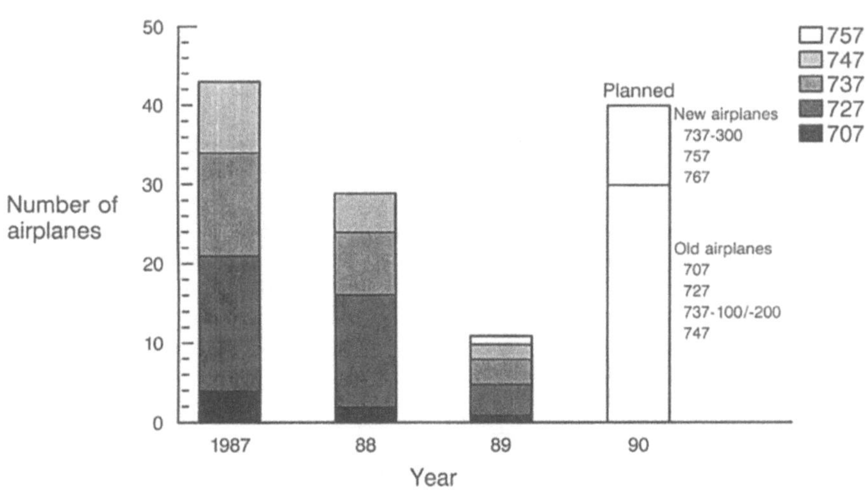

Fig. 4. Airplane Fleet Survey Status

MAINTENANCE ACTIONS—AGING FLEET

Recent aging fleet concerns have resulted in both specific Boeing initiatives and joint industry, airline, and airworthiness authority actions. Boeing formed a special Corrosion Task Force. In addition, meetings were held with airline chief executives as a result of the aging fleet survey findings. A conference on aging airplanes was held in Washington, D.C., in June 1988 that resulted in several recommendations, listed in Reference 1, that include establishing task forces from the airlines, manufacturing industry, FAA, and NASA to continue the work begun at this conference.

Working groups were formed in August 1988 with the charter to foster a consistent approach to the aging fleet concerns for each major manufacturer of commercial airplanes. The task groups for each Boeing model comprise about 15 airline representatives and Boeing structural specialists. The airline members were selected to provide a good representation of fleet experience with a particular model. These groups have been meeting to—

- Select service bulletins for high-time airplane modifications based on potential safety problems, probability of occurrence, and inspectability.
- Develop directed corrosion inspection and prevention programs based on fleet data.
- Review adequacy of SSID programs in terms of candidate fleet coverage and inclusion/deletion of significant structural items.
- Develop comprehensive maintenance recommendations for older airplanes in the fleet.
- Assess structural repair quality relative to long-term operation.

Service Bulletin Reviews

As airplanes age, the incidence of fatigue increases and corrosion becomes more widespread. Problems are often addressed in isolation during the early service use of airplanes. With age, two or more problems in an area may degrade airplane structural fail-safe capability. This increases the need to incorporate preventive modifications in areas with known problems.

A candidate list of service bulletins was established by Boeing as a baseline after a thorough review of service bulletins for long-term operation applicability. These service bulletins were reviewed by the respective working groups for selection of those recommended for terminating actions. The thresholds for these mandated repairs/modifications were typically selected as the life goal objectives in flight cycles for fatigue-related problems. Earlier thresholds may be necessary for items driven by corrosion or stress corrosion considerations. The resulting service bulletins for which mandatory modifications are recommended are shown in Figure 5. These selections were guided by a rating system developed by working group members to reflect their own experience. Summary documents of affected service bulletins have been issued for each Boeing model. These serve as a reference for airworthiness directive action by airworthiness authorities. The impact of the mandated modification program is different for each model. An example giving the estimated man-hours for the modification program on the 747 model is provided in Figure 6. Newer airplanes have the intent of many of the service bulletins incorporated during production.

Corrosion Prevention and Control Requirements

Boeing aging fleet evaluations have shown that some operators do not utilize a proven corrosion prevention and control program. This leads to unacceptable degradation of structural integrity and, in an extreme instance, results in loss of an airplane. The Boeing aging fleet

Model	Candidate bulletins		Mandatory modification recommended	Other recommended action	No new recommendation
707	Wing	94	72	13	9
727		18	12	3	3
737		21	15	5	1
747		25	8	7	10
707	Body	68*	54*	3*	11*
727		51	45	4	2
737		37	25	10	2
747		37	13	14	10
727	Doors	9	8	0	1
737		9	8	1	0
747		3	0	1	2
707	Land gear	5	1	1	3
727		16	1	0	15
737		3	2	1	0
747		4	4	0	0
707	Prop	8	1	3	4
727		2	1	1	0
737		0	0	0	0
747		9	5	2	2
707	Emp	21	13	4	4
727		17	7	8	2
737		10	8	0	2
747		5	1	3	1
707	Total	196	141**	24	31
727		113	74	16	23
737		80	58	17	5
747		83	31	27	25

* Including doors. ** Covers 82 structural areas.

Fig. 5. Service Bulletins Recommended for High-Time Airplanes

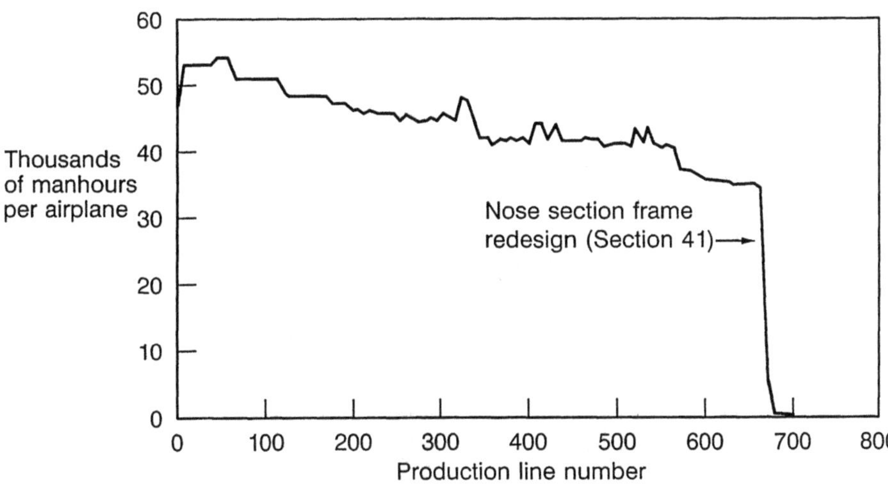

Fig. 6. Impact of Service Bulletin Modifications Program on Model 747

working group charter therefore included the development of a mandatory corrosion-directed inspection and prevention program for each airplane model.

Following review, an industry approach has been developed. This consists of a minimum baseline program, including an implementation plan, thresholds, and repeat inspection intervals for basic corrosion control/prevention tasks. Full details of the tasks will include inspection, repair, application of prevention standards, and reporting. The thresholds and repeat intervals are defined in calendar periods (i.e., independent of flight cycles or flight-hours). Operators who continue to experience significant corrosion must make appropriate adjustments to their program to prevent or control the problem. Priority for implementing the program and alternative means of compliance will be subject to negotiation between operators and appropriate airworthiness authorities.

Although this corrosion control and prevention program addresses the aging fleet, it is imperative that it be implemented early due to the insidious nature of corrosion. This means thresholds of 5 to 6 years for some structure, Figure 7. These programs define minimum requirements for continuing airworthiness. However, additional and more frequent tasks may be required for an economically balanced program. Less stringent requirements may apply to operators with established corrosion prevention and control programs that have proved to be successful.

General area		707/720		727		737		747	
		Threshold	Repeat	Threshold	Repeat	Threshold	Repeat	Threshold	Repeat
Wing	Outer-external	10	4	10	5	8	4	10	2
	Leading edge interior	8	2/4	10	5	8	4	6	1.5
	Outer-main box-interior	10	8	10	10	10	10	20	10
	Trailing edge interior	8	2/4	10	5	8	2	10	2
	Center section interior	10	8	10	8	10	8	20	10
Fuselage	External (including doors and landing gear bays)	6	2	6	1.5	5/8	1.5/2	10 upper 5 lower	5 upper 2 lower
	Flightcrew compartment	10	8	10	8	10	8	15	8
	Upper lobe interior	8	8	10	8	8	8	15	8
	Lower lobe interior (except bilge)	6	6	6	6	6	5	6	6
	Lower lobe - bilge	6	3	6	3	6	2/4	6	4
	Section 48 interior	10	5	10	5	8	4	10	5
V/H Stabilizer	External surfaces	10	2/4	10	2	10	2	10	5
	Leading edges	10	8	10	8	10	8	15	8
	Main box interiors	10	8	10	8	10	5	15	8
	Trailing edges	10	8	10	4	10	5	10	5
	Center section	10	5	10	5	8	4	10	8
	Center engine inlet duct	-	-	10	8	-	-	-	-
Nose and main landing gear		Landing gear overhaul		Landing gear overhaul		Landing gear overhaul		Landing gear overhaul	
Powerplant and strut		4	2	5	2	5	5	7/15	3/15

Note: Some specific areas/items within the general areas have independent thresholds and repeat intervals.

Fig. 7. Corrosion Inspection Thresholds and Inspection Interval (Years)

Supplemental Structural Inspection Program (SSIP) Reviews

The major issues of the SSIP reviews planned for mid-1989 were—

- Adequacy of the present candidate fleet for fleet leader sampling.
- Inclusion/deletion of significant structural items.

The candidate fleet for models 727, 737, and 747 were those airplanes that were fleet leaders at the time of initial SSID releases in 1983. Boeing periodically reviews the candidate airplane list for any significant changes in fleet distribution, composition, or utilization. To date, minor changes only have occurred in the active candidate airplanes subject to SSID compliance. However, a few high flight cycle airplanes will be added to the candidate fleet to account for airplanes retired from service.

Some SSIs were not incorporated in the SSIDs based on obvious damage containment capability providing damage detection without directed inspections. Concerns on the extent of corrosion that has been observed and the probable interaction with fatigue damage warranted a detailed review to determine if any additional SSIs should be included in SSID updates. Thin-gauge fuselage lap splices are an example of this category of structure.

Another class of SSI is selected on the basis that the primary crack origin is in the most difficult area of structure to inspect, for example, internal structure. Cracking is assumed to spread into adjacent, external structural elements that are more readily inspectable. Detection of a crack in the adjacent element should lead to a thorough investigation of the extent of damage, and as a result, the primary crack origin should be detected. One such example is fuselage frame cracking with secondary skin cracking. However, experience has shown that multiple frame cracking can occur prior to any significant external secondary skin cracking.

Inspection for multiple site damage (MSD) is a key consideration to ensure structural integrity of areas with numerous identical details operating at similar stress levels. Full-scale fatigue testing of aging airplane structures has demonstrated that rapid linkup in local areas may occur and analysis criteria based on this behavior, developed since original SSID releases, have been used as part of a review of SSI coverage.

Old Airplane Maintenance Programs

Comprehensive maintenance program guidelines acceptable to the airlines, airworthiness authorities, and Boeing do not exist that properly address older airplanes in the fleet. The Supplemental Structural Inspection Program, for example, addresses one issue—fatigue cracking—in isolation. It is also perceived that maintenance intervals on older airplanes should reduce as a function of age. Some sampling inspection prgrams should be revised or there should be a reversion to full fleet inspections. An activity is in progress to review existing recommendations with a view to resolving these deficiencies.

Structural Repair Quality

Major repairs are performed and validated based on existing FAA-approved data such as service bulletins and structural repair manuals or require new substantiating data approval by Designated Engineering Representatives (DER), repair approval authority (SFAR 36), or other means. Aging fleet concerns have been raised about the quality of repairs with regard to damage tolerance relative to long-term operations. Repair evaluations will be focused on damage tolerance analyses of typical major repairs of significant structure, particularly involving patches or splices. These evaluations will provide recommended mandatory inspection requirements in terms of methods, thresholds, and repeat intervals. The airplane operators will complement this activity by documenting major repairs, based on visual inspections and airplane records, within a specified time period for external repairs and when areas are accessible for any reason for internal repairs. These repairs will also be verified to be in compliance with FAA-approved data, or have DER or SFAR36 approval on an individual basis.

A parallel activity to provide guidance material to classify major repairs has been addressed by a joint industry/operator task force. Guidelines in the form of a logic system will be provided in Air Transport Association report (51-10-01). Application of the guidelines will exempt air carriers from the criteria for major repairs (FAR Part 43 Appendix A).

SUMMARY

The Boeing Company is dedicated to the design and manufacture of safe commercial jet transports. The successful discharge of this responsibility over the last three decades, Figure 8, has contributed significantly to a position of industry leadership and reflects the top priority given to safety. This paper illustrates that the structural integrity of commercial airplane structures is a very serious and disciplined process. High standards must be maintained to ensure the safety of aging airplanes until economics dictate their retirement.

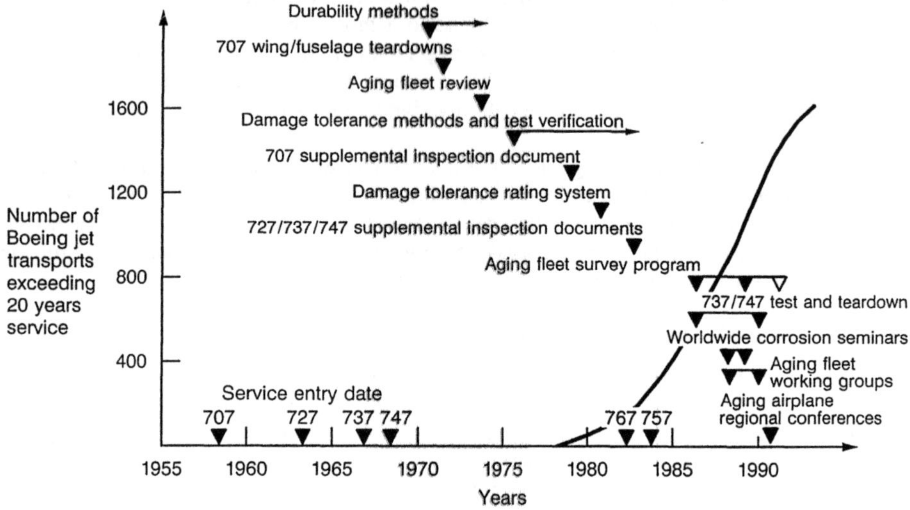

Fig. 8. Boeing Fleet Support Action

REFERENCE

1. Goranson, U. G. and M. Miller, "Aging Jet Transport Structural Evaluation Programs." Presented at the 15th Symposium of the International Committee on Aeronautical Fatigue (ICAF) held in Jerusalem, Israel, June 21-23, 1989. Ed. A. Berkovits, Published by Engineering Materials Advisory Services, Ltd., West Midlands, U.K.

10. NASA Airframe Structural Integrity Program

Charles E. Harris, Head
Mechanics of Materials Branch
NASA Langley Research Center
Hampton, VA 23665

ABSTRACT

NASA has initiated a research program with the long-term objective of supporting the aerospace industry in addressing issues related to the aging commercial transport fleet. The interdisciplinary program combines advanced fatigue crack growth prediction methodology with innovative nondestructive examination technology with the focus on multi-site damage (MSD) at riveted connections. A fracture mechanics evaluation of the concept of pressure proof testing the fuselage to screen for MSD has been completed. Also, a successful laboratory demonstration of the ability of the thermal flux method to detect disbonds at riveted lap splice joints has been conducted. All long-term program elements have been initiated and the plans for the methodology verification program are being coordinated with the airframe manufacturers.

INTRODUCTION

Basic research related to the fatigue and fracture of metals, computational fracture mechanics, structural analysis methods, and nondestructive examination (NDE) methods for material defect characterization have been ongoing at NASA Langley for many years. All of these disciplines have been brought to bear on the problems facing the aging commercial transport fleet. After participating in the two previous International Aging Aircraft Conferences and conducting a tour of the U. S. commercial transport manufacturers, NASA has identified several areas of concern where our ongoing research offers a potential solution. Therefore, in coordination with the FAA and the airframe manufacturers, NASA has developed a specialized research program that is an outgrowth of our base research program with the broad objectives of methodology specialization, verification, and technology transfer. The two key program elements are advanced fatigue crack growth prediction methodology applicable to complex built-up structure and innovative NDE technologies to detect fatigue cracks, corrosion, and disbonds in adhesively bonded joints. The focus of the program is multi-site damage (MSD) in unbonded lap splice joints. However, the research is generic in nature and the developed methodology is expected to be applicable to many other structural components that may be fracture critical. The purpose of this paper is to provide a brief overview of the relevant

Springer Series in Computational Mechanics
Atluri, Sampath, Tong (Eds.)
Structural Integrity of Aging Airplanes

ongoing research at NASA Langley and to present the plans of the specialized Airframe Structural Integrity Program (ASIP).

FATIGUE CRACK GROWTH PREDICTION METHODOLOGY

The concept of crack closure to explain crack growth acceleration and retardation was pioneered at NASA Langley almost two decades ago. A plasticity-induced closure model [1] employing fracture mechanics principles was shown to be quite accurate in predicting the fatigue crack growth in aluminum alloys for a number of basic crack configurations for both constant amplitude and spectrum loadings. Furthermore, the closure model has been successfully used to explain the small-crack phenomenon exhibited by many aluminum alloys. The crack closure concept is illustrated schematically in Figure 1. Residual plastic deformations produced by the crack-tip plastic zone causes residual compressive stresses in the wake of the advancing crack at the minimum load. These compressive stresses must be overcome before the crack is fully open on each loading cycle. Therefore, the full cyclic stress range is not effective in extending the crack. The crack growth rate data must be correlated with the effective stress-intensity factor range rather than the full range to yield meaningful predictions of total crack growth. Figure 2 illustrates the growth rate data for 2024-T3 which spans 8 orders of magnitude and couples the threshold behavior with large-scale stable tearing. The successful application of the closure model to predict small-crack growth rate data has resulted in a total life prediction methodology which treats initiation by predicting the growth of micron size cracks initiating at inclusion particles in the microstructure [2].

This type of methodology is necessary to predict the fatigue crack growth of small cracks initiating at a rivet hole before they grow to a detectable size. Furthermore, this methodology may be used to predict the necessary inspection intervals to monitor crack growth before critical sizes are reached and link-up of adjacent cracks occur. Therefore, one of the objectives of the ASIP is to specialize the closure methodology for riveted splice joint crack configurations.

A rigorous fracture mechanics treatment of cracks initiating at rivet holes and MSD will require the generation of stress intensity factor solutions to several basic crack configurations. NASA has developed several computational methods for computing stress intensity factor solutions to complex crack configurations. The boundary force method (BFM) [3], an indirect boundary element method, is well suited to two dimensional (2-D) problems such as the crack configuration illustrated in Figure 3a. For more complex problems, the finite element method with the force method or the virtual crack closure technique (VCCT) [4] have been successfully employed to obtain solutions to three dimensional (3-D) crack configurations. An example of this capability is illustrated in Figure 3b for a surface crack emanating from a semi-circular notch. These techniques will be used to generate the solutions to the two crack configurations shown in Figure 4 with the BFM being used to obtain 2-D solutions for MSD. Because MSD linkup is likely to be governed by net section yielding, an elastic-plastic analysis of the loaded rivet hole and a typical splice joint will also be required.

NASA has recently completed an evaluation of the concept of pressure proof testing the fuselage of aging airplanes as an augmentation to NDE. The evaluation was conducted from the fracture mechanics viewpoint by employing both the crack closure prediction methodology and flat panel test results. The results revealed that the remaining fatigue life of a structure with near critical cracks with the proof test is longer than without the proof test because of the effect of crack growth retardation. However, when repeated proof tests are conducted, the proof test interval required to insure failures during the proof test rather than in-flight will shorten the overall life of the structure. The required proof test interval to screen critical MSD in riveted splice joints would be every 275 cycles for a maximum proof test load of 1.33 times the normal cabin pressure. Complete details of this evaluation may be found in reference 5.

GLOBAL/LOCAL STRUCTURAL ANALYSIS METHODOLOGY

The behavior of large cracks in fuselage structures such as mid-bay cracks or splice joint cracks after MSD linkup are strongly influenced by the stiffening effects of the circumferential frames and longitudinal stiffeners. It is not practical to model all of the structural details in a finite element analysis. Greater efficiency can be achieved by exploiting a global/local strategy where local details that produce stress gradients can be treated in a companion analysis to the global structural analysis. Figure 5 shows an example stress analysis of a circular cutout in a blade-stiffened panel where quite accurate results were obtained by the global/local analysis methodology [6]. To accurately predict the behavior of a growing crack in a stiffened shell structure, the global/local methodology must be extended to include an adaptive mesh concept so that the local refined mesh can change in a manner dictated by the growing crack. This concept is illustrated schematically in Figure 6. The structural analysis methodology will have to account for geometric nonlinear behavior as well as large deformation behavior. This is necessary to predict the crack growth direction and crack opening of large cracks that may result in a rapid decompression rather than a catastrophic in-flight failure.

The fracture mechanics and global/local analysis methodology must be verified by an experimental test program. As shown in Figure 7, there are various levels of testing required to achieve a full verification of a structural analysis methodology. The goal of the ASIP program is to achieve verification through the curved panel and subscale barrel test article level. Tests of the coupons and flat panels shown schematically in Figure 8 required to verify the fracture mechanics solutions are already underway. However, we are investigating the possibility of using existing data from the damage tolerance tests previously conducted by the airframers as benchmarks for the structural analysis methodology in lieu of conducting new tests for the sole purpose of methodology verification.

QUANTITATIVE NONDESTRUCTIVE EXAMINATION TECHNOLOGY

A wide range of nondestructive measurements science research is ongoing at NASA Langley including defect characterization, stress and strain field mapping, and smart sensors for materials and structures. A number of NDE techniques under development at NASA are relevant to detecting fatigue cracks in aging aircraft. These techniques include thermography, ultrasonics, magnetics, and acoustic emission.

Perhaps the most promising technique in the near-term is an advanced thermal method which is based on heat flux rather than absolute temperature [7]. The thermal flux method has been demonstrated to be successful in detecting disbonds in lap splice joints as illustrated in Figure 9. This technique shows such great promise that NASA and the FAA have developed a technology transfer program and are tentatively planning a full scale field demonstration on an airplane at an inspection depot within the next year.

It is well known that the physical measurements from NDE techniques must be coupled with computational models such as those obtained from elasticity and fracture mechanics to provide useful quantitative defect characterization information. One of the goals of the ASIP program is to couple the disciplines of physics and mechanics to obtain a more thorough understanding of the crack growth process, more reliable experimental maps of the stress gradient fields produced by a crack, and more quantitatively useful NDE techniques for field inspection applications. The coupon and panel tests planned to verify the fracture mechanics solutions will rely on NDE measurements as well as provide a cross-reference on the utility of the NDE techniques. For example, the high resolution thermal method shown schematically in Figure 10 will be used to provide maps of the local stress gradient fields produced by a crack. This technique relies on a mathematical algorithm derived from thermoelasticity that relates the stress field to the temperature gradients measured by the infrared radiation emitted by the elastic body undergoing cyclic deformation. This information will provide a useful benchmark to the stress fields predicted by the fracture mechanics analyses. Also, a special facility shown schematically in Figure 11 has an x-ray cat scanner system mounted to a servo-hydraulic loading frame that is anticipated to provide 3-D images of microcracks in the 10-50 micron size. The 3-D image is constructed by processing x-ray exposures taken from various locations around the specimen. This technique will allow for the determination of both the orientation and crack growth increments of small cracks during testing rather than inferring this information from an examination of the fracture morphology after the specimen fails.

Finally, the long-range goal of the ASIP program is to develop user friendly, economical, and quantitatively reliable inspection techniques for field applications. In addition to the thermal flux method previously discussed, an ultrasonic phased array technique [8] currently under development shows promise as a candidate for field applications. The technique is similar to phased array RADAR and provides high resolution detection, image aberration correction, and quantitative characterization through signal processing. As illustrated in Figure 12, this technique has the potential to detect fatigue cracks and corrosion by scanning a relatively large field-of-vision during the inspection.

SUMMARY

NASA has initiated a research program with the long-term objective of supporting the aerospace industry in addressing issues related to the aging commercial transport fleet. The interdisciplinary program combines advanced fatigue crack growth prediction methodology with innovative nondestructive examination (NDE) technology with the focus on multi-site damage (MSD) at riveted connections. The fatigue crack growth prediction methodology is centered around the plasticity-induced closure model developed at NASA which will be specialized for MSD and implemented into a global-local analysis scheme for analyzing a complex built-up structure such as a stiffened fuselage. The NDE technology will focus on advanced thermal, magnetics, and ultrasonics methods. Interdisciplinary laboratory research will focus on acquiring a more fundamental understanding of the physics of fatigue crack growth and will employ thermal methods for stress gradient field mapping and 3-D x-ray radiography for crack imaging. A methodology verification program will be coordinated with the airframe manufacturers to enhance our ability to achieve meaningful technology transfer.

All program elements have been initiated. A fracture mechanics evaluation of the concept of pressure proof testing the fuselage of aging airplanes to screen for MSD has been completed. The results of this evaluation indicate that the required proof test interval would have to be about 275 cycles to be considered as an augmentation to NDE. Current activities are focused on generating meaningful fracture mechanics solutions to 3-D crack configurations at single rivet holes and 2-D MSD configurations. The thermal flux method has been demonstrated to be capable of detecting a lap splice joint disbond in the laboratory. The near-term NDE plans are to conduct a field test demonstration of the thermal method at an inspection depot on an actual airplane.

REFERENCES

1. Newman, J. C., Jr., "A Crack Closure Model for Predicting Fatigue Crack Growth Under Aircraft Spectrum Loading," Methods and Models for Predicting Fatigue Crack Growth Under Random Loading, ASTM STP 748, American Society for Testing and Material, 1981, pp. 53-84.
2. Phillips, E. P. and Newman, J. C., Jr., "Impact of Small-crack Effects on Design-Life Calculations," Experimental Mechanics, Vol 29, No 2, June 1989, pp. 221-225.
3. Tan, P. W., Raju, I. S. and Newman, J. C., Jr., "Boundary Force Method for Analyzing Two-Dimensional Cracked Bodies," NASA TM-87725, May 1986.
4. Shivakumar, K. N., Tan, P. W. and Newman, J. C., Jr., "A Virtual Crack-Closure Technique for Calculating stress Intensity Factors for Cracked Three Dimensional Bodies," International Journal of Fracture, Vol. 36, 1988, pp. R43-R50.
5. Dawicke, D. S.; Poe, C. C., Jr.; Newman, J. C., Jr.; and Harris, C. E.; "An Evaluation of the Pressure Proof Test Concept for Thin Sheet 2024-T3," NASA TM-101675, 1990.
6. Ransom, J. B. and Knight, N. F., Jr., "Global/Local Stress analysis of Composite Panels," NASA TM-101622, 1989.
7. Winfree, W. P.; Welch, C. S.; James, P. H .; and Ciamer, E.; "Thermographic Detection of Delaminations in Laminated Structures," Review of Progress in Quantitative Nondestructive Evaluation, Edited by D. O. Thompson and D. E. Chimenti, Vol. 8, pp. 1657-1662, 1989.
8. Madaras, Eric I.; "Ultrasonic Phased Array Test Bed System (UPATS) with LaRC," Solicitation No. 1-33-1250-0258.

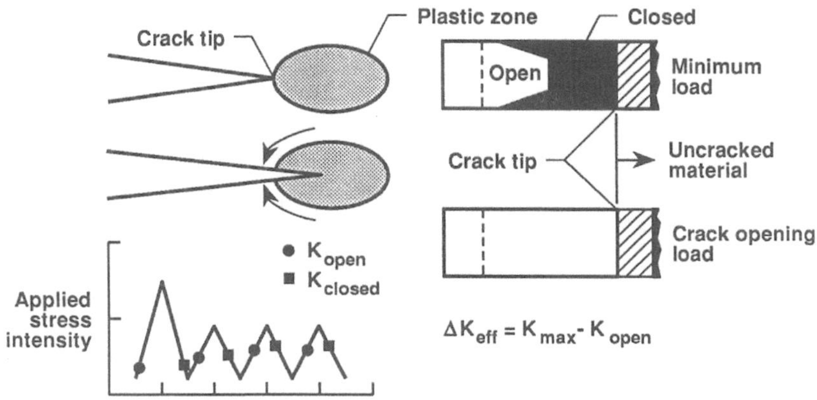

Fig. 1. Fatigue crack growth controlled by closure mechanism

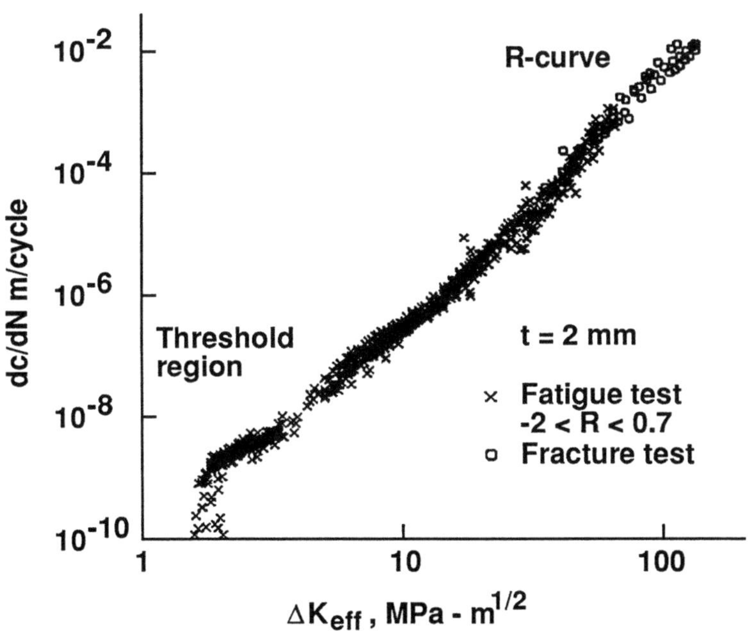

Fig. 2. Crack growth behavior of thin sheet 2024-T3

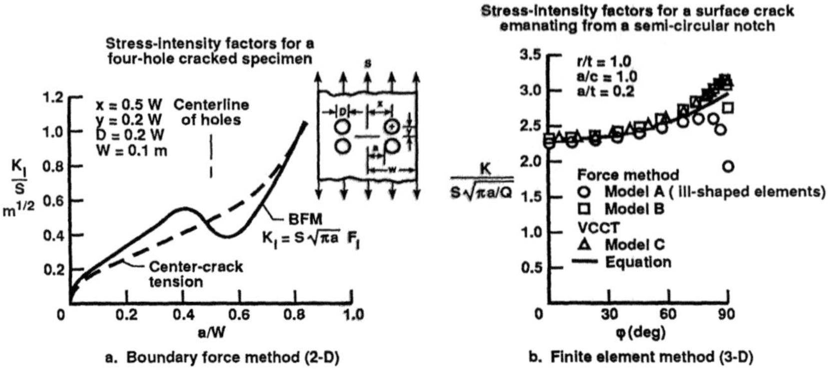

Fig. 3. Fracture mechanics analyses to 2-D and 3-D crack geometries

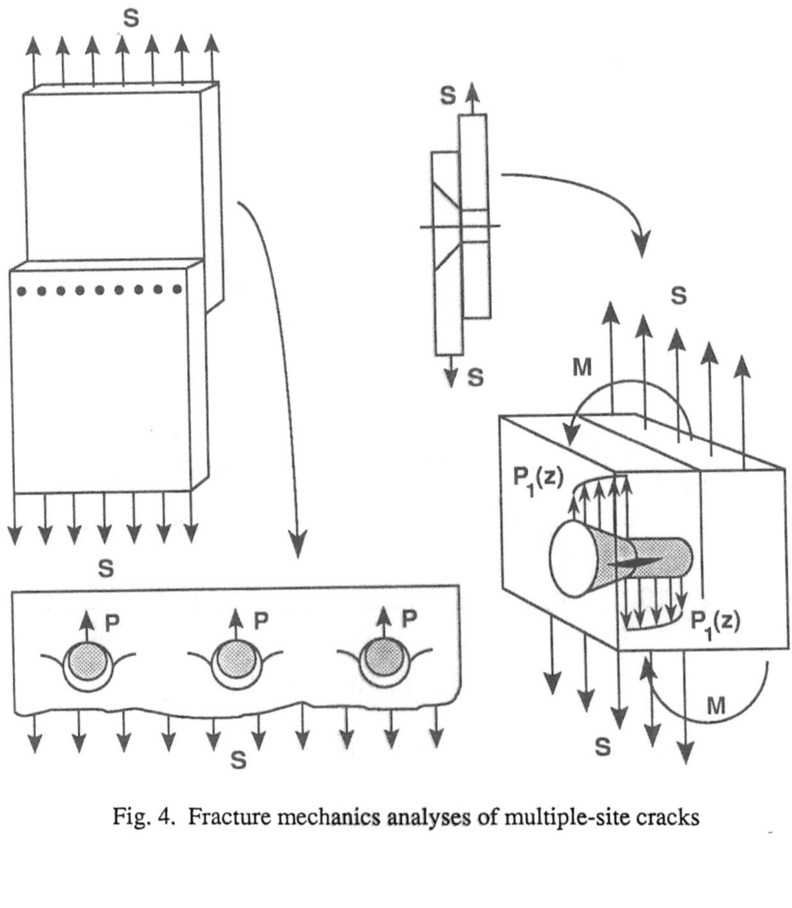

Fig. 4. Fracture mechanics analyses of multiple-site cracks

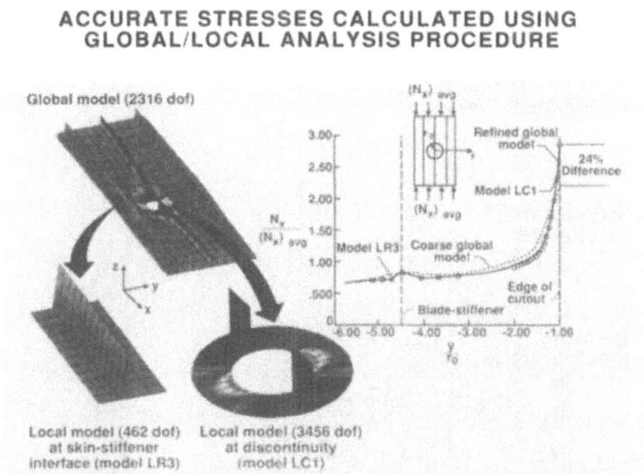

Fig. 5. Glocal/local analysis methodology for an airframe

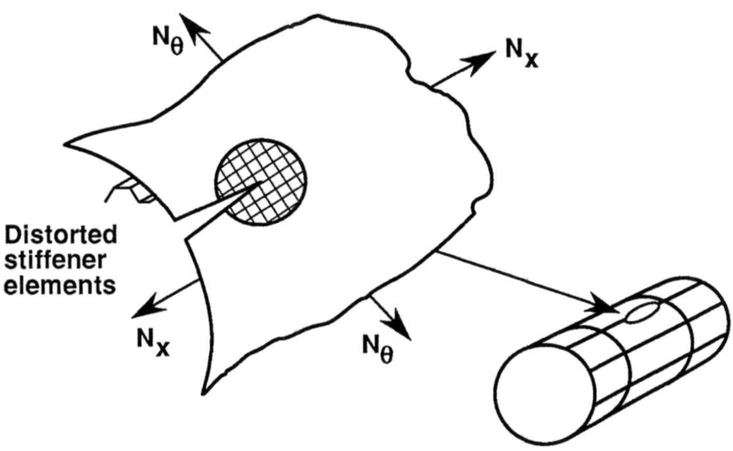

Fig. 6. Global/local analysis with adaptive mesh strategy

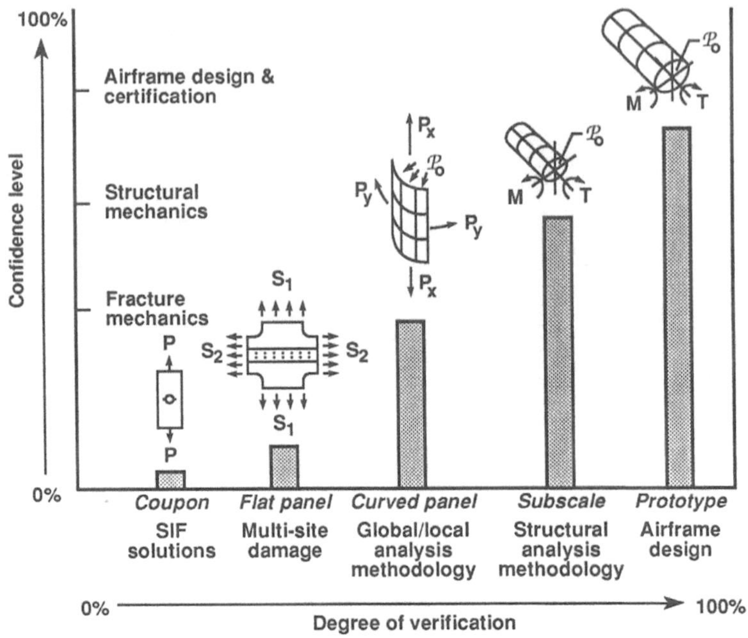

Fig. 7. Prediction methodology verification program

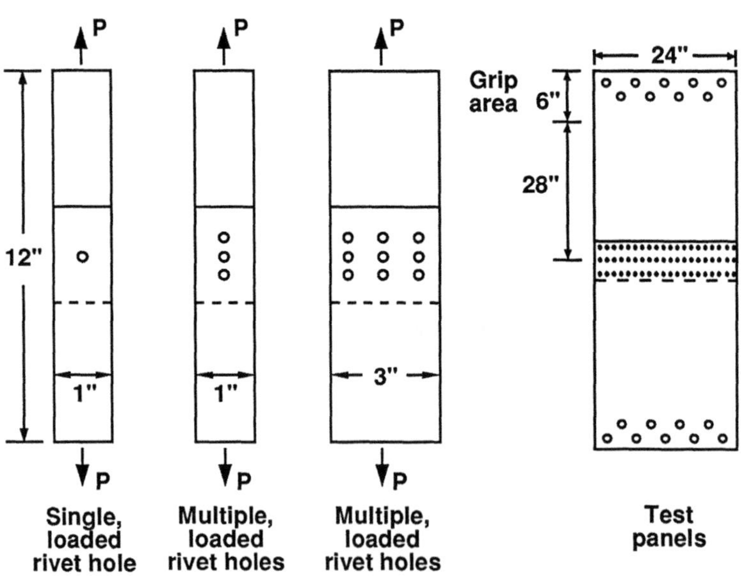

Fig. 8. Coupon and panel test specimens

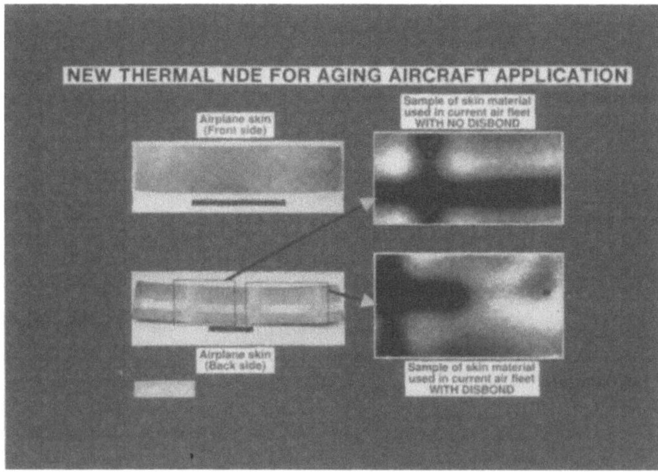

Fig. 9. Application of thermal flux method to aging aircraft

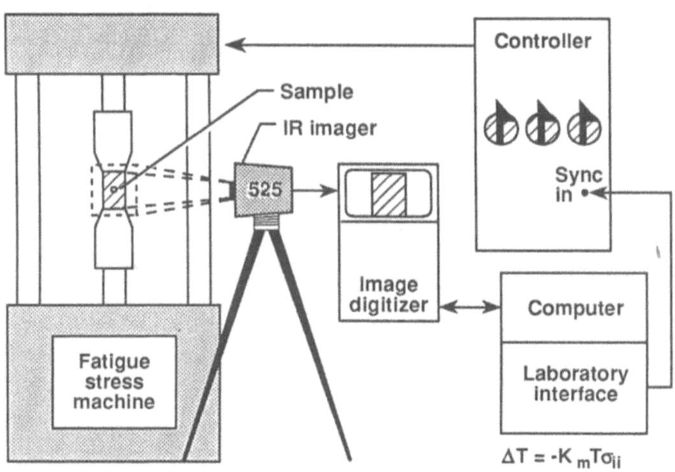

Fig. 10. Laboratory setup for stress field resolution by thermoelasticity

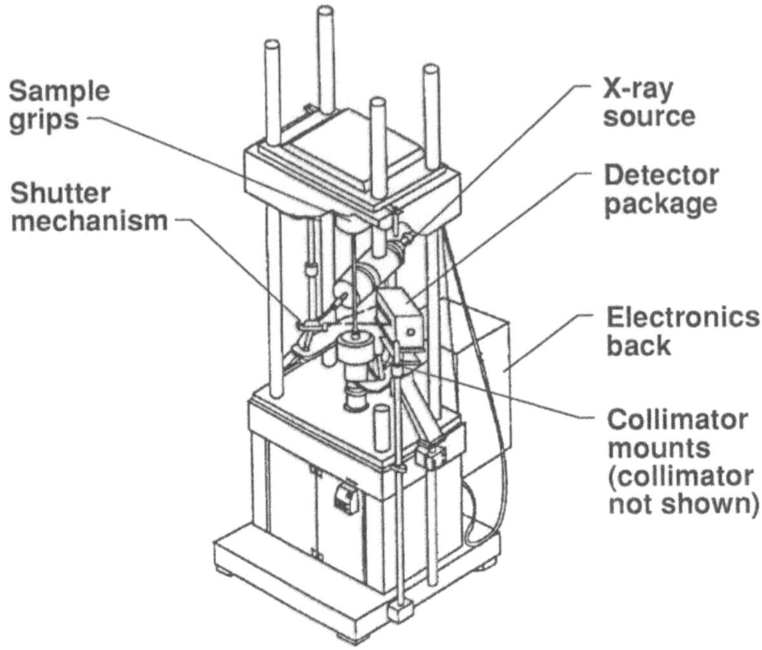

Fig. 11. Quantitative experimental stress tomography system (Quest-1)

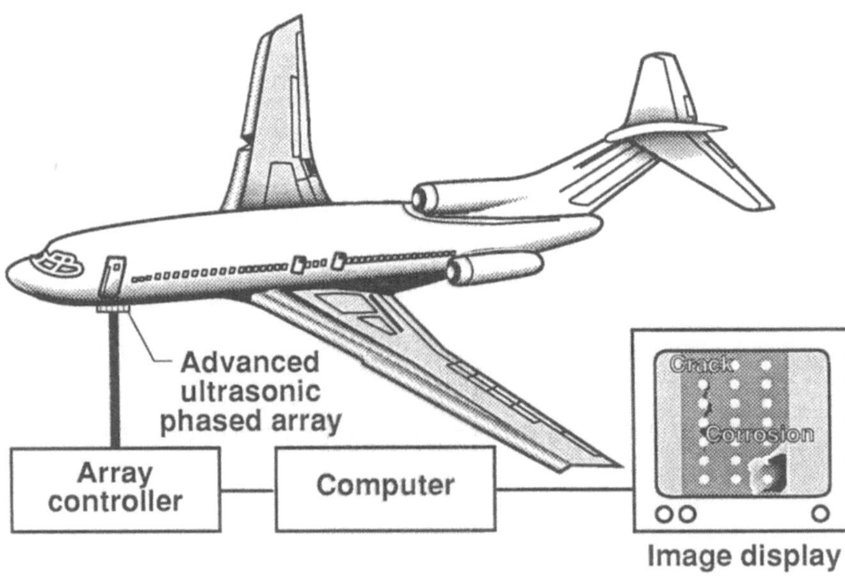

Fig. 12. Application of ultrasonic phased array imaging to aging aircraft

11. The Aloha Airlines Accident – A New Era for Aging Aircraft

Presentation by
William R. Hendricks
Director of Accident Investigation
Federal Aviation Administration
to
International Symposium on Structural Integrity
of Aging Airplanes
March 21, 1990
Atlanta, Georgia

This report concerns some of the factors relating to the Aloha Airlines accident which occurred on April 28, 1988, in the Hawaiian Islands. What is different about this occurrence is that in the past it has usually been the high fatality spectacular catastrophic accidents that have worked as the catalyst for change. High visibility aviation accidents with great loss of life would serve as the trigger for stampeding the government and industry into action. This phenomena is sometimes referred to as the "tombstone effect." Examples of this premise are: (1) the Eastern Airlines flight 66, B-727 approach accident that occurred at the JFK International Airport on June 24, 1975, killing 113 passengers and crew. Although thunderstorm/down draft accidents, certainly, were not a new phenomena at that time, this accident because of the high visibility focused on it precipitated the first full scale efforts by the FAA and the industry to develop and implement windshear detection and training programs, and (2) the spectacular PSA mid-air collision accident over the city of San Diego on September 25, 1978, which claimed 144 lives. While, certainly, not the first mid-air collision accident, resulted

Springer Series in Computational Mechanics
Atluri, Sampath, Tong (Eds.)
Structural Integrity of Aging Airplanes

in massive changes to the National Airspace and Air Traffic Control System including a new requirement for a network of terminal control areas which are still in place today.

While the Aloha Airlines accident wasn't as dramatic or have the high numbers of fatalities as the above referenced accident, it did garner considerable media attention and public interest. More importantly it sparked the interest of people who count when it comes to effecting change in the aviation system, namely, the aviation technical community. Thus, it is believed that this accident will be remembered as the precipitant event that launched a new era featuring advanced technology, improved procedures and higher standards for maintaining the world's aging aircraft.

In examining this event we will first review the in-service structural history of the B-737 and the interrelationships between the FAA, the manufacturer, and the operator as they relate to the causal factors involved in the accident; we will then evoke at some of the specific activities and recommendations for corrective action that have been initiated since this occurrence.

Aloha Airlines Flight 243, a Boeing 737-200 was on a regularly scheduled passenger flight from the Island of Hilo to Honolulu, Hawaii. Shortly after leveling the airplane at flight level 240 with the first officer at the controls the crew heard a loud "clap" or "whooshing" sound followed by rushing air behind them. They observed that the cockpit door was gone and noticed "blue sky" where the first class cabin ceiling had been. The captain immediately took control of the airplane and began a 4000 feet per minute emergency decent using the speed brakes while maintaining an indicated airspeed of 280-290 knots. ATC was advised of the emergency and approved a diversion to Maui. The captain stated that during this time, they were experiencing severe vibration and that the airplane handled very rough. He slowed the airplane to about 200 knots as the flight descended through 10,000 feet. The flaps were set to 5 degrees but when 15 degrees flaps were tried the airplane became less controllable so they were returned to the 5 degrees position. The captain attempted to slow the aircraft to the approach speed range but as it decelerated below 170 knots the airplane again became less controllable so the speed was maintained at 170 knots throughout the approach. It was also noted that they had lost power on the No. 1 engine. Flight 243 landed on runway 2 at the Kahului Airport at 1358. A normal touchdown and landing was accomplished using the brakes and the No. 2 engine thrust reverser to stop on the runway. The entire episode from the time of the structural failure at FL 240 to touchdown on Maui lasted approximately 11 minutes.

There were 95 persons on board--89 passengers, 2 flightcrew, 3 flight attendants, and an FAA air traffic controller in the jumpseat. The senior flight attendant in the first class cabin was lost during the explosive decompression. Fifty-nine passengers were taken to the hospital with injuries ranging from minor to serious.

All factors considered, it was a miraculous escape for all but one of those persons on board flight 243 and is indeed a testimony to the skill of the flightcrew and to the residual strength of the B-737.

Investigation disclosed that the aircraft sustained substantial airframe damage to the fuselage upper skin structure from the bulkhead aft of the passenger entry door to the fuselage circumferential structural frame just forward of the leading edge of the wings. The entire fuselage structure from above floor level beginning at fuselage station 360 to 540 (180 inches) was missing. All fractures associated with the perimeter of the missing structure were typical of static overload separations.

Several of the support beams beneath the floor were cracked and the floor panels had lifted up into the forward cabin area. The underfloor damage was subsequently found to have caused movement of the left engine fuel control cable to the shut off position, and then the cable broke. The passenger oxygen lines from the oxygen bottle to the masks were broken when the roof separated, thus disabling the passenger oxygen system.

This was a high-time, high-use airplane. It was manufactured in 1969 and at the time of the accident had accumulated 35,493 flight hours. Of more significance, it had 89,090 cycles which equates to about 13 flights or pressurization cycles a day for 19 years. This was the second highest number of flight cycles in the worldwide B-737 fleet. A sister ship, also owned by Aloha Airlines, has accumulated 90,051 cycles, the highest number for any B-737. That aircraft was in the Aloha hangar for maintenance at the time of the accident.

It is fairly well understood that the safety and integrity of the system relies on the interdependent relationship between the FAA, the manufacturer, and the airline in properly carrying out their individual responsibilities in the design, production, certification, tracking maintenance and surveillance of all aircraft in the system. Notwithstanding the facts of this accident, the contention is that the system almost worked.

Very early on in the service history of the B-737, the manufacturer issued several service bulletins pertaining to disbonding, corrosion, and fatigue cracking around the fuselage lap joints.

The fuselage is divided into sections, with sections 41, 43, and 46 comprising the majority of the pressure vessel. These sections, along with section 48, are butt joined at circumferential frames to form the entire fuselage. Section 43 forms the forward cabin area from BS 360 to BS 540, the area of separation on the accident airplane. The skin panels in the upper lobe of section 43 are the length of the entire section--about 18 feet. Each panel is constructed by bonding together two pieces of .036 aluminum sheets. The tear straps are formed by an acid milling process which takes away one layer of the skin on the prepared surfaces. The result is a lighter waffle-like surface. The panels are joined longitudinally by overlapping the edges of the individual panels and fastening them with rivets and a bonding process. One row of rivets secures the lap joint to a stringer underneath the skin. Through line no. 291, single thickness (.036") sheets at the 3-inch overlap area, or lap joint, were joined together with a "cold bonding" process. This cold bonding process used an epoxy impregnated woven cloth the width of the joint to join together the longitudinal edges of the skin panels. The joint was then bonded and mechanically fastened together with three longitudinal rows of countersunk rivets, the middle row of which attached the lap joint to the underlying stringer. This bond cured at room temperature after assembly. After line no. 291 (1972), because of delaminating at the lap joints Boeing discontinued the cold bonding process and used a smooth close fitting "fay" surface sealed lap joint with increased joint thickness. The top overlapping panel was retained at the double panel thickness (.072) which formed a three layer lap joint. This additional thickness also alleviated the "knife edge" problem caused by the countersunk rivets in the single layer lap joints. This redesigned lap joint became necessary as problems of delamination of the cold bonded lap joint were encountered.

The first service bulletin which concerned lap joint delamination, corrosion and repair was issued in 1972. Boeing revised this service bulletin three times, elevating it to alert status with the third revision on August 20, 1987. This upgraded status followed several reports from operators of fatigue cracking along the fastener locations of the fuselage lap joints. An inspection and repair procedure was included in the service bulletin.

Following this third revision of the service bulletin by Boeing along with confirming reports of multiple site fatigue cracking along the upper row fasteners, the FAA acted by issuing an Airworthiness Directive in October 1987 which required a visual inspection of the fasteners along the No. 4 stringer (4L and 4R) from fuselage station (FS) 360 to 1016 on all B-737's with 30,000 or more cycles. Any cracks detected would have to be repaired in accordance with instructions contained therein. It also specified that if any cracks were found on the visual inspection, an eddy-current inspection was to be performed along the entire stringer.

In accordance with the requirements of the AD issued by the FAA in November 1987, Aloha inspected its B-737 fleet and on the visual examination of the accident airplane, at least two areas of corrosion were found along the No. 4 stringer. These areas were properly repaired and, according to Aloha, the eddy-current inspection along the entire stringer, as required by the AD, was also performed. According to the maintenance records, no other defects were found. However, after the accident, the airplane was again inspected by the NTSB and many other areas of detectable multi-site cracking were found along the No. 4 stringer as well as along many other areas of the fuselage lap joints. Aloha's maintenance records did not show evidence of an eddy-current inspection although the record of the repair was shown.

Also in the fall of 1987, prior to the FAA's issuance of the AD, Boeing, as part of its aging fleet research program, sent a team of engineers to Aloha to collect data and to inspect some of the high time 737's. Although the accident airplane was one of four Aloha 737's with over 65,000 cycles, it was not one of the airplanes examined during this visit. However, in general discussions with Aloha following the visit and in its report delivered to the operator on April 14, 1988 (2 weeks before the accident), the Boeing team did express concern about the condition of the higher time airplanes as well as adequacy of the maintenance program to cope with the existing corrosion problems in this highly corrosive operational environment.

In theory the system was working. All of the responsible parties were attuned to the lap joint problem and had, in fact, taken some form of action to correct a deficiency. Obviously, the actions that were taken were insufficient and, in this case, the system failed and an accident occurred.

Congressman Oberstar (Chairman of the Subcommittee on Aviation; House Public Works Committee) summed up the episode at the First Aging Aircraft Conference in June of 1988. He said, in part:

-- What concerns us--is that most of the right steps were taken but the system still failed.

-- Boeing should have issued a service bulletin to inspect cracks along the Nos. 4 and 10 stringers--and they did.

-- The FAA should have issued an AD requiring an inspection of the stringer #4 lap joint--and they did, about 6 months before the accident.

-- Aloha should have acted on these warnings and they did--but they didn't find the problem.

Some of the findings in the NTSB report were directed at these issues:

-- Were the service bulletins issued by Boeing explicit enough or early enough?

  The NTSB stated in their report that Boeing should have issued an alert bulletin in 1984 after the first reports of cracking were received. The regular service bulletin that was issued was upgraded to Alert Status in 1987.

-- Was the Airworthiness Directive issued by FAA as complete, comprehensive or timely as it should have been?

The Board stated that the FAA should have mandated the inspection of all lap joints instead of limiting the inspection to only the No. 4 stringer lap joints.

-- Were Aloha's inspection personnel sufficiently trained for the task? Were they skilled and experienced enough to recognize the problem when they encountered it? Was the inspection as vigorous and meticulous as it should have been to discover cracks?

   The answer was no to all questions. The NTSB stated that a proper eddy-current inspection should have detected those cracks that were found along the No. 4 lap joint after the accident. Either the inspection was not performed or it was ineffectively performed.

-- Was the FAA's surveillance of Aloha's maintenance program adequate?

   The Board stated that the FAA failed to evaluate properly the Aloha Airlines maintenance program and to access the airlines inspection and quality control deficiencies.

A number of other issues were also identified in the aftermath of this accident including, the adequacy of current maintenance practices for high time aircraft, the adequacy of aviation technology to detect early signs of impending failures of aircraft structure, the adequacy of current corrosion control techniques for high time aircraft and the human factors realities of repetitive visual and NDI inspections.

A good many actions have already been completed and many more are actively underway. With regard to the B-737, the FAA did issue an AD last year that required or will require, depending on the number of airplane cycles, replacement of all upper row rivets along the lap joints for aircraft number 291 and below. In addition, in early March, based on recommendations made by an industry/FAA task force, the FAA adopted regulations that will require extensive structural modifications of older 727's, 737's, and 747's when they reach their "economic design goal." The AD's are effective on April 17, 1990 and will have an immediate impact on 737's and 747's that are 20 years old with 75,000 cycles, and 727's that are 20 years old with 60,000 cycles. A total of 115 U.S. airplanes are in the immediate category, and it is estimated that it will cost 142 million to modify them over a 4 year period. Additional aircraft will be covered by the AD as they reach the age and cycle threshold of their "economic design goal."

The requirements of these AD's represent a significant change in FAA policy regarding the maintenance and safety of aging airplanes in that it sets a life limit on individual parts considered to be critical structure. This differs from the longstanding practice of repetitive inspections and necessary repairs to assure safety throughout the life of an aircraft.

The FAA is also adopting many of the Task Force recommendations on corrosion that will result in a comprehensive industry-wide corrosion prevention/corrosion control program. This effort will have several elements, including corrosion level definitions, program implementation, repeat intervals and a mandatory reporting system for corrosion levels found above the level allowed by FAA.

In view of these new requirements the industry is facing some fairly substantial expenditures which will be required to maintain older airplanes.

The General Accounting Office (GAO) report on aging aircraft points out that airlines must now weigh the price and availability of new aircraft against the operating efficiency and costs to maintain the older planes in a safe operating condition.

The report states, "In the coming years, airline decisions to sell or maintain older planes will be made more frequently. About 31 percent of the U.S. fleet now exceeds the economic design goals originally set by the manufacturer and, by the year 2000, 64 percent of the worldwide fleet of U.S.-manufactured aircraft will be 20 or more years old..."

The reported average cost for a new airplane is about 55 million dollars, while a major overhaul for an older aircraft can run over 2 million dollars for a Boeing 727 and, depending on how much work needs to be done, between 4 and 20 million dollars for a Boeing 747. Based on these comparisons, it is clearly understandable why we are seeing an increase in the aging aircraft fleet.

In conclusion, while the loss of one life in the Aloha accident is a tragedy in itself, we are all thankful that it didn't take a major catastrophe or a "tombstone effect" to launch the many initiatives now underway to improve the safety of aging aircraft. These activities will ultimately affect several thousand transport airplanes worldwide, will involve millions of dollars worth of structural repairs, and will require considerable input and resources from the FAA, manufacturers, and the operators. In addition to these efforts, it is of great benefit to the aviation industry to have the best technical experts in the field, meet together to discuss and resolve some of those challenging issues. The entire industry will look forward to the results of this conference.

Thank you.

12. Fuselage Longitudinal Splice Design

AMOS W. HOGGARD

Douglas Aircraft Company
McDonnell Douglas Corporation
Long Beach, California

Summary

In the past few years there has been much discussion about the design of fuselage longitudinal skin splices. This paper presents one manufacturer's approach to the design of those splices. As such, this paper is not meant to be an authoritative source on the design of such details, but should be viewed as a case history of one way that has proven successful. This paper also details where research is needed to improve the design processes and the subsequent in-service inspection requirements for fuselage structure.

The Design Process

The design process for fuselage longitudinal splices begins by establishing the request design goals for the structure. Nominally there are four basic design criteria that must be developed for splice design (see Figure 1): ultimate strength, fatigue resistance, damage tolerance characteristics, and corrosion prevention. These four design criteria can be quantified by establishing exactly two design goals: the overall structural loadings and the design life of the aircraft.

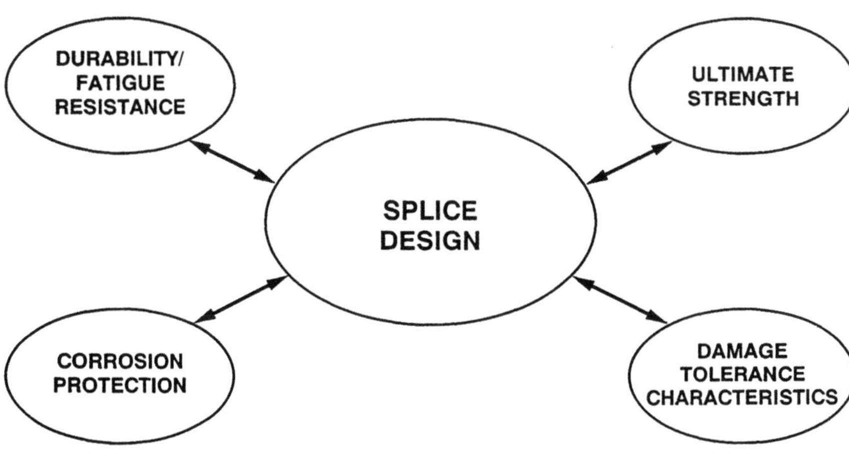

FIGURE 1. DESIGN CONSIDERATIONS FOR SPLICES

Springer Series in Computational Mechanics
Atluri, Sampath, Tong (Eds.)
Structural Integrity of Aging Airplanes

Aircraft design starts with the definition of all design conditions and design goals for the structure. These conditions not only specify the extreme structural design requirements for the aircraft. but also the length of time (years. flight hours. and landings) the aircraft should operate without suffering degradation due to fatigue and corrosion. The design parameters for a modern transport aircraft represent extreme loading conditions, which occur infrequently in the life of the aircraft. Figure 2 presents some examples of the extreme conditions, including encounters with exceptionally high levels of turbulence in a thunderstorm. abrupt maneuvers to avoid collision. landings at descent rates far above normal, and a dive pullout following an inadvertent upset. On a statistical basis, these extreme events will occur less than once in the life of an aircraft. In the last decade, the probability of such encounters has decreased because of refinements in our ability to predict adverse weather conditions and the use of predetermined and controlled route structures.

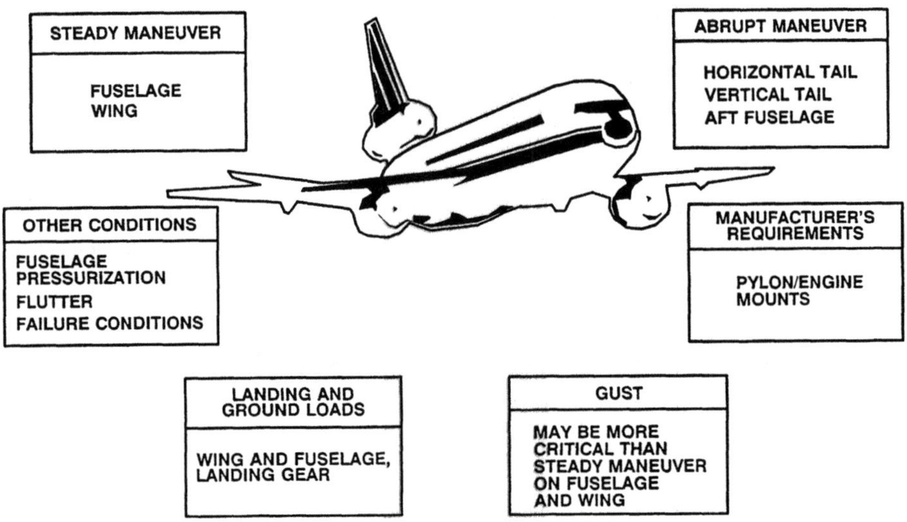

FIGURE 2. CRITICAL-DESIGN CASES

The design conditions for fatigue and corrosion are the routine loading and environmental conditions that an aircraft sees on a day-to-day basis. Normal loading conditions, experienced by each aircraft, cover all variations in ground and flight loads. These include loads due to routine taxi, takeoff, climb, descent, and landing. Figure 3 illustrates a representative internal load time history that results from normal operation.

The extreme design conditions are dictated by regulatory authorities (FAA, CAA, JAA, etc.) throughout the world. The authorities publish regulations similar to the Federal Aviation Administration (FAA) Federal Aviation Regulations (FAR). In certain instances, Douglas not only adheres to the authorities' recommendations but also goes beyond the requirements in order to meet established goals.

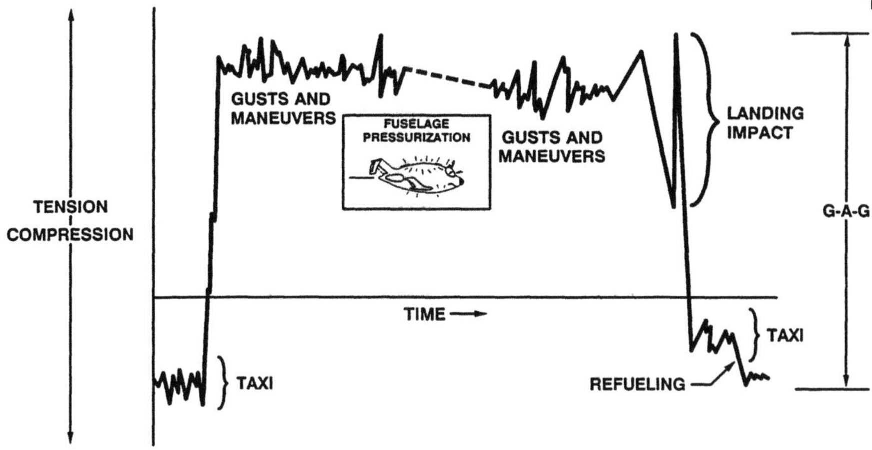

FIGURE 3. AIRCRAFT STRUCTURE IS DESIGNED FOR LONG LIFE DUE TO REPEATED CONDITIONS (FAR 25.571)

Douglas' design goals are initially determined for a life of 20 years. In the past, this has been established on both a financial and a technical basis. In the first case, the customer expects an amortization period of at least 20 years, and in the second case, it is reasonable to expect that breakthrough technologies will be introduced on the average of every 20 years. These breakthrough technologies generally tend to make the older technology aircraft less economical to operate. In order to meet the 20-year requirement, Douglas designs and builds products for a minimum average useful life of 40 years. This provides the customer with a high statistical probability that the aircraft will be relatively free from fatigue or corrosion problems for 20 years, providing the routine maintenance tasks are performed.

Design of primary load-carrying structure, including splices, is a balance between (1) the amount of material required to prevent structural failure from one of the extreme loading conditions described above (ultimate strength), (2) the amount of material required to sustain structural integrity with certain levels of hidden or undiscovered damage (damage tolerance or fail-safe), and (3) the amount of material required to delay the onset of widespread cracking until the aircraft has reached its useful life (fatigue resistance). This is done through a process of selecting design-limit stress levels, materials, and design features that accomplish all three requisite conditions.

There are two basic fuselage longitudinal splice designs — the lap splice and the butt splice (Figure 4). Typically, Douglas splices are adaptations of the ones shown in Figure 4, and are developed as a result of the design process shown in Figure 5. The process is highly dependant on empirical results from testing. To date, a reliable analytical technique to accurately predict splice fatigue life has yet to be fully developed. This is due in part to the many variables that affect the life

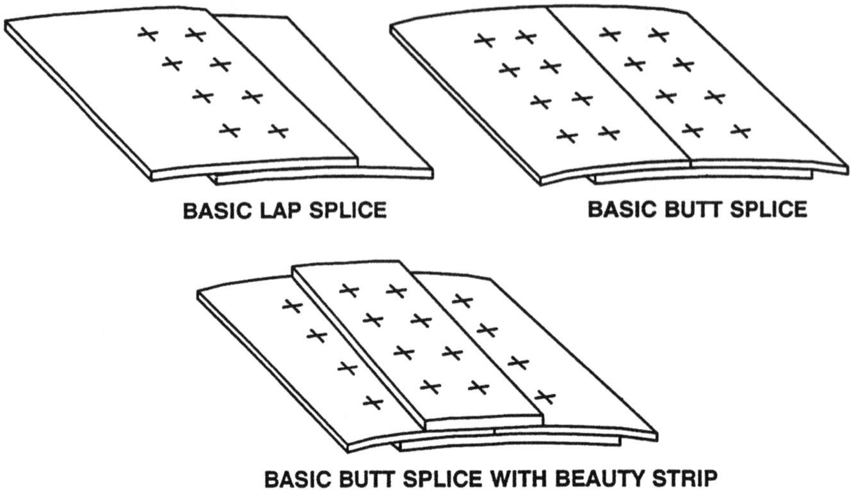

FIGURE 4. BASIC SPLICE DESIGN CONCEPTS

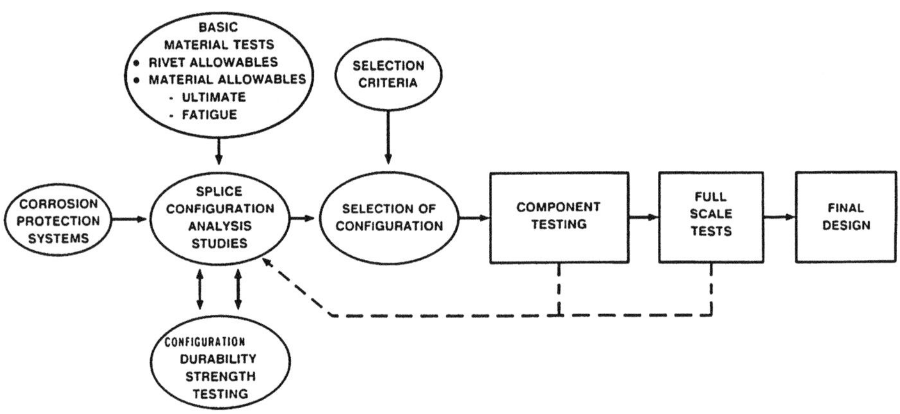

FIGURE 5. THE SPLICE DESIGN PROCESS

of the splice. Figure 6 details some of the more significant variables that affect the fatigue life of a splice. Hole preparation technique, if not standardized and closely controlled, could lead to a variety of conditions, including knife edges and open holes, which shorten the life of the splice. Adequate detail design of the splice, including appropriate ratios of bearing-to-bypass loads, has a very significant effect on splice fatigue life. In-plane and out-of-plane bending loads, including the basic splice offset and the nonlinear effects of fuselage pressure pillowing, together with rivet bending, also have a significant effect on splice life. These particular effects are worsened by the presence of a faying surface corrosion protection system. Appropriate rivet sizing for the material thicknesses being joined can materially affect the hole-filling and clamp-up capability of the fastener, which again affects fatigue life.

HOLE PREPARATION
- **SIZING**
- **COUNTERSINK**
- **DEBURRING**

LOAD TRANSFER
- **BEARING/BYPASS LOAD**
- **BYPASS STRESS**
- **BEARING STRESS**

INPLANE/OUT-OF-PLANE BENDING
- **FAYING SEAL THICKNESS**

FASTENER CLAMP UP

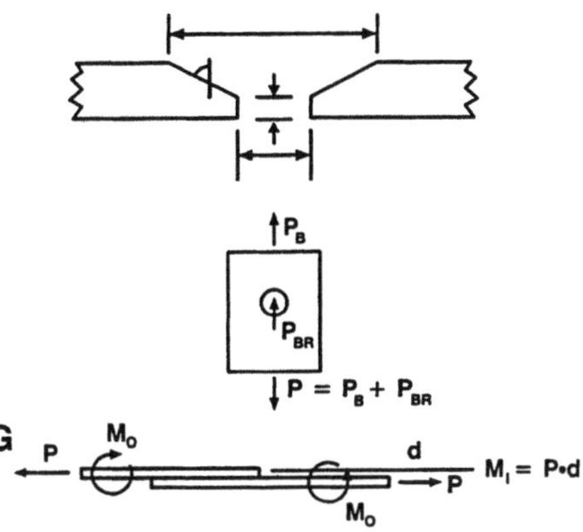

FIGURE 6. FACTORS AFFECTING SPLICE FATIGUE LIFE

The basic fuselage longitudinal splice designs used on Douglas products are shown in Figure 7. Unique to the Douglas product line is the use of finger or scallop doublers to enhance fatigue life and crack detectability. Of interest, the DC-9 longitudinal splice is the only lap splice design that Douglas employs. As a general policy, Douglas does not use any type of structural bonding techniques for primary structure.

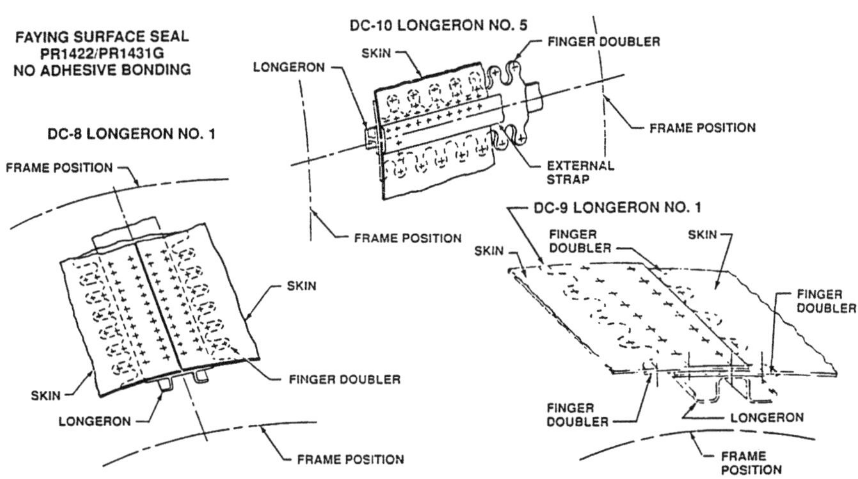

FIGURE 7. TYPICAL DOUGLAS SPLICE DESIGNS (LONGITUDINAL SKIN SPLICE)

Design for Ultimate Strength

Design for ultimate strength involves establishing various material allowables and comparing them to the requisite design conditions that may exist in a given area of the splice. Normally, the design is based on the worst possible conditions of splice location and loading. In doing this, the splice configurations are minimized.

Various tests are normally run to establish the material allowables of the fastening systems with the materials used. These tests use the basic materials and methods used in making the splice. Various material thicknesses and attachment sizes, assembled using the hole preparation techniques for the actual splice configuration, are tested to establish the material allowables. A multitude of tests are run to reduce the risk of statistical intangibility and arrive at material allowables that are meaningful and reliable.

Design for Fatigue Resistance

Designing the splice for structural strength is a relatively simple procedure compared to designing for fatigue resistance and damage tolerance. At Douglas, design for fatigue resistance implies that the design will not be the limiting factor in determining the overall life of the fuselage structure. This criterion stipulates several things: first, that the splice will have a long fatigue life with a high degree of confidence; second, that fatigue cracking in other design features will precede the onset of cracking in the splice; and third, that when cracking does occur in the splice, the cracks will be easily detected.

Design for a long and reliable life begins with selection of an appropriate fatigue life goal for the structure being designed. The difference between the design life goal mentioned earlier and the fatigue life goal is defined by establishing scatter factors that ensure that the fleet will reach a certain life without a major fatigue problem. This statistical concept is illustrated in Figure 8. These factors are derived from the Douglas structural life policy, which is based upon the two-parameter lognormal distribution. The policy says that for any particular fleet of aircraft there will be a characteristic median life and standard deviation. The median life is also called the

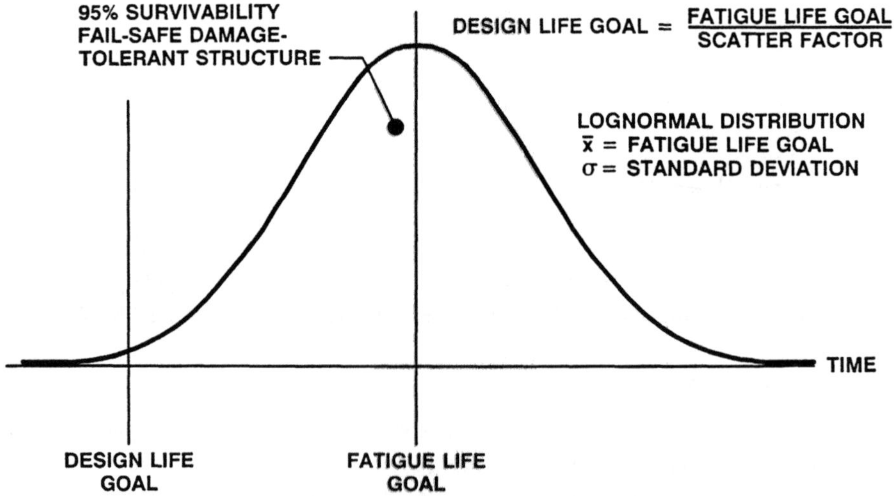

FIGURE 8. STRUCTURAL LIFE CONCEPT

fatigue life goal. In other words, the fatigue life goal is the life at which 50 percent of the fleet will survive the onset of widespread cracking for a particular design detail. The standard deviation is based on the material characteristics and is modified to account for usage variations. For aluminum structure, the standard deviation is 0.20.

Fuselage structure is considered to be fail-safe structure; i.e., the failure of one isolated component is not immediately a safety concern for the aircraft. For the purposes of design, Douglas uses a survivability rate for fail-safe structure of 95 percent. For aluminum structure, this equates to a scatter factor of approximately two between the fatigue life goal and the design life goal. The scatter factor of two also approximates a fleet failure rate equivalent of 1.0×10^{-9} when considered in conjunction with the probability of encountering limit load.

Even though the splice is designed for a high degree of fatigue resistance, it must also be designed for ease of inspection. This generally means that the fatigue characteristics must be such that the predominate mode of failure occurs in a structural component visible from the outside of the aircraft. This concept is shown in Figure 9. In this particular example of a DC-9 lap splice, the predominate fatigue failure is expected in the skin of Fastener Row B. Failures in Row A or D are also acceptable, but not in Row C. In the past, the predominate mode of failure has been determined by fatigue testing.

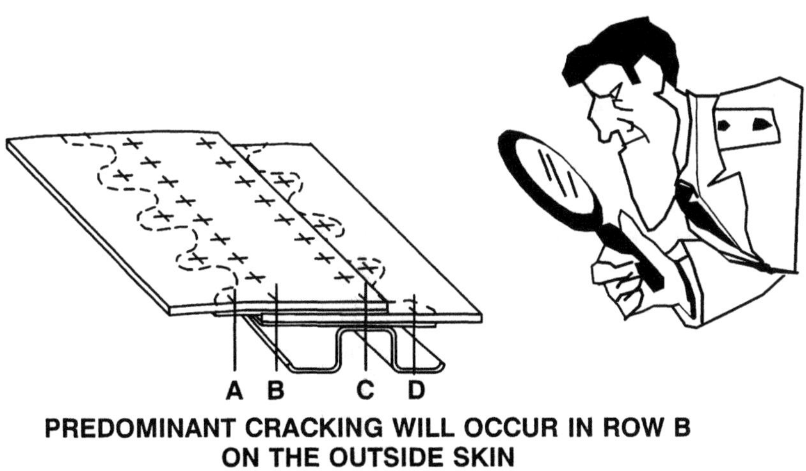

FIGURE 9. SPLICE INSPECTABILITY

The design process illustrated in Figure 5 is an iterative process for developing a splice configuration that both has a fatigue life greater than the fatigue life goal and develops detectable cracks. For a given design stress level and ultimate strength considerations (e.g., material/rivet allowables), many design configurations are selected as possible candidates for the final design. The variables considered include variation in doubler configurations, rivet type and placement, rivet installation technique, and faying surface treatments. Several small 9- to 18-inch-wide specimens similar to the one shown in Figure 10 are constructed for each candidate configuration and are then tested to establish the fatigue life and failure mode. Figure 11 details the total number of test coupons run to establish the final design configurations for the DC-8, DC-9, and DC-10 aircraft.

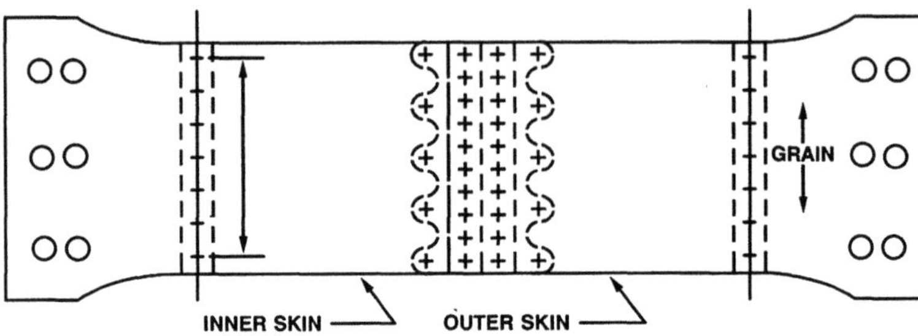

TEST SPECIMEN

FIGURE 10. TYPICAL FUSELAGE LONGITUDINAL SPLICE TEST COUPONS

	COUPONS	COMPONENTS	FULL-SCALE TESTS
DC-8	200+	21	1
DC-9	280	8	3
DC-10	200	6	1

FIGURE 11. FUSELAGE SPLICE DEVELOPMENT TESTS

As shown in Figure 12, the testing to establish design configuration for the splice does not end at the coupon test stage but continues on into component tests and finally the full-scale two-lifetime fatigue test. This is significant because the interaction of design features and more closely represented loadings of the component and full-scale tests may vitally affect the fatigue performance of the best design. Such a situation occurred in the DC-10 development test program, in which a splice design selected in the coupon testing stage did not exhibit detectable fatigue cracks in the curved panel component tests. The discovery of this caused a significant delay in the DC-10 fatigue test program.

Design for Damage Tolerance

If the design for fatigue resistance can be thought of as one end of a very long scale, the design for damage tolerance is logically on the other end of that scale. Such a concept is shown in Figure 13.

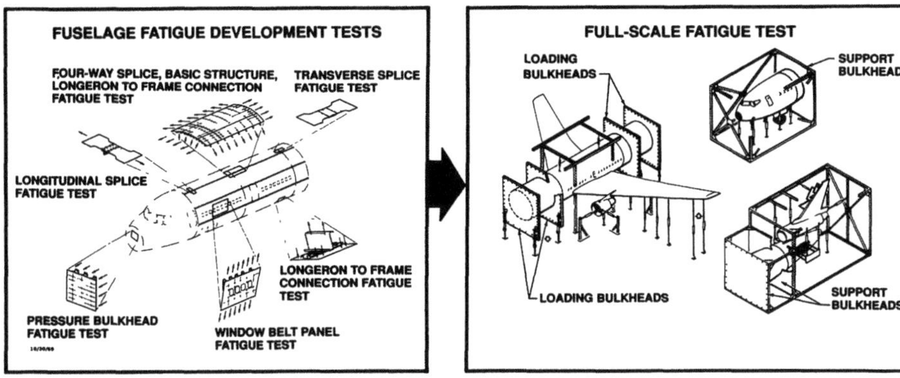

FIGURE 12. THE TESTING PROCESS – DC-10

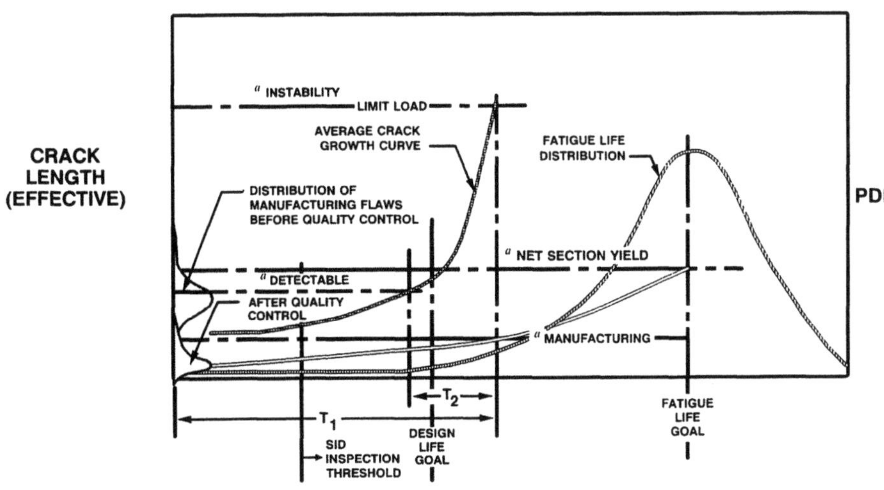

FIGURE 13. THE INSPECTION REQUIREMENT

Damage tolerance concerns cause one to consider the potential growth of a single isolated crack in the design of the splice. A single small crack in itself is not critical to the integrity of the splice. It is the subsequent growth of the crack and loss of residual strength that are of concern. In performing the damage tolerance analysis, the generally accepted practice has been to assume the existence of the largest flaw size that would be missed during quality control inspection on the manufacturing line. For the purpose of the analysis, the flaw is orientated perpendicular to the most significant principal stress direction. The evaluation considers average crack growth characteristics of the material and a realistic fatigue loading spectrum. The results of the analysis produce both the total time for the flaw to grow from the initial flaw size to instability (T1) and the time from detectable flaw size to instability (T2).

Also shown in Figure 13 is the relationship of the damage tolerance considerations to the considerations of fatigue. Even with the most exacting quality control procedures, not all flaws are eliminated. Earlier, in the discussion about fatigue life, the concept of a standard deviation was introduced that accounts, in part, for the inherent flaws within the material. The implication is that the fatigue problem is just a generalization of the damage tolerance analysis but only at a smaller crack size. Repeating the analysis for that smaller flaw size, if properly done (accounting for all loadings and the presence of other cracks), will approximate the fatigue life goal established earlier. For the special case of splices, in which the smaller flaws may be considered to be located at every hole, the instability flaw size is reduced to that of net section yield.

The point of this whole discussion is that the cyclic nature of the loading an aircraft experiences will most likely cause a crack(s) in the structure to grow and eventually weaken the local structure to a point at which it cannot withstand the extreme design conditions. This condition must be prevented. This is accomplished by designing a splice with reasonably slow crack growth characteristics and providing adjacent stiffening members with crack stopping capabilities. Ultimately, protection is provided by ensuring an adequate routine maintenance program augmented by special inspection programs offered by the Supplemental Inspection Documents (SID) as shown in Figure 14.

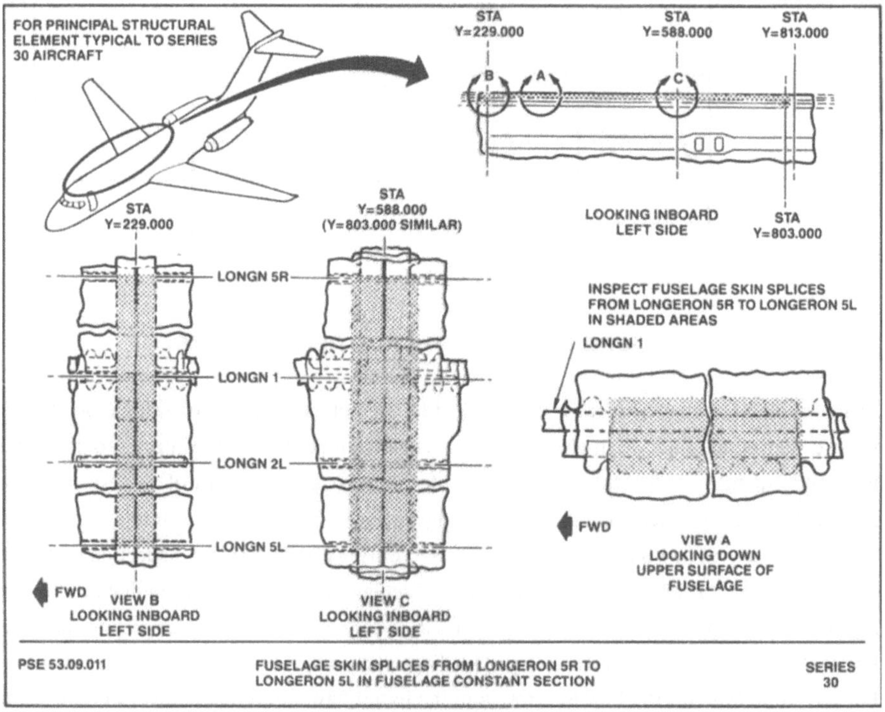

FIGURE 14. DC-9 PRINCIPAL STRUCTURAL ELEMENTS

The procedures used to perform the supplemental inspection of the longitudinal splice joints for cracks employ an eddy current reflectance probe. Figure 15 shows this probe performing an inspection of a splice on a DC-9. This probe and associated equipment provide a means by which the condition of a splice can be rapidly assessed. While the assessment is fast, the detectable flaw size is not as fine as with the eddy current pencil probe. Additional research is needed in this area that both optimizes speed of inspection and detectable flaw size and minimizes inspector fatigue.

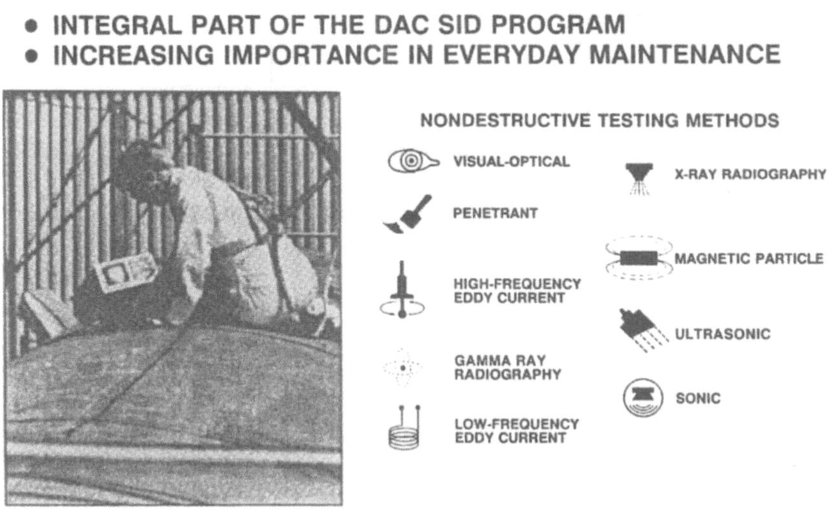

FIGURE 15. NONDESTRUCTIVE TESTING

Design for Corrosion Prevention

Faying surface corrosion can ruin the most carefully designed and tested splice and do it essentially without warning. Any splice contains several faying surfaces, and those surfaces form an area where liquids can collect and cause corrosion. The five-element DC-9 lap splice shown in Figure 16 is a good example of this. Douglas has found that the only successful means to prevent moisture entrapment in the faying surface is to use a faying surface sealant to exclude the moisture. If this is not done, any small leak paths at fasteners or at part edges (Figure 17) will allow the void to become a vacuum at altitude, drawing in moist air during descent, which causes condensation inside the joint. Faying surface corrosion can start when corrosive liquids are present in the joint, most especially when dissimilar metals are involved. Figure 18 details the corrosion-prevention systems employed for the DC-9 lap splice. They adhere to the IATA standards for corrosion prevention. As indicated, Douglas uses a concept of zoned prevention, which puts the maximum protection where it is most needed.

A Case History

The DC-9 was certified and entered service some 25 years ago this year. The aircraft was designed for a short-haul operation with an average flight length of three-quarters of an hour and a design

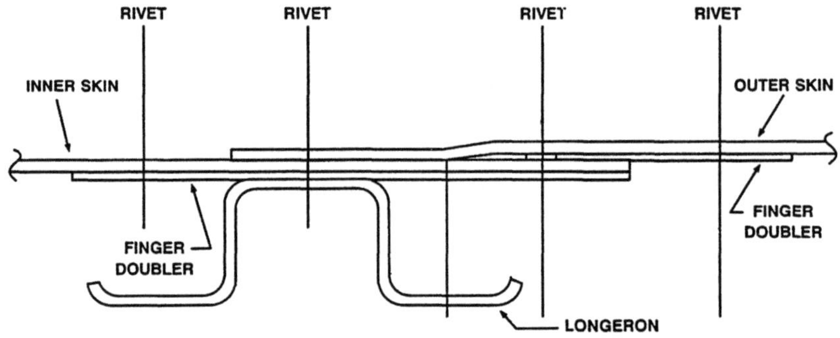

FIVE-ELEMENT SPLICE

FIGURE 16. DC-9 LONGERON NO. 1 SPLICE

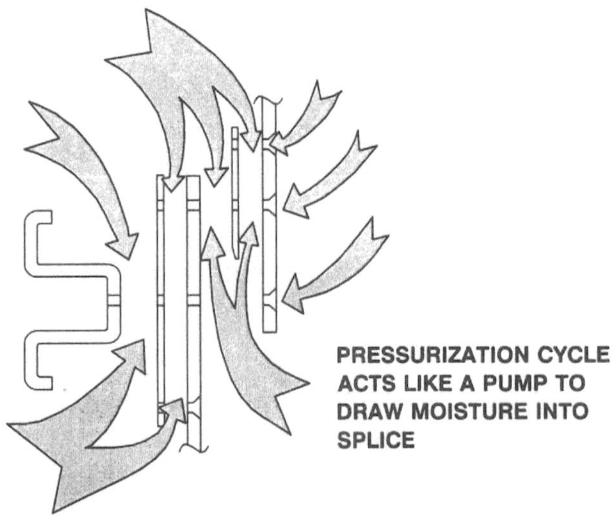

FIGURE 17. MOISTURE INTRUSION PATHS

life goal of 30,000 flight hours/40,000 landings. Figure 19 gives the current status of the DC-9 fleet along with other Douglas commercial products. As it will be quickly noted, well over half of the active DC-9 fleet exceeds every design service-life goal.

When it became evident that the DC-9 aircraft would be flown well beyond test-supported life, Douglas purchased one of the original production models out of regular airline service to begin the fatigue test again (Figure 20). The purpose for this was to reveal new problem areas on the airframe that have not developed in service so that they could be dealt with in a timely and

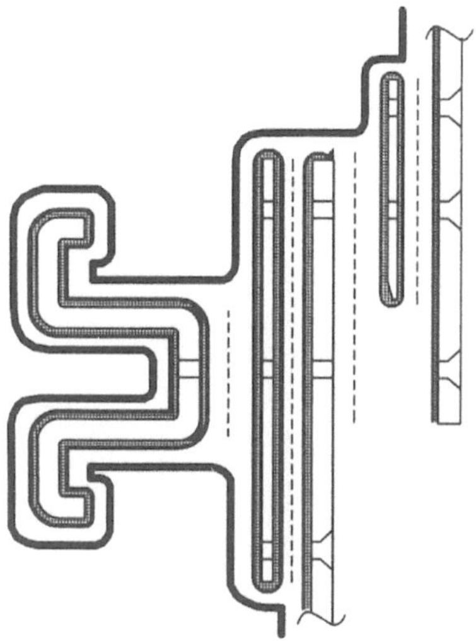

FIGURE 18. SPLICE CORROSION PROTECTION

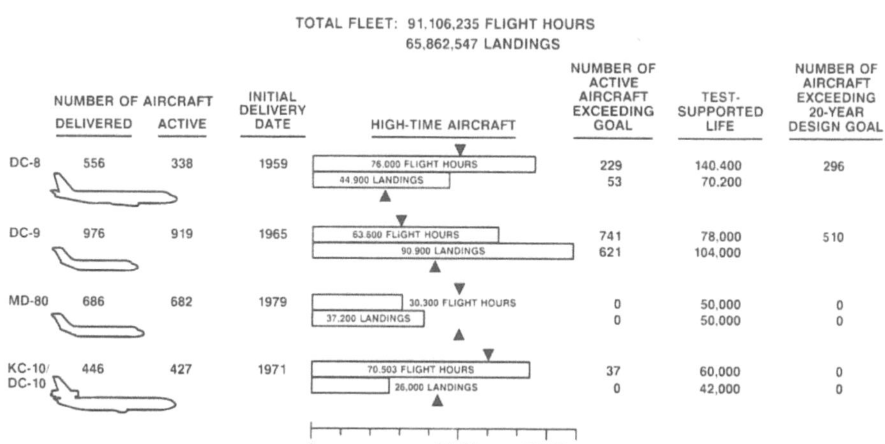

FIGURE 19. DOUGLAS JET INVENTORY STATUS – JANUARY 1990

cost-effective manner for the airlines and evaluate the splice joints for susceptibility to the development of multiple site fatigue damage (Figure 21).

When Douglas purchased DC-9 Fuselage No. 3 in 1981, it was one of the five high-time aircraft and had accumulated over 66,500 landings. It was in regular airline service and was accruing landings at the rate of 3,500 landings a year. At the time, it was scheduled for a major structural maintenance check (D-check) within a month. The aircraft was stripped of all interior items, and the

FIGURE 20. EXTENDED FATIGUE TEST PROGRAMS – DC-9 FUSELAGE NO. 3

MULTIPLE SMALL CRACKS EMANATING FROM EACH FASTENER HOLE IN A LINE OF FASTENERS

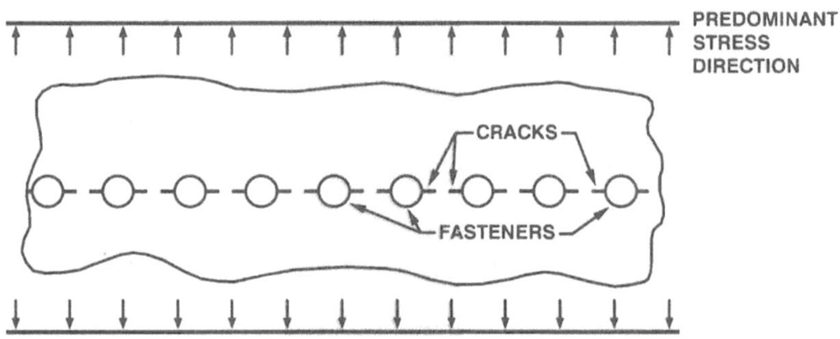

STRENGTH CAPABILITY OF THE STRUCTURE IS DRASTICALLY REDUCED AT VERY SHORT CRACK LENGTHS

MODE OF FAILURE - NET SECTION YIELD FOR ALUMINUM STRUCTURE

FIGURE 21. MULTIPLE-SITE FATIGUE DAMAGE

structure was thoroughly inspected. After a total of 15 years of service, the structure was found to be remarkably free of corrosion and fatigue cracks. The problem areas found were either repaired or tagged for careful evaluation as the test progressed. Certain service bulletins were selected for incorporation before the test began.

The testing extended the full-scale tested life on the DC-9 fuselage structure from 125,000 flights to more than 208,000 flights. At the end of the test in 1986, an extensive evaluation of the structure was conducted, including the teardown of more than 22 feet of fuselage longitudinal lap splice (Figure 22). More than 8,000 individual rivet holes were checked for fatigue cracks with no findings reported. In addition, the examination showed that, 20 years after assembly of the fuselage, there was no corrosion in the faying surfaces.

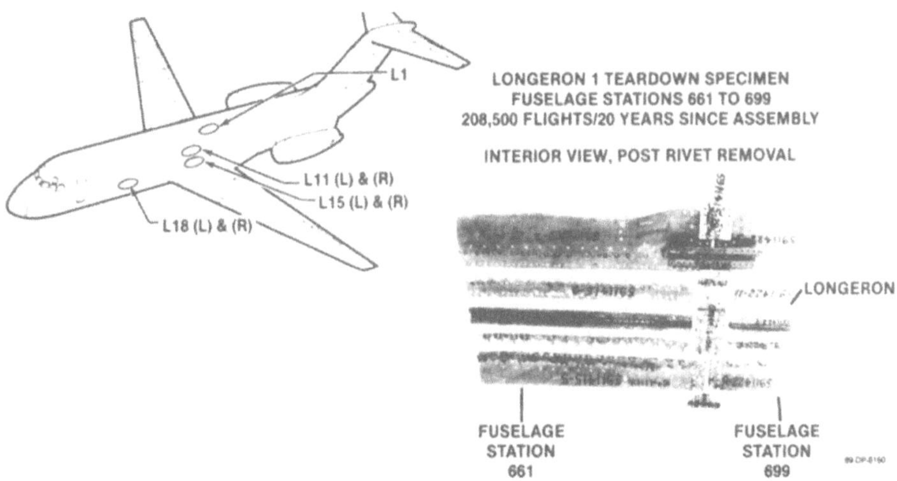

FIGURE 22. DC-9 POSTTEST TEARDOWN OF FUSELAGE SPLICES

Conclusion

This paper has outlined one manufacturer's method for developing reliable and long-lived fuselage longitudinal splices. The procedure, largely empirical, requires a significant investment in development testing. This is due to the current lack of sufficiently sophisticated analytical techniques to evaluate the complex nonlinear behavior of stiffened fuselage structure under combined loadings. Douglas would welcome the development of low-cost analytical alternatives to the current proven methods that would reliably predict the behavior of a longitudinal fuselage splice. We would also welcome the development of more innovative and reliable techniques for performing required inspections aimed at discovering and correcting problems before they become safety issues.

Acknowledgements

The author wishes to thank Mervin Danforth and Donald Hagemaier for their invaluable assistance in preparing this paper.

13. Representation and Probability Issues in the Simulation of Multi-Site Damage

A. R. Ingraffea and M. D. Grigoriu

School of Civil and Environmental Engineering
Cornell University, Ithaca, New York 14853-3501

D. V. Swenson

Department of Mechanical Engineering
Kansas State University, Manhattan, Kansas 66506-7191

Summary

The types of multi-site damage which are observed in fuselage skins preclude use of common simplifications employed in simulating crack growth. Interaction of cracks with each other and with structural features causes arbitrary growth patterns. Representation of a number of arbitrarily growing cracks in a finite element model is a problem which can be addressed with new codes built on a topological data structure and employing automatic remeshing. Examples of such codes applied to problems of interacting cracks, crack interaction with rivet holes, and cracking beneath a boron/epoxy patch are presented. A technique for incorporating probability methods into simulations of arbitrary crack growth is also presented and discussed via an example problem.

Introduction

The types of multi-site damage which are observed in fuselage skins preclude the use of common simplifications, such as symmetry, self-similar propagation, and non-interaction of cracks, usually employed in simulating crack growth. Interaction of cracks with each other and with structural features causes arbitrary growth patterns. In the context of finite or boundary element modeling, this causes a problem of *representation*. By representation we mean the accommodation of an arbitrary number of cracks of arbitrary, evolutionary trajectory in an accurate numerical model of an airframe component. Representation is a problem which can be addressed with new codes built on a topological data structure and employing automatic remeshing.

Observations which further exacerbate the problem of multi-site damage are an apparently extreme sensitivity of crack trajectories to small changes in initial conditions of crack location, and a requirement for very accurate trajectory predictions in the presence of large field gradients around structural discontinuities such as rivet holes. These observations indicate that probability-based simulations may be required in some cases.

This is a survey paper of R&D being conducted at Cornell and Kansas State Universities in the areas of representation and probability as applied to

Springer Series in Computational Mechanics
Atluri, Sampath, Tong (Eds.)
Structural Integrity of Aging Airplanes

multi-site damage. In the following sections, the focus will be specific example applications of a new class of finite element codes to fatigue problems in aging aircraft. Details concerning the theoretical bases and implementation of these codes are contained in the cited references.

Interaction of Cracks from Adjacent Rivet Holes

A fundamental problem in multi-site damage is growth and coalescence of fatigue cracks emanating from adjacent rivet holes. In an example, preliminary simulation of this problem, the FRacture ANalysis Code (FRANC) [1,2,3,4] will be used. Figure 1a shows an initial mesh of a section of lap joint under biaxial tension. In Figure 1b, two cracks of unequal length have propagated from adjacent rivet holes. In order to show the ability to accommodate curvilinear growth and crack interaction, the cracks are allowed to continue to propagate here under the rules of LEFM until the state shown in Figures 1c and 1d is reached. Figure 1c shows the overall trajectories, while 1d is a close-up view of the region of interaction. Of course it is likely that at this state LEFM would no longer be valid and a plasticity-based coalescence criterion would be needed to complete the simulation; however, the trajectories shown in this figure are still qualitatively very similar to those observed in MSD.

This simulation was performed interactively on an engineering workstation. All remeshing was performed automatically, and at each step of propagation stress-intensity-factors are computed, their values refined by automatic local rezoning, and stored with associated crack lengths so that a history of K versus a is always available.

Crack Propagation Under a Boron/Epoxy Patch

As another example of representational difficulties, we present an analysis of a crack repaired with boron/epoxy composite patch, a promising new repair technology [5, 6]. The goal of the analysis is to calculate the stress intensities as the crack grows under the composite patch. This problem is a specific application in a general class of problems in layered structures, another similar problem is cracks propagating in riveted lap joints. A common characteristic of these problems is that load is transferred from one layer to another due to either rivets or adhesives between the layers. This load transfer changes the stress intensity at the crack tip.

To analyze these problems, we use separate two dimensional models of individual layers. The finite elements for each layer can overlap in space and can be connected either with rivet elements or by an adhesive layer. The rivets and adhesive act as springs between the layers, transmitting shear

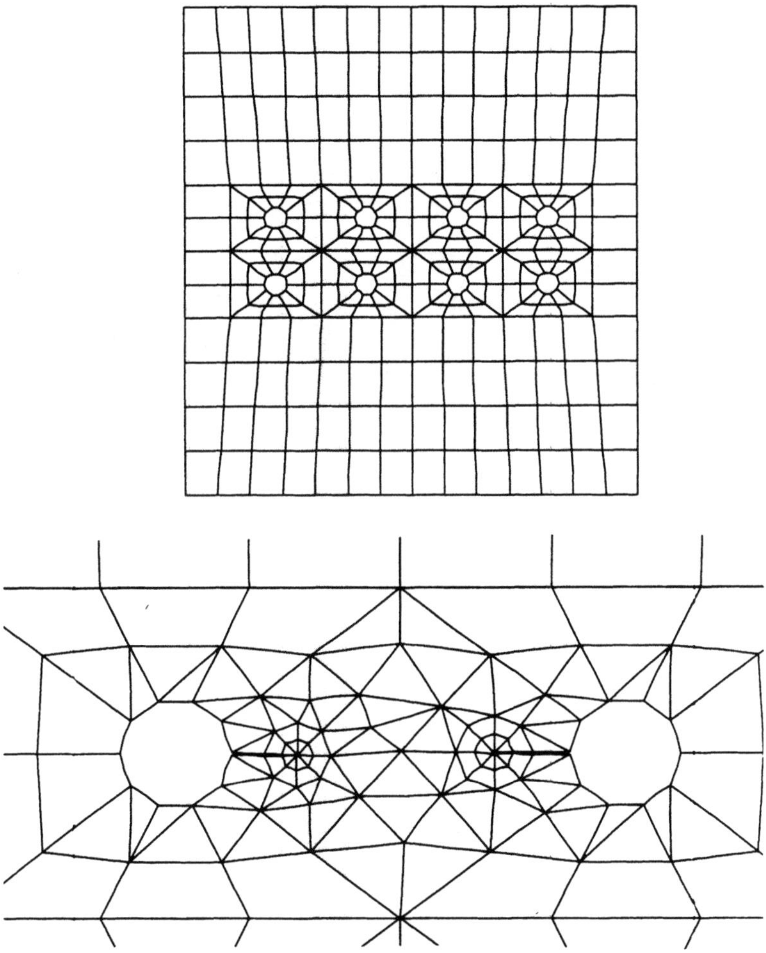

(a)

(b)

Fig. 1 (a). Initial Mesh for Section of Lap Joint.
(b). Expanded View of Two Un-equal Length Cracks.

forces which are a function of the relative displacements of the two layers. With this approach, the user can introduce a crack in one layer, follow the propagating crack, and predict crack growth and load transfer to the other layers. For more details of this model see [7].

Finite Element Model

Crack propagation in the layered structure is modeled using a finite element fracture analysis code called CRACKER [8], originally developed to model dynamic fracture propagation. The finite element approximation uses six-

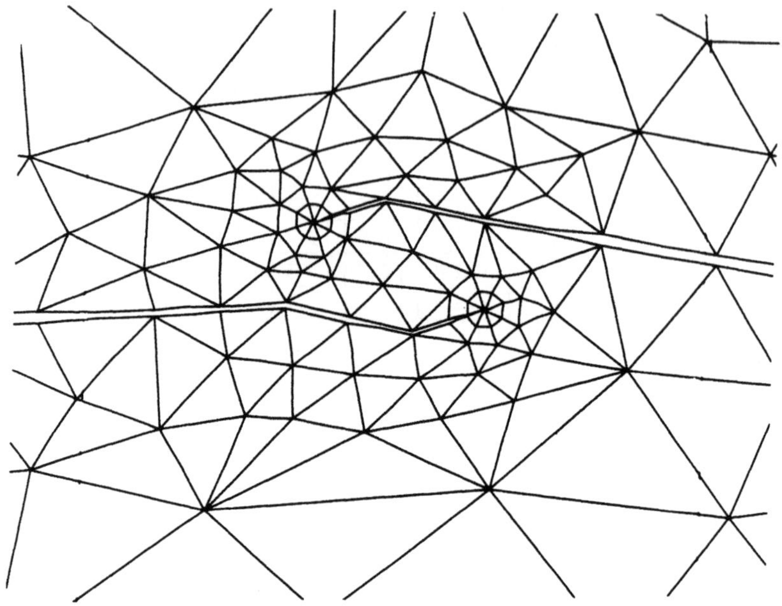

(c)

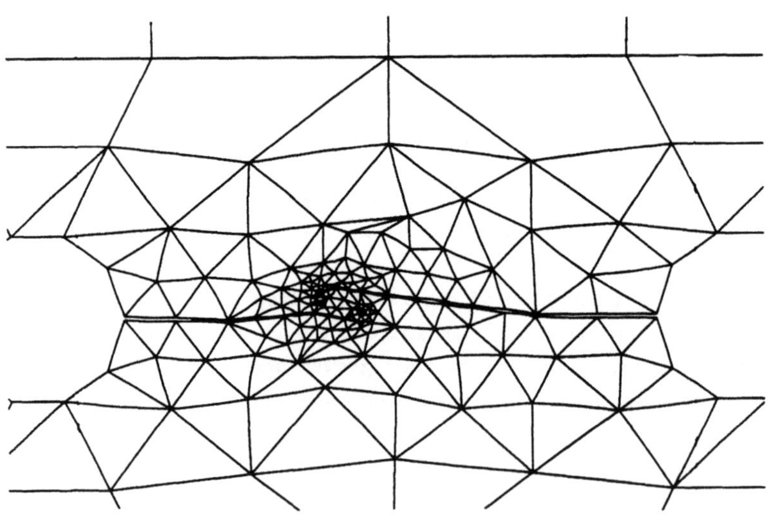

(d)

Fig. 1(c). Interaction of Cracks Shown in (b).

(d). Detail of Crack Over-Lap Region in (c).

noded triangular elements with quadratic shape functions. These elements were chosen because they make possible relatively simple and robust automatic remesh algorithms [9]. In addition, shifting the side nodes to the quarter-points results in the desired stress singularity at the crack tip [10]. Automatic remeshing is performed by deleting elements in the vicinity of the crack tip, moving the crack tip, and then inserting new elements that connect the new crack to the existing mesh.

An analysis consists of a series of solutions. After each solution, the stress intensities are calculated. The maximum circumferential stress criterion is used to determine the direction of crack propagation. The crack tip is then moved one element length in this direction, remeshing is performed, and a new solution calculated. The final results are the predicted crack path and the stress intensities along the path. This information can be integrated using a growth model to predict the number of cycles for propagation.

The rivets act as springs between the layers, with the adhesive problem being a generalization of the rivet problem. We assume that the thickness of the adhesive is small compared to the thickness of the adherents. Thus the stresses in the adhesive are functions of the shear modulus and relative displacements of the two layers. The stresses in the adhesive are applied as surface tractions on the layer meshes.

In general, it is not necessary that the adhesive and adherent meshes be the same. However, special care must be taken along the crack face, since the adhesive stresses are discontinuous across the face, but smooth on either side of the crack face. In order to assure accurate calculation of these stresses and accurate integration of the resulting surface tractions, it is necessary that the crack be modeled in the adhesive and that element boundaries in the patch lie on the crack face. We have met this condition by using identical meshes in adhesively bonded layers. Automatic remeshing is performed on all layers near the crack as the crack moves.

Experiment

Boeing Military Airplanes tested a patch repair on a typical lap joint using the test specimen shown in Figure 2. A crack was grown in fatigue from a rivet hole until it reached a length of 2 inches. The specimen was then patched on its cracked surface with 0.01 inch thick FM-73 adhesive and boron/epoxy patch material with fibers running perpendicular to the direction of the crack propagation. The repair is shown in Figure 3a. The patched panel was then cyclically loaded ($R = 0.05$) to 16 ksi. Essentially, the crack was arrested after application of the patch.

Analysis

The finite element model consists of four layers (cracked panel, uncracked panel, adhesive, and patch), with each mesh overlapping in space. The cracked and uncracked layers are connected by rivets. The cracked and patch layers are connected by adhesive. The cracked and uncracked layers use an isotropic material, while the patch is represented by an orthotropic material. The mesh outline and finite element mesh for the cracked panel is shown in Figure 3b.

Results

Figure 4 shows a series of displacement results for the cracked panel as the crack propagates. Figure 5 gives the calculated stress intensities

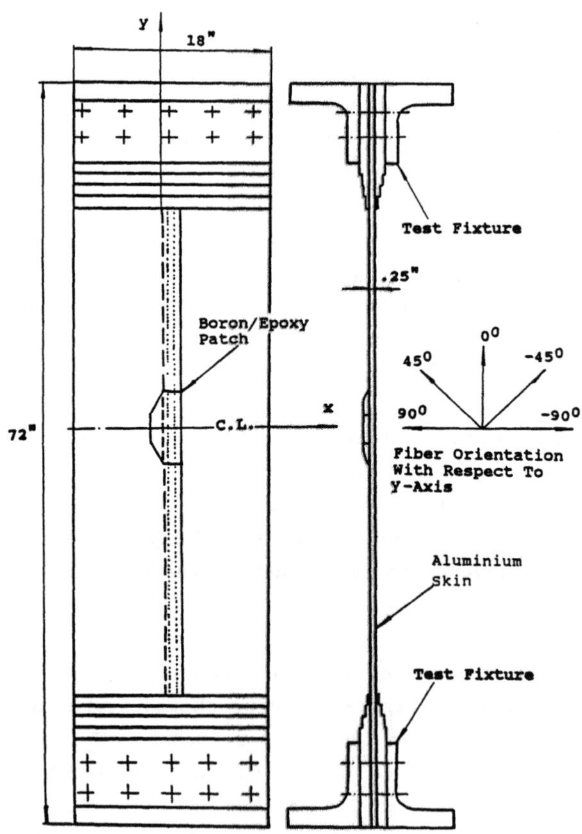

Fig. 2. Boron/Epoxy Patch Lap Joint Specimen.

(a) 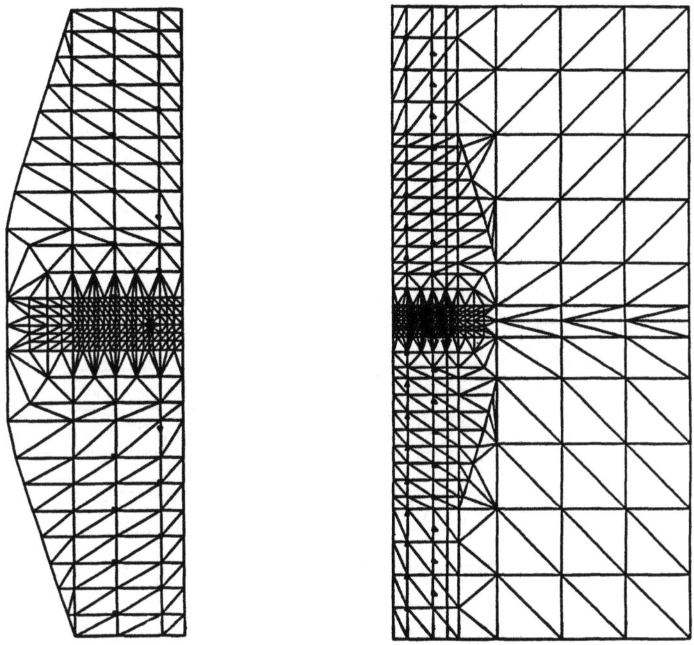 (b)

Fig. 3(a). Mesh Used for Boron/Epoxy Patch.

(b). Mesh used for Cracked Panel.

during crack propagation. From this comparison, we can see that the boron/epoxy patch significantly reduces the stress intensity. In addition, the stress intensity is approximately constant (reflecting the bridging of the patch over the crack). Using a Paris model, there is approximately a two orders of magnitude reduction in crack growth rate compared to the unpatched crack. This is consistent with Boeing data where cracks essentially arrested after patching.

Observations

This analysis demonstrates that the linear elastic fracture mechanics approach has been successfully used to predict crack growth in patched structures. The main factor to be considered in the model is the load transfer occurring between the cracked and uncracked layers and the patch layer. Care must be taken to ensure that the discontinuous adhesive stresses are accurately integrated to obtain accurate loads on the adherents.

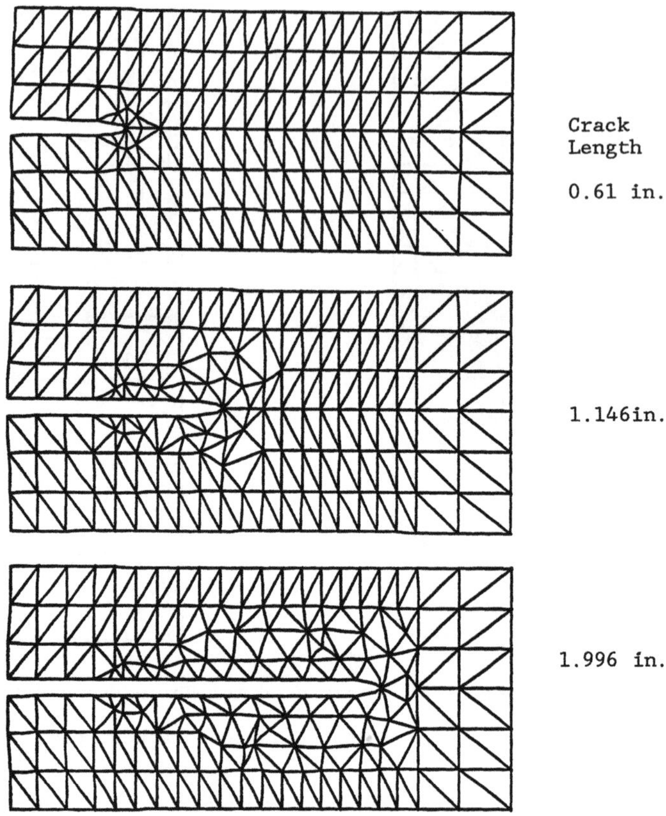

Fig. 4. Predicted Crack-Opening-Displacements (X150).

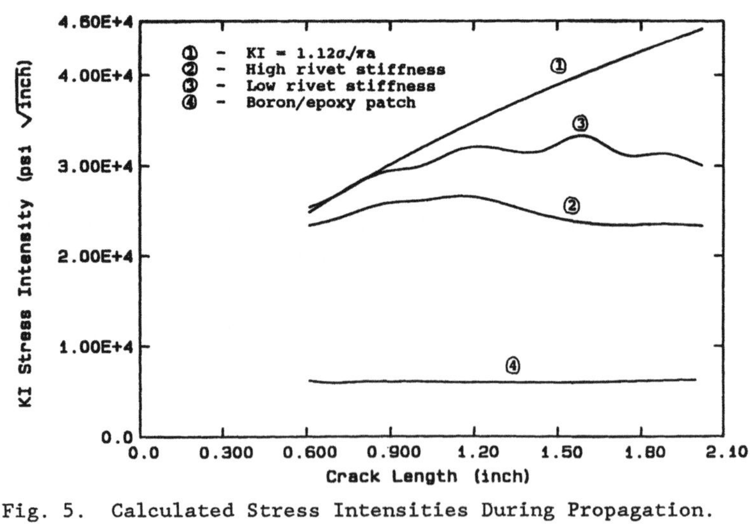

Fig. 5. Calculated Stress Intensities During Propagation.

Future work should be directed to implementing the same concepts in three dimensions, since many aircraft structures are constructed using thin sheets. These sheets are then riveted and/or adhesively bounded together. Cracks could then be grown in one of these layers and the crack growth rate could be accurately predicted.

<u>Probability Issues in Crack Propagation</u>

Recent experiments at Cornell have further indicated the necessity to approach simulation of interaction of cracking with structural discontinuities on a probabilistic basis. Figure 6 shows the initial finite element mesh used to predict the crack propagation response of plexiglas beams containing simulated rivet holes. The results of two tests and simulations thereof are discussed here. The two tests differ only in a small change, one-quarter of an inch, in the location of the starter crack.

Figure 7 shows the observed and simulated results for these tests. With respect to the observed trajectories, it may be seen that substantially different response occurred. Figure 7a shows the crack following a tortuous path between two holes and arresting on the far side of one of them. In Figure 7b the crack is drawn into the near side of the first hole encountered. Such sensitivity to relatively small changes in initial conditions is at least suggestive of chaotic response.

With respect to the comparison between predictions, made by FRANC, and observations, Figure 7a shows that FRANC, running in auto-mode, ie. without human interaction, tracked the crack closely until its approach to the second hole. There the small difference between the trajectories allowed the prediction to be perturbed by, but not attracted into this hole. Figure 7b shows the predicted and observed trajectories to be quite close. The observation here is that small inaccuracies in predicted trajectory, in the presence of sharp field gradients created by structural discontinuities, may lead to gross inaccuracies in overall behavior. These observations strongly suggest that reliable life predictions may in some cases need to be based on probabilistic fracture mechanics.

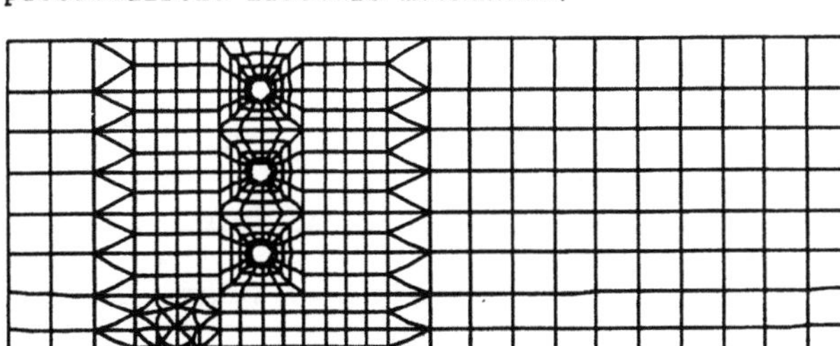

Fig. 6. Initial Mesh for Beam With Holes.

Probabilistic Fracture Mechanics

Deterministic models are characterized by a precise description of loading conditions and system initial state. As a result, there is a unique crack trajectory for a specified structure, boundary conditions, and load history. However, predictions of deterministic models are inconsistent with experiments showing variations of crack trajectories and loads at fracture initiation corresponding to nominally identical specimens. This consistency poses significant difficulties in the validation of deterministic codes for fracture mechanics.

In contrast, probabilistic models can account for all sources of uncertainty, e.g., loads, system geometry and state, and material mechanical properties. The output of the analysis consists of families of crack trajectories, loads at fracture initiation, and other fracture mechanics descriptors that can be characterized probabilistically. Results of

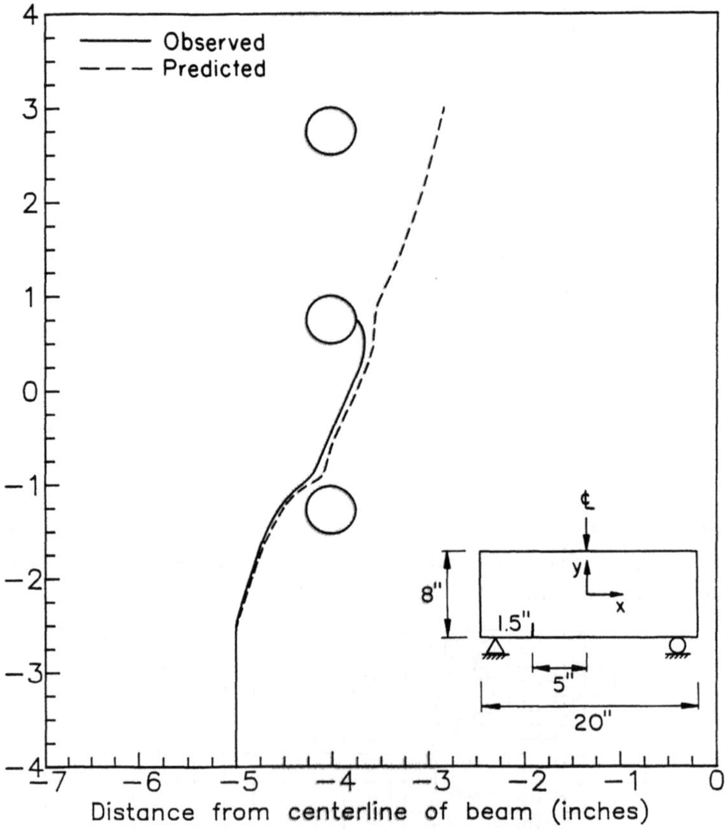

Fig. 7(a). Prediction and Observation for Offset of 5 in.

probabilistic analyses can be used to validate theoretical and numerical approaches based on experimental results and statistical tests.

Consider for illustration a homogeneous and isotropic plate of complex geometry with a crack length 2a that is subject to a n-dimensional far field random stress $\underline{S}$. The plate material has a random fracture toughness K_{Ic}. Due to the uncertainty in $\underline{S}$ and K_{Ic} deterministic fracture mechanics codes cannot be used directly and fracture descriptors, such as initial direction of crack extension θ and load at fracture initiation, become random quantities. Methods for finding probabilistic characteristics of θ and determining the probability of fracture initiation are discussed in Ref. [11] The methods involve two phases. First, stress intensity factor vectors $\{k_{Ii}, k_{IIi}\}$, $i = 1, \ldots, n$, are determined for loading conditions $S_i = 1$ and $S_j = 0$, $j \neq i$, based on a deterministic code for fracture mechanics analysis, such as FRANC. The random stress intensity factor vector $\underline{K}^T = \{K_I, K_{II}\}$ can

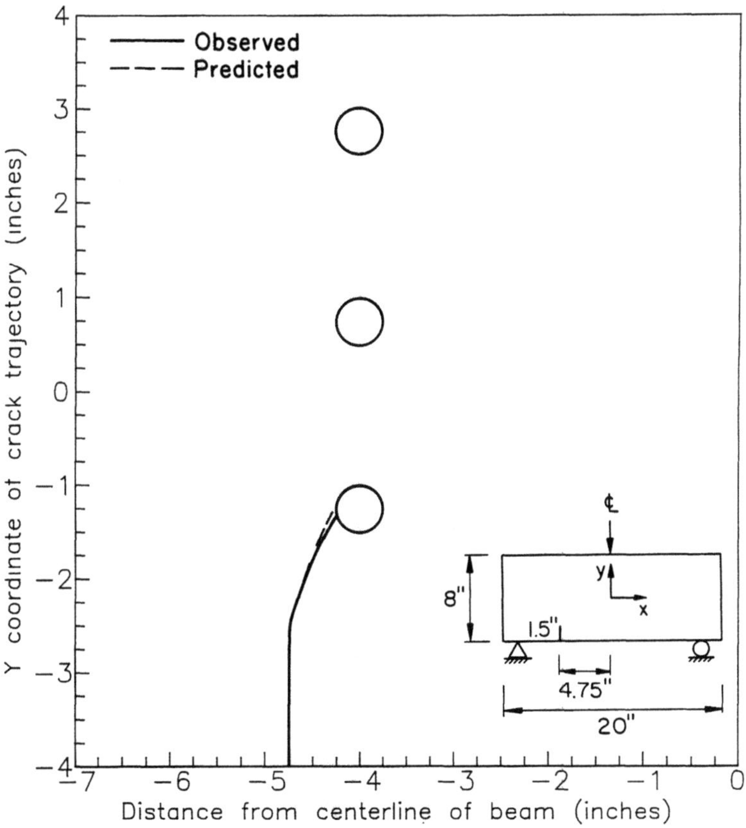

Fig. 7(b). Prediction and Observation for Offset of 4.75 in.

be obtained from

$$\underline{K} = \underline{k}\ \underline{S} \tag{1}$$

in which the transfer matrix

$$\underline{k} = \begin{Bmatrix} ---,\ k_{Ii},\ --- \\ ---,\ k_{II,i},\ --- \end{Bmatrix} \tag{2}$$

has dimension (2, n). In the second phase, probabilistic characteristics can be determined for $\underline{K}$ in Eq. 1 and displayed in a workstation window. These characteristics can be employed to calculate the probability of fracture initiation and determine the distribution $F(\theta)$ of initial direction of crack extension θ. Confidence intervals (θ_1, θ_2) can also be obtained for θ from the equations $F(\theta_1) = p$ and $F(\theta_2) = 1 - p$, where p is a prescribed probability.

Experimental results reported in Ref. [12] are used to validate prediction of a probabilistic fracture analysis code, PROFRANC, under development at Cornell. The code is based on several theories of linear fracture mechanics and probabilistic techniques outlined in the previous paragraph. Figure 8 shows the specimen used in analysis and Table 1 the statistics of its geometry and material properties. Figure 9 shows prediction of the probability density of the load at fracture initiation for the 800 and 900 series of specimens when (a) only K_{Ic} is random and (b) K_{Ic} and plate geometry are uncertain. Observed values of P are also shown. The figure shows qualitatively that probabilistic predictions of PROFRANC are consistent with observations.

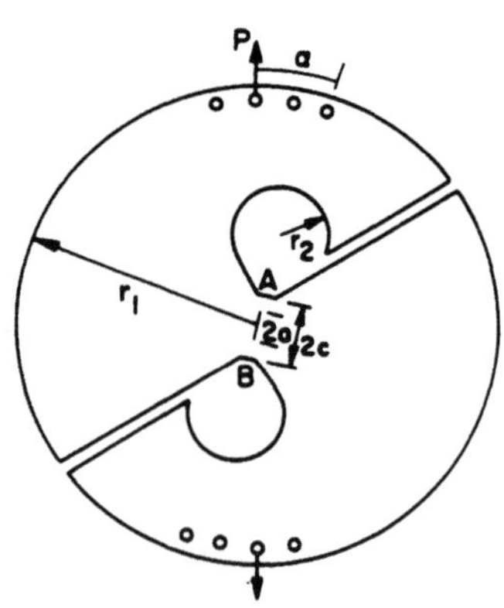

Fig. 8. Test Specimen Used for Probability Study. From [11].

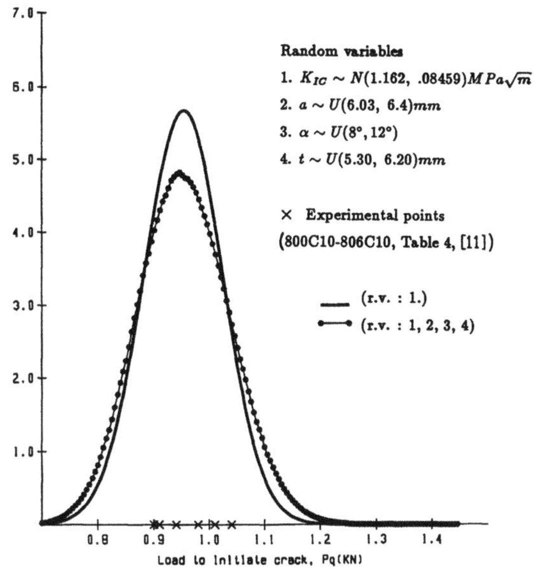

(a)

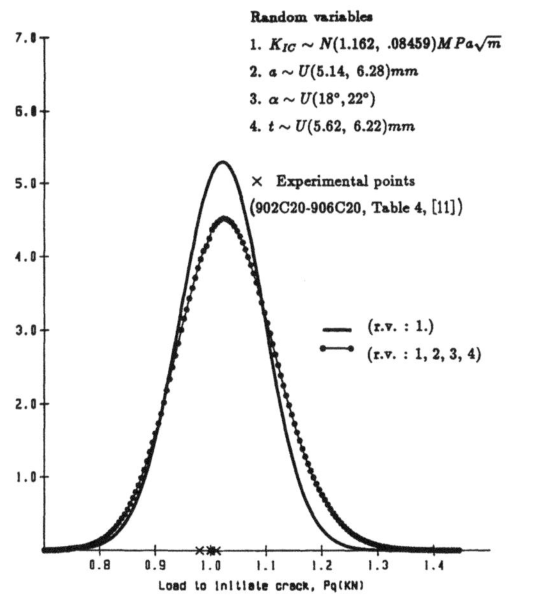

(b)

Fig. 9(a). Probability Density Curve of the Load to Initiate Fracture for the 800 Series Tests.

(b). Probability Density Curve of the Load to Initiate Fracture for the 900 Series Tests.

Table 1. Geometrical and Material Characteristics.

Probabilistic Descriptors	Toughness, $MPam^{\frac{1}{2}}$	Half Crack Length, a (mm)		Thickness t (mm)		Loading Angle alpha (degrees)	
		S-800	S-900	S-800	S-900	S-800	S-900
Mean	1.162	6.225	5.770	5.745	5.910	10	20
C.O.V	.00715	.0114	.1083	.0675	.03	1.33	1.33
Distribution	Normal			Uniform	Probability		

Conclusions

The types of multi-site damage which are observed in fuselage skins preclude use of simplifications commonly employed in simulating crack growth. No symmetry is generally available, and crack growth is usually curvilinear. Further, interaction of cracks with each other and with structural features causes arbitrary growth patterns. Representation of a number of arbitrary growing cracks in a finite element model is a problem which can be addressed with new codes built on a topological data structure and employing automatic remeshing. Examples of such codes applied to problems of interacting cracks, cracks interacting with rivet holes, and cracking beneath a boron/ epoxy patch were presented.

Through comparisons between observations of curvilinear crack growth and deterministic predictions, it was also observed that simulations may be very sensitive to initial conditions. Further, small errors in predicted trajectory may, in the presence on strong field gradients prevalent in lap joints, lead to gross inaccuracies in predicted life. These observations strongly suggest that reliable life predictions should be based on probabilistic fracture mechanics.

Acknowledgements

The work on simulation of effects of patching is being funded by Boeing Military Airplanes and KTEC. We would like to particularly thank Tom Derber, Mark Hoffman, Leonard Wright, and Gary Cassatt for their support. The research in probabilistic fracture mechanics is being supported by the Air Force Office of Scientific Research, Project FQ8671-8900185. We would like to thank the project monitor, Dr. Spencer Wu, for his support. All simulations at Cornell were performed in the laboratory of the Program of Computer Graphics.

References

1. Wawrzynek, P.; Ingraffea, A.R.: Interactive finite element analysis of fracture processes: an integrated approach. Theo & Appl. Frac. Mech. 8 (1987) 137-150.

2. Wawrzynek, P.; Ingraffea, A.R.: An edge - based data structure for two-dimensional finite element analysis. Engrg. with Comp. 3 (1987) 13-20.

3. Ingraffea, A.R.: Case studies of simulation of fracture in concrete dams. Eng. Frac. Mech. 35 (1990) 553-564.

4. Wawrzynek, P.A.; Ingraffea, A.R.: An interactive approach to local remeshing around a propagating crack. Finite Elements in Anal. and Design 5 (1989) 87-96.

5. Jones, R.; Callinan, R.J.: Bonded repairs to surface flaws. Theo. & Appl. Frac. Mech. 2 (1984) 17-25.

6. Baker, A.A.: Fiber composite repair of cracked metallic aircraft components - practical and basic aspects. AGARD, Reprint from Conference Proceedings No. 402.

7. Chih-Chien, C.: A finite element model of fracture in layered structures. Engineering Experiment Station Report 220, Kansas State University, Manhattan, KS (1990).

8. Swenson, D.V.; Ingraffea, A.R.: Modeling mixed-mode dynamic crack propagation using finite elements: theory and applications. Comp. Mech. 3, (1988) 381-397.

9. Shaw, R.D.; Pitchen, R.G.: Modifications to the Suhara-Fukuda method of network generation. Int. J. for Numer. Meth. in Eng. 12 (1978) 93-99.

10. Henshell, R.D.; Shaw, K.G.: Crack tip finite elements are unnecessary. Int. J. Numer. Meth. in Eng. 12 (1975) 93-99.

11. Banks-Sills, L.; Bortman, Y.: A mixed-mode fracture specimen: analysis and testing. I. J. Fracture 30 (1986) 181-201.

12. Grigoriu, M.; Saif, M.T.A.; El Borgi, S.; Ingraffea, A.R.: Mixed mode fracture initiation and trajectory prediction under random stresses. I. J. Fracture (in pren.).

14. Bonded Repair of Multi-Site Damage

R. Jones, N. Bridgford, G. Wallace and L. Molent

Aircraft Structures Division, Aeronautical Research Laboratory Melbourne, Australia

Summary

Repairs using bonded composites have numerous advantages over mechanically fastened repairs. Adhesive bonding does not result in stress concentrations due to additional fastener holes. Composites are readily formed into complex shapes, permitting the repair of irregular components. In service damage monitoring is possible, with the appropriate fibre matrix system, by direct inspection through the repair using eddy current methods or by thermal emission measurements. This paper presents a bonded repair for fuselage lap joints containing multi-site damage. The effectiveness of this repair is confirmed by the results of a laboratory test program.

1. Introduction

The high acquisition costs associated with the purchase of modern transport and military aircraft has resulted in a greater utilization of existing aircraft fleets. This is illustrated by the percentage of Boeing aircraft, as of March 1988, exceeding their economic design life objective.

Aircraft Designation	Percentage of World Fleet Exceeding Design Life (flight hours)
707	22
720	43
727	22
737	3
747	11

This trend, of operating aircraft approaching or exceeding their original design life, has been reflected in an increased number of structurally significant defects. Recent examples of this include:

(1) Corrosion damage and subsequent fatigue cracking in fuselage lap joints.
(2) Corrosion damage and subsequent fatigue cracking in pressure bulk-heads.
(3) Corrosion damage and subsequent fatigue cracking in control surfaces.

Springer Series in Computational Mechanics
Atluri, Sampath, Tong (Eds.)
Structural Integrity of Aging Airplanes

This has led to several joint airline, industry and airworthiness actions. A recent conference on this problem, in Washington, U.S.A, in June 1988 resulted in a number of recommendations which included:

(1) The development of better ways for finding and assessing structural defects.
(2) The need to develop improved NDT methods.
(3) The development of improved repair methods.

At present there is concern that, when repairing multi-site damage, the close proximity of a large number of mechanically fastened repairs may lead to a compromise in the damage tolerance of the structure even though, in isolation, each repair may be satisfactory. The challenge is to develop alternative repairs to such multi-site damage. Externally bonded repairs using high modulus unidirectional fibre composites provide one possible alternative approach. They are ideally suited to aero-space applications as they can be fabricated to reproduce complex shapes and curvatures and do not inhibit current inpection processes.

The present paper describes one possible externally bonded repair. Constant amplitude fatigue tests are performed to illustrate its ability to significantly extend the fatigue life of fuselage lap joints containing multi-site damage. A detailed discussion of the relevant design criteria, application technology and the extent of current utilization is given in [1].

2. Background to Adhesively Bonded Repairs

In the military scene the Aeronautical Research Laboratory, Australia, has pioneered the use of bonded repairs to fatigue and stress corrosion cracking, see [1,2,3] and Table 1. This method has proved to be effective and highly cost efficient as compared with conventional repairs. To date the repair to the stiffener runout region of the F111 wing pivot fitting (WPF) is by far the most ambitious of current repairs, see [3]. In this case failure of the WPF occurs due to a small defect in the stiffener. The repair consists of two separate external doublers, each with more than 120 plies. Each doubler consists of two segments. The lower segment is bonded to the D6ac steel WPF to provide a bridge level with the aluminium skin step and the upper segment bonded to the aluminium wing skin and the lower segment. This results in an alternative load path allowing the load to by-pass the critical region. Indeed this resulted in a strain reduction, in the critical region, of up to 30 percent. This design, which provides an alternative load path by-passing the critical defect, is applicable to a wide range of aero-space problems and represents a possible method for repairing fuselage lap joints containing multi-site damage.

Table 1: Current Bonded Repairs/ Reinforcements

Cracking	Material	Component	Aircraft	Comment
Stress-Corrosion	7075-T6	Wing Plank[a]	Hercules	> 300 repairs
Fatigue	Mg alloy	Landing Wheel[a]	Macchi	Life doulbed
Fatigue	2024-T3	Fin Skin	Mirage	In-service since 1978
Fatigue	AU4SG	Lower Wing	Mirage	> 500 repairs
Fatigue	2024-T3	Upper Wing Skin	Nomad	> 105900 simulated Flying hours
Fatigue	2024-T3	Door Frame	Nomad	> 106619 simulated Flying hours
Stress-corrosion	7075-T6	Console Truss	F-111	In-service since 1982
Lightening Burn	2024-T3	Fuselage Skin	Orion	In-service since 1986
Static Failure	D6ac	Wing Pivot	F-111	Recent repair
Stress-corrosion	7075-T6	Wing Plank	Orion	Recent repair

a - Now standard RAAF procedure

Repairs using bonded composites have numerous advantages over mechanically fastened repairs. Adhesive bonding does not result in stress concentrations due to additional fastener holes. Composites are readily formed into complex shapes, permitting the repair of irregular components. In service damage monitoring is possible, with the appropriate fibre matrix system, by direct inspection through the repair using eddy current methods or by thermal emission measurements. The high degree of anisotropy has the advantage of eliminating the parasitic stiffening of the structure in other than the required direction.

For a cracked structure a bonded repair significantly reduces the stress intensity factor and as a result dramatically reduces crack growth. In many cases crack growth is completely eliminated, see [1] pages 155-157. This reduction in the stress intensity factor is achieved by two mechanisms. The first is a general reduction in the net-sectional stresses resulting from the application of the repair. The second mechanism is due to the high modulus fibres attempting to hold together the adjacent crack faces, i.e. preventing opening.

Prior to the advent of bonded repair technology the procedure most commonly used involved a mechanically fastened repair which achieves its objective primarily by reducing the net section stresses. The significant difference between the two repair methods lies in the behaviour of the repair region adjacent to the crack. In a mechanical repair the free length between the first set of fasteners and the crack surface is strained allowing opening of the crack faces. For a bonded repair, as the crack opens the adhesive near the crack is subjected to shear resulting in transfer of load into the repair. This load transfer reduces the crack opening displacement. As the transfer length associated with a bonded repair is very much smaller than the spacing of the fasteners, the crack opening displacement is considerably smaller than that achieved by a mechanical repair and the crack growth rate is much less.

Whilst bonded repair technology was originally developed for the repair of thin metallic wing skins, e.g. the boron epoxy repair to Mirage aircraft in service with the Royal Australian Airforce (RAAF) see [2], and for which crack growth can be eliminated, see [1] page 172, recent laboratory tests have shown that a correctly designed repair can be used to eliminate crack growth in thick structural components [4].

3. Repair of Multi-Site Damage

The Boeing 737 is currently one of the best selling jet liners in the world. However, corrosion induced cracking at the fuselage skin lap joint led to the, now extensively documented, failure of the Aloha 737 in Hawaii on April 28, 1988. (The Boeing 727, 707 and 747 also have variants of this joint.) The failure was essentially due to the linking, into one large crack, of numerous small cracks at a number of fasteners. This problem is common to a wide range of mechanically fastened fuselage lap joints and is generally initiated by corrosion damage.

Current repair methods for this area involve removing the damaged material and the use of a rivetted scab patch, introducing even further stress concentrations. The objective of this investigation is to evaluate a possible bonded repair for mechanically fastened fuselage lap joints containing multi-site damage.

3.1 Specimen geometry

The basic specimen used in this investigation consists of two 2024-T3 clad aluminium alloy sheets 1.016mm thick, fastened with three rows of BACR15CE 100° shear head counter-sunk rivets, 3.968mm (5/32 inch) in diameter, see Figure 1. The width of the specimen was chosen to coincide with the distance between the tear straps on Boeing 737 aircraft. The upper row of rivet holes contained initiation sites, on either side, nominally 1.2mm long. This length was chosen so that the defect was obscured by the fastener head. The specimens were manufactured by Australian Airlines, from material supplied by them, to aircraft standard. This joint is representative of those found in 'wide bodied' aircraft fuselage skins, in particular in the 737 fuselage. As a worst case scenario the joints were not bonded or treated with sealant.

The repair is a multi-segment boron/epoxy laminate, fabricated in a similar fashion to the recent repair to F-111C aircraft [3] and designed to retard, and if possible, eliminate crack growth. This laminate is bonded to the joint and its main purpose is to provide an alternative load path for each row of rivets, see Figure 2. The repair is applied after the specimens have been pre-cracked. To investigate the failsafe nature of this repair two specimens were repaired following complete failure of the upper row of rivets. Since the amount of bending, in a typical aircraft fuselage joint, was unknown the local bending was optimized by testing two specimens bonded to either side, i.e. back to back, of a 12.5 mm aluminium honeycomb core, see Figure 3. In this configuration strain gauge results indicated no global bending and no parasitic stiffening due to the honeycomb.

3.2 Applied loads

The specimens were tested in an INSTRON servo-hydraulic test machine. The loads applied to the specimen were chosen to give a remote peak stress range of 13.4 ksi, which is typical of fuselage hoop stresses. For the unrepaired specimens the tests were terminated when the upper row of fasteners had completely failed, i.e. all of the cracks had joined. Crack growth was monitored optically using a travelling microscope, with a magnification factor of forty to one. For the repaired specimens, which did not fail in the test section, eddy current techniques were used to check for crack growth.

3.3 Fatigue test results

To date twelve specimens have been tested, see Table 2. In the initial process used to bond the specimens to the honeycomb core, i.e. specimens A1-A4, adhesive flowed between the two aluminium sheets and across the upper row of fasteners. This resulted in a large number of cycles being required for crack initiation. In the case of specimens A1 and A2 final failure did not result until approximately 140,000 cycles after initiation, see Table 2. Indeed with the exception of A2 all specimens failed due to failure of the top skin and in the upper row of rivets, see Figure 4. However, specimen A2 failed at the rivet line in the lower skin, after the failure of A1. This is particularly important since in operational aircraft this location is very difficult to inspect. The fatigue lives of the unrepaired specimens varied from 25,000 to 140,000 cycles with a mean of approximately 75,000 cycles. This estimate of the fatigue life compares favourably with that provided by Boeing [5] for the 737 fuselage.

The variation in the fatigue lives was partially due to the nature of the cracking. In general, a short life occured when the largest cracks in the specimen grew towards each other from adjacent rivet holes. A long life occured when failure was due to crack initiation and growth from widely separated rivet holes. The crack growth data for the longest cracks in specimens A5, 6, 9, 10 and 12, up to the point of initial failure, is shown in Figure 5.

In each case the number of cycles required for two cracks to join, which we will term initial failure, does not significantly differ from the number of cycles to final failure. This difference varies from a few hundred cyles for specimens A5 and A6 to 4400 cycles for specimen A10 and 6400 cycles for specimen A12 and has a mean value of 2970 cycles, i.e. approximately four percent of the mean life.

Whilst a detailed analytical or finite element analysis of the failure process has not yet been performed the experimental results indicate that the initial failure is essential one of plastic overload in the remaining ligament. Initial failure occurs when the distance between the crack tips is approximately 2mm.

To date six repaired specimens have been tested, viz: A3, A4, A5, A6, A7 and A8. This repair has been found to increase the fatigue life by more than an order of magnitude, see Table 2. Whilst specimen A5 failed in the grips no failures occured in the test section. In the case of the repairs to specimens A5 and A6 the specimens had previously been tested, in the unrepaired state, to failure and the repairs subsequently applied in-order to evaluate the failsafe nature of the repair. To investigate the damage tolerant nature of the repair, these specimens included teflon inserts to simulate delaminations, 25.4mm by 12.7mm at the step in the lap joint, this was to represent the effect of contaminants at this location, as well as teflon inserts 12.7mm by 12.7mm at the ends of both the upper and lower doublers. These defects did not result in failure by delamination growth. To further investigate the damage tolerance specimens A5 and A6 were also statically loaded to 18.8 ksi without failure.

None of the repaired specimens have experienced failure in the test section. Given the large number of cycles experienced this implies that the repair also sufficiently supresses the failure mode in the lower skin.

4. Conclusion

The present test program has shed light on the phenomena of multi-site damage in pressurized fuselage lap joints. Some useful crack propagation data have been generated. Although failure consistently occurred in the upper row of fasteners, it was also demonstrated that crack growth can occur in the inner skin at the lower row of rivets. Consequently any proposed repair should be applicable to both locations and preferably increase the required inspection interval for the rivet holes.

It has been demonstrated that an externally bonded composite repair will significantly increase the fatigue life of fuselage lap joints containing multi-site damage. The repair was demonstrated to be damage tolerant and carry in excess of full operational load, even after complete failure of the aluminium skin.

Whilst this work is of a generic nature the results should be directly applicable to current operational aircraft. Further work is required for bi-axial stress states. However, based on current results a repair has recently been applied to a lap joint on the lower fuselage of an Australian Airlines 727 aircraft, see Figure 6.

Whilst this paper has concentrated on the repair of fuselage lap joints, work is also in progress on other aspects of multi-site damage, such as the recent repair to stress-corrosion damage, at numerous sites, to an Ansett Airlines B-767 keel beam, see Figure 7.

Acknowledgements

The authors wish to acknowledge the advice and assistance given by Mr R. Frazer of Australian Airlines and to Dr A. A. Baker, Mr. I. Fry and Mr. J. Roberts of Aircraft Materials Division, A.R.L., for advice on materials aspects and for fabrication and application of the repairs.

References

1. Baker A.A. and Jones R., Bonded repair of aircraft structures, Martinus Nijhoff Publishers, The Netherlands, 1988.
2. Baker A.A., Callinan R.J., Davis M.J., Jones R. and Williams J.G. , Aplication of BFRP crack patching to Mirage III aircraft, J. Theoretical and Applied Fracture Mechanics **2**, 1-5, (1984).
3. Molent L., Callinan R.J. and Jones R., Design of an all boron epoxy doubler reinforcement for the F-111C wing pivot fitting: Structural design aspects, J. Composite Structures **11** (1989) pp57-83.
4. Heller M., Williams J.F., Hill T. and Jones R., Increasing the fatigue life of cracked fastener holes using bonded repairs, J. Theoretical and Applied Fracture Mechanics **11, 1** (1989)
5. R. M. Bader, A review of research on aeronautical fatigue in the United States 1987-1989, ICAF National Reviews, Compiled by A. Berkovits, ICAF Document No 1681, 1989.

Table 2: Fuselage Lap Joint Fatigue Test Results

SPECIMEN	STATUS	PRECRACK CYCLES AT 17.8 ksi	CYCLES TO FAILURE AT 13.4 ksi	LIGAMENTS JOINED (TOP ROW)	LONGEST CRACK mm
*A1**	Unreinforced	-	977,600[a]	Yes	-
*A2**	Unreinforced	-	977,600[a]	*No*[b]	9.0
*A3**	Reinforced	178,400	$>$ 1,300,000[ϕ]	-	4.5
*A4**	Reinforced	178,400	$>$ 1,300,000[ϕ]	-	-
A5	Unreinforced	5,000	-	No	1.8
A6	Unreinforced	5,000	25,000	Yes	-
A7	Reinforced	5,000	$>$ 4,001,000[ϕ]	-	1.0
A8	Reinforced	5,000	$>$ 4,001,000[ϕ]	-	1.0
A9	Unreinforced	-	-	No	7.1
A10	Unreinforced	-	105,700	Yes	-
A11	Unreinforced	-	-	No	2.4
A12	Unreinforced	-	67,000	Yes	-
A5[+]	Unreinforced	n/a	41,400	Yes	-
A6/2[@]	Reinforced	n/a	see below	prior	-
A5/2[@]	Reinforced	n/a	$>$ 1,300,000	prior	-
A6/2[@]	Reinforced	n/a	$>$ 1,316,400	prior	-

* - Adhesive seepage occurred across top fastener row (enhanced fatigue life)

a - No crack detected at 837,000 cycles

b - Failed lower fastener row

\+ - Continuation of specimen test

@ - Repaired after complete failure, using room temperature curing adhesive

ϕ - Test in progress

NOTE: All specimens contained spark eroded notches under the fasteners at the top row

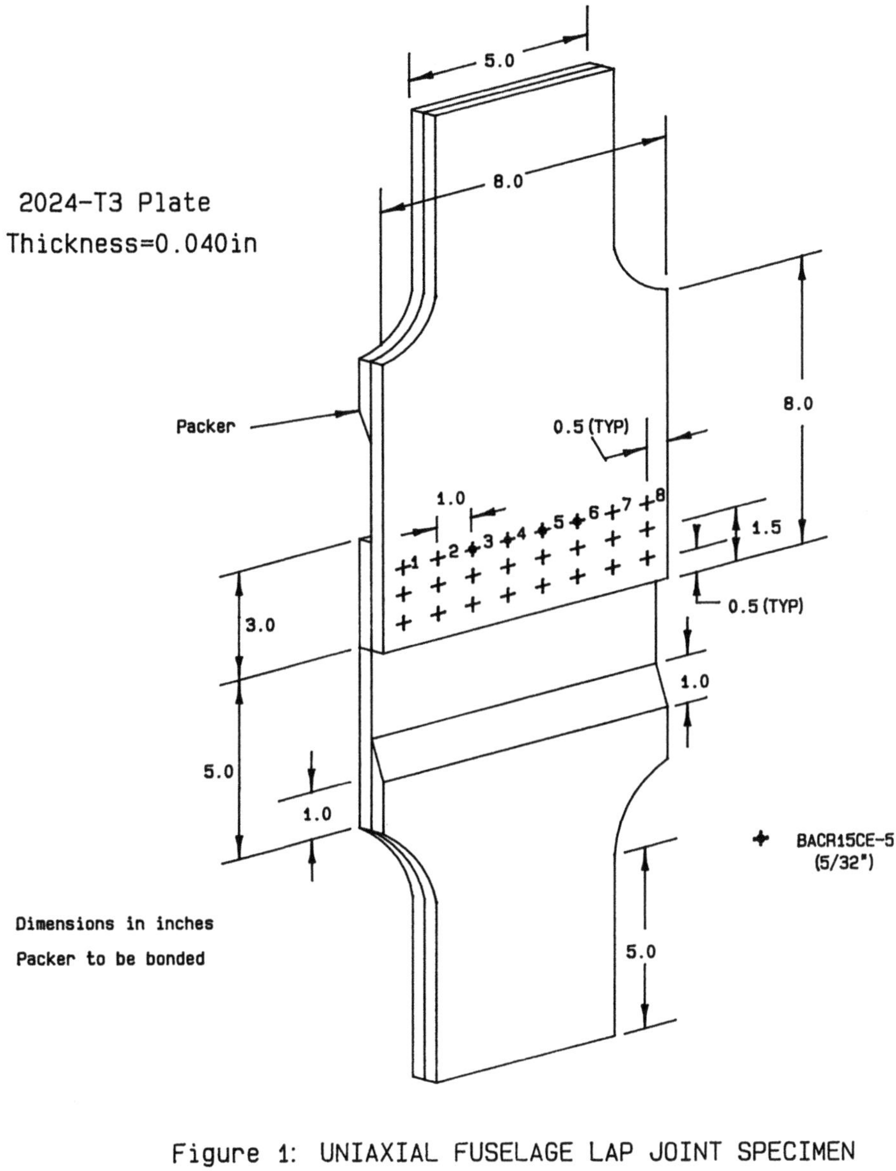

Figure 1: UNIAXIAL FUSELAGE LAP JOINT SPECIMEN

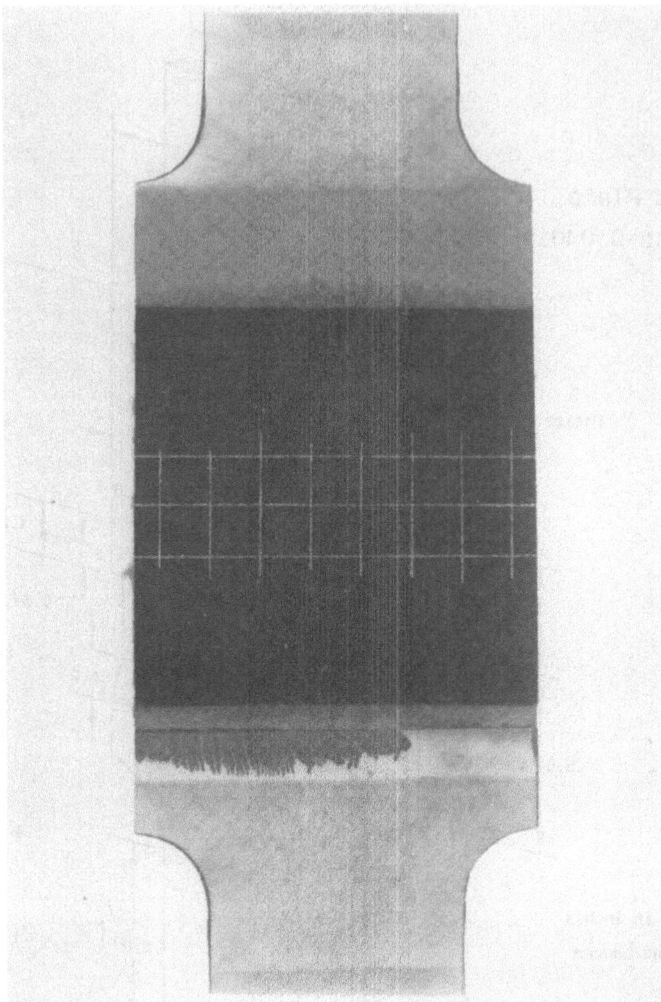

Figure 2: REINFORCED LAP JOINT SPECIMEN

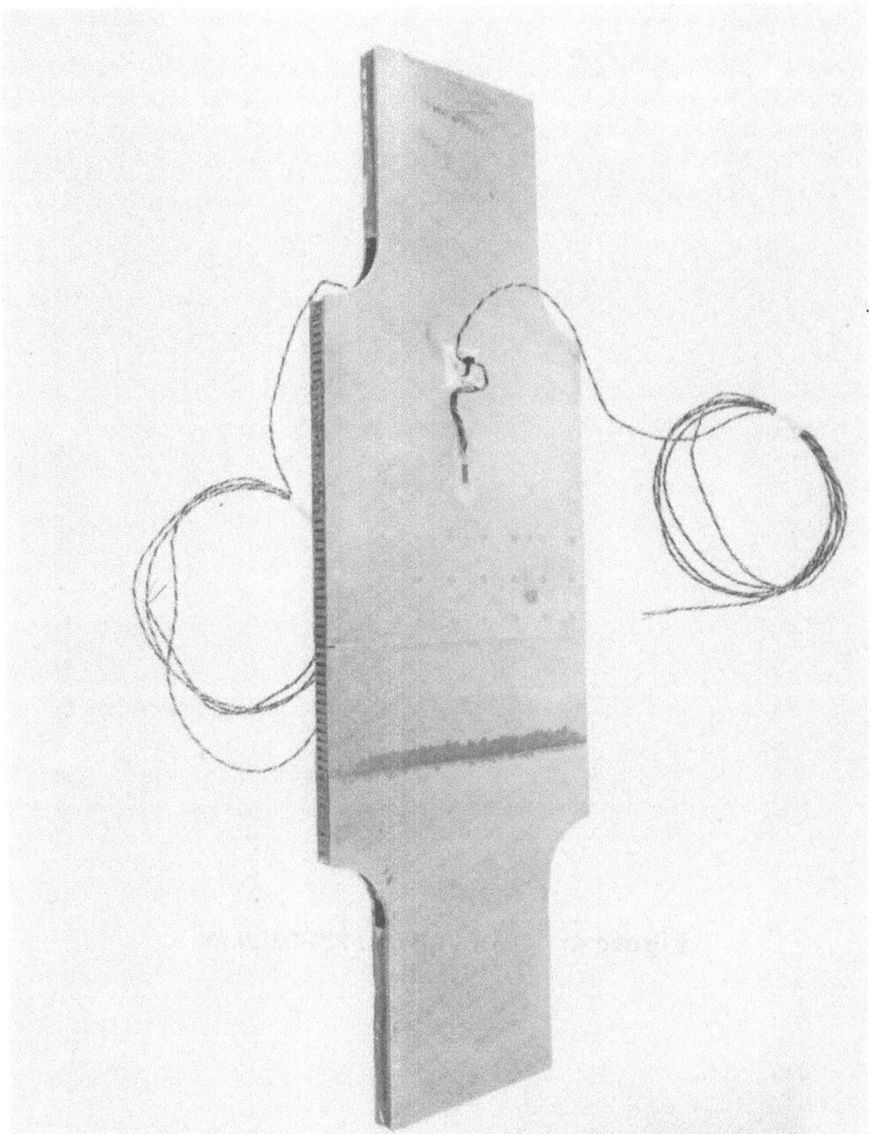

Figure 3: VIEW OF SPECIMENS BONDED TO HONEYCOMB BACKING

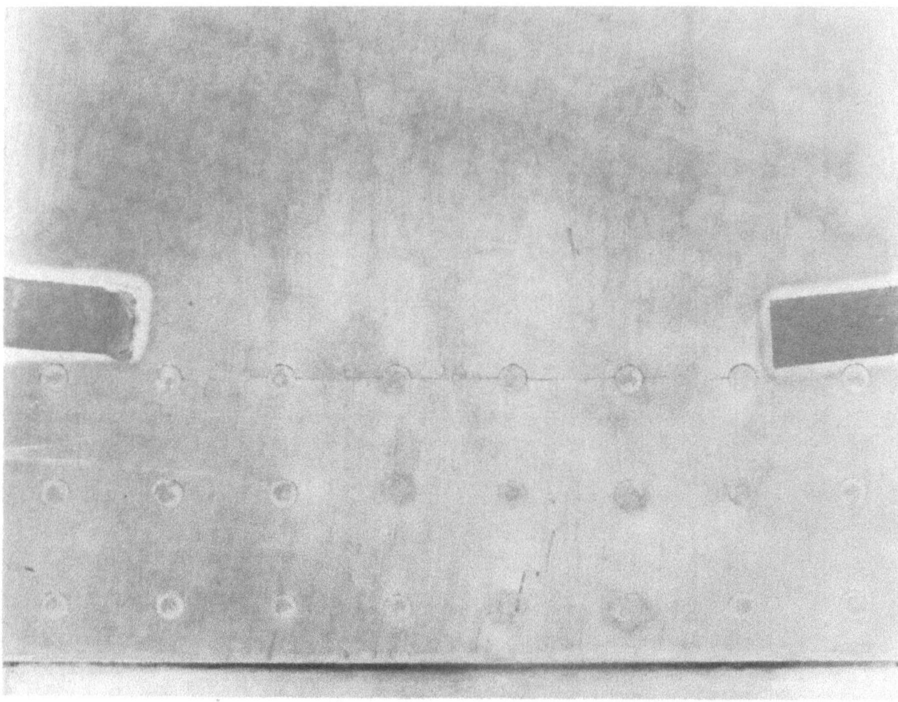

Figure 4: FAILURE OF SPECIMEN A6

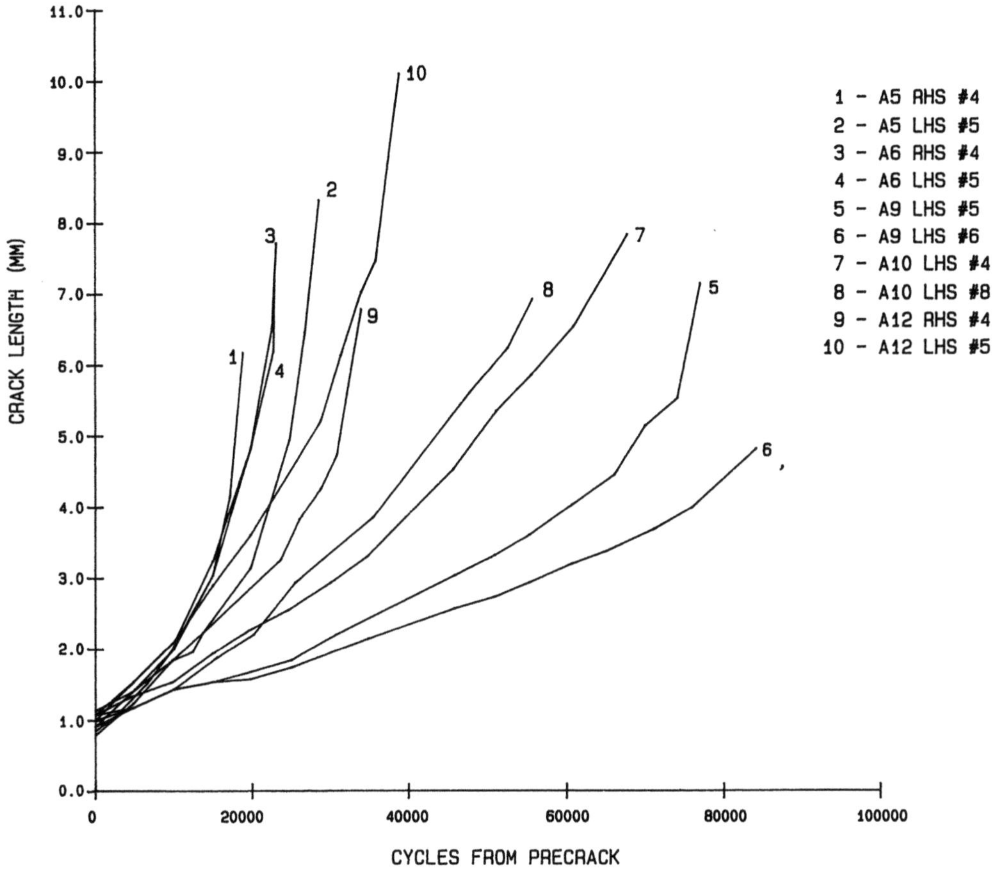

Figure 5: COMPARISON OF SPECIMEN CRACK GROWTH RATES

Figure 5: LAP JOINT REPAIR TO A B-727

Figure 6: B-767 MULTI-SITE DAMAGE

15. Applications of Advanced Fracture Mechanics to Fuselage

M. F. KANNINEN, P. E. O'DONOGHUE, S. T. GREEN,
C. P. LEUNG, S. ROY, AND O. H. BURNSIDE

Engineering and Materials Sciences Division
Southwest Research Institute
San Antonio, Texas

Summary

Multi-site damage (MSD) in the form of cracking at rivet holes in lap splice joints has been identified as a serious threat to the integrity of commercial aircraft nearing their design life targets. Consequently, to assure the safety of aircraft that have accumulated large numbers of flights, flight hours and years in service requires inspection procedures that are based on the possibility that MSD may be present. For inspections of aircraft components to be properly focused on the defect sizes that are critical for structural integrity, fracture mechanics analyses are needed. The current methods are essentially those of linear elastic fracture mechanics (LEFM) which are strictly valid only for cracks that extend in a quasi-static manner under small-scale crack tip plasticity conditions. While LEFM is very likely to be appropriate for subcritical crack growth, quantifying the conditions for fracture instability and subsequent propagation may require advanced fracture mechanics techniques. The specific focus in this paper was to identify the conditions in which inelastic-dynamic effects occur in (1) the linking up of local damage in a lap splice joint to form a major crack, and (2) large-scale fuselage failure by a rapidly occurring fluid/structure interaction process .

Introduction

A recent rash of airplane accidents has drawn considerable public attention to what is now widely known as the aging aircraft issue. A variety of failure modes has been experienced in the accidents that have focused attention on this issue. However, undoubtedly because of the Aloha Airlines aircraft canopy separation, large-scale fuselage fracture is likely the best known. This failure was abetted by the presence of wide-spread cracking along rows of rivet holes in the lap splice joints in a fuselage skin - a phenomenon generally called multi-site damage (MSD). Of greatest technical concern, the Aloha Airlines incident revealed that the crack arrest capability of the design can be defeated when MSD exists.

Considering the great number of older aircraft that are presently in service - indeed, to determine what "older" really means - some type of quantitative examination is clearly mandated. While details may differ, all such approaches must conform to the "damage tolerance" philosophy in which periodic proof tests and/or nondestructive inspections (NDI) are combined with fracture mechanics analyses for the rate of growth to criticality of the most severe crack that would go undetected. For most situations, ordinary linear elastic fracture mechanics (LEFM) will suffice for this purpose. However, the presence of MSD engenders a number of complications that might lie beyond the limits of LEFM. The objective of this paper is to quantify the role of inelastic and dynamic effects that are embodied in an advanced fracture mechanics approach.

Springer Series in Computational Mechanics
Atluri, Sampath, Tong (Eds.)
Structural Integrity of Aging Airplanes

Because the complexities arising from crack growth analyses are superposed on the already awesomely complex frame/stringer/doubler/splice construction of a fuselage [1], unless it is unequivocally warranted, the aircraft industry should not be presented with the further demands of a more complex fracture mechanics analysis methodology. Accordingly, this paper seeks to identify just those few areas in which advanced fracture mechanics techniques need to be used. By way of illustration, preliminary results are presented herein for the elucidation of inelastic and dynamic effects associated with MSD in fuselage lap splice joints.

Background Discussion

The possibility of catastrophic large-scale crack propagation in an aircraft first came to light in connection with the DeHavilland Comet, a BOAC aircraft, in 1954 [1]. Following the loss of two of those aircraft due to large-scale fracture, most manufacturers incorporated some type of crack arrester device in their fuselage designs. For example, Boeing bonded a rectangular grid of skin onto the inside surface of the fuselages of their 737 aircraft, a process that doubles the thickness of the skin locally [2]. The spacing of these "tear straps" was determined such that an axially propagating crack in the fuselage (the most likely direction) would be turned into the circumferential direction and arrested, leaving a manageable rate of cabin air loss and a minimal decrease of structural integrity.

This approach, known as "safe decompression", was validated experimentally by Boeing for suddenly injected cracks (e.g., from a thrown engine fan blade) of up to forty-inches in length. Although the details are different, a generally consistent approach is used by Douglas [3]. Indeed, several of the recent incidents have been of this type, but with large-scale crack propagation only taking place in one of them. That several airplanes have landed safely with a ten-inch square "flap" in the fuselage lends credence to the safe decompression concept. However, the Aloha airline incident has given rise to concern that this approach will invariably be successful, particularly when MSD has developed.

The Aloha airplane failure clearly was aided by the premature debonding of the riveted skin overlap sections, stress concentrators in the design, and the severe operating conditions to which the Aloha fleet is subjected; e.g., salt air environment, many short flights, long service times [4]. Nevertheless, as it is widely agreed that other aging commercial aircraft could eventually be expected to deteriorate in this manner, MSD surely is a key issue in the structural integrity of aging airplanes. As a working definition, MSD consists of a group of small cracks that appear in an airframe at about the same time and originate from similar details located in a common area. To a large extent, LEFM procedures are useful in analyses of potential fuselage failures involving MSD. Nevertheless, some aspects are likely to extend beyond LEFM methodology in several ways. In terms of a chronology that starts from a relatively small fatigue crack, these might be as follows.

First, the thin skin material used in most aircraft is ductile which requires consideration of stable tearing and an *elastic-plastic analysis*. Second, the small cracks that constitute the MSD can be misaligned such that stable tearing (i.e., via R-Curve behavior) will occur in a *non-collinear* manner. Third, as a long crack in a thin-shell under internal pressure will bulge outwards to a significant extent, a *large deformation* analysis may be needed. Fourth, when local instabilities occur, *dynamic effects* could be present that could enhance the crack driving force for cracks at adjacent stress concentrators while possibly reducing the resistance to crack growth. Fifth, if the crack is sufficiently open to allow significant air leakage from the fuselage, the local stress state will be a result of a *fluid/structure interaction*. Sixth, when a large-scale instability occurs (e.g., by the linking up of a number of MSD sites), the direction of *rapid crack propagation* can change abruptly from axial to circumferential, and it may or may not arrest.

To examine these items, a computational capability such as that developed and verified in a SwRI internal research project is needed; see Table 1. The code that was developed, known as PFRAC, was found to be in very good agreement with full-scale experimental results for unreinforced monolithic gas transmission pipes. These validations provide confidence for translating the computational procedure to conditions appropriate for aircraft fuselages; i.e., much larger diameters, very thin walls, much lower pressure differentials, smaller relative crack lengths, reinforced structures. This code has been used in the following for a preliminary judgment on the importance of various nonlinear effects using the nominal structural geometry for a Boeing 737 aircraft with material properties of 2024-T3 aluminum.

TABLE 1
SWRI INTEGRATED FLUID/STRUCTURE/FRACTURE COMPUTATIONAL CAPABILITY

STRUCTURAL MECHANICS UNIT
• DYNAMIC FINITE ELEMENT SHELL CODE • ELASTIC-PLASTIC MATERIAL BEHAVIOR • GEOMETRICALLY NONLINEAR DEFORMATION • EFFICIENT EXPLICIT TIME INTEGRATION
FLUID MECHANICS UNIT
• THREE-DIMENSIONAL FINITE DIFFERENCE CODE • NON-STEADY GAS DYNAMICS • TWO-PHASE FLOW BEHAVIOR • MOVING OBSTACLE FEATURE
FRACTURE MECHANICS UNIT
• DYNAMIC CRACK PROPAGATION FEATURE • NODE RELEASE ALGORITHM • ENERGY RELEASE RATE COMPUTATION (DYNAMIC J) • CRACK TIP REFINEMENT VIA SUBSTRUCTURING

Analysis of Stationary Cracks in Fuselage Structures

As the first step, a computation was made to determine the local bulging of the skin due to an axial crack in an unreinforced skin. An example result, in which the fuselage is simply treated as a circular cylindrical shell, is shown in Figure 1. Figure 2 shows the corresponding stress intensity factor values computed for a range of axial crack lengths for unreinforced skins. The definition $K = (EJ)^{1/2}$ has been used for convenience where E denotes the elastic modulus and, as fully described in reference [5], J is a measure of the crack driving force for elastic-plastic conditions.

Computational results like those in Figure 2 are meaningful only in relative terms. While one benchmark is the fracture toughness, these values can vary over a wide range; NB, for 2024-T3 aluminum, K_c ranges from 75 ksi in$^{1/2}$ up to 150 ksi in$^{1/2}$. Another basis of comparison would be "standard" LEFM expressions for the stress intensity factor. The best known of those appropriate for this application would be that developed for axial cracks in internally pressurized cylinders by Folias; see reference [5]. This is:

$$K = \sigma\sqrt{\pi a}\left[1 + 3.22\frac{a^2}{Dh}\right]^{1/2} \tag{1}$$

where σ denotes the hoop stres, $2a$ is the crack length, D is the cylinder diameter and h is the wall thickness. The conventional "flat plate" solution is recovered from Eq. (1) by simply setting $a << \sqrt{Dh}$. Now, referring to Figure 2, it can clearly be seen that the computed solutions differ markedly from these two reference cases. But, of more importance, a distinct difference can be seen between the geometrically nonlinear elastic solution and the elastic-plastic geometrically nonlinear solutions obtained from the enhanced PFRAC code.

On the basis of the results shown in Figure 2, it is evident that elastic plastic deformation can be significant in thin-skin aluminum 2024-T3 materials. Accordingly, the further computations that were carried out to examine the effects of stiffening used this approach. The preliminary results are shown in Figure 3 where periodic tear straps have been introduced into the analysis. It can be seen that this mode of reinforcement provides only a modest reduction in the stress intensity factor; likely because it is the nonlinear material behavior that dominates.

The significant differences between the PFRAC solutions accounting for the geometric nonlinear deformation and the flat plate and Folias solutions that are revealed in Figure 2, particularly evident for crack lengths greater than the tear strap spacing, are not entirely unexpected. What may be more surprising are the further differences that result from the imposition of elastic-plastic material behavior. Because the neglect of elastic-plastic behavior is anti-conservative, it would appear that the geometrically nonlinear elastic-plastic approach should be adopted for the consideration of long cracks (i.e., crack lengths greater than the tear strap spacing) in fuselage structures.

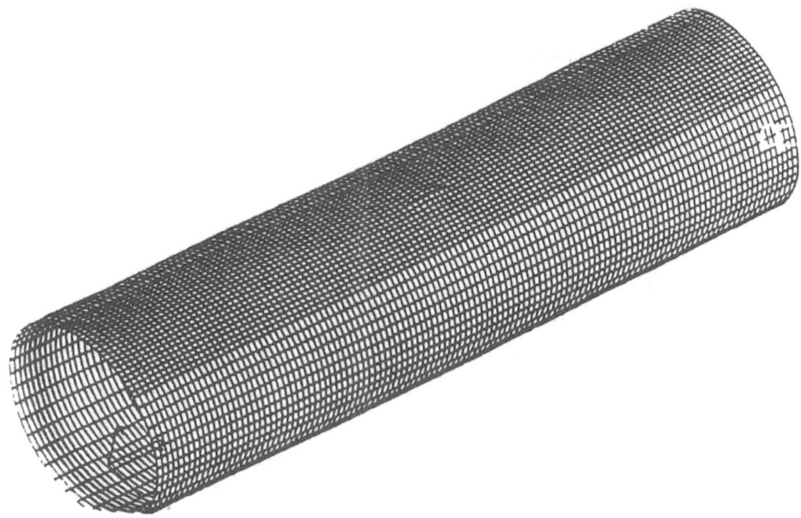

(a) Overall View of Cracked Fuselage

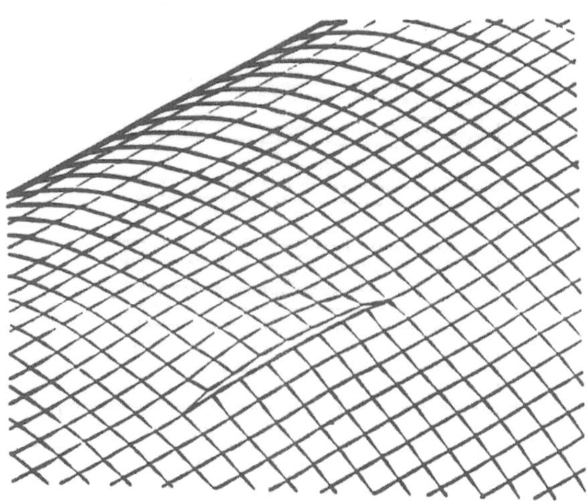

(b) Magnified View of Cracked Region

Fig. 1. Preliminary Coupled Fluid/Structure Computation of Aircraft Fuselage Deformation and Decompression Due to a 24-inch Axial Crack

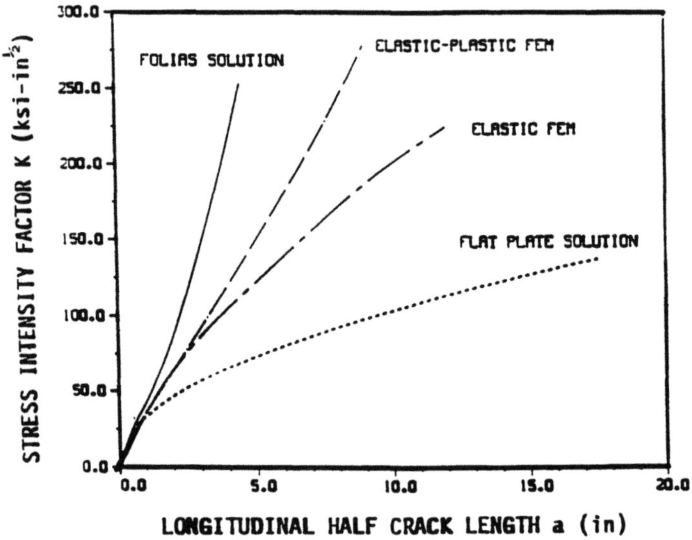

Fig. 2. Comparison of Computed Stress Intensity Factors for Different Geometric and Material Behavior Modeling Assumptions

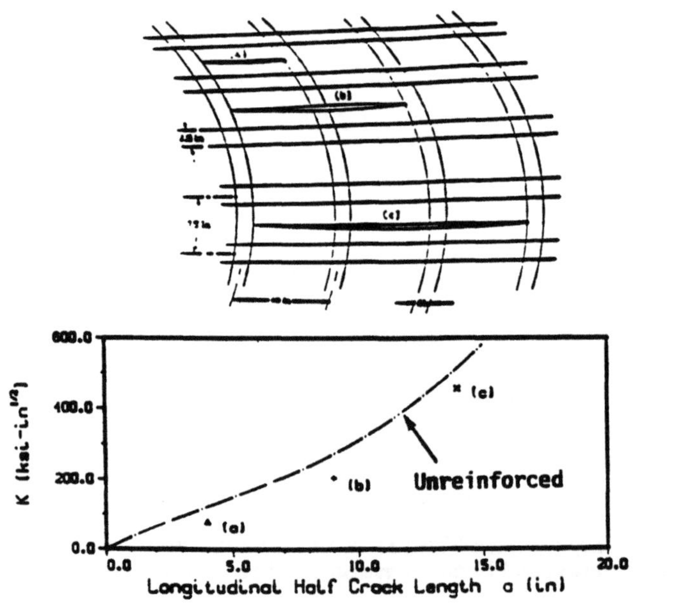

Fig. 3. Comparison of Elastic-Plastic Computational Results for Unreinforced and Reinforced Aircraft Fuselages

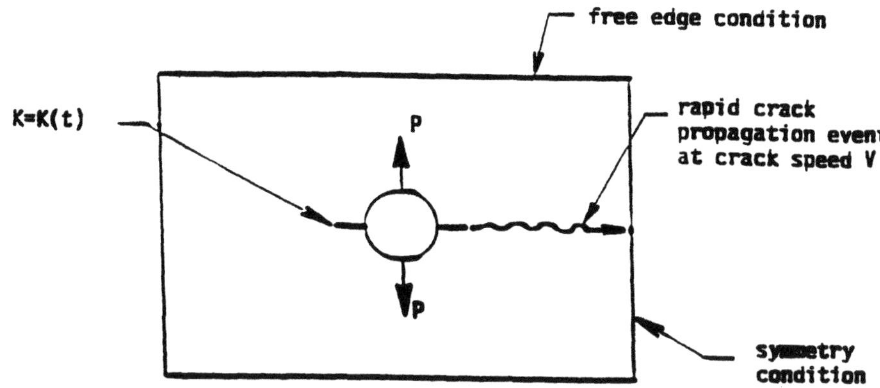

Fig. 4. Computational Model for the Determination of Dynamic Stress Enhancement Arising from Rapid Load Transfer at MSD Sites

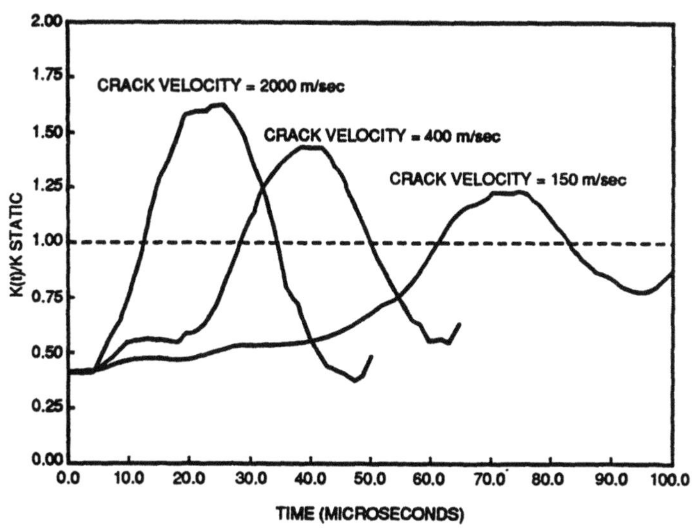

Fig. 5. Dynamic Fracture Mechanics Computational Results for the Configuration Shown in Figure 4

Analysis of Dynamic Effects in Fuselage Structures

The process through which cracks at near-by rivet holes link-up to form a large crack is one to which dynamic effects could contribute. In particular, because inertia effects generally act to enhance the stress state while rapid motion tends to reduce the material's resistance to crack advance, the neglect of dynamic effects could be anti-conservative.

Rather than attempt to simulate actual aircraft fuselage geometries, some relatively simple computations were performed for a preliminary assessment of the dynamic enhancement of the crack driving forces resulting from the rapid link-up of MSD sites. The configuration, which roughly represents the dimensions of typical skin lap joint configurations in Boeing 737 and Douglas DC-9 aircraft, is shown in Figure 4. The crack on the right was allowed to propagate rapidly to the edge of the specimen, while the increase in the stress intensity factor for the stationary crack on the left was calculated. Elastodynamic computational results obtained from this model using the SwRI code VISCRK are shown in Figure 5.

Figure 5 represents three separate simulations, each having a different constant crack speed. As expected, the higher the crack speed, the greater the increase in the stress intensity factor. Here, the increase ranges from 25% for a crack speed of 125 m/s to 80% for a crack speed of 2000 m/s; the latter representing an upper bound for the crack speed.

In addition to local dynamic stress enhancement arising in MSD-related crack growth, dynamic effects associated with the possible subsequent global crack growth event can also arise. To examine these, elastodynamic computations under conditions geometrically similar to those of Figures 2 and 3 were performed; i.e., a pristine structure without MSD. Because dynamic fracture toughness data for 2024-T3 aluminum are currently lacking, these computations simply considered rapid axial crack propagation at assigned constant speeds. Figure 6 shows the resulting dynamic stress inensity factor values for a fuselage structure reinforced by tear straps and frames.

It can be seen in Figure 6 that the dynamic K values are relatively independent of crack length, unlike those obtained for quasi-static growth (cf, Figure 2). That these K values are lower than those obtained for quasi-static growth should be viewed in light of the fact that, generally, dynamic fracture propagation toughness values are less than static initiation toughness values. Note that these results were obtained by assuming that the cabin pressure is unaffected by the crack propagation event. This is equivalent to assuming that the insulation blankets remain intact so that, while carrying no significant portion of the presure load, they effectively prevent the egress of cabin air. Clearly, this will not be possible when the crack has propagated far enough whereupon a coupled fluid/structure interaction will take place. That the resulting decompression will significantly reduce the crack driving force is shown in Figure 7.

Finally, to set the stage for examining the safe decompression concept in a fuselage structure with MSD, illustrative computations were performed to examine the effects of a crack propagating rapidly in an axial direction that is turned at a tear strap and subsequently arrested. It was assumed that the loads acting on the fuselage are due to the cabin air pressure only, even though it is well known that the aircraft weight gives rise to axial bending stresses. In addition, external aerodynamic effects have also been neglected at this stage of the effort, even though these may well be significant. The computations were initiated from a relatively small critical crack along the rivet line midway between the tear straps by stable tearing. Subsequently, the dynamic event transpired for which the arrested state is shown in Figure 8.

Discussion

As it is not feasible without considerably disrupting the public and incurring severe economic consequences to simply to retire all older aircraft now operating, damage tolerance procedures must be developed to ensure their integrity. However, given that any aircraft could expect some degree of MSD, the use of the damage tolerance approach that is mandated becomes much more complex. This is due both to the smaller crack sizes that can be critical in MSD (i.e., in comparison to a single dominant crack), and to the possible reduction in the resistance to fracture propagation arising from MSD ahead of a crack. Whether or not NDI or periodic proof testing is combined with fracture mechanics, it is clear that the analyses must confront the possibility of MSD.

The objective of the work reported in this paper was to quantify the importance of non-LEFM effects in applications of fracture mechanics to aircraft fuselage structures with MSD. These effects included (1) dynamic processes, (2) inelastic deformation, and (3) geometric nonlinearities. To focus the work, the computations were carried out for fuselage lap splice joints in terms of the instability of linked-up cracks, and the rapid propagation/arrest of long cracks. The progress that was made did not provide a complete description of these processes; in part, because of the lack of appropriate material dynamic-inelastic fracture propagation data. Nonetheless, the work has set the stage for the computational procedure that is needed for attacking key issues such as that illustrated in Figure 9.

In addressing the type of conditions exemplified by Figure 9, particularly in view of uncertainties in material properties and service conditions, it is clear that those having the responsibility for aircraft structural integrity will be best served by technical approaches that balance validity against excessive complexity. Accordingly, the replacement of LEFM procedures by advanced fracture mechanics techniques should be made cautiously lest the analysis process be unduly burdened for the sake of marginal increases in accuracy. The approach taken in this paper was therefore a heuristic one, addressed to quantifying nonlinear and dynamic effects in order to determine the pay-off for the complexity thereby introduced. It is felt that, while the work is still in a preliminary stage, several opportunities were clearly identified.

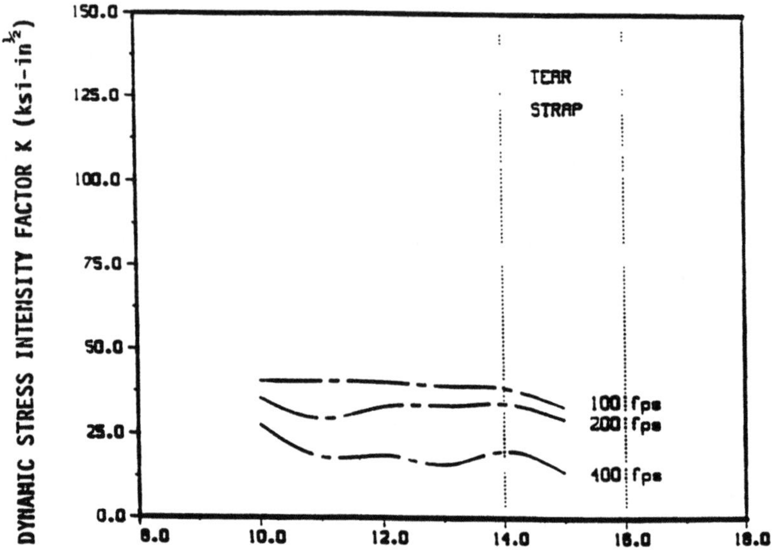

Fig. 6. Dynamic Stress Intensity Factors Computed for Rapid Axial Crack Propagation in an Aircraft Fuselage Structure Reinforced with Tear Straps and Frames

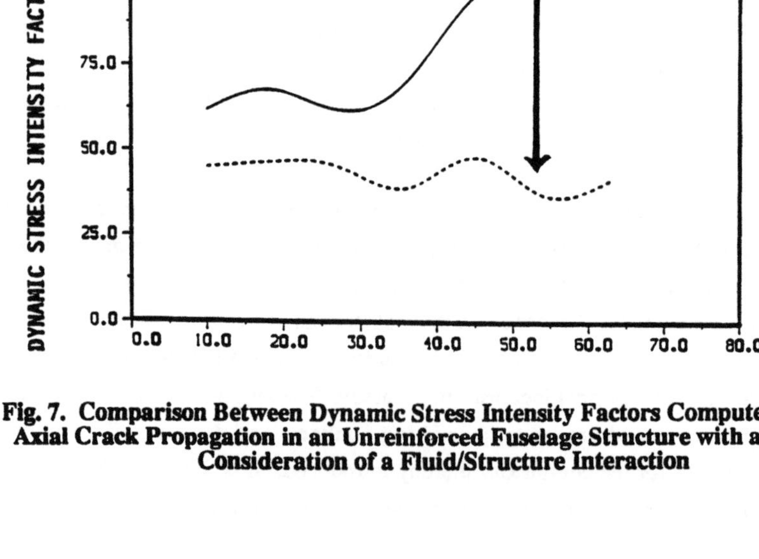

Fig. 7. Comparison Between Dynamic Stress Intensity Factors Computed for Rapid Axial Crack Propagation in an Unreinforced Fuselage Structure with and without Consideration of a Fluid/Structure Interaction

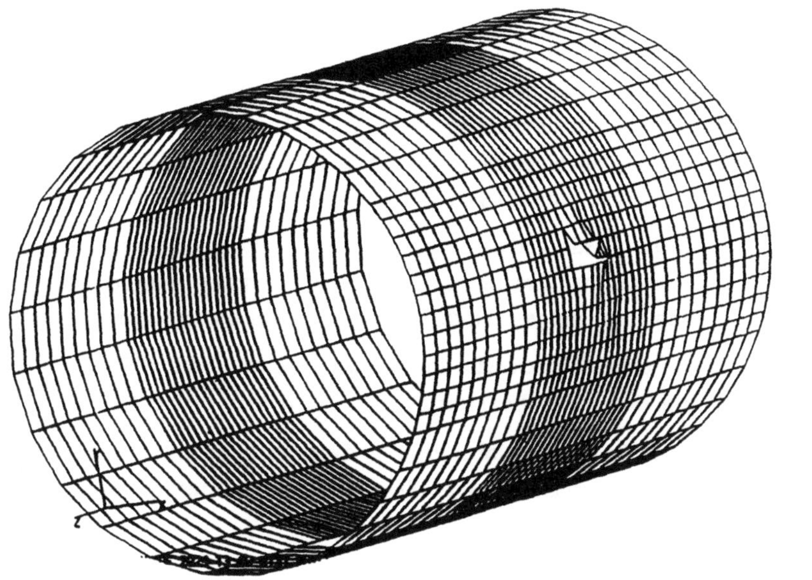

Fig. 8. Elastoplastic-Dynamic, Large Deformation, Fluid/Structure Analysis Computational Result for an Arrested Crack in an Aircraft Fuselage

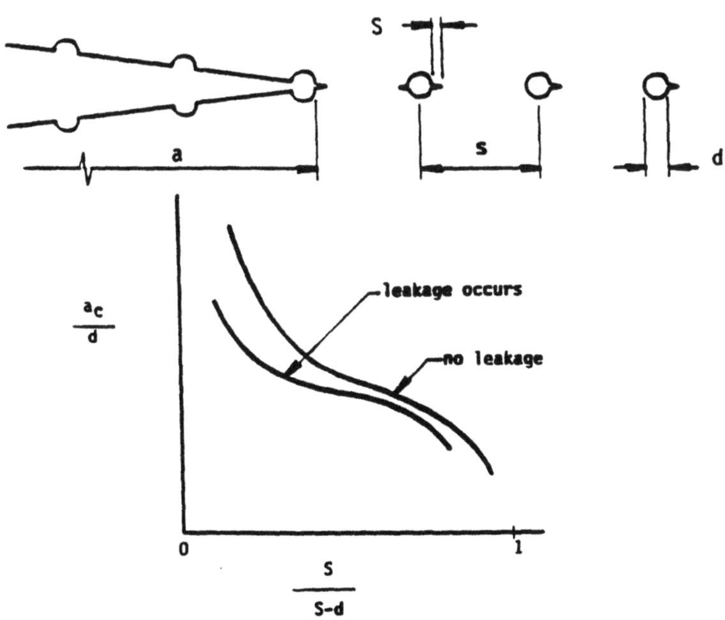

Fig. 9. A Goal for Further Work: The Quantification of Critical Crack Sizes in Aircraft Fuselage Structures with Multi-Site Damage

Conclusions

The computational results provided in this paper were developed in the course of a heuristic study aimed at identifying those particular areas in which the complexity of nonlinear and dynamic fracture mechanics treatments are warranted for aircraft fuselage structures where multi-site damage (MSD) might be present. While the work is far from complete, some preliminary conclusions have been reached, as follows:

1. Elastic-plastic geometrically nonlinear analyses of long cracks in aircraft fuselage structures (i.e., greater than the tear strap spacing) differ significantly from linear and geometrically nonlinear elastic fracture mechanics predictions.

2. The rapid load transfer in fuselage structures with MSD can engender a time-dependent stress enhancement due to local dynamic effects.

In addition, it has been shown that global inelastic-dynamic, fluid/structure analyses of "safe decompression" in fuselage structures with MSD can be made in a realistic manner.

Acknowledgements

Portions of this work were supported by the DOT Transportation Systems Center Aging Aircraft Program through Purchase Order No. DTRS57-90-P-80689. The authors would like to express their gratitude to Dr. S. G. Sampath, TSC, for making their participation possible. Other portions of the work reported herein were conducted through a Southwest Research Institute internal developmental project for which the authors would like to express their appreciation to the SwRI Advisory Committee for Research. The assistance of Mr. D. Rhia in post-processing many of the computational results is also acknowledged. Finally, informal discussions with individuals knowledgeable in aircraft structures that enabled the authors' fracture mechanics analysis capabilities to be much better focused on aging aircraft issues were very generously provided by T. A. Swift, the Federal Aviation Administration; U. Goranson and M. Miller, Boeing Commerical Airplane Company; and A. W. Hoggard, the Douglas Aircraft Company.

References

1. Swift, T.: Damage Tolerance in Pressurized Fuselages. 14th Symp. Int'l. Comm. Aeronautical Fatigue. Ottawa, Canada: 10-12 June 1987.

2. Goranson, U. G.; Miller, M.: Aging Jet Transport Structural Evaluation Programs. 15th Symp. Int'l. Comm. Aeronautical Fatigue. Jerusalem: June 1989.

3. McGrew, J. A.; Hoggard, A. W.: Douglas Aging Aircraft Programs. SAE Aerospace Technology Conference. Anaheim, California: September 1989.

4. Anonomous: National Transportation Safety Board Aircraft Accident Report. NTSB/AAR-89/03 June 1989.

5. Kanninen, M. F.; Popelar, C. H.: Advanced Fracture Mechanics. New York: Oxford University Press 1985.

16. Axial Crack Propagation and Arrest in Pressurized Fuselage

M. KOSAI AND A. S. KOBAYASHI

Department of Mechanical Engineering
University of Washington
Seattle, Washington 98195
USA

Summary

The rapid crack propagation, crack curving and arrest mechanisms associated with a pressurized, thin-walled ductile steel tubes are used to develop a model of axial rupture of an aircraft fuselage. This model is used to replicate axial crack propagation along a line of multi-site damage (MSD) and crack curving and arrest near a tear strap of an idealized fuselage.

Introduction

Static linear elastic fracture mechanics (LEFM) has been used to predict a monotonically increasing stress intensity factor with increasing crack length of an axial crack in a pressurized pipe [1,2] as well as to model dynamic crack branching in a pressurized steel pipe [3]. Plasticity effect has also been incorporated to this LEFM analysis using semi-empirical adjustments [4,5], or by kinematic modelling of the yielded pipe [6]. Many of these analyses, however, do not consider the dynamic effects generated by the propagating axial crack as well as the influence of the large plastic deformation of the crack flaps.

Under the sponsorship of the US Department of Transportation (DOT), one of the author and his colleagues studied axial crack propagation, curving and arrest in a small scale model of a line pipe [7,8]. These experimental results were then used to modify a numerical model, which was previously developed for analyzing bursting thick-walled steam pipes. This study showed that the main driving force of the axial crack in a bursting pipe was the gas pressure acting on the crack flaps which literally tore the pipe apart. It was also found that the axial stretch of the crack flap, not only served as an energy dissipation mechanism, but also imposed an axial tensile stress, which is larger than the hoop stress, as much as five pipe diameters ahead of the crack tip thus leading to crack curving and crack arrest.

While an aircraft fuselage is an infinitely more complicated structure than a monolithic line pipe, the crack propagation, curving and arrest mechanisms of the latter could be of use in a postmortem analysis of a fuselage failure as well as for estimating the crack arrest capability of a tear strap. In the following, a brief

Springer Series in Computational Mechanics
Atluri, Sampath, Tong (Eds.)
Structural Integrity of Aging Airplanes

review of the cogent results generated in the DOT pipe fracture study as well as the crack curving and crack branching criteria, which were developed under a Office Naval Research (ONR), contract are presented. These results are then used to assess the crack arrest capability of a hypothetical fuselage with tear straps.

Analysis of Axial Rupture of Subsize Gas Transmission Lines

The experimental phase of the DOT studies consisted of measuring the dynamic depressurization rate, crack extension rate, static and the dynamic axial and circumferential strains along the axial crack in a 2-in. diameter, schedule 10, carbon steel pipe. These pipes were pressurized to approximately 80 percent of the yield strength in hoop stress in a series of static and dynamic pipe rupture experiments. The test section of the pipe was pre-grooved to a depth of approximately 20 percent of the pipe wall thickness to simulate a locally embrittled zone for continuous and controlled axial crack propagation in the ductile pipe wall. Such grooving simulated multi-site damage (MSD) in a fuselage. The grooves were carefully machined to avoid local elevation of the ultimate strength of the pipe wall material due to strain hardening, and thus the pipe wall fractured with a 100% shear lip. The propagating crack always arrested upon entering the ungrooved section, which simulated the fuselage section with a tear strap of the pressurized pipe. The rupturing profiles, i.e., the pipe flap motions, were recorded by a high speed framing camera. Figure 1 shows typical sequential photographs of a rupturing pipe where the crack bifucated upon entering the ungrooved section of the pipe.

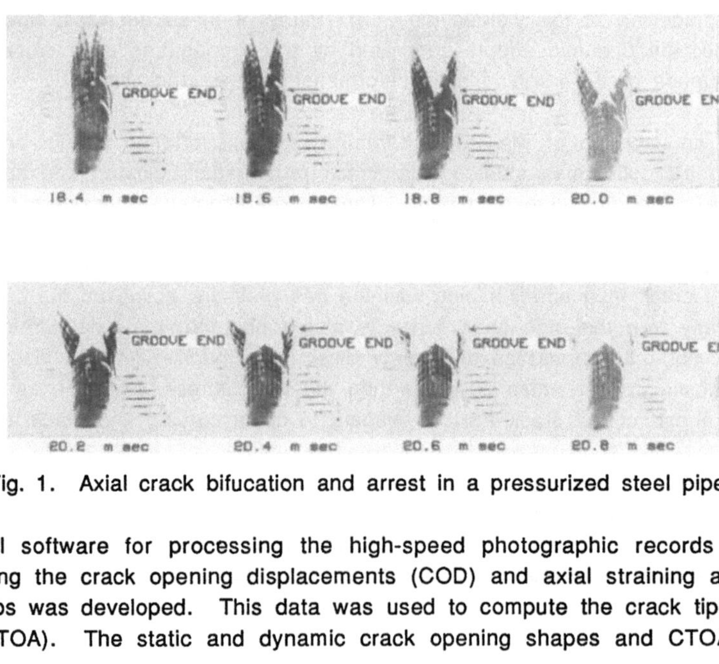

Fig. 1. Axial crack bifucation and arrest in a pressurized steel pipe

A special software for processing the high-speed photographic records and for determining the crack opening displacements (COD) and axial straining along the crack flaps was developed. This data was used to compute the crack tip opening angle (CTOA). The static and dynamic crack opening shapes and CTOA's were similar and the CTOA remained constant over a crack velocity range of 600 to 1000 fps but varied with the groove depth. When the groove depth was decreased to simulate a tougher pipe material, the CTOA increased and the crack velocity decreased.

Records of the strain gage rosettes, which were placed along the crack path in the dynamic pipe rupture tests, showed that the maximum strain exceeded the yield strain and changed from the circumferential to the axial directions approximately 5 to 9 pipe diameters ahead of the propagating crack tip, as shown in Figures 2 and 3. These results indicated that a long and narrow axial-stretch zone preceded the propagating crack tip. This large axial strain, which is much larger than those reported previously, is consistent with the observations of Shoemaker et al. [9] and Urednicek [10]. The narrow axial-stretch zone is also the essential feature of the thin shell model of a bursting pipe by Freund et al. [11] who used this model to predict crack kinking at arrest due to a mechanical crack arrester [12].

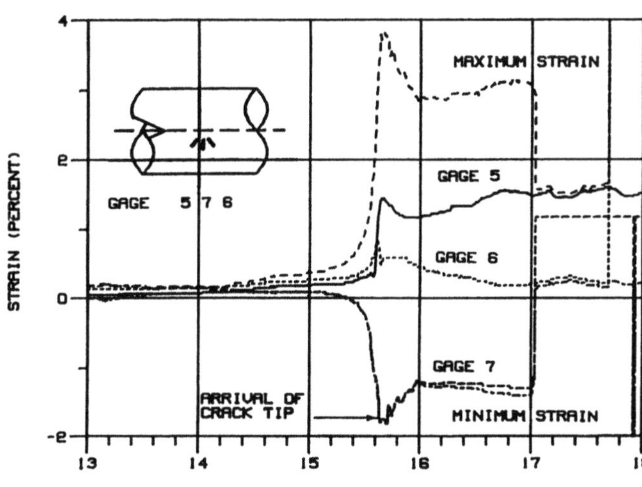

Fig. 2. Strain rosette readings adjacent to a propagating axial crack in a pressurized steel pipe

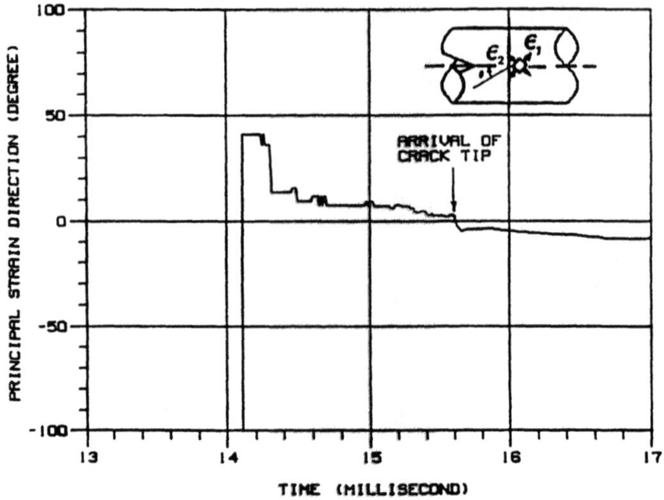

Fig. 3. Variation in principal strain direction derived from data of Fig. 2

The paint, which was sprayed as a background to highlight the painted grid lines, acted as a brittle coating in this test. The presence of large axial strains preceding the propagating crack tip and the alignment of the principal strain direction with the axial direction of the crack in the region of the crack tip was graphically demonstrated by the brittle coating crack pattern in Figure 4 where the crack arrested in the ungrooved section of the pipe.

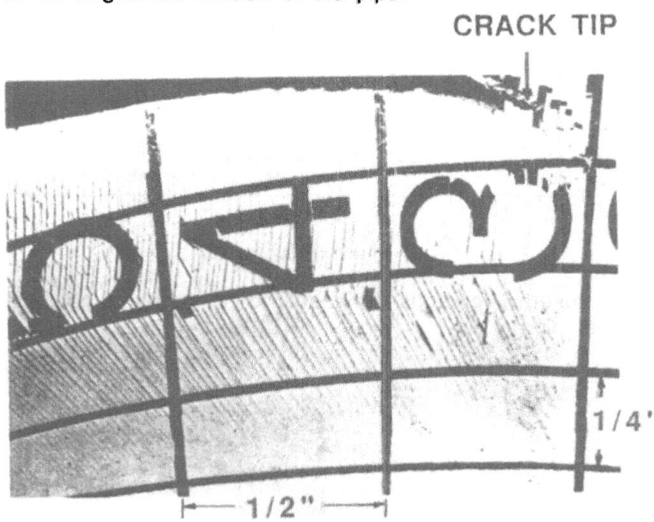

Figure 4. Brittle coating cracking pattern along crack path

Figure 4 shows the axial principal strain orientation and the region of large straining which extended on both sides of the crack path and at least one inch ahead of the crack tip.

The strain at the loci of the crack ends of the brittle coating pattern, i.e., along the isoentatics, is estimated to be larger than two percent. Also to be noted is the change in the direction of the cracking pattern approximately 1/2 in. away from the axial crack path in the upstream region. The 30° change in crack pattern indicates that the principal strain direction abruptly rotated upstream of the crack tip with the opening and stretching of the flaps and is consistent with Freund's shell model of axial pipe fracture [11].

The width of the brittle coating crack pattern and the lengthening of the painted grid lines indicated that a large axial plastic straining occurred in a narrow region surrounding the crack tip. The brittle coating pattern thus provided a qualitative confirmation of the strain gage measurements and its direction as well as the axial stretch model of the pipe flaps.

The CTOA of a crack, which ran out of the groove and arrested in the ungrooved section, is almost triple that of a running crack. This large CTOA suggests that the crack flap motion continued unabated for a short duration after the crack had slowed to an arrest.

Dynamic Crack Curving and Branching

The function of a tear strap in a pressurized fuselage is not so much to arrest a propagating axial crack by reducing the circumferential stress in the crack path, but to deflect the crack in the circumferential direction. A larger opening generated by the deflected crack will then result in a controlled depressurization of the fuselage and hence eliminate the driving force.

The mechanics of elastic crack curving as well as crack branching was studied by one of the authors and his colleague under an ONR contract [13, 14]. The dynamic crack curving criterion, which is a dynamic extension of the static criterion by Streit and Finnie [15], postulates that the micro-cracks ahead of a crack tip dictates the direction of crack propagation. The crack curving criterion assumes that when the circumferential stress within a prescribed crack tip region attains a maximum value off the axis of a self-similar crack extension, crack curving will occur. This maximum condition, which is based on linear elastic fracture mechanics (LEFM), results in a characteristic crack tip distance, r_o, in which the propagating crack will deviate from its axis by an angle of θ_c. For a static crack, these values are

$$r_o = \frac{9}{128\pi}\left[\frac{K_I}{\sigma_{ox}}\right]^2$$

and the crack curving angle, θ_c, is

$$\theta_c = \cos^{-1}\left[\frac{1 \pm \sqrt{1+\frac{1024\pi}{9} r_c \left(\frac{\sigma_{ox}}{K_I}\right)^2}}{\frac{512\pi}{9} r_c \left(\frac{\sigma_{ox}}{K_I}\right)^2}\right]$$

where K_I and σ_{ox} are the mode I stress intensity factor and the remote stress component or the first nonsingular term in the elastic crack tip stress field, respectively. The elastic crack curving criterion requires that $r_o < r_c$ for the crack to curve an angle, θ_c, away from its axis where r_c is a material constant which specifies the characteristic crack tip region in which the off axis micro-cracks are triggered and connected to the main crack tip.

In the presence of a large driving force, or a large dynamic stress intensity factor, the crack will bifurcate in order to shed the excess driving force and thus results in crack branching while the crack branching angle is governed by the crack curving criterion. This crack branching criterion was used successfully to correlated the predicted and measured crack branching angle, $2\theta_c$, and the estimated crack branching stress intensity factor, K_{IB}, in a bursting steel pipe [16]

Finite Element Modeling of Axial Rupture of Aircraft Fuselage

Axial rupture of an aircraft fuselage differs significantly with the rupture of a line pipe in that the ratio of the half crack length, a, to the fuselage radius, R, is one to two orders of magnitude smaller than that of the line pipe. Also the crack flaps of the fuselage are initially loaded statically by the insulation, which acts as a bladder to maintain the cabin pressure on the flaps, and later by the aerodynamic windage as the flaps peel open. Figure 5 shows the axial and circumferential strain variations with crack extension from the static tests conducted in the DOT pipe rupture studies where the pipe pressure was maintained by a neoplene bladder reinforced by a Kevlar insert. The small flap openings in these tests, together with static pressure acting on the flaps simulated the initial phase of an axial rupture of a fuselage. As shown in Figure 5, the deforming flap in the static line pipe test also generated an axial strain which exceeds the circumferential strain at an distance approximately one half of the pipe radius ahead of the crack tip. This recurring theme in both the static and dynamic axial crack studies of the line pipe is the large axial strain which precedes the crack tip. This axial strain, which is three to five times larger than the circumferential strain, is the cause of crack curving once the propagating crack runs out of the groove in the line pipe.

In order to estimate the axial strain ahead of the crack tip in a rupturing aircraft, a finite element model of a two bay crack in an idealized fuselage, as shown in Figure 6, was studied. This quasi-static analysis accounts for the deforming flap geometry but ignores the stress wave effects. The quasi-static analysis was justified since the measure crack velocities in the DOT pipe rupture studies [7,8], full-scale steel pipe rupture studies [17] and dynamic fracture testing of single-edge notched, 7075-T6 and 7178-T6 aluminum specimens [18] all recorded crack velocities less than ten percent of the dilatational wave velocities.

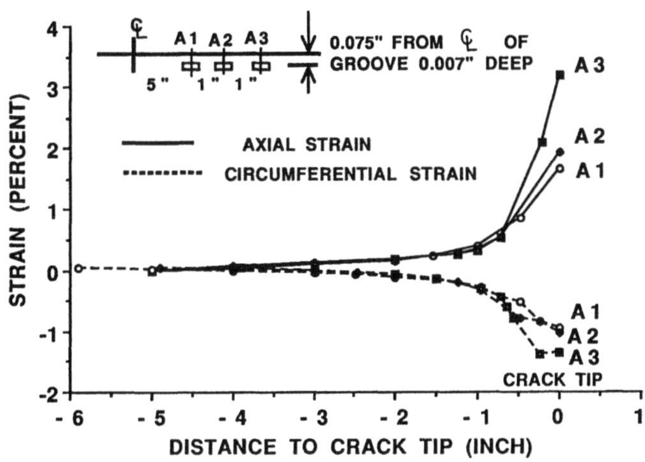

Fig. 5. Variations in strains. Crack length 16 in. Groove depth 0.007 in. Test No. S17.

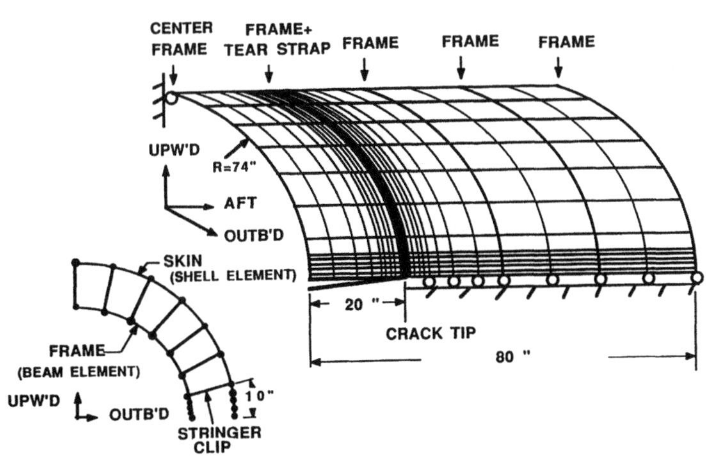

Fig. 6 Finite element model of an idealized fuselage with an axial crack

Table 1 shows some of the structural details and the material properties of the idealized fuselage considered in this study. The fuselage was assumed to be a pressurized cylindrical tube with an uniform axial tensile stress of pR/2t. While structural details, such as butt and overlap joints were ignored, the frame stiffness was incorporated by the beam elements as shown in Figure 6. Windage loading on the crack flaps was not considered due to the lack of measured pressure distribution on a fuselage flap of a moving airplane.

Parameter		Symbol	Value	Unit
Skin and Tear Strap	Material	2024-T3 Clad Sheet		
	Thickness	t	0.036	Inches
	Ultimate Strength	F_{tu}	62	ksi
	Yield Strength	F_{ty}	47	ksi
	Young's Modulus	E	10.5×10^3	ksi
	Fracture Toughness	Kc	91 *	ksi $\sqrt{in}$
	Fuselage Diameter	D	148	Inches
	Tear Strap Spacing	S	10	Inches
	Tear Strap Width	W	2	Inches
	Frame Spacing	b	20	Inches
	Crack Length	2a	40	Inches
	Limit Pressure Differential	P_{limit}	7.5	psi
	Fail-Safe Pressure Differential	P_{fs}	9.0	psi

* Ref. MCIC-HB-01R, "Damage Tolerant Design Handbook"

Table 1. Structural and material property data for aircraft fuselage analysis

Elastic and elastic-plastic static analyses, both based on large deformation shell theory, were conducted. Both analyses were conducted for a stationary axial crack of half crack length of a = 15 to 20 inches at one inch intervals. As such, this analysis does not account for the history of crack extension where the crack would grow from an initial crack length of the order of 1 in. to a = 15 - 20 inches. Thus the effect of the residual stresses in the unloaded crack flaps is also ignored. In all analyses, the center frame, as shown in Figure 7 was assumed to be intact and the riveted skin was assumed to have torn off 10 in. away from the crack. The results of this analysis is discussed in the following.

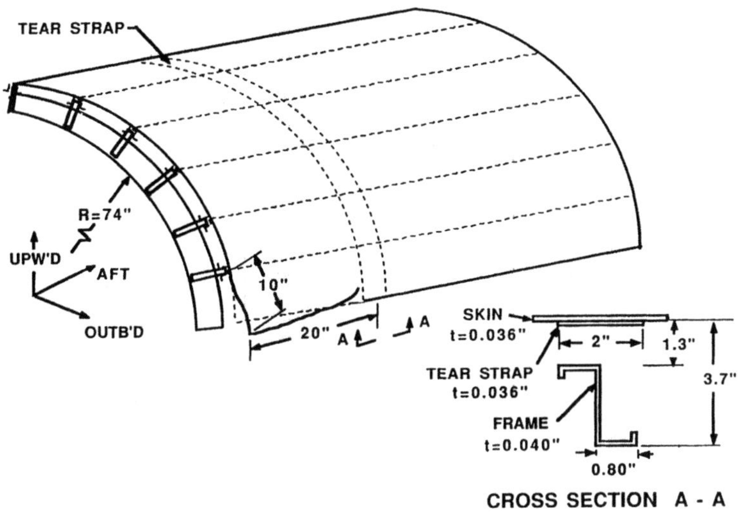

Fig. 7. Axial rupture of an idealized fuselage

Elastic Analysis

As the bursting pipe study showed, large plastic deformation precedes the propagating crack tip which is followed by extensive plastic unloading along the edges of the flaps. Although the elastic analysis in this section does not model the actual flap deformation process accurately, this zeroth order approximation provides an insight to the problem at hand.

Figure 8 shows the elastic stress distribution in a cracked fuselage, which is pressurized to a gage pressure of 7.5 psi and an axial tensile stress of 7.7 ksi, with a crack length of a = 16 in. The numbers with an asterisk are the computed axial stresses in each finite element. The lower numbers represent the axial to circumferential stress ratios. The coarse finite element mesh of Figure 6 was calibrated against a known solution, i.e. a central crack in an infinite plate under uniaxial tension in this case, such that the stress intensity factor and the remote stress component, σ_{ox}, could be extracted from the circumferential and axial stresses at the center of the forward crack tip element. This calibration scheme was essential for extracting meaningful fracture data from the coarse grid finite element analysis.

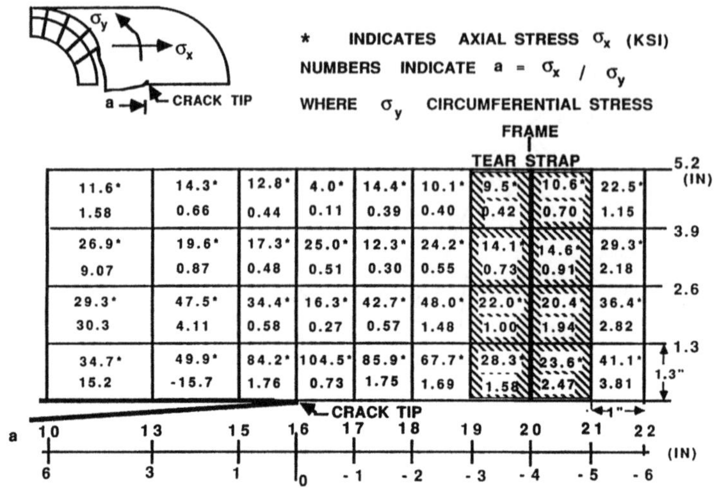

Fig. 8. Elastic stress distribution in a cracked pressurized fuselage

Table 2 shows the mode I stress intensity factor, K, and the characteristic distance, r_o, for half crack lengths ranging from a = 15 through 20 in. where the crack was assumed to have penetrated halfway through the tear strap. Also shown is the crack curving angle, θ_c, which was predicted by using the crack curving criterion as described in the previous section, for these crack lengths. The critical material property for crack curving, r_c, was assumed to be 1.0 ~ 1.5 mm following similar analysis for steel pipes [16]. Note that the high axial stress for

p (psi)	a (in)	K $(ksi\sqrt{in})$	σ_{ox} (ksi)	r_o (mm)	θ_c (deg) r_c =1mm	θ_c (deg) r_c =1.5mm
	15	144.	73.0	2.2	-	-
	16	149.	88.3	1.6	-	-
4.5	17	151.	109.	1.1	-	35.6
	18	157.	130.	0.8	27.6	46.9
	19	110.	52.3	2.5	-	-
	20	113.	62.3	1.9	-	-
	15	220.	98.0	2.9	-	-
	16	225.	120.	2.0	-	-
7.5	17	228.	148.	1.3	-	21.8
	18	235.	178.	1.0	2.8	40.0
	19	166.	68.2	3.4	-	-
	20	172.	79.5	2.7	-	-

Table 2. Results of elastic analysis

much of the crack length, a, would have caused the crack to curve if the fuselage was not weakened by stress concentrations and multiple site damages along its path. Thus, similar to the 2-in. pipe rupture tests, the axial crack in the fuselage propagated straight ahead only in the presence of the MSD's which acted as a crack guide.

Exception to the tendency for crack curving can be seen in longest crack which was assumed to have penetrated into the tear strap. In this case, $r_o > r_c$ and the necessary condition for crack curving is violated, thus suggesting that the crack will propagate axially once it has fractured the tear strap.

Elastic-Plastic Analysis

Figures 9, 10 and 11 show the axial stresses and the axial-to-circumferential stress ratios in the finite elements surrounding the crack tip for half crack length of a = 15, 18 and 20 in. Figures 9 and 10 represent the states of crack tip stress as the crack approaches the tear strap and Figure 11 represents the state of stress when the crack penetrated the tear strap. Note that the axial-to-circumferential stress ratio exceeds unity for a = 18 in. and is a low 0.6 when the crack had penetrated the tear strap.

The crack curving and crack branching criteria described above obviously is not applicable to the elastic-plastic states of stress of Figures 9, 10 and 11. If, however, the state of stress governs the fracture process, then the crack curving and crack branching criteria can be used to estimate the onset of crack curving as well as the crack curving angle. The rational for using such crude approximation is analogous to the use of plasticity corrected stress intensity factor in LEFM.

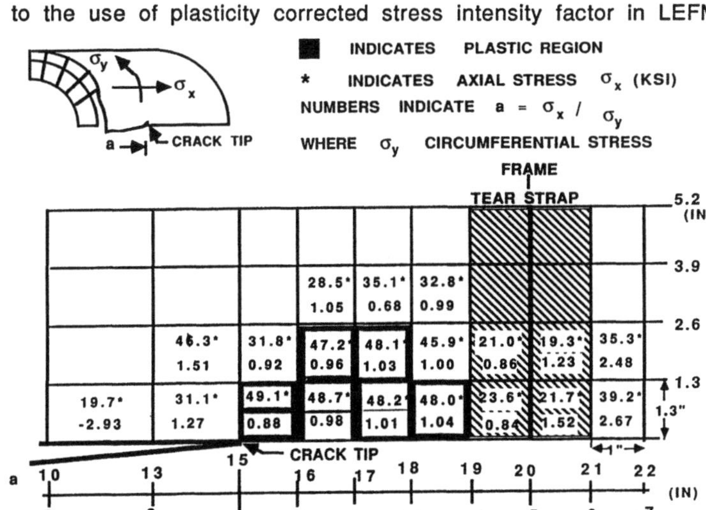

Fig. 9. Elastic-plastic stress distribution in a cracked pressurized fuselage

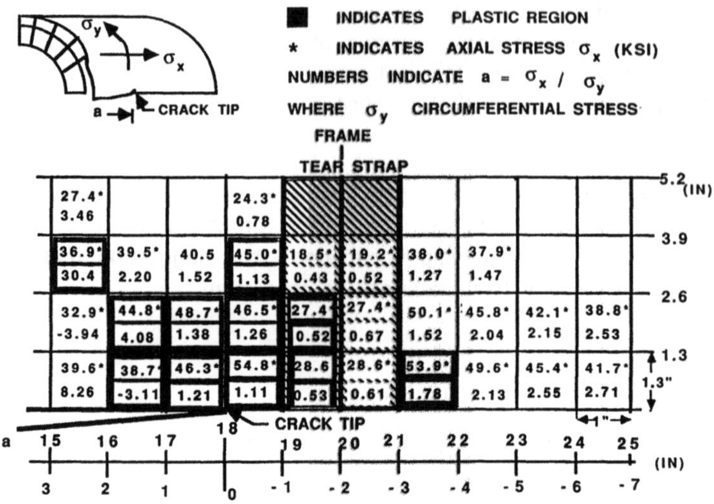

Fig. 10. Elastic-plastic stress distribution in a cracked pressurized fuselage

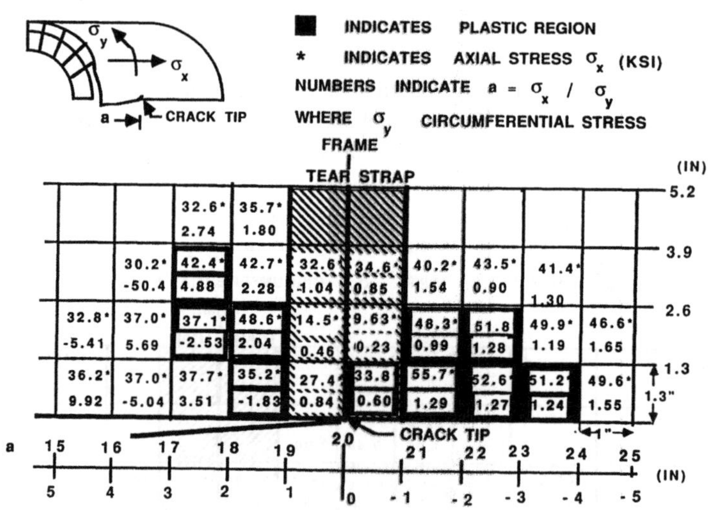

Fig. 11. Elastic-plastic stress distribution in a cracked pressurized fuselage

Table 3 shows the plasticity corrected mode I stress intensity factor, K, the remote stress component, σ_{ox}, the characteristic distance, r_o, and the crack curving angle, θ_c for two fuselage pressures of p = 4.5 and 7.5 psi. The results show that crack curving is inevitable in the skin before the crack arrived at the tear strap.

p (psi)	a (in)	K $(ksi\sqrt{in})$	σ_{ox} (ksi)	r_o (mm)	θ_c (deg) r_c =1mm	θ_c (deg) r_c =1.5mm
	15	75.5	62	0.8	26.7	46.5
	16	78.1	62	0.9	20.9	44.0
4.5	17	79.2	62	0.9	17.9	42.9
	18	73.0	62	0.8	31.2	48.7
	19	85.3	28.4	5.1	-	-
	20	83.9	39.1	2.6	-	-
	15	86.2	61.7	1.1	-	34.9
	16	88.3	62	1.1	-	32.7
7.5	17	85.2	62	1.1	-	36.6
	18	78.5	62	0.9	19.9	43.6
	19	90.2	24.7	7.6	-	-
	20	90.8	36.6	3.5	-	-

Table 3. Results of elastic-plastic analysis

The fuselage in this analysis is a shell of uniform thickness without rivets. The stress concentrations at the row of rivet holes in the actual fuselage will locally increase the circumferential stress and simulate the groove in the line-pipe experiment. Thus an axial crack in the actual fuselage will continue to propagate axially, particularly in the presence of a multiple site damage. As the axial crack in such damaged fuselage approaches the tear strap, however, crack curving is inevitable if the integrity of the tear strap is maintained as shown by Figure 10. If the axial crack penetrates the tear strap, as shown in Figure 11, then it will continue to propagate axially without curving through the tear strap.

Conclusions

Large deformation elastic-plastic finite element analysis of a pressurized fuselage showed that a large axial stress component exists ahead of an axial crack. The computed crack tip stress, when combined with a previously developed crack curving criterion, predicted crack curving throughout much of the skin.

Discussions

The numerical analysis and results presented in this paper should be considered a first order approximation to the complex state of stress in an actual fuselage. The analysis could be easily upgraded by incorporating the history effect of plastic deformation by incrementally advancing the axial crack tip.

A more difficult task is to develop a crack curving and crack branching criteria in the presence of large plastic deformation. If, for example, a maximum strain criterion is used for crack curving, then crack curving will not occur until the crack tip is close to the tear strap as indicated by the axial-to-circumferential stress ratios in Figures 9, 10 and 11.

Acknowledgement

The results on axial cracking of line-pipe was obtained under DOT contract DTRS - 5683-C-0009. The results on crack curving and crack branching as well as the analysis reported in this paper was obtained under the sponsorship of ONR contract N00014-89-J-1276. The authors wish to express their gratitude to Dr. Y. Rajapakse for his continuous support of this investigation.

References

1. Folias, E.S.; A Finite Line Crack in a Pressurized Cylindrical Shell. Intl. J. of Fracture Mechanics. 1 (Jun. 1965) 104-113.

2. Erdogan, F.; Kibler, J.J.; Cylindrical and Spherical Shells with Cracks. Intl. J. of Fracture Mechanics. 5 (Sept. 1969) 229-236.

3. Ramulu, M.; Kobayashi, A.S.; Kang, B.S.-J.; Dynamic Crack Curving and Branching in Line-Pipe. ASME J. of Pressure Vessel Technology. 104 (1982) 317-322.

4. Hahn, G.T.; Sarrate, M.; Rosenfield, A.R.; Criteria for Crack Extension in Cylindrical Pressure Vessels. Intl. J. of Fracture Mechanics. 5 (1969) 187-210.

5. Maxey, W.A.; Dynamic Crack Propagation in Line Pipe. Proc. of Intl. Conf. on Analytical and Experimental Fracture Mechanics. Sijthoff and Noordhoff (1980) 109-223.

6. Freund, L.B.; Parks, D.M.; Analytical Interpretation of Running Ductile Fracture Experiments in Gas Pressurized Linepipe. Crack Arrest Methodology and Applications. Hahn/Kanninen eds. ASTM STP 711 (1980) 359-378.

7. Kobayashi, A.S.; Emery, A.F.; Love, W.J.; Chao, Y.H.; Johannson, O.; Crack Bifurcation and Arrest in Pressurized Pipe. Fracture Mechanics: Nineteenth Symposium. ASTM STP 969 (1988) 441-465.

8. Kobayashi, A.S.; Emery, A.F.; Love, W.J.; Chao, Y.H.; Subsize Experiments and Numerical Modelling of Axial Rupture of Gas Transmission Lines. ASME Pressure Vessel Technology. 110 (May 1988) 155-160.

9. Shoemaker, A.K.; McCartney, R.F.; Displacement Consideration for a Ductile Propagating Fracture in a Line Pipe. ASME Journal of Engineering Materials Technology. 96 ((1974) 318-322.

10. Urednicek, M.; Control of Ductile Fracture Propagation in Large Diameter Gas Transmission Pipelines. ASME Winter Annual Meeting Preprint. 83-WA/PVP-11 (November 1983).

11. Freund, L.G.; Parks, D.M.; Rice, J.R.; Running Ductile Fracture in a Pressurized line Pipe. Mechanics of Crack Growth. ASTM STP 590 (1976) 243-262.

12. Freund, L.B.; Li, V.C.F.; Parks, D.M.; An Analysis of a Wire-Wrapped Mechanical Crack Arrester for Pressurized Pipelines. ASME J. of Pressure Vessel Technology. 101 (Feb. 1979) 51-58.

13. Ramulu, M.; Kobayashi, A.S.; Dynamic Crack Curving - A Photoelastic Evaluation. Experimental Mechanics. 23 (March 1983) 1-9.

14. Ramulu, M.; Kobayashi, A.S.; Kang, B.S.-J.; Dynamic Crack Branching - A Photoelastic Evaluation. Fracture Mechanics: Fifteen Symposium. Ed. by R. J. Sanford. ASTM STP 833 (1984) 130-148.

15. Streit, R: Finnie, I.; An Experimental Investigation of crack Path Directional Stability. Experimental Mechanics. 20 (January 1980) 17-23.

16. Ramulu, M.; Kobayashi, A.S.; Kang, B.S.-J.; Dynamic Crack Curving and Branching in Line-Pipe. ASME Journal of Pressure Vessel Technology. 104 (November 1982) 317-322.

17. Bromo, F.; Bramante, M.; Spedaletti, M.; Ductile Fracture Propagation in Pipelines: Results of Instrumented Full-Scale Burst Tests on 48" and 56" Diametral Pipes for gas Transmission. Analytical and Experimental Fracture Mechanics. G.C. Sih and M. Mirahik eds. Sijthoff and Noordhoff (1980) 567-578.

18. Kobayashi, A.S.; Engstrom, W.L.; Transient Analysis in Fracturing Aluminum Plate. Proc. of JSME 1967 Semi-International Symposium (1987) 172-182.

ASK/cm

17. Structural Integrity of Aging Airplanes: A Perspective

James W. Mar

Massachusetts Institute of Technology
Cambridge, MA 02139

This meeting is being held because of an accident which occurred nearly two years ago. The topics to be covered addresses issues highlighted by that accident.

In this country, the featured players in the commercial transport industry are the FAA, the regulatory agency, the designers and manufacturers of the aircraft, and the operators who run the airlines for the public's benefit (See Figure 1). The interconnects between the featured players, represented by the lines, are of many different types. Some are mandated by regulation but many are voluntary and are just as important as the legislated ones. For example, the ATA/AIA/FAA Aging Aircraft Task

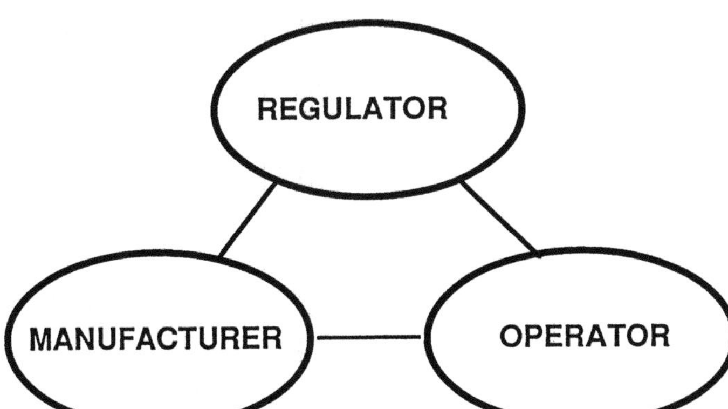

FIGURE 1
THE PRINCIPAL PLAYERS

Springer Series in Computational Mechanics
Atluri, Sampath, Tong (Eds.)
Structural Integrity of Aging Airplanes

Force is addressing issues such as corrosion, human factors and NDE in a much more timely and cooperative fashion than is possible within the regulations.

Additionally, there is a large supporting cast composed of people in NASA, the NTSB, Universities and DoD (see Figure 2). The interconnects between the supporting cast and the featured players are profuse and are not shown. All segments of the supporting cast are present in the audience and most, if not all, are represented on the program.

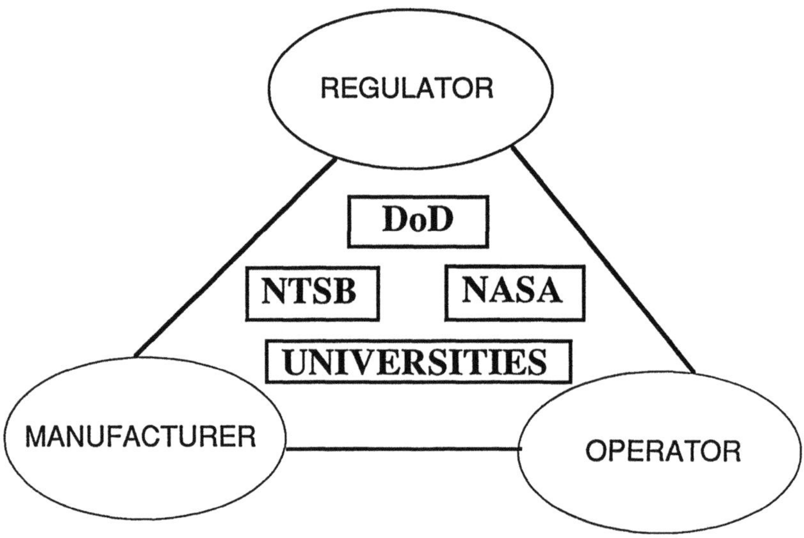

FIGURE 2
THE SUPPORTING CAST

The aerospace industry has always responded to accidents and service experiences with positive actions which further enhance safety. This meeting is one of those actions which will further improve an already remarkable record of safety and service. The title of this symposium is "Structural Integrity of Aging Airplanes". As a background for the subsequent presentations, it may be informative to relate some of the underlying structural issues and the concomitant underlying philosophies.

The following sentence is the second sentence in a chapter of a book on "Aeroplane Structures" by Pippard and Pritchard entitled "The Partially Disabled Airplane" published in 1919:

> **Any airplane should be capable of reaching the ground safely even after certain parts of it have failed in the air.**

This chapter deals primarily with the transference of loads to the incidence wires in biplane wings after the flying wires have failed. The concerns for these kinds of failures grew out of the experiences of fighter aircraft in WWI. Your lay friends should be comforted by this statement.

It should be noted and remembered that fail safety has been a guiding design philosophy for airplane structures because the consequences of major failures are catastrophic. Aerospace engineers are acutely aware much wisdom is required to properly specify the parts which are to "fail safely". Service experiences have played an important role in the rules which now govern structural integrity. This sentence from Pippard and Pritchard's book merits being highlighted in the introductory chapter of books dealing with the design of aircraft (curiously enough, it is not to be found in the second edition of "Aeroplane Structures").

The theme of this Symposium is "Structural Integrity of Aging Aircraft". Fatigue is a dominant concern for geriatric aircraft.

However, it should be remembered that the design stress levels also must satisfy many other requirements. These other requirements, which are depicted in the chart (see figure 3), have gradually been added onto the list of design requirements to be addressed by the designer.

There were a number of early transport airplane accidents which were caused by fatigue. A Martin 202 in 1948, two DC-6's in the 50's and a B-52 in Wolfboro, North Carolina. Valuable lessons were learned from each of these service experiences.

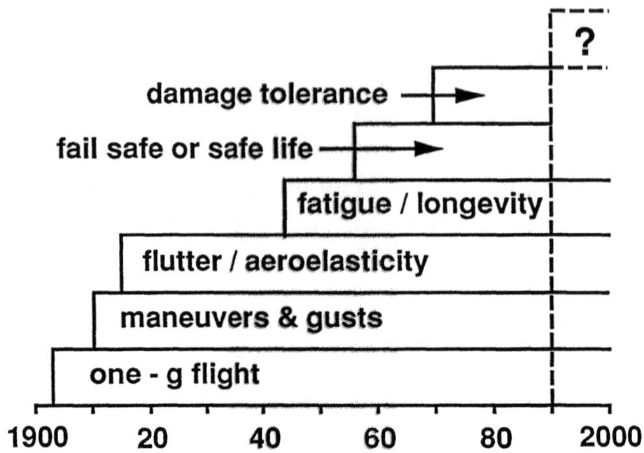

FIGURE 3
INTEGRITY REQUIREMENTS

There was an early awareness of airplane fatigue even among the non-engineering public. Neville Shute wrote a novel about airplane fatigue, titled "No Highway" a few years before the Comet accidents and it was later made into a movie called "No Highway in the Sky" starring Jimmy Stewart and Marlene Dietrich. Accidents and service experience shaped the design philosophies adopted by designers and also the government mandated rules. It might be informative to summarize a few of the key experiences which have helped mold the present rules.

Finally, two suggestions for possible rule changes sharpened by the Aloha incident will be discussed.

The following words appeared as a rule in CAR 4-b in 1956:

> **It shall be shown by analysis and/or test that catastrophic failure ..... are not probable after fatigue failure or obvious partial failure of a single principal structural element.**

These words enunciate a fundamental basic design philosophy for all the current jet transports. The phrase "catastrophic failure

... are not probable" means loss of the aircraft is to be highly improbable; "obvious partial failure" implies the failure is to be easily detectable; and "single principle structural element" means the structural configuration shall be fail safe. The residual strength was specified to .8 of limit load. According to Bob Allen, recently retired from the FAA, the .8 factor was based upon analyses of the safe landings of two DC-6 airplanes after in-flight partial failures of their wings.

Many different types of aircraft have accumulated flight hours, flight cycles and chronological age well beyond their original structural and economic life goals. There is a widely held belief by the designers of these aircraft (those which met the requirements of CAR 4-b) that, with proper inspection, maintenance and repair, the life of an air frame could be unlimited. Thus inspections, maintenance and repair are as important as good design for the safe operation of the commercial transports. There are examples where lax attention to inspections, improper maintenance and faulty repair have led to serious accidents.

The Aloha incident suggests that the then-present inspection, maintenance and repair procedures as well as the understanding of the potential consequences of old age needed to be re-evaluated and bolstered. For example, there needs to be an evaluation of what constitutes "proper inspection and maintenance" in terms of frequencies and sophistication for the aging aircraft. Another factor is that by law, the FAA is required to address the economic impact of airworthiness directives. It may be that the economic burden of repeated inspections and repair will lead to the retirement of aged aircraft.

Three events which played prominent roles in shaping the present damage tolerance philosophy and rules will be briefly discussed. The first of these concerns the US Air Force F-111 airplane. The contract for the development and production was awarded to the General Dynamics Corporation in October of 1964. The longevity of the airplane was specified to be 4000 flight hours and 4000 landings. Demonstration of the life was to be verified a full scale fatigue test by cyclic loading to four lifetimes; in the parlance of fatigue design, the scatter factor is 4. Fatigue analysis was based upon Miner's Rule. The structural design was based upon a safe life philosophy, i.e., to demonstrate by analyses and tests that there would be no fatigue failures within the contractually designated life of the aircraft. At that time, the weight penalties associated with a fail safe design were deemed unacceptable in a fighter-type airplane. An important feature of the design was the capability to vary the sweep of the wing while the airplane was in flight. The movable part of the wing was attached to the carry-through structure by means of the wing pivot fitting (see Figure 4). It was fabricated of D6-ac steel heat treated to 220 ksi. In the time period of the design of the F-111, the concepts of fracture toughness and crack growth were not a part of the specifications. The wing pivot fitting was welded together from several sub-components to form a monolithic structure. Once assembled, inspection of the interior was not feasible.

Just over twenty years ago there was an in-flight structural wing failure of a newly introduced U.S. Air Force airplane. The accident had occurred on 22 Dec., 1969 when F-111 #94 crashed while doing a low level bombing run. The cause of the crash was the separation of the left wing at the left wing pivot fitting (see Figure 4). Aircraft #94 had only a total of 107 total flight

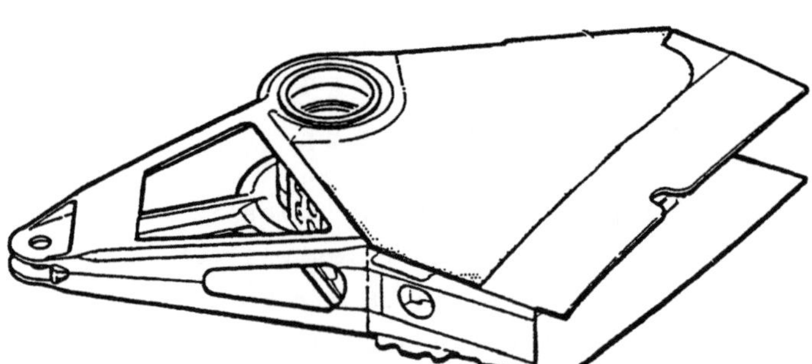

FIGURE 4
F-111 WING PIVOT FITTING

hours. It had been delivered to USAF on 26 Nov. 1968. At the time of the wing failure, the aircraft was determined to have been pulling only about 3.5 g's, less than half of the design limit load factor.

The cause of the fracture was obvious to the naked eye even at the accident crash site (see Figure 5). A photograph taken closer to the wing more clearly reveals the flaw which was the direct cause of the failure (see Figure 6). The fracture mechanics specialists can see that the shape of the flaw is like the moon on a thumbnail. A close-up of the fracture face is shown in Figure 7. The very dark part of the crack was present when the airplane was delivered to the Air Force. The bright narrow band is the result of the crack growth over the 107 flights.

FIGURE 5
F-111 #94 WING ON DESERT FLOOR

The F-111 airplane eventually passed all all of its contractual static strength and fatigue tests to four lifetimes. However, the sensitivity of the structure to small flaws made it necessary to run a cold proof test in order to assure sufficient integrity for safe operations. The cold proof test was and is conducted in a

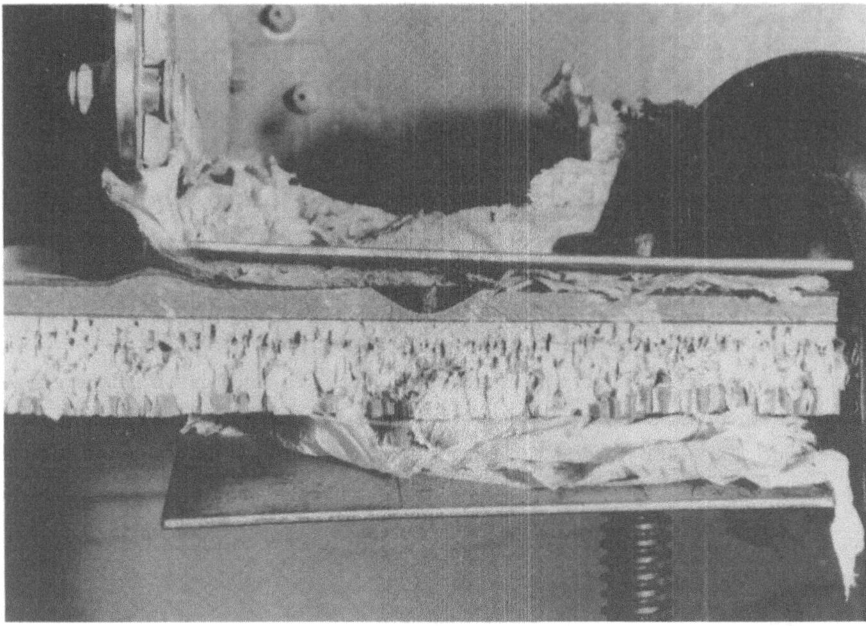

FIGURE 6
F-111 #94 CRACK

FIGURE 7
F-111 #94 FRACTURE FACE

chamber in which the metal temperature is cooled down to about -40 degrees F. At this temperature, the fracture toughness of the D6-ac steel is lower than at the normal operating temperature of the airplane. This means the critical crack sizes are larger during normal operation of the airplane and this additional size as well as a reduction in the permissible flight load factors leads to a safe operating flight period. Positive and negative limit loads are applied. Figure 8 is a time lapse photograph of such a test showing the wing in its limit positive and negative deflected positions. The efficacy of the cold proof is dependent upon faith in the discipline of fracture mechanics.

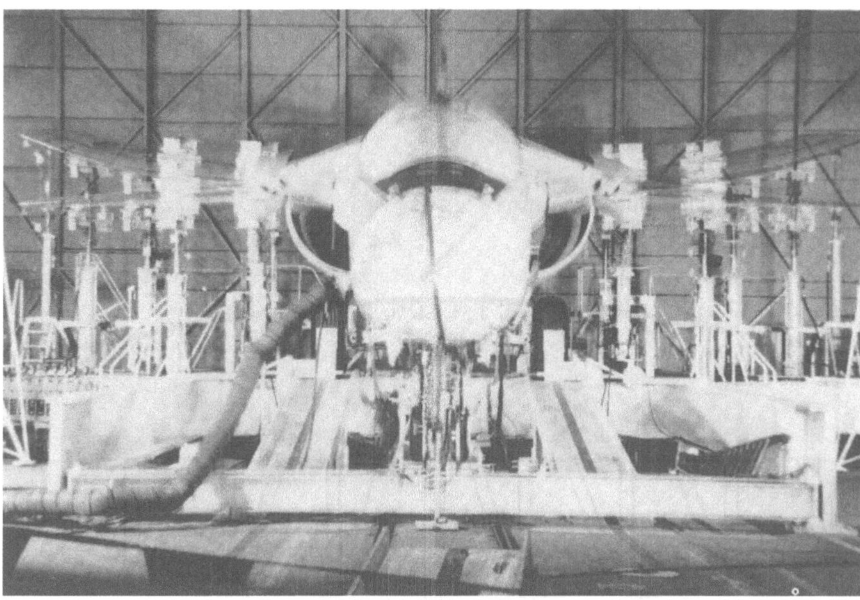

FIGURE 8
F-111 COLD PROOF TEST

The second event which will be related concerns the C-5A cargo transport. Lockheed was awarded the contract to design and build the C-5A cargo transport in Oct of 1965. The longevity of this design was to be 30,000 flight hours and 5950 pressure cycles for the fuselage. It was Lockheed's hope to also develop a variation which could be used for the civilian market. Thus, in addition to the military specifications, Lockheed also imposed the civil rules which during the competition was CAR 4b and which then became the FAR Part 25 rules.

To show compliance with the longevity requirements, the US Air Force required a full scale fatigue test to four lifetimes. This resulted in probably the most complex full scale fatigue test ever conducted on a subsonic airplane. The test article consisted of a wing attached to a fuselage which was immersed in a tank of water for the pressurization cycles. Cycles were applied to the wing through a whiffle-tree system.The fatigue analysis was based upon Miner's Rule.

The wing configuration consisted of planks made of 7075-T6 extrusions joined together by spanwise rows of taperlok fasteners (see figure 9). The principal structural elements were the planks and the wing was considered by Lockheed to be fail safe because the wing could carry the FAR Part 25 loads with one of the planks failed.

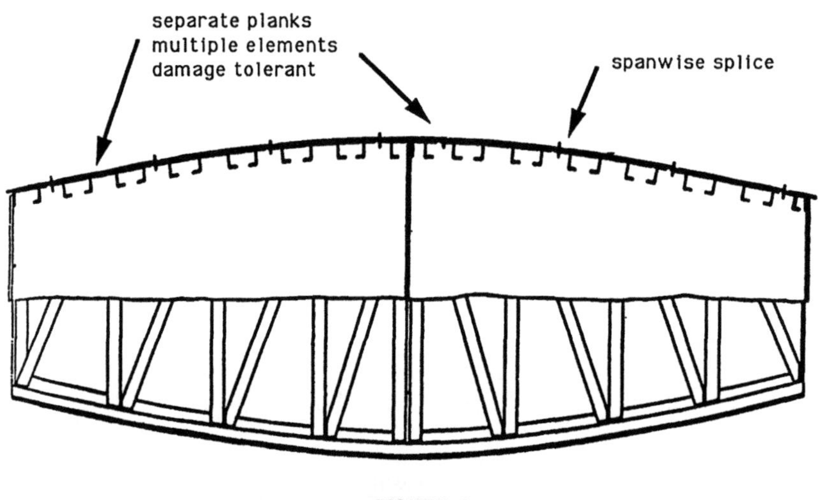

FIGURE 9
C5-A WING BOX

Many cracks developed in the spanwise splices of the wing very early in the projected 4 lifetimes of testing (see figure 10). Cracking also occurred in adjacent planks and in today's terminology, the C5-A wing developed MSD, multiple site damage. As can be see in the plot, the threshold for MSD occurred before the completion of one lifetime of testing.

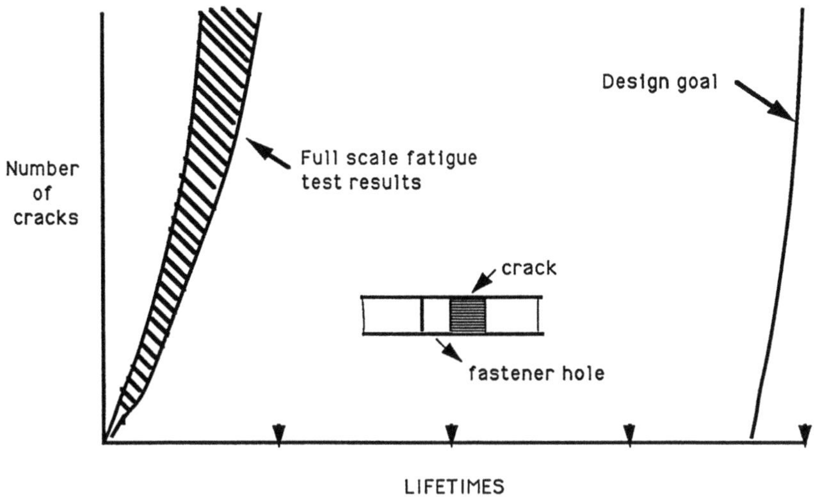

FIGURE 10
C-5A WIDESPREAD CRACKING

The time required to conduct the cyclic testing of the original fatigue test article was very long because both fuselage pressure cycles and wing loading cycles had to be sequenced. An additional wing fatigue test article consisting of just a semi-span was added to the test program. The test spectrum was truncated according to Miner's Rule. It was found that crack initiation did not occur in the two different test articles at the equivalent amount of damage as predicted by Miners Rule.

A lesson for the present aging aircraft programs is that the C5-A wing could not be considered to be fail safe beyond the threshold of MSD, i.e., if adjacent panels contained cracks.

The original wing design of the C5-A airplane could not meet the US Air Force requirements. The structural boxes of the center, inner and outer wing (see Figure 11) were replaced by a new design incorporating new materials, much lower stress levels and the notions of damage tolerance.

The problems of the F-111 and the C5-A airplanes led the US Air Force in July,1974 to mandate a new design philosophy contained in Mil Std 83444 "Airplane Damage Tolerance Requirements".

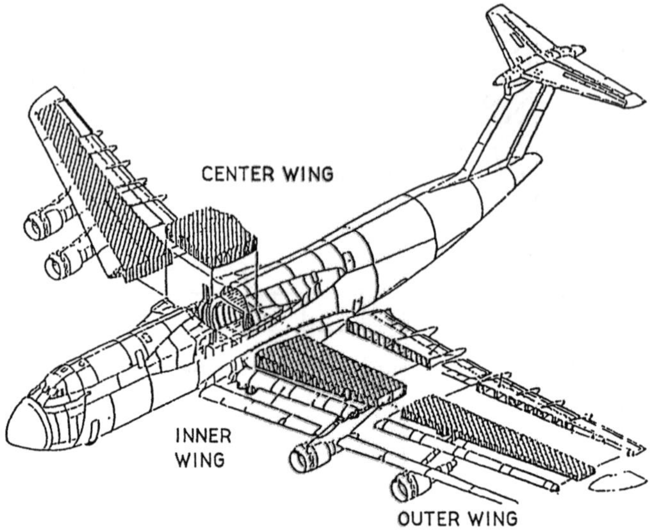

FIGURE 11
C5-A NEW WING

The damage tolerance philosophy is made quantifiable by the discipline of fracture mechanics. Prior to fracture mechanics, "tough" had a nice ring to it but engineers could not relate "tough" to the imposed stresses. "Per-cent elongation" as measured from a simple tension test was used as an indicator of toughness but "per- cent elongation" also cannot be used to size a structural member. The notion of "stress intensity" which is a measure of the stress at the tip of a crack has been developed. "Fracture toughness" is now a measured material characteristic equally important as tensile ultimate strength, tensile yield strength and other such measurable parameters. It permits the determination of critical crack sizes which may lead to catastrophic failure. Additionally, crack growth is proportional to change in stress intensity as the stress at the tip of the crack is cycled by the change in loads.

Many aerodynamicists find it difficult to accept fracture toughness and stress intensity as important parameters because they have the dimensions of stress times the square root of length. They believe structures people should deal with dimensionless numbers. Of course these are the same people who insist upon countersunk fasteners in fuselages in order to reduce the drag.

The third event to be discussed concerns an accident involving the B707 horizontal stabilizer which is a single spar structure. This design is an excellent illustration of the concept of fail safety. As can be seen (see Figure 12), the spar has an extra mass of material in the middle of the web. Under normal operations, this extra material does not participate in the carrying of the bending moment. However, if either the upper or lower cap fails, the middle mass becomes a spar "cap" and the spar is now a smaller member capable of carrying the prescribed fail safe loads. Thus, fail safety is achieved in this design by a mass of a material which becomes a "spar cap" after failure of either of the two normal spar caps.

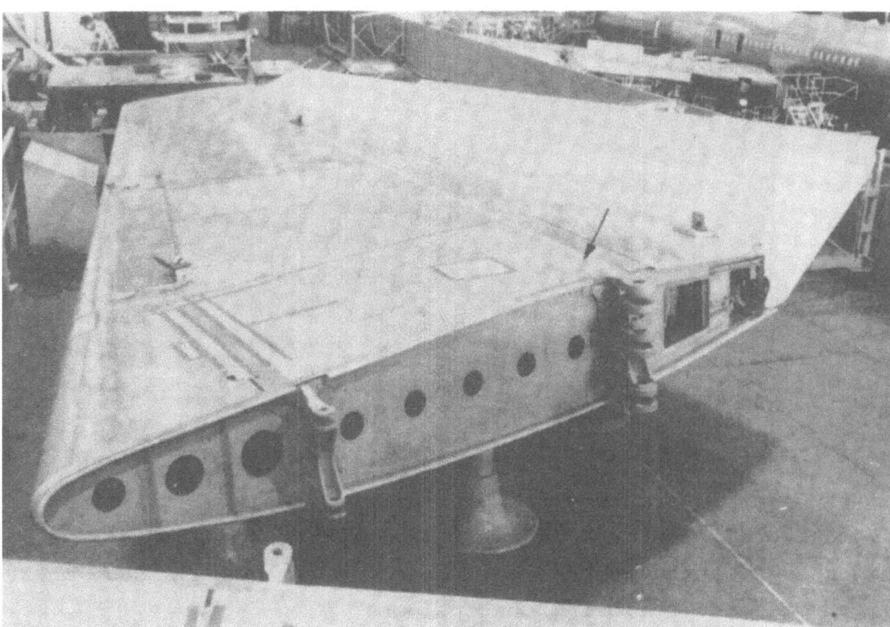

FIGURE 12
B707 HORIZONTAL STABILIZER

A 707 operating as a freighter crashed short of the runway at Lusaka Airport, Zambia on 14, May 1977. The cause of the accident was the failure of the horizontal tail because the crack caused by fatigue reached critical size under the loads associated with the landing. Total airframe flight hours was 47,621 and the total number of landings was 16,723. The airplane had been manufactured in 1963. The design life goals for the 707 were 60,000 flight hours and 20 years. By today's standards the airplane was past mid-life and may even to be considered to be aging. As can be seen the crack (see Figure 13) was of a visually detectable size for a long period of time.

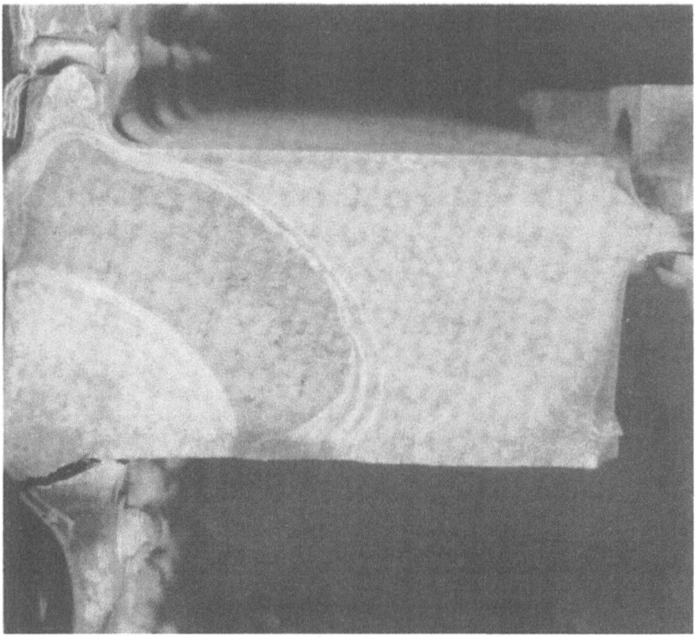

FIGURE 13
LUSAKA FATIGUE FRACTURE

The accident report ("Boeing 707 321C.G-BEBP, Report on the accident near Lusaka International Airport, Zambia, on 14 May 1977", Aircraft Accident Report 9/78, Department of Trade, Accidents Investigation Branch, London, Her Majesty's Stationery Office) made the following conclusions:

- **Following the failure of the stabilizer rear spar top chord the structure could not sustain the flight loads imposed upon it long enough to enable the failure to be detected.**
- **It therefore cannot be classified as fail safe.**

The Lusaka accident report brings forth that a fail safe design concept by itself does not result in a fail safe design. Inspectability is equally important. The words "long enough to enable the failure to be detected" clearly states that fail safety exists only if the obvious partial failure can be found in a timely fashion. Inspections and design must be combined before fail safety is achieved. Another aspect of this accident is that the reporting of discrepancies noted in fleet service of the 707 had not been given proper attention.

Prior to the Lusaka accident, the principal players (see Figure 1) had been engaged in the writing of what is now familiarly referred to as "25.571". It is rule 25.571 of the Federal Aviation Regulations, Part 25, "Airworthiness Standards: Transport Category Airplanes". This rule contains the following words:

> **......catastrophic failure due to fatigue, corrosion, or accidental damage, will be avoided throughout the operational life of the airplane.**

We have been conditioned by recent events to the potential hazards of MSD. But in 1978, the writers of the rule foresaw MSD as shown by this excerpt from rule 25.571:

> **......*Damage* at *multiple sites* due to prior fatigue exposure must be included where the design is such that this type of damage can be expected to occur**

In principle, safe life is still permitted as a design approach. The following words can be found in rule 25.571:

> **Fatigue (safe life) evaluation.**
> **....damage tolerance is not required if .....impractical. .....structure must withstand the repeated loads ..... during its service life *without detectable cracks*.**

Practically, the only major airframe component which is still permitted to be designed according to safe life is the landing gear. Implicit in the safe life philosophy is the belief that a "detectable crack" marks the end of life.

With the preceding as a background, the primary technical issues to be addressed for the aging aircraft fleets are as follows:

MSD, Multiple Site Damage
Corrosion
NDI, Non-Destructive Inspection
Structural Repairs
Terminating Actions

MSD is the undesirable condition caused by wide spread cracking of the structure. Neither multiple load path nor crack arrest features can be depended upon to provide sufficient integrity after the structure has reached the onset of MSD.

Corrosion will decrease the size of structural members. It can also have bad synergisms with factors which lead to early cracking. These synergistic effects are not well understood and are not considered in the design process. Structurally, corrosion should not be permitted to occur.

NDI is the key to assessing the health of the aging fleet. NDI needs to be able to detect the small crack sizes which are characteristic of MSD.

Structural repairs, which are more prevalent in the aging fleet, must also satisfy damage tolerance and inspectability requirements.

Terminating actions is the FAA language for actions to overcome potential structural problems before they occur. The terminating actions create a new structure for which life and inspection intervals must be determined.

Before old age occurs, the following is a possible scenario under which structural integrity is maintained:

- **Cracks initiate at holes along splices and link up to form cracks long enough to be detected visually.**
- **Linked cracks are detected during scheduled checks before catastrophic failure occurs.**
- **Critical crack lengths are very long, giving many opportunities for detection prior to reaching critical sizes.**

Old age in this context is after the age which marks the onset of MSD. The manufacturer, through the medium of service bulletins to the operators, provides information on where, when and how to find possible damage. Compliance with the service bulletins is voluntary. The FAA can give the service bulletin the strength of a rule by issuing an airworthiness directive.

After old age, the following is a possible scenario:

- **Widespread cracking may ensue leading to the condition termed MSD, Multiple Site Damage.**
- **Critical crack sizes for catastrophic failure is greatly reduced by the presence of MSD.**

Fatigue initiated damage is the primary concern of aging air craft. In an aircraft which is properly inspected and maintained, corrosion and accidental damage will be controlled and understood long before the design life is reached. Fatigue initiated damage is a phenomena where the probability of its existence at any point increases with time, i.e., with flight cycles. The number of points with possible damage will also increase with flight cycles. As the number of damage sites increase, there is an increasing probability of not detecting at least one point which does not have limit load capability. At this stage, the airplane can no longer be considered airworthy.

Before the onset of MSD, the fuselages of U.S. designed aircraft can carry the design operating pressures even if there is a longitudinal crack which spans two bays (approximately 40 inches) with the intervening frame completely broken. However, this capability is greatly reduced if the tear straps and frames contain cracks.

A row of fastener holes each of which has emanating cracks of approximately the same sizes is a simple depiction of one form of MSD (see figure 14). These cracks may not have even grown out from under the head of the fastener represents a crack configuration which makes detection difficult even with NDI methods. Visual detection of cracks have grown to sizes larger than the fastener heads cannot be relied upon. A linked crack, e.g. of several inches, in bay next to one wherein MSD exists would have a much lower critical stress.

The potential problem for aging aircraft which are in the regime of possible MSD are two fold:

- **Degradation of damage tolerance.**
- **Negation of fail safety.**

In this context, damage tolerance means the ability of the structure to carry loads to permit a safe landing after discrete events such as bird strikes, uncontained engine failures and other kinds of FOD. Fail safety means the ability of the structure to carry limit loads after the failure of a principal structural element.

Once, an airplane fleet has passed the age which MSD is probable, the following actions are no longer adequate nor sufficient to maintain structural integrity:

- **Visual inspections**
- **Sampling**
- **Service experience**

Visual inspections cannot reliably detect the small cracks which mark the threshold of MSD. "Small" means cracks which have just emerged from under the head of the fastener or which may still be hidden from direct viewing by the fastener head. Thus, the electronically driven NDI techniques such as high frequency eddy current, HFEC, must be used to maintain airworthiness.

Sampling is most useful for the purpose of determining a condition such a local hot spot, i.e., where cracking is likely to initiate because of a poor design detail. This information is then used to direction inspections of the rest of the fleet if the poor design detail is characteristic of the entire fleet. In these situations, the cracking at these local hot spots is still contained from catastrophic propagation by the damage tolerance characteristics of the remaining structure. Sampling thus is a a means to protect the fleet by gaining knowledge in a timely fashion. However, damage tolerance will have been degraded once the threshold of MSD has been exceeded. In these situations, the existence or non-existence of MSD must be determined for every aircraft in the fleet.

Service experience is always useful. But in a fleet wherein the potential for MSD has been established, the absence of cracking at a particular location as determined by previous service experience should not be used to limit the scope of the inspections.

An example of the schedule for terminating actions is the Air-worthiness Directive, AD, imposed on the Boeing Model 737 airplanes (Docket No. 88-NM-160-AD; Amdt. 39-6184):

Flight Cycles	**Modify with next**
70,000 or more	**6 months**
60,000 to 69,999	**12 months**
50,000 to 59,999	**18 months**
40,000 to 49,999	**24 months**
less than 40,000	**36 months**

The effective date of the AD was May 8, 1989. The terminating actions require the replacement of approximately 70000 countersunk fasteners with protruding head fasteners one size larger than the existing ones. Under this schedule, none of the affected aircraft (except for a handful) will accumulate the design life goal of flights before the modifications are completed. This reflects the FAA policy which is not to rely indefinitely on inspections to maintain safety of aging aircraft.

The longitudinal splices of the first two hundred and ninety-one 737 airplanes used a cold bond process and three rows of countersunk fasteners. Service experience indicates that some of the bonds have deteriorated so that the loads are transferred to the countersunk fasteners. The countersink at the upper row leads to a high stress concentration. Once the bond has failed, wide spread cracking may ensue. It has been determined that the threshold for MSD in these 737 fuselages is approximately 30,000 cycles after the failure of the bond.

The question mark at the top of the chart shown in Figure 3 represents ongoing discussions regarding new rules and philosophy. These are motivated by the current concerns regarding aging aircraft. The FAA and the Industry have always responded to lessons learned from service experiences and from accidents. It is likely that the Aloha incidents will also lead to the promulgation of new rules or lead the earlier adoption of rules or changes to existing rules which have been under consideration. Rule making is not a static process but rather is an ongoing search for means to provide greater integrity. Two proposals have been suggested:

- **Manufacturers of aircraft certified to the current 25.571 should be required to establish the life cycles for the onset of MSD and to provide the validating data and analyses.**
- **Amend 25.571 to require proof by analyses and tests that the structure will not be subject to MSD within a safe operational life to be established at certification.**

Answers to issues raised concerning the present generations of jet aircraft would have been greatly facilitated if the onset of MSD had been established. Presently, a rigorous definition of MSD does not exist.

It seems likely that aircraft will be operation near or beyond their original longevity goals. The events of the the past few years suggest that actions to insure continuing airworthiness would be greatly facilitated if MSD is addressed at the time of certification.

I would like to conclude these remarks with a few non-technical comments. My friends, the principal players, in the three constituencies shown in Figure 3 have been using the three-legged stool to describe how important it is for them to work together (see Figure 14). And they have. But I believe there is a serious unbalance which needs to be redressed. The generations who went into the FAA after WWII and in the 50's and 60's are now retiring. It is important that these people be replaced with the best.

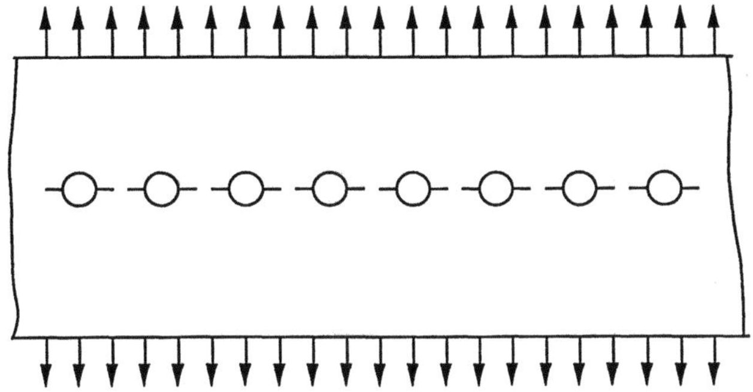

FIGURE 14
MULTIPLE DAMAGE

This cannot happen unless FAA salaries are increased. It's easy for us to blame the state of affairs on Congress; after all, they are blamed for almost everything else. But who elects Congress. We do. Engineers don't seek public office maybe because there are no formulae for success.

I also believe two of these three legs bear more of the load than the third. I believe the FAA is seriously understaffed in carrying out its assigned mission. We have got to find the means to strengthen the FAA. Ours is a capitalistic society and until FAA salaries are competitive with the private sector, the FAA will find it difficult to attract the best people.

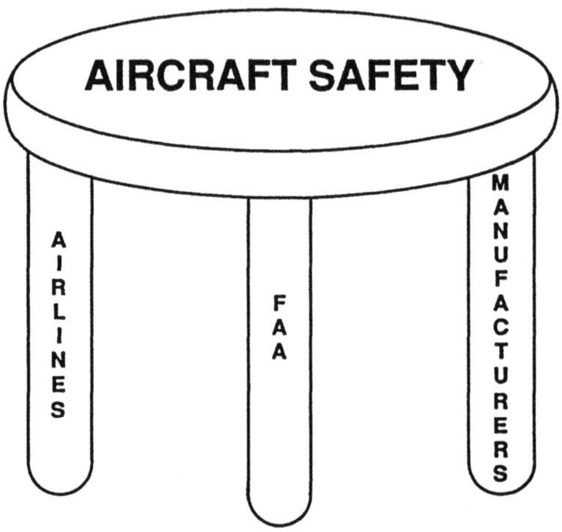

FIGURE 15
THE THREE-LEGGED STOOL

18. A Laboratory Study of Fracture in the Presence of Lap Splice Multiple Site Damage

Ronald A. Mayville and Thomas J. Warren
Arthur D. Little, Inc.
Cambridge, MA 02140

Abstract

Flat coupons were tested in the laboratory to determine a fracture criterion for link-up of fuselage lap splice multiple site damage at adjacent rivet holes. Experiments were performed on 0.040 inch (1 mm) thick 2024-T3 clad aluminum sheet. Continuous and riveted lap splice coupons were tested with simulated uniform (equal crack lengths) and nonuniform MSD, and the effects of notch sharpness were also studied. A net section stress criterion was found to provide excellent predictions of fracture for uniform MSD and uniform stress distributions. This same criterion provides conservative predictions for nonuniform MSD in uniform stress fields. An overload/cyclic stress experiment was also conducted to explore the pressurized proof test scenario of ensuring structural integrity.

Key Words: aircraft, fuselage, 2024-T3 aluminum, multiple site damage, fracture.

Introduction

The phenomenon of fuselage lap splice Multiple Site Damage (MSD) is currently of great interest in the study of aging aircraft structural integrity. This type of degradation, in which fatigue cracks emanate from the edges of several adjacent and colinear rivet holes, was found to contribute significantly to the Aloha Airlines accident in April 1988 [1]. Qualitative analysis of this type of damage suggests that it can progress from small, perhaps difficult to detect, indications to a size covering one or more bays in a relatively short time.

To date, all evidence suggests that the transition from small rivet hole cracks to the large fuselage crack necessary to result in unstable fracture requires enough cycles (flights) to provide adequate opportunity for discovery. Full scale, pressurized fuselage tests by Boeing show that lap splice MSD, if left without repair, progresses from: (a) a series of small adjacent rivet hole cracks in each of two or more neighboring bays to; (b) two or more

Springer Series in Computational Mechanics
Atluri, Sampath, Tong (Eds.)
Structural Integrity of Aging Airplanes

adjacent cracks whose lengths are approximately equal to the bay width, typically 10 inches to; (c) a very large crack, say 36 inches and finally to; (d) a large flapping type failure. Figure 1 shows an unusually severe example of stage (a) MSD. It was found during regular inspection in a B727 fuselage lap splice with approximately 43,000 flights. Crack lengths are given by bar heights for each side of each rivet hole. This figure shows that cracks of approximately equal length cluster in between the frames, which are regions of higher stress.

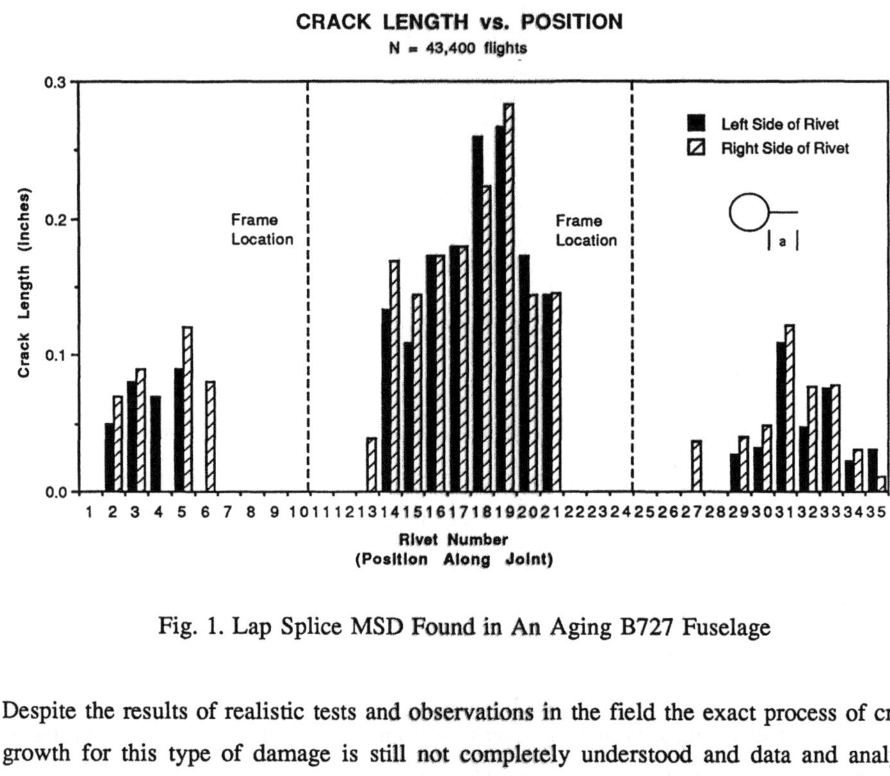

Fig. 1. Lap Splice MSD Found in An Aging B727 Fuselage

Despite the results of realistic tests and observations in the field the exact process of crack growth for this type of damage is still not completely understood and data and analysis techniques are required to predict initiation, fatigue crack propagation, link-up, and final fracture as part of the lap splice structural integrity assessment.

The objective of the investigation reported here is to determine a criterion for the part of the crack growth process corresponding to the link-up of small lap splice MSD cracks.

It is generally considered that link-up of adjacent rivet hole cracks occurs when the net section stress attains some critical value. However, there appear to be no data to support this belief or to show what the critical stress is. The approach taken in this investigation was an experimental one in which fuselage lap splice MSD and loading were simulated in flat laboratory coupons. Several parameters were investigated including: crack length, crack sharpness, nonuniform vs. uniform damage, and the effect of an overload on subsequent fracture behavior under cyclic loading.

The results confirm the validity of the net section stress criterion for prediction of link-up of MSD and that the critical stress level is between the material yield and tensile strengths.

Experiments

The basic coupon geometry tested in this study is shown in Figure 2. It is fabricated from 0.040 inch (1 mm) thick 2024-T3 clad aluminum sheet with the following tensile properties: yield strength = 50 ksi (345 MPa), tensile strength = 67 ksi (462 MPa). The longitudinal axis

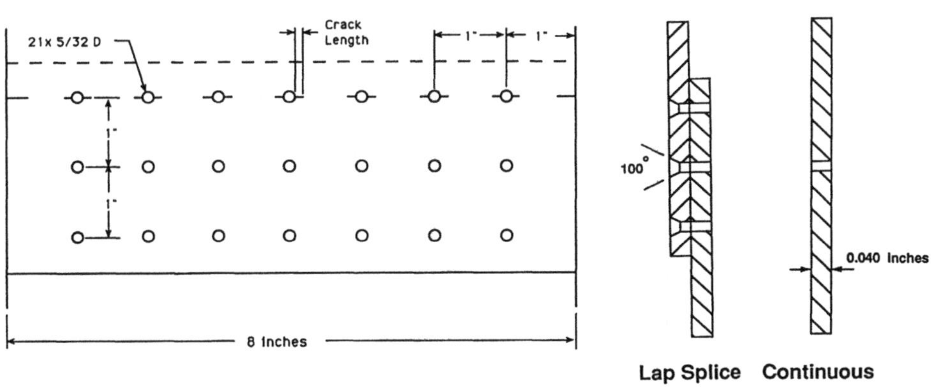

Fig. 2. Test Coupon Configuration

of the coupon corresponds to the rolling direction of the sheet for all tests. The coupon width was 8 inches (200 mm) and included continuous and riveted lap splice configurations. Continuous coupons contained a single row of equally spaced 0.156 inch (4.0 mm) diameter holes and riveted coupons contained three rows of 0.156 inch (4.0 mm) holes, countersunk approximately 80% of the thickness with a 100 degree included angle. Hole spacing was 1 inch (25 mm) in both the transverse and longitudinal direction.

Multiple site damage was artificially induced using a jewelers saw and only in the top row of holes in riveted coupons. The notch width was 0.014 inch (0.36 mm). In general, equal length notches were cut into each side of each hole to provide uniform MSD, but for a few coupons these notches were made of unequal length to simulate nonuniform MSD. Notches were inserted in the top row of riveted coupons prior to the riveting step. In one case, the notch was sharpened by fatigue using a maximum cyclic load that was always less than 60% of the final fracture load.

Table 1 lists the various configurations tested in monotonic tension to fracture. A single experiment, on a riveted coupon, was performed under cyclic loading after an overload. The

Table 1. Coupon Configurations Tested: 2024-T3 Clad Aluminum, 0.040-inch (1mm) thick sheet; 8-inch (200mm) width

Specimen	Average Crack Length (inches)	Joint Type	Notch Sharpness	Type of MSD	Nominal Stress at Fracture (ksi)
1	0.0	Continuous	--	--	52.6
2	0.10	Continuous	Sawcut	Uniform	37.3
3	0.23	Continuous	Sawcut	Uniform	24.8
4	0.30	Riveted	Sawcut	Uniform	15.2
5	0.31	Riveted	Fatigue	Uniform	15.3
6	0.23	Riveted	Sawcut	Nonuniform 4 at 0.3 12 at 0.2	20.3
7	0.25	Riveted	Sawcut	Nonuniform 6 at 0.3 10 at 0.2	19.1

nominal (far field) stress corresponding to the overload was 20 ksi (138 MPa) and the maximum cyclic stress was 15 ksi (103 MPa) with an R-Ratio of 0.1. This corresponds to an overload-to-cyclic load equal to 1.33, which is the value that has been considered in proof testing scenarios.

Original crack lengths were measured on the fracture surface using an optical microscope after complete fracture of the specimen had occurred. Some crack tip tunneling was observed for the fatigue-sharpened notches and the average length of the tunneled portion was included in the crack length estimate. Small amounts of stable crack extension from the notches were observed but, in general, fracture of the entire specimen was very sudden.

A few experiments were performed to determine the net section stress at fracture for 2024-T3 removed from an aging aircraft. One inch (25 mm) wide specimens were fabricated from aluminum sheet removed from an aging Boeing 727 aircraft that had experienced 41,000 flights. Specimens were tested in both the longitudinal and transverse orientations with and without a simulated rivet hole of diameter 0.156 inches (4.0 mm).

Results

Figure 3 shows the results from the coupons tested under monotonic tension. The figure is a plot of nominal (skin) stress at complete fracture vs. average original crack length from the minimum (0.156 inch) hole diameter. Data are included for both continuous and lap-spliced coupons, for uniform and nonuniform MSD, and plain and fatigue sharpened notches. Also included are lines corresponding to net section stress fracture criteria for two values of stress: 60 ksi (414 MPa) and 55 ksi (379 MPa). The data for uniform MSD, including the specimen with no cracks emanating from the holes, are seen to agree quite closely with the 60 ksi value for continuous and lap spliced coupons and plain and fatigue sharpened notches. On the other hand, the net section stress at fracture for the two coupons with nonuniform MSD is significantly lower: approximately 55 ksi (379 MPa).

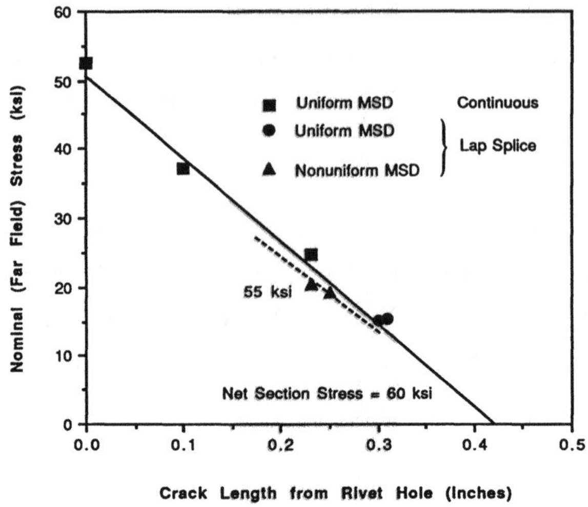

Fig. 3. Simulated Fuselage Lap Splice MSD Fracture Data

Table 2 lists the net section stress values at fracture for the 1 inch (25 mm) wide specimens removed from the aging aircraft. These data also support a 60 ksi (414 MPa) net section stress criterion for the specimens containing holes.

Table 2. Fracture Test Results for 1-Inch (25mm) Wide 2024-T3 Aluminum Specimens Removed from The Fuselage Skin of An Aging Aircraft; Skin Thickness = 0.048 Inch (1.2mm)

Numer of Specimens	Orientation*	Hole Diameter (inches)	Net Section Fracture Stress (ksi)
3	Axial	0.156	61.0
1	Axial	0	66.7
2	Circumferential	0.156	59.4
1	Circumferential	0	66.3

*Relative to fuselage

Fracture of the cyclically loaded coupon occurred 550 cycles after the application of the overload. The overload for this experiment was chosen after the 60 ksi (414 MPa) net section fracture stress criterion had been determined from the other tests. The crack lengths were measured on the surface after prefatiguing and then a nominal stress was applied equal to approximately 95% of the predicted fracture stress. This situation simulates conditions for which the damage present was the maximum size that would escape detection by proof loading. A small amount of stable crack extension from the MSD cracks was observed upon the application of this stress. Subsequent cyclic loading at the lower stress level resulted in sudden fracture of the entire coupon in 550 cycles.

Discussion

Uniform MSD

The success of a net section stress criterion for predicting fracture conditions in the coupons tested in this investigation is consistent with the high stresses and plastic deformation associated with fracture. The fact that the critical net section stress, 60 ksi (414 MPa), is less than the ultimate tensile strength, 67 ksi (462 MPa), is indicative of some degree of notch sensitivity for this aluminum alloy. However, the notch radius below which fracture stress becomes constant is evidently relatively large, since the same net section stress at fracture is obtained for specimens with 0.156 inch (4.0 mm) diameter holes, 0.014 inch wide notches, and even fatigue sharpened cracks. This may be explained by the relatively large crack tip opening displacement at fracture for the 2024-T3. Valid, linear elastic plane stress fracture toughness values for this alloy, obtained from wide panels, are on the order of 140 ksi $\sqrt{in}$ (154 MPa$\sqrt{m}$) [2]. Using this value in the relation between stress intensity and CTOD gives:

$$CTOD = \frac{K_I^2}{E\sigma_o} = \frac{(140)^2}{(10^4)(50)} = 0.040 \; inches (1mm)$$

where E = Young's Modulus (10^4 ksi)
σ_o = yield strength (50 ksi).

The CTOD is larger than the width of the jewelers saw used to cut the notches in the specimens and about 25% of, but the same order of magnitude as, the diameter of the rivet holes. Tests from a sheet of aluminum removed from the fuselage of an aging aircraft suggest that the same critical net section stress obtained in this study is applicable to aging fuselage structures.

The results of this study also show that fracture in riveted lap splices and continuous sheets with simulated MSD occurs at the same net section stress. This indicates that either the rivet bearing load has a negligible effect on fracture or that the bearing load is shed to the next row of rivets, thereby subjecting the top row of holes to pure bypass stress as in the continuous coupons.

Nonuniform MSD

In a real aircraft structure it is likely that the multiple site damage will be nonuniform, as shown in Figure 1. What is the criterion for link-up of defects for this more realistic situation? The limited data of this study suggest that if MSD is characterized as average crack size, then a lower critical net section stress must be used than for uniform MSD, Figure 3. However, it is unclear what net section stress value should be used for other distributions of nonuniform MSD and over what distance along the joint the crack lengths should be averaged. Combine these difficulties with the nonuniform stress distribution in an aircraft and one is led to find a simple criterion in which conservative engineering predictions can be made for the conditions under which link-up occurs.

One such criterion is that link-up occurs when any crack length satisfies the net section stress fracture criterion based on uniform MSD. In other words, fracture occurs when

$$\sigma_{nom} Wt = \sigma_f \, nlt, \qquad (1)$$

$$\sigma_{nom} = \sigma_f \left(\frac{D-2a-d}{D}\right),$$

where σ_{nom} = nominal (far field) skin stress
σ_f = critical net section stress
W = width of the region of interest
t = skin thickness
n = number of ligaments in the region of interest
l = ligament length
a = crack size from the rivet hole
d = rivet hole diameter
D = rivet spacing.

Applying this criterion to the coupons with nonuniform MSD, for which the fracture stress was approximately 20 ksi (138 MPa), the predicted critical crack size using a net section stress of 60 ksi (414 MPa) is 0.255 inches (6.5 mm), compared to the maximum crack size of 0.30 inches (7.6 mm). As another example, consider the fatigue and fracture testing of a 20 inch (508 mm) wide panel from another investigation [3]. This panel had a riveted lap splice identical to that used in the coupons of this study and was loaded in fatigue with a maximum stress of 16 ksi (110 MPa). Multiple site damage occurred in this panel with the distribution of crack sizes shown in Figure 4, which was measured approximately 400 cycles prior to complete fracture. The dashed line in this figure shows the critical crack size calculated according to equation (1). Although the maximum crack size exceeds the critical value at N = 26,110, fracture occurred relatively soon afterwards, showing that the criterion is conservative by approximately 2% of the total life. These two examples show that the criterion given by equation (1) is conservative and provides more accurate predictions as the MSD becomes uniform. It is likely that this criterion will not be applicable to the case in which only one or two rivet cracks are present in a joint, but this situation is not considered as MSD for our purposes.

In cases of a nonuniform stress distribution the maximum stress could be used in place of the nominal stress in equation (1). This will undoubtedly provide additional margin of conservatism to the predictions and ongoing experiments in a current study are being performed to explore the applicability of the criterion. If one assumes that the maximum stress in the aircraft fuselage skin depicted in Figure 1 is 14 ksi (97 MPa) then the critical

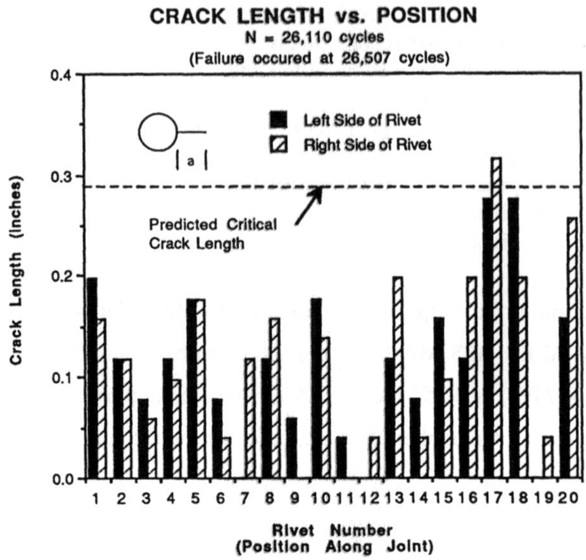

Fig. 4. MSD Generated in A Laboratory Tested 20-Inch Wide Panel

crack size by equation (1) is 0.282 inches (7.2 mm). Only one crack has achieved this length and link-up has not occurred, indicating that the criterion is conservative (again, assuming a nominal stress of 14 ksi.)

One of the applications of the criterion given by equation (1) is to calculate the number of cycles (flights) required to grow the largest crack that is just detectable to the critical size. While such a calculation is outside the scope of this study, we note that the number of cycles to grow one of the cracks in the 20 inch wide panel referred to earlier from 0.080 to 0.310 inches was 10,600.

Finally, the single experiment designed to study the benefit of a proof test to identify critical MSD cracks showed that fatigue life subsequent to the overload was only 550 cycles. This is a relatively small number of cycles, thereby indicating that proof testing alone is not a viable approach to ensuring structural integrity of the fuselage lap splices.

Conclusions

- The results of this study indicate that a net section stress criterion can be used to estimate critical crack size for the link up of cracks between rivet holes in fuselage lap splices containing MSD.

- It is proposed that a net section stress of 60 ksi (414 MPa) be used to calculate critical crack size with the assumption that MSD is uniform. This should lead to conservative (smaller critical crack size) predictions in the case of nonuniform MSD.

- The same net section stress criterion applies for continuous and riveted lap splice coupons, notched and fatigue cracked coupons, and for new and aging aircraft material.

- A single test performed to explore the benefits of pressurized proof testing showed that only 550 cycles of life remained after the simulated proof lead.

Acknowledgements

The author is grateful for the technical and financial support of the FAA Technical Center and the Transportation Systems Center, Department of Transportation, Contract No. DTRS-57-88-C-00019, under the supervision of Dr. Oscar Orringer. The assistance in technical review by Dr. Peter D. Hilton, Arthur D. Little, Inc. is also gratefully acknowledged.

References

1. National Transportation Safety Board Report PB89-910404 "Aircraft Accident Report; Aloha Airlines Flight 243 Boeing 737-200 N73711 Near Maui Hawaii", April 28, 1988 (June 14, 1989) also: "Aloha Airlines Probe Raises Questions about FAA Surveillance of Maintenance", Aviation Week and Space Technology (Aug. 28, 1989) pg. 77.

2. Swift, T., "Damage Tolerance in Pressurized Fuselages,", presented at the 14th Symposium of the Int. Committee on Aeronautical Fatigue (ICAF) New Materials and Fatigue Resistant Aircraft Design, Ottawa, Canada (June 10-12, 1987).

3. "Causative Factors Relating to MSD in Aircraft Structures", Contract DTRS-57-89-C-00007 with the Transportation Systems Center, DOT.

19. How Likely is Multiple Site Damage?

Oscar Orringer

U.S. Department of Transportation
Transportation Systems Center
Cambridge, MA 02142

Summary

One way to avoid multiple site damage in airframes of the future is to screen design details by means of coupon tests. A simple conceptual model for estimating the risk of multiple site damage is presented. The risk is expressed as the fraction of similar details which can be expected to form fatigue cracks at times close enough to each other to allow a fracture cascade after the damage has propagated. Examples based on an approximate representation of typical transport category fuselage skin properties and stress environments show that the risk becomes significant for details which exhibit much less fatigue life scatter than is observed in plain fatigue tests and for which the characteristic life is of the order of the service life. Results for confidence limits are also presented to show that at least 40 replicate details must be tested to obtain adequate estimates of scatter and characteristic life.

Background

The question which heads this paper was posed by Mr. Thomas Swift, FAA NRS for structural integrity, in the wake of the Aloha 737 accident and subsequent inspection findings of multiple site damage (MSD) in other B-737 and DC-9 fuselages. The answer will have a strong effect on FAA policies and industry actions to protect aging transport aircraft against in-flight structural failures from MSD. On one hand, MSD might be no more than an isolated phenomenon, which is best dealt with via airworthiness directives addressing specific cases. For example, the first 291 B-737's, which have cold-bonded fuselage lap splices and include the Aloha 737, have already been addressed by an AD. On the other hand, MSD might be a phase which any airframe should be expected to enter if it flies long enough. In that case, FAR 25.571 and/or the special supplemental inspection program criteria for aging fleets might require modifications.

Springer Series in Computational Mechanics
Atluri, Sampath, Tong (Eds.)
Structural Integrity of Aging Airplanes

In either case, MSD is now recognized as a situation with fatigue crack growth and fracture characteristics significantly different from the characteristics of the isolated cracks on which current damage tolerance requirements are based. The most important differences are inspectability and time available for detection (Figure 1), which may require more frequent D checks and improved nondestructive inspection methods to control MSD.

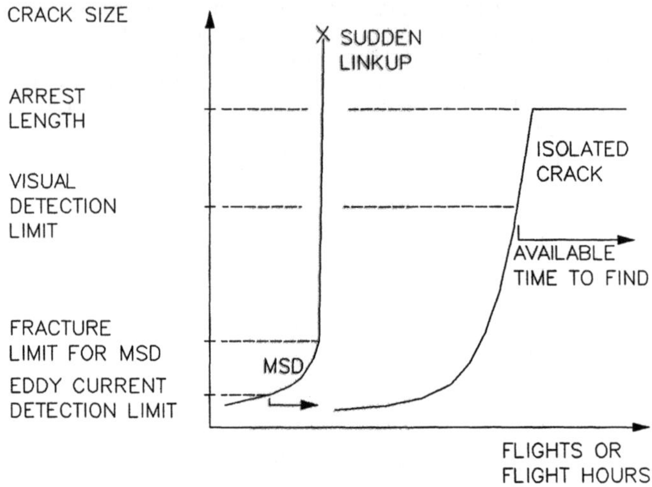

Fig. 1. Comparison of MSD and isolated crack

The MSD observed in the Aloha 737 post-accident investigation and subsequent inspections of other airframes typically consists of large numbers of small skin cracks originating at the holes along a fastener row, usually part of a longitudinal skin splice in fuselage structures. To have the potential for sudden linkup which would effectively bypass the isolated crack arrest capability of the structure, the MSD crack sizes must be spread over no more than a moderate range. This characteristic can be used as a basis to develop an answer to the question about the likelihood of MSD.

<u>Estimation of MSD Likelihood</u>

Figure 2 illustrates a conceptual framework for estimating the likelihood of MSD which has the potential to precipitate a sudden linkup. A cascade scenario is postulated for the linkup phenomenon, i.e., the fracture is assumed to start at one critical crack of length A_{CR} and then to continue as separate fractures are precipitated by overloads at adjacent critical cracks of shorter length, down to some minimum length A_{ACR} .

An average fatigue crack growth curve is used to trace back to the times at which the cracks would have had to form in order to have reached their present lengths. The interval between these times, ΔT , identifies a segment of the crack formation life distribution curve. The area under this segment (hatched area in Figure 2) represents the fraction of structural details expected to have formed cracks in the interval ΔT . This interval is the same as the interval required for a crack of length A_{ACR} to grow to length A_{CR} , provided that all of the cracks are assumed to grow at the same rate. Cracks which form earlier are presumed to have been visually detected and repaired already, and cracks which form later are presumed to be too small to participate in the fracture cascade.

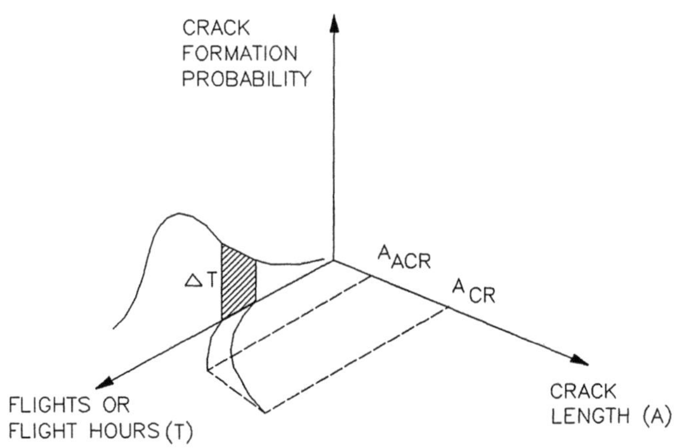

Fig. 2. MSD likelihood model concept

Hypothetical Example

A numerical example will serve to show how the likelihood of MSD can be estimated. The following example is based partly on empirical data from recent findings and partly on order-of-magnitude estimates for material properties. The numbers are intended to roughly represent MSD cracking along a row of fastener holes in a longitudinal splice of an aluminum-skin fuselage. However, the example is intentionally simplified to focus on the overview rather than the details.

Critical Crack Sizes

The critical cracks lengths A_{ACR} and A_{CR} can in principle be estimated from a stability analysis based on fracture mechanics or net section failure theory. In the present case, A_{CR} will be based on the recent findings from striation-count tracings of fatigue crack growth in B-737 fuselage splices. The findings reported by Boeing suggest that the total length (two cracks plus hole diameter) for a critical crack is about 0.64 inch. The individual crack lengths were not reported, but will be assumed equal for the purpose of the example, i.e., A_{CR} = 0.24 inch for one side (Figure 3).

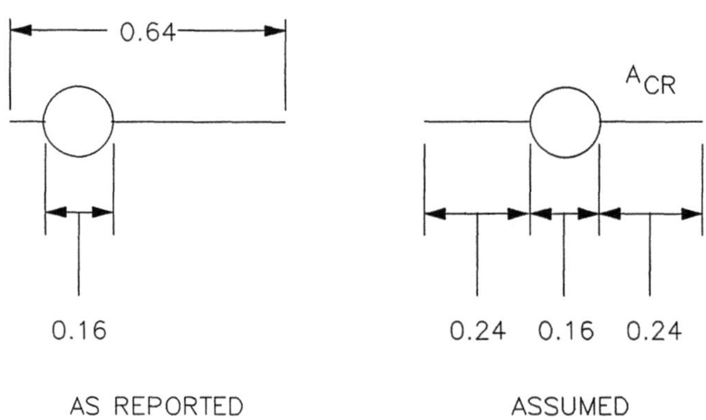

Fig. 3. Critical crack length for fracture initiation

A scenario for MSD crack locations must also be specified. Figure 4 illustrates a simple assumption, in which the first critical crack breaks a ligament between two fastener holes, thus overloading the adjacent ligaments, one of which is assumed to contain a crack of length A_{ACR}. The rivets immediately adjacent to the fracture are assumed to be unable to transfer load across the broken ligament, and an overload factor of 3/2 is then assigned to the adjacent ligaments. If basic net section failure theory is applied, the adjacent critical crack length is 2/3 of the first critical crack length, i.e., A_{ACR} = 0.16 inch.

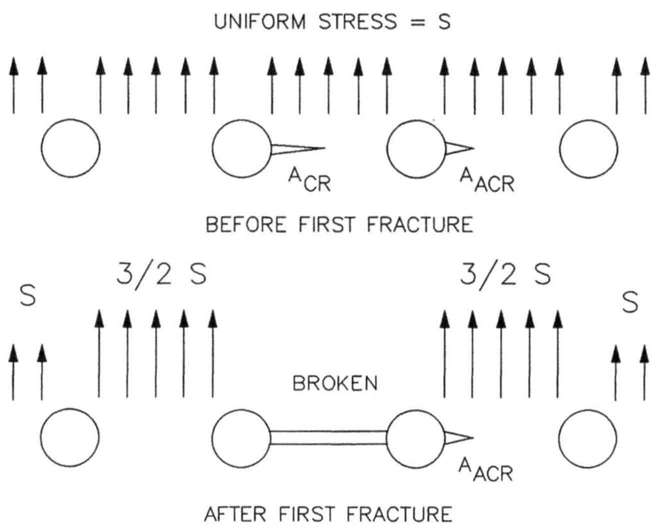

Fig. 4. Scenario for MSD fracture cascade

<u>Crack Growth Time</u>

For aluminum skin, assume a Paris crack growth rate equation:

$$\frac{dA}{dT} \cong 10^{-10}(\Delta K)^4 \qquad (1)$$

where the crack growth rate is in inches/cycle, T represents flights (one flight is equivalent to one GAG stress cycle at R

= 0), and ΔK is the stress intensity factor range in ksi-in.$^{1/2}$. For the purpose of the example, assume an approximate stress intensity factor range:

$$\Delta K \sim 3\Delta S\sqrt{\pi A} \tag{2}$$

where ΔS is the GAG stress cycle (hoop stress for zero to service pressure). Assume that ΔS = 10 ksi.[1]

Substituting eq. (2) into (1) with $(3\sqrt{\pi})^4 \sim 10^3$ and integrating then gives the following result for the crack growth time:

$$\Delta T \sim \frac{10^7}{(\Delta S)^4}\left[\frac{1}{A_{ACR}} - \frac{1}{A_{CR}}\right] \tag{3}$$

With A_{ACR} = 0.16 and A_{CR} = 0.24, eq. (3) gives 2,083 flights, which will be rounded to 2,000 flights for convenience.

Crack Formation Life Distribution Model

A brief digression is needed at this point to introduce and explain a life distribution model for crack formation. The Weibull probability distribution [1] is often used to correlate laboratory fatigue test data and will be used here to model the distribution of crack occurrence times at fastener holes in the hypothetical skin splice. The cumulative and density functions of the Weibull distribution are given by:

$$P(T) = 1 - e^{-(T/\beta)^\alpha} \tag{4}$$

$$p(T) = \frac{dP}{dT} = \frac{\alpha T^{\alpha-1}}{\beta^\alpha} e^{-(T/\beta)^\alpha} \tag{5}$$

1 This stress corresponds to two thirds of a 15 ksi nominal hoop stress to account for the stress reduction effect of a doubler or lap. The nominal stress is approximately in the middle of the range for transport category fuselages.

where T is the total number of flights accumulated and α , β are the distribution parameters which must be determined. The parameter α is the shape factor, which characterizes the spread of the distribution; β is the scale factor, which characterizes the average.[2]

The cumulative function $P(T)$ represents the fraction of the fastener hole population that has formed cracks at any time up to T . The density function $p(T)$ represents the instantaneous rate of crack formation. The fraction, F , of holes that have formed cracks in the interval ΔT is needed in the present case and can be expressed as either a difference of cumulative fractions or the product of rate and interval:

$$F = P(T+\Delta T)-P(T) = e^{-(T/\beta)^{\alpha}} - e^{-[(T+\Delta T)/\beta]^{\alpha}} \qquad (6)$$

$$F \cong p(T)\Delta T = \frac{\alpha T^{\alpha-1}\Delta T}{\beta^{\alpha}} e^{-(T/\beta)^{\alpha}} \qquad (7)$$

Equation (7) is a useful approximation when ΔT is small compared to T . In particular, it is easy to show that the density function assumes its maximum value when $T = \beta$, and therefore:

$$F_{max} \cong \frac{\alpha \Delta T}{\beta e} \qquad (8)$$

Equation (8) is a useful tool for quick screening of MSD risks.[3]

The distribution parameters must be estimated from test or service data. They can be treated as material properties when they are derived from well controlled plain fatigue tests in the laboratory. For example, Figure 5 is a schematic illustration of typical results of a series of laboratory fatigue tests at different

---

2 Strictly speaking, 63 percent of the lives in the distribution are less than or equal to β .

3 If F_{max} is small, then the risk of MSD can never be high, and the case can be dismissed. A large value indicates a potential for high risk, but the actual risk may be either high or low, depending on the ratio of service life to the fatigue life scale parameter.

values of constant-amplitude stress. The fatigue life distribution generally depends on the stress level, trending toward greater scatter (lower α) and longer life (greater β) as the stress amplitude is reduced.

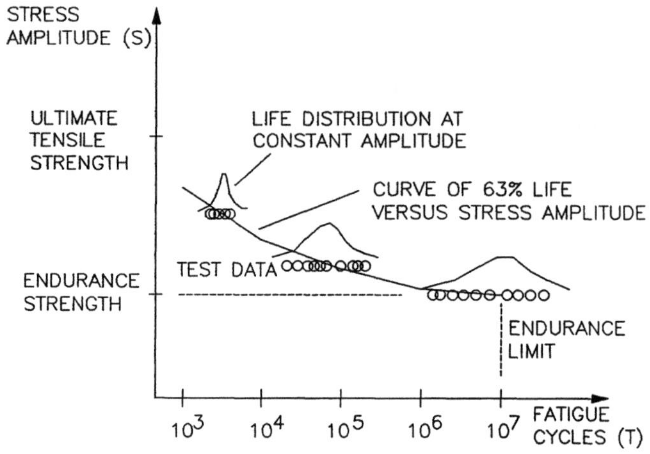

Fig. 5. Probabilistic description of fatigue life

Convenient measures of the tendency to scatter can be constructed by taking ratios of long to short life. For example, one may take the ratio of the 90th to 10th percentile lives, $T_{0.9}/T_{0.1}$, where $P(T_{0.x}) \doteq 0.x$. The value of the 90/10 scatter ratio is easily obtained from eq. (6):

$$\ln\left[\frac{T_{0.9}}{T_{0.1}}\right] = \frac{\ln(-\ln 0.1)-\ln(-\ln 0.9)}{\alpha} \tag{9}$$

or

$$\frac{T_{0.9}}{T_{0.1}} \cong e^{3.08/\alpha} \tag{10}$$

Figure 6 plots the 90/10 scatter ratio versus the shape factor.

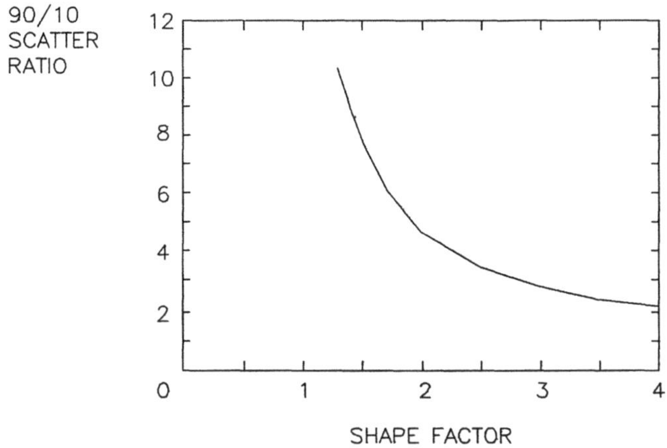

Fig. 6. Life scatter ratio versus shape factor

In plain fatigue tests like those depicted in Figure 5, the 90/10 life scatter ratio is typically found to range from 2 to 10 as the stress amplitude is decreased from about 80 percent of the material's ultimate tensile strength to levels near the endurance strength. Thus, one would expect the shape factor to decrease from about 4 at the high stress to 1.3 at the low stress.

<u>Distribution Parameters for MSD Example</u>

Parameters for the MSD example cannot be estimated from plain fatigue test data because the physical process of fatigue at a fastener hole is different from the plain fatigue process. Rough estimates for the scale factor can be made from the results of lap-splice coupon tests reported by Boeing, but there has not been enough data reported to allow such estimates for the shape factor.

The worst case reported from the Boeing tests was an unbonded coupon, believed to represent the conditions found in an inspection of a VASP 737 which had badly degraded cold bonds from the beginning of service. This coupon had a fatigue life of about 45,000 flights. At the other extreme, coupons which were fabricated with improved hot-bond procedures had fatigue lives as long as 370,000 flights. Therefore, values of 40,000 and 400,000 flights will be assumed for β to represent these extremes.

In the absence of scatter data, rough estimates for α can be based on general experience and practice in the airframe industry. A scatter factor of 4 has often been used in the past as a contractual requirement for demonstrating that military airframes meet a fatigue life guarantee. Such contracts require a ground test of the airframe to four service lives, without widespread cracking, to demonstrate one service life capability for the fleet. However, the meaning of the scatter factor is subject to different interpretations. The test airframe might be either the best airframe in the fleet, in which case the scatter ratio should also be 4, or it might be an average airframe, in which case the scatter ratio might be 8 (see Figure 7).

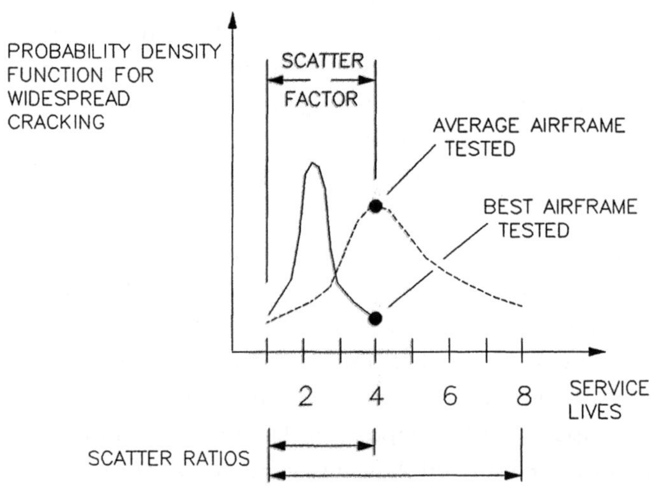

Fig. 7. Interpretations of scatter factor

Figure 7 suggests scatter ratios from 4 to 8 (α from about 2.2 to 1.5), but the cited experience does not involve any tests or service in which MSD was encountered. A scatter ratio of about

2 (α of about 4) might better represent MSD situations, in which effects analogous to those caused by the high stress amplitude of a plain fatigue test may be suspected.

Example Results

Table 1 summarizes the estimates of the input data for the example MSD risk analysis. Table 2 summarizes the results obtained when these values are entered in eq. (8) to screen for MSD risk. These results suggest isolated cracking rather than MSD.

Discussion

The situation just described is hardly MSD, yet most of the numbers in the example should reflect skin splice environments like those in fuselages which have recently experienced MSD. Therefore, it is appropriate to identify shortcomings in the model and/or input assumptions. Three possibilities are discussed below.

The first possibility is an error in the assumption for the shape factor of the fatigue life distribution. This is the only input quantity for which no MSD data is available. Although the extrapolation discussed earlier seems reasonable, it is appropriate to turn the analysis around by asking what scatter ratio would be needed to produce MSD, e.g., half the details with cracks simultaneously in the critical range. For the unbonded and best bond cases solution of eq. (8) with F_{max} = 0.5 gives α = 27.2 and 272, respectively. The corresponding 90/10 life scatter ratios obtained from eq. (10) are 1.12 and 1.01, respectively. Instead of fatigue effects in real structures, such ratios bring to mind Mr. Justice Holmes' story of the wonderful one-horse shay. Therefore, it appears that other sources of error must be sought.

The estimate for the fatigue life scale factor β is a second possible error source. For example, the life of 45,000 simulated flights reported for the unbonded condition was based on a coupon fatigue test, and the final state of the test specimen was not reported. Coupons are believed to have longer lives than skin splices subjected to the full effects of pressure loads.

Table 1. Summary of inputs for hypothetical example

Quantity	Units	Value
Critical crack length, A_{CR}	in.	0.24
Adjacent critical crack length, A_{ACR}	in.	0.16
Critical time interval, ΔT	flights	2,000
Shape factor	---	
for scatter ratio of 7.8		1.5
for scatter ratio of 4.1		2.2
for scatter ratio of 2.2		4.0
Scale factor	flights	
unbonded		40,000
best bond		400,000

Table 2. Results for F_{max}

Bond condition \ Scatter ratio	7.8	4.1	2.2 (MSD)
Unbonded	0.03	0.04	0.07
Best bond	0.003	0.004	0.007

The time to the end of the test may also have included some crack growth time. In the VASP 737 MSD case mentioned earlier, the aircraft had accumulated about 40,000 flights, but fatigue crack growth was traced back via striation counts for about half the service. Hence, one could argue that β = 20,000 flights should be used to represent the unbonded condition. The example analysis would still predict isolated cracking in this case, or it would require a slightly greater 90/10 scatter ratio of 1.25 ($\alpha = 13.6$) to predict half the fastener hole population cracked.

The third possible source of error is the cascade scenario, which might be unrealistically strict. As posed in the example, this scenario implies that all adjacent cracks must be critical at an overload factor of 3/2, which represents the state of the adjacent ligaments after the first crack has fractured one ligament. However, larger overloads and smaller critical sizes for adjacent cracks are possible after two or more ligaments have been broken. For example, assume that N ligaments of length L have broken, that the fasteners along the fracture are ineffective, and that the stress in the adjacent ligament can be approximated by averaging the first two terms of the stress field near the tip of a sharp crack:

$$S(x) = \left[1 + \sqrt{\frac{NL}{2x}}\right] S \tag{11}$$

where x is the distance ahead of the crack tip. The average overload stress is then given by:

$$S_{OL} = \frac{1}{L}\int_0^L S(x)dx = (1+\sqrt{2N})S \tag{12}$$

Now suppose that the structure's arrest capability is limited to a two-bay crack (say 20 inches, or about 18 ligaments). One can then use eq. (12) with $N = 18$ together with:

$$\Delta K(S_{OL}\ ,\ A_{ACR}) = K_C \tag{13}$$

where K_C is the effective toughness for the material and skin gage under consideration, to estimate the adjacent critical crack length which, as MSD, would nullify the arrest capability. Using eq. (2) for ΔK and a typical value of K_C = 95 ksi-in.$^{1/2}$ for thin-gage aluminum, for example, gives A_{ACR} = 0.07 inch.[4] Re-calculation of the critical time interval from eq. (3) then gives ΔT = 10,000 flights.

Tables 3 and 4 summarize the revised inputs and results, respectively. The model now predicts a risk of MSD failure for the lowest assumed scatter ratio of 2.2 (α = 4).

Table 3. Comparison of original and revised inputs

Quantity	Units	Values	
		original	revised
Adjacent critical length, A_{ACR}	in.	0.16	0.07
Critical time interval, ΔT	flights	2,000	10,000
Life scale factor, unbonded	flights	40,000	20,000

Table 4. Revised results for F_{max}

90/10 Life Scatter Ratio	7.8	4.1	2.2 (MSD)
F_{max}	0.28	0.37	0.74

4 Strictly speaking, one must assume the adjacent cracks to be arranged in order of decreasing length from the middle to the ends of the bays. Such an assumption is consistent with the bulging effect on stress distributions in semi-monocoque fuselage skins.

The foregoing results highlight the fact that the judgement of MSD risk is sensitive to the fatigue life parameter values. Since these parameters can only be estimated from tests, however, it is important to establish the number of replicate tests required to obtain suitably accurate estimates.

For example, suppose that n nominally identical details tested under the same conditions have individual fatigue lives T_i $(i = 1, 2, ..., n)$ and that estimates for the parameters are calculated from the maximum likelihood formulas for the Weibull distribution [2, 3]:

$$\hat{\alpha} = \frac{n \sum_{i=1}^{n} T_i^{\hat{\alpha}}}{n \sum_{i=1}^{n} (T_i^{\hat{\alpha}} \ln T_i) - \left(\sum_{i=1}^{n} T_i^{\hat{\alpha}} \right) \left(\sum_{i=1}^{n} \ln T_i \right)} \tag{14}$$

$$\hat{\beta} = \left[\frac{1}{n} \sum_{i=1}^{n} T_i^{\hat{\alpha}} \right]^{1/\hat{\alpha}} \tag{15}$$

The simultaneous $100\gamma\%$ confidence limits on the true values α, β are then given by:

$$Pr\left\{ \hat{\alpha} - \frac{0.78\delta\hat{\alpha}}{\sqrt{n}} \leq \alpha \leq \hat{\alpha} + \frac{0.78\delta\hat{\alpha}}{\sqrt{n}} \right\} \geq 1 - \frac{1}{\delta^2} = \gamma \tag{16}$$

$$Pr\left\{ \hat{\beta} - \frac{1.05\delta\hat{\beta}}{\hat{\alpha}\sqrt{n}} \leq \beta \leq \hat{\beta} + \frac{1.05\delta\hat{\beta}}{\hat{\alpha}\sqrt{n}} \right\} \geq 1 - \frac{1}{\delta^2} = \gamma \tag{17}$$

These limits are illustrated in Figures 8 and 9, respectively, as the bounds A and B, where:

$$\alpha = \left(1 \pm \frac{A}{100} \right) \hat{\alpha} \tag{18}$$

$$\beta = \left(1 \pm \frac{B}{100} \right) \hat{\beta} \tag{19}$$

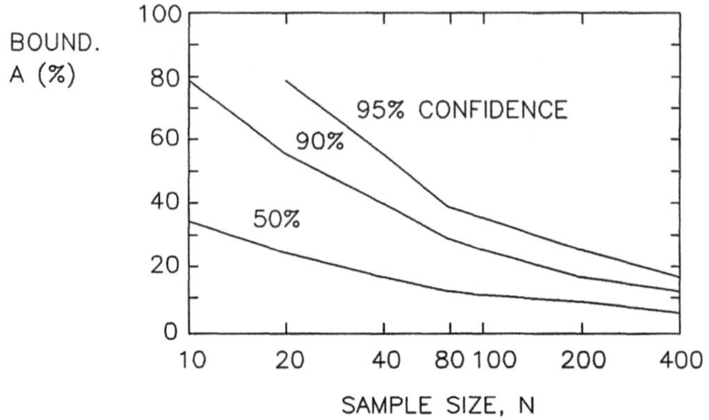

Fig. 8. Bounds on Weibull shape factor

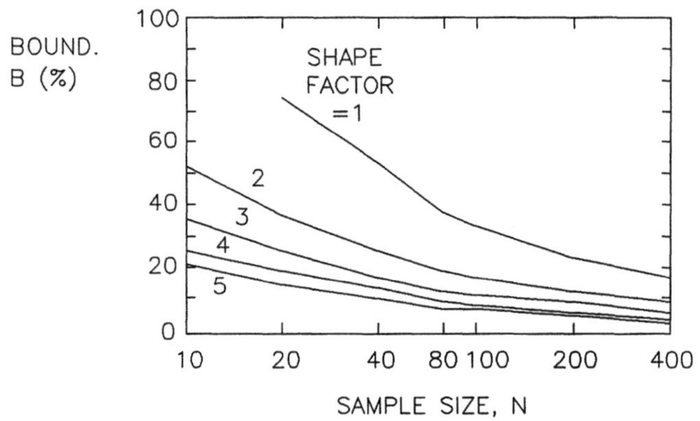

Fig. 9. Bounds on Weibull scale factor (90% confidence)

Based on these confidence limits, a sample size of n = 40 can be judged adequate to screen for MSD risk. If $\hat{\alpha} = 4$ is estimated from the test series, then the 50% confidence limits on the true shape factor are $3.3 \leq \alpha \leq 4.7$. While a 50% confidence level is not a strong statement, it is sufficient to justify a conservative judgement about the tendency of cracks to form in a group. Also, for $\hat{\alpha} = 4$ the 90% confidence limits on the scale factor are $0.87\hat{\beta} \leq \beta \leq 1.13\hat{\beta}$, i.e., the true scale factor is quite likely to be within the risk range if the estimate implies high risk.

Conclusions

The MSD likelihood model concept presented in this paper appears to be a practical framework for finding quantitative answers to the question: how likely is multiple site damage? Applications to actual cases can be based on the actual fatigue, crack growth rate, and critical crack length properties determined for actual structure details.

The existing methods of fracture mechanics and net section failure theory are adequate for making estimates of crack growth times and critical crack lengths, provided that stress intensity factor formulae, crack growth rate equations, and service stress spectra appropriate to the specific structure are employed. Alternatively, design development type test specimens can be used to establish estimates for crack growth times and critical crack lengths.

Basic material fatigue properties should not be relied upon to characterize fatigue life distributions in structures because the fatigue processes in the plain material and the structure are likely to be different. Enough fatigue data to estimate life scatter as well as average life is essential in order to assess the risk of multiple site damage which could cause an in-flight failure. Obtaining such data from fleet service difficulty reports will pose the difficult statistical problem of extrapolating average behavior from short-life data.

Coupon tests of representative details are a practical alternative for screening during design. The example results suggest that low life scatter ratio and average fatigue life of the order of service life are the principal attributes of high-risk MSD. These attributes can serve as indicators of MSD risk in screening programs based on design development or coupon type tests. The necessary sample size of 40 details is large but not impractical for coupon or flat panel tests. In order to derive the fatigue life scale factor from such a program, some of the test specimens should be fractographically examined to trace back to the scale factor for the crack formation time.

The example results also suggest that the model predictions are sensitive to the crack cascade assumption. This assumption can be improved if cascade scenarios can be based on typical crack location distributions as observed in airframes.

References

1. Weibull, W.: Fatigue Testing and Analysis of Results. New York; London: Pergamon Press 1961.
2. Hahn, G.J.; Shapiro, S.S.: Statistical Models in Engineering. New York: Wiley 1967.
3. Breiman, L.: Statistics: With a View Toward Applications. Boston: Houghton Mifflin Co. 1973.

20. Fractographic Analysis of Initiation and Growth of Fatigue Cracks at Rivet Holes

R. Pelloux, A. Warren, and J. O'Grady

Massachusetts Institute of Technology
Department of Materials Science & Engineering
Cambridge, MA 02139

Abstract

A series of fatigue tests were performed on riveted panels of clad 2024-T3 without epoxy bonds. Fatigue crack initiation occurred at the apex of the rivet hole chamfers. Transgranular fatigue crack growth by ductile striation formation occurred through the sheet. The fracture features at low, medium and high growth rates were examined with the SEM. Microscopic crack propagation rates as measured by fatigue striation spacings correlate with macroscopic crack growth rates observed. The fatigue crack growth rate is fairly constant over a length of 6 mm (0.25 in.) from the edge of the rivet hole, due to the fact that the stress intensity range is approximately constant in this region. Transition to fast fracture and unstable crack propagation is readily identified due to marked yielding of the cladding material.

Introduction

It is common practice in failure analysis [1] to investigate the fracture surface features at low and high magnifications, in order to identify the modes of crack initiation and crack growth. The fracture path and fracture features provide information on the local fracture mechanisms. The fracture path can be transgranular due to cleavage, microvoid coalescence, or mechanisms associated with a series of crack front progression marks such as ductile fatigue striations or stress corrosion striations. The fracture path can also be intergranular indicating grain decohesion due to stress corrosion cracking, creep deformation, or fatigue cracking.

Springer Series in Computational Mechanics
Atluri, Sampath, Tong (Eds.)
Structural Integrity of Aging Airplanes

The fracture features which characterize the crack initiation process are more difficult to identify because they are typically at the scale of one or two times the grain size. Crack initiation may be due to a cluster of brittle inclusions or to local damage created by repeated microplasticity as in fatigue. A localized corrosion pit, associated with enhanced corrosion activity, is also a very common initiation site.

In all cases the microscopic features of the fracture surfaces, if they are not destroyed by post fracture damage, provide extensive information regarding the processes of crack initiation and the rates of crack growth. Slow crack propagation may be due to a number of time dependent processes including fatigue, stress corrosion cracking, and creep crack growth.

The microscopic fracture features should be considered in the context of continuum fracture mechanics. The orientation of the fracture plane with respect to the tensile axis should be noted. The fracture may be either plane strain (fracture plane normal to the maximum tensile stress axis) or plane stress (typical of a shear fracture). The size of the plastic zone and the approximate stress intensity factor can be estimated by examination of the shear lip, the near surface fracture region which is characterized by planes of shear fracture at $\pm 45°$ to the plane of the sheet. Because the plane stress plastic zone is approximately equal to the width of the shear lip, the plastic zone size can be observed directly on the fracture surface. In the case of a slowly growing crack, the size of the plastic zone and/or the shear lip width are related to the stress field at the crack tip which is described by the linear elastic stress intensity factor, K (or stress intensity range, ΔK).

The Multi-Site Damage (MSD) problem associated with the initiation and propagation of small fatigue cracks from rivet holes in the fuselage of ageing airplanes presents an ideal case study for the application of fractography to failure analysis. In the case of MSD, fatigue cracks are reported to initiate at rivet holes in the fuselage lap joints after the epoxy bond failed as a result of corrosion in high humidity environments. The fatigue cracks initiate from the sharp chamfer edges of the rivet hole (diameter 0.156 in.). The fatigue cracks propagate lengthwise i.e., normal to the direction of maximum hoop stress, through the tapered section to the full width of the sheet (diameter 0.250 in.). Typically, two cracks are present, one extending from each side of the fastener hole. The growing cracks are initially hidden under the rivet head. Once a crack is visible on the outside surface of the sheet, it has already propagated over a distance in excess of 1 mm (0.040"). The cracks grow to a length of approximately 6 to 8 mm (0.25 in. to 0.3 in.) on each side of the rivet, before fracture by tensile instability. Note that the rivets are spaced an inch apart center to center. Crack growth in service has been reported to occur over 20,000 to 40,000 cycles.

The main goal of MSD failure analysis is to report the rates of crack propagation, da/dN (measured by the spacing of fatigue striations), as a function of crack length, a. If the data is plotted as $\frac{1}{da/dN}$ versus a, the integration of the area under the curve gives the number of cycles for crack propagation from $a_{initial}$ to a_{final}. Work of this type has been reported by the Boeing Company [2] and data has been used to produce curves of tip-to-tip crack length (2a) versus number of cycles. The crack propagation data reported by Boeing for different airplanes indicates fairly uniform crack growth rates over a distance of 6 mm (0.25 in.) on each side of each rivet hole. Depending upon the type of airplane and the service conditions, the crack growth rates range from 0.13 µm to 0.5 µm ($5 \cdot x 10^{-6}$ in. to $2 x 10^{-5}$ in.) per cycle.

In view of the fact that service cracks were not readily available for a fractographic analysis, we attempted to duplicate the conditions of MSD fatigue fracture in the laboratory. These experiments were performed to characterize the rates of fatigue crack propagation of cracks initiated at rivet holes and establish, by fractography, the characteristic features of MSD fatigue fractures. This work was performed under ideal conditions i.e., there was no corrosion or no post fracture damage of the fracture surfaces. One difference between the service conditions, and the laboratory experiments is the cyclic frequency. The test frequency was 5 Hz as compared to 1 cycle/hour in the field. Mechanical properties of the alloy are not expected to be altered by the frequency, but results are expected to show differences in behavior due to environmental effects.

Experimental Procedures

Lap joint specimens were made from sheets of clad and bare 2024-T3 aluminum. The sheets were 1 mm (0.040 in.) thick. The clad layers were typically of the order of 0.05 mm (0.002 in.) on each side of the clad sheet. The specimens prepared measured 76 mm by 355 mm (3 in. by 14 in.) and 102 mm by 355 mm (4 in. by 14 in.). Lap joints were created by overlapping 50 mm (2 in.) of the sheets. Joints were fastened with two rows of 3 rivets with 25 mm (1 in.) spacing within the row and between the rows. The hole dimensions and the rivets used are identical to those currently used on Boeing 737 aircraft. The rivet has a 100° shear head and requires an equivalent countersink to be drilled around the hole. The specimens were drilled and riveted using pneumatic equipment and alignment jigs at East Coast Aerotech.

The fatigue crack growth tests were performed on a servohydraulic fatigue machine equipped with hydraulic grips. The specimens were tested using a sinusoidal waveform with a frequency of 5.0 Hz in tension-tension with a stress ratio, R = 0.1. The typical maximum stress level was 100 MPa (15 ksi). The crack length of a single crack growing from the edge of a rivet was measured optically using a travelling microscope with an accuracy of 0.010 mm (4×10^{-4} in.) measurements were made every 2000 cycles. During crack length measurements, the test frequency was lowered to 1 Hz to facilitate observation of the crack tip.

Proof tests were performed on specimens containing cracks of 1 to 2 mm from the edge of a rivet. A maximum stress of 20 ksi, representing a 33% overload, was applied for one cycle. The tests were then continued at a maximum stress of 15 ksi until failure.

The fracture surfaces were removed from the failed specimens, cleaned, and gold coated for examination using the scanning electron microscope (SEM) at magnifications ranging from 25x to 10,000x.

Metallographic sections through rivets as-installed were made to examine the geometry of the riveted joint.

Results

Table 1 summarizes the test conditions and life data for both the continuous loading tests and the overload tests.

Table 1: Testing data and cycles to failure for lap joint specimens.

Specimen	Test Type	Max Stress MPa (ksi)	Cycles to Failure	Test Frequency
2	simple	77 (11.25)	130,000	9.5
4	simple	107 (15.6)	100,000	5.0
9	simple	103 (15.0)	42,670	5.0
11	simple	103 (15.0)	61,100	5.0
12	simple	103 (15.0)	45,000	5.0
19	simple	103 (15.0)	47,710	5.0
10P	proof test	103 (15.0)	35,170	5.0
13P	proof test	103 (15.0)	81,000	5.0
16P	proof test	103 (15.0)	40,740	5.0

17P	proof test	103 (15.0)	28,000	5.0
18P	proof test	103 (15.0)	42,270	5.0
14BH	button head	103 (15.0)	155,000	5.0
15BH	button head	103 (15.0)	95,410	5.0

Typical plots of crack length (a) versus the number of cycles (N) are reported in Figures 1 and 2. Figure 1 shows the a versus N data for tests conducted with constant load amplitudes. Figure 2 shows the retardation associated with the application of a single overload cycle.

Because of the limited number of data points collected, the da/dN data reported is calculated using da/dN = $\frac{a_{i+1} - a_i}{\Delta N}$. The da/dN data shows little scatter.

From the a versus N curve and from da/dN versus a data, it is found that the crack growth rates do not change markedly from the edge of the fastener hole to a crack length of the order of 6 mm (0.25 in.).

The calculation of K versus a was performed using the equation given by Broek [3]. This equation

$$K = \frac{1}{2} \sigma \left(\sqrt{\pi a} + \frac{1}{\sqrt{\pi a}} \right)$$

assumes uniform loading on one side of the panel and wedge force loading due to the rivet on the opposite side of the panel, shown schematically in Figure 3. The variation of K versus a for a stress of 100 MPa (15 ksi) is given in Figure 3. The a/w correction was neglected in this calculation. It is important to note that K remains approximately constant versus crack length over a wide range of crack lengths. The stress intensity is approximately constant for cracks which just appear from the edge of the rivet to a crack length of the order of 6 mm (0.250 in,), following an initial transient at low values of a.

Cracks coming out of the rivet holes are not normal to the load axis of the panel, but are slanted in the direction of the load bearing rivet due to the rivet wedge loads. A detailed analysis of the crack orientation near the rivet would give a measurement of the relative importance of the wedge force and the loss of the epoxy bond.

Large amounts of scatter exists in the time it takes a crack to appear at the edge of the rivet head. Note that this first observation of a crack corresponds to a crack which has grown at least 1 mm (0.040 in.) from the sharp edge of the chamfer under the rivet head. This variability in crack initiation is attributed to the variability in the interface stresses between the rivets and the sheets of clad aluminum alloys, and the variability in the amount of damage created during installation of the rivets. Sections thru the assembled joints show that microcracks exist in the clad layer, and cavities were observed at interfaces between the two sheets and the expanded rivet shank.

Fractography

Figures 4, 5, 6, and 7 show the typical features observed on fatigue fracture surfaces. Figure 4 is a low magnification view of the fracture surface including the edge of the chamfer. The edge of the chamfer is very often damaged during riveting. Furthermore, there is usually excessive fretting corrosion damage in the form a black hydrated aluminum oxide present near the origin. Fatigue striations are very difficult to resolve within the first 500 μm from the tip of the chamfer. Once the crack propagates outside of the chamfer, the crack front is fairly straight and plane strain conditions exist at the crack tip. The crack tip stresses change slowly from plane strain to plane stress as the cracks get longer. Figure 5 shows at low magnification the typical interaction between voids around inclusions or second phase particles and the small plateaus typical of fatigue crack propagation in aluminum alloys. The direction of crack propagation is clearly defined by the ridges of the plateau. At higher magnification fatigue striations are clearly resolved as shown in Figure 6. Extensive secondary cracking parallel to the fatigue striations exists. It is important not to confuse the secondary cracks with fatigue striations. The spacing of fatigue striations is typically an order of magnitude smaller than the spacing of the secondary crack. When the fracture surfaces are corroded, the secondary crack spacing may be mistaken for the fatigue striation spacings. Near the surfaces of the sheet in the cladding region, where the pure aluminum is very ductile the fatigue striations are clearly defined and are shown in Figure 7. The orientation of the striations observed indicates that the crack front is leading in the center of the sheet and trailing near the surface. Measurements of striation spacings are easier near the free surfaces. A micrograph of fine fatigue striations with an average spacing of 0.2 μm is shown in Figure 8.

Measurements of the striation spacing is usually performed in a selected fracture area by taking an average of 5 to 10 measurements. Each measurement reports the average spacing of 10 fatigue striations. The striation spacing varies from 0.2 to 0.5 μm in the range of crack growth measured. Good correlation of the experimental data with the da/dN versus ΔK data published by Schijve [4] for clad 2024-T3 is observed. The macroscopic law of crack propagation can be written as

$$da/dN \text{ (m/cycle)} = 10^{-11}\,\Delta K^4. \quad (\Delta K \text{ in units of } MNm^{-3/2}).$$

The crack growth rate is 0.1 μm at 10 $MNm^{-3/2}$ and 0.5 μm at 15 $MNm^{-3/2}$. This is within the range of crack growth rates observed in most of the tests performed here.

It has been shown by Schijve that in this ΔK range there is very good correlation between the microscopic crack growth rates, measured by the striation spacing, and the macroscopic crack growth rates, determined from crack length measurements.

At the transition between fatigue crack propagation and fast fracture, the rate of crack propagation reaches 1 μm. The fast fracture regions are clearly identified on all the fracture surfaces so that the remaining ligament length can be determined.

In the tests where the cladding was not damaged during handling and specimen assembly, well defined slip lines are observed in the pure aluminum surface grains. The sharp slip lines indicate the extent of plasticity at the surface near the fracture front. Near the origin, many of the slip lines have led to microcracks which could serve as secondary initiation sites. Careful elimination of the clad surface may give an indication of the plastic zone size and in turn this measurement could be used to estimate the stress intensity factor. Further work remains to be done to establish those correlations.

Conclusions

1. MSD cracks were duplicated in the laboratory by fatigue testing of 2024-T3 clad aluminum panels with riveted lap joints. Fatigue crack growth tests were performed at 5 Hz with a typical σ_{max} of 100 MPa (15 ksi) and stress ratio of R = 0.1.

2. Crack growth rates observed did not vary markedly over the range of crack length examined. The da/dN values ranged from 0.2 μm/cycle to 0.5 μm/cycle. The lack of variation in the crack propagation rate is attributed to the fact that the stress intensity range (ΔK) remains fairly constant in the range of crack lengths observed.

3. Fractography shows that fatigue striations are the typical fracture features. The crack propagation regime is typical of the Paris Law regime where $da/dN = 10^{-11} (\Delta K)^4$ m/cycle.

4. Further work will require the investigation of hold time effect and corrosion fatigue on the rates of crack initiation and crack growth.

5. The large scatter in crack initiation life is a result of the variability in the local stresses, and the degree of damage at the riveted joints.

References

1. Metals Handbook 8th Edition Volume 9, Fractography and Atlas of Fractographs, ASM, Metals Park, OH, 1961.

2. U. G. Goranson, M. Miller, "Ageing Jet Transport Structural Evaluation Program," in Proceedings 15th Symposium of the ICAF, Israel, June 1989.

3. D. Broek, Elementary Engineering Fracture Mechanics, Fourth Revised Edition, Martinus Nijhoff Publishers,Boston, 1986, p. 87.

4. J. Schijve, F.A. Jacobs, P.J. Tromp, Fatigue Crack Growth in Aluminum Alloy Sheet Material Under FlightSimulation Loading. Effects of Design Stress Level and Loading Frequency, NLR-TR 72018, Amsterdam, 1972.

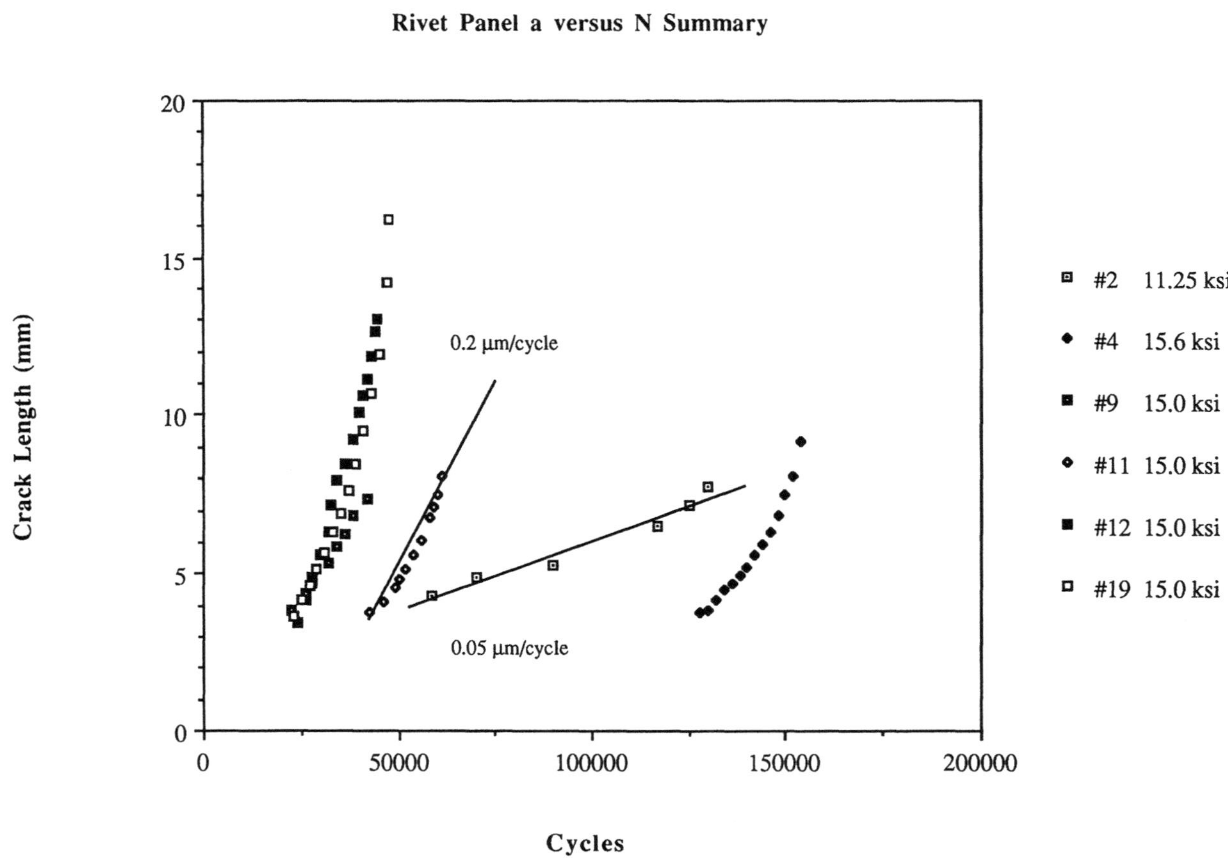

Fig. 1. Crack length (a) vs. N cycles for different tests. Note that da/dn is fairly constant.

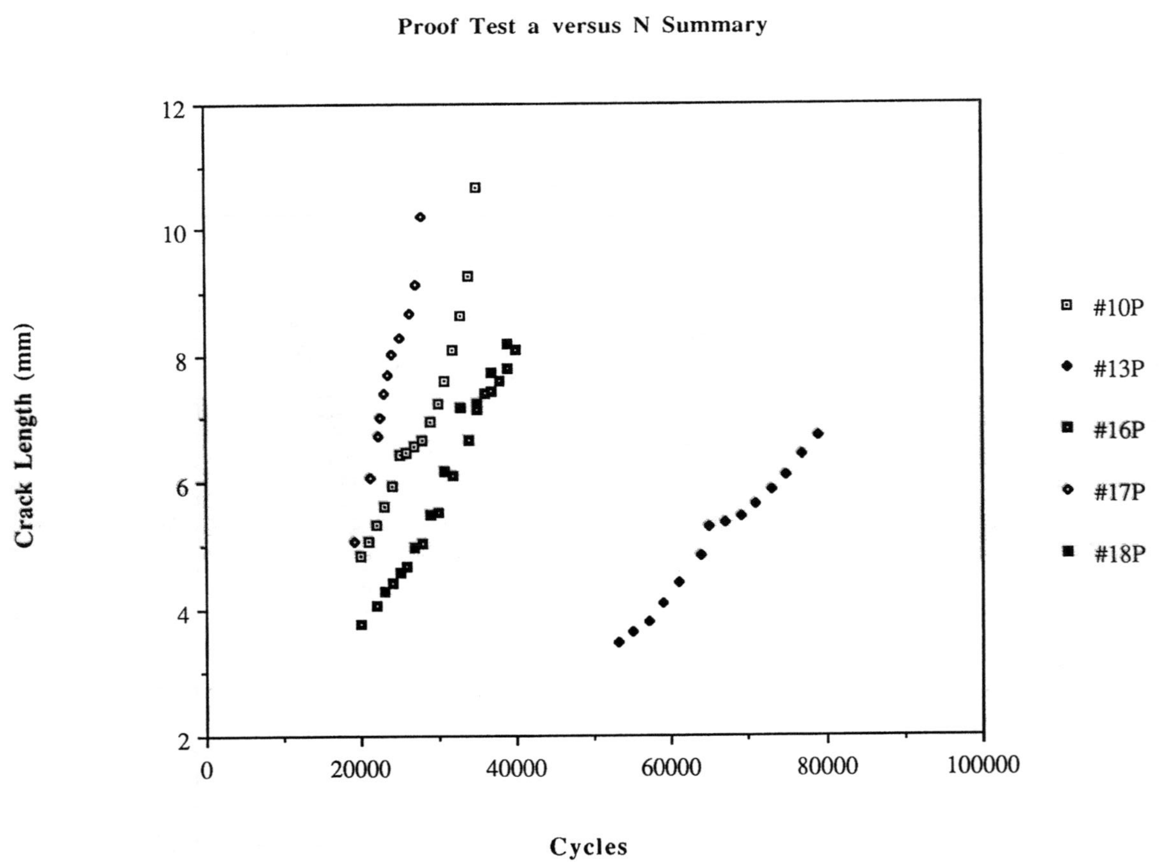

Fig. 2. Crack length (a) vs. N cycles for tests with one overload (proof) cycle.

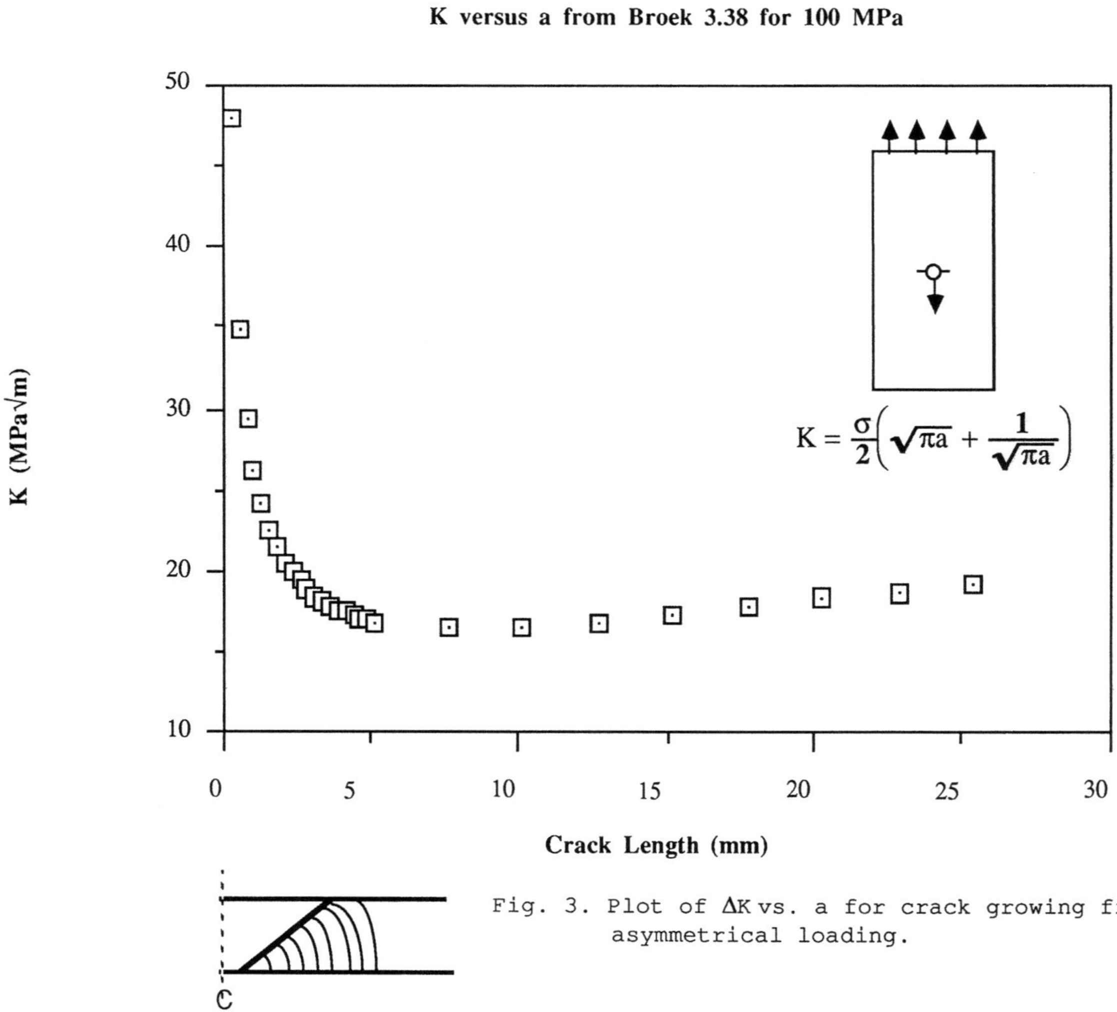

Fig. 3. Plot of ΔK vs. a for crack growing from hole with asymmetrical loading.

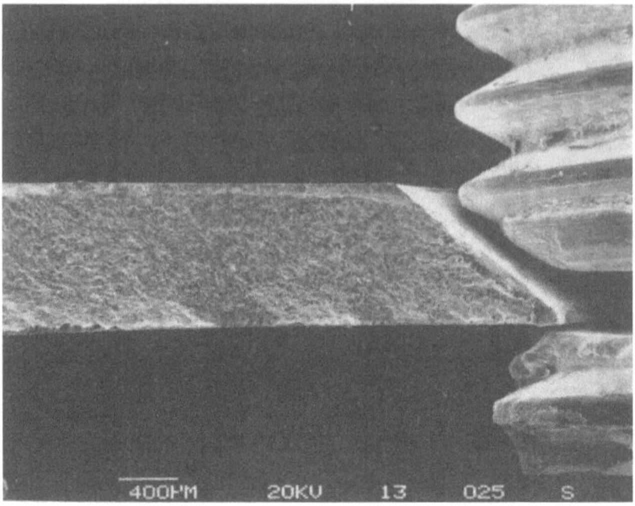

Fig. 4. Overall view of fatigue fracture surface from edge of hole.

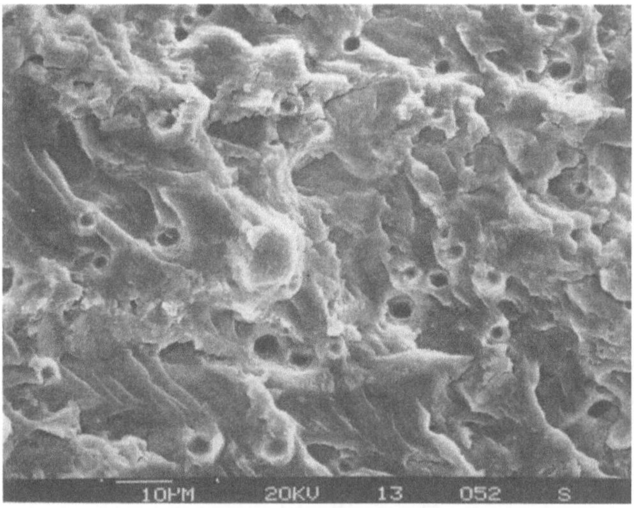

Fig. 5. Fracture features typical of fatigue crack growth at K = 15 MPa $\sqrt{m}$.

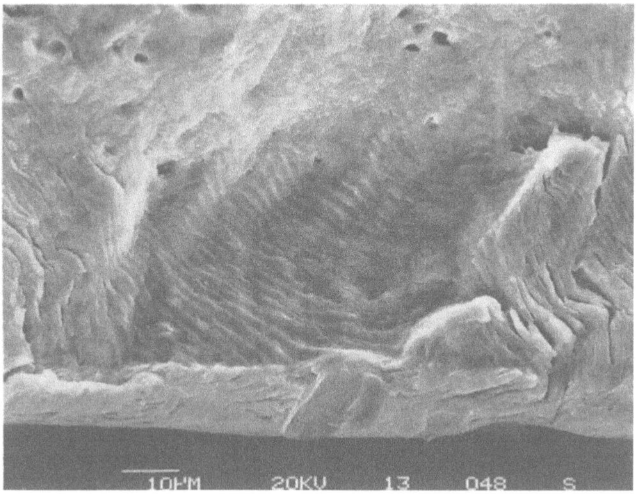

Fig. 6. Ductile fatigue striations in the clad layer.

Fig. 7. Typical fatigue striations with a spacing of 0.2 microns (.8 x 10^{-6} inch).

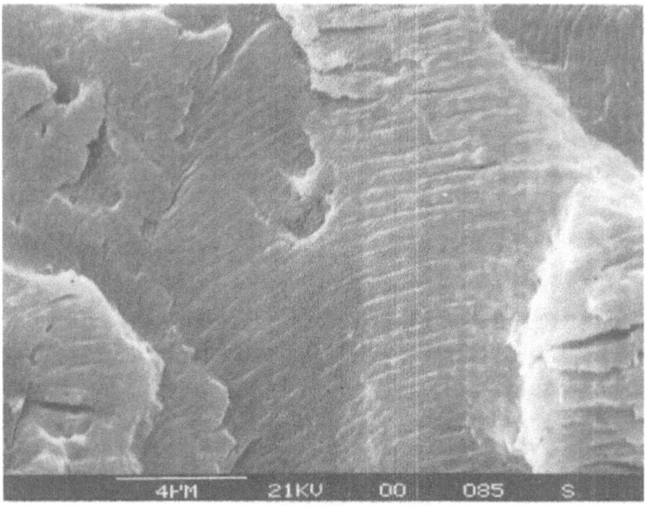

Fig. 8. Fatigue striations and secondary cracks (da/dn = 1 micron).

21. Results of Review of Fokker F 28 "Fellowship" Maintenance Program

R. Roll, A. van Dalen, A.A. Jongebreur

Fokker Aircraft Company
Schiphol-oost, The Netherlands

ABSTRACT

The philosophy as developed by Fokker with regard to the structural maintenance is presented. The maintenance schedules have always included a part with a mandatory character: the Structural Integrity Program (SIP). In this document the stress corrosion aspect is also covered.
As a result of the first FAA aging aircraft conference a Fokker/Piedmont-USAir Task Group was formed for the F28, resulting in a maintenance review activity. The first meeting of the Task Group was in March 1989.
The objectives of the Task Group are presented.
The review is now in the final stage. The last meeting is scheduled in May 1990.
The paper discusses the results of the review as far as these are available and closes with a number of conclusions.

C O N T E N T S

1. HISTORICAL REVIEW OF F28 STRUCTURAL MAINTENANCE

1.1. INTRODUCTION

The aircraft this presentation is focused on is the Fokker F28 "Fellowship". It is a shorthaul jet, originally designed for 65 passengers. The first flight was in May, 1967 and the first

Springer Series in Computational Mechanics
Atluri, Sampath, Tong (Eds.)
Structural Integrity of Aging Airplanes

delivery in February, 1969 (fig.1). There have been some developments in fuselage lengths, wing span and design weights and it can be seen in the same figure that the maximum take-off weight grew from 65,000 to 73,000 lbs. The version shown in figure 2 is the last development,

<u>FOKKER F28 FELLOWSHIP - GENERAL DATA</u>

FIRST FLIGHT: **9 May, 1967**
TYPE CERTIFICATE AND FIRST DELIVERY: **February, 1969**

Developments:

Mark no	Certification date	MTOW (lbs)	MLW (lbs)	MZFW (lbs)	Nbr of pax
Full Scale Test	- - -	56700	54000	46650	- -
1000	February, 1969	66500	59000	54500	65
2000	August, 1972	65000	59000	54500	79
3000	July, 1978	73000	64000	56000	65
4000	December, 1976	73000	69500	62000	85

NOTE: Weights specified represent the last developments of the respective mark no's.

Figure 1

Figure 2

the Mk4000, having a stretched fuselage to accommodate 85 passengers and an increased wing span.
The latest distribution of lives in flying hours and flight cycles is given in the diagram in figure 3. In the meantime, the leading F28 aircraft are approaching 90,000 flight cycles.

F28 DISTRIBUTION OF LIVES (JANUARY 1990)

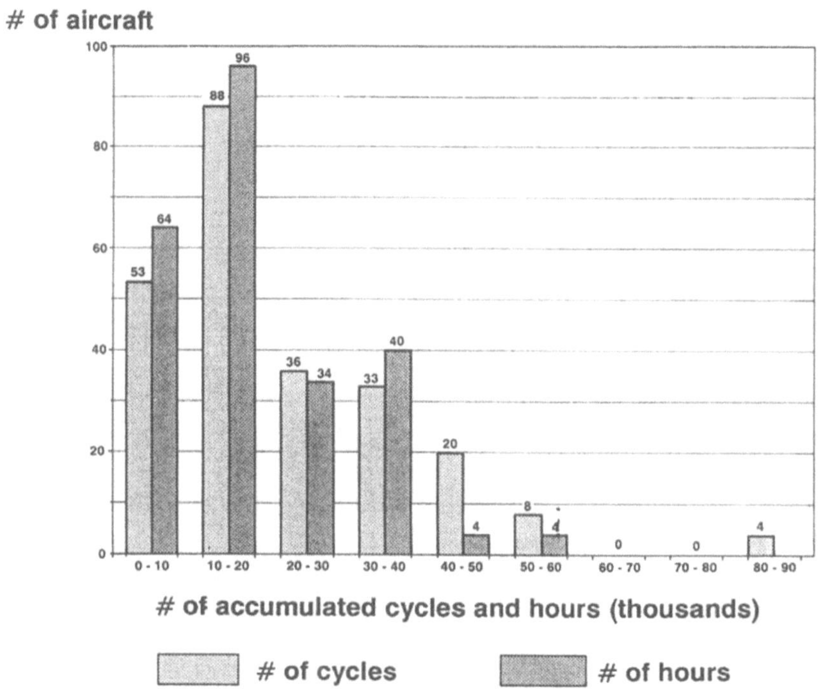

Figure 3

1.2. DESIGN PHILOSOPHY

Some aspects of the Fokker structural design philosophy are relevant to be mentioned when discussing aging aircraft (fig.4).

* Fokker mainly applies laminated, adhesive-bonded structures, including all skin doublers and stringers, which provide optimum durability for fatigue.
* Excellent environmental protection is obtained by application of etching and anodizing, combined with bonding primer, sealing and a Skydrol-resistant paint scheme.
* Full-scale testing of all major components and detail testing as required, combined with analysis, have identified the main problem areas, resulting in modifications or mandatory

FOKKER DESIGN PHILOSOPHY FOR STRUCTURE (HIGHLIGHTS)

. Structural Concept for Optimum Durability (related to fatigue and corrosion)

. Laminated, adhesive bonded structure (skin, doublers, stringers) resulting in excellent fatigue properties.
. Etching and anodizing, combined with bonding primer provide optimum adhesion and environmental protection.
. Skydrol resistent paint scheme and sealing for further perfection of the environmental protection.

. Maintenance Aspects

. Fatigue and Stress Corrosion requirements published in a separate document.
. This document, the Structural Integrity Program, is mandated by the RLD.

Figure 4

inspections. The derivative versions have all been certified analytically, based on original test evidence and analysis of the Mk1000. Modifications were introduced wherever stress levels approached or exceeded the allowable values.

* Fatigue inspections and retirement lives are published in separate, mandatory, well-identified documents, now named the "Structural Integrity Program" (SIP). This principle has been adhered to from the late fifties, when the Fokker F27 was launched, although the format and title of the document were changed twice. Based on new service experience, fatigue load spectrum revisions, continued reevaluation and, recently, Fokker 100 test results, the SIP is regularly revised and amended.

1.3. FATIGUE DOCUMENT

The reasons for a separate, mandatory, fatigue inspection document are summarized in figure 5. The main points are that anticipated defects are involved, which have a direct bearing on safety and for which fairly reliable inspection thresholds are available. It is noteworthy that, rather than flying hours, flight cycles are almost exclusively used as the unit for inspection periods, being the only exact parameter for fatigue.
Stress corrosion, which was experienced after a number of years of operation, was also included in the same document, for similar reasons as fatigue.
It has to be admitted that the title "Structural Integrity Program" may cause confusion in relation to the existing denominations "Structural Inspection Document" and "Supplemental Structural Inspection Program". There are several similarities, but also principal differences. See figure 6.
The SIP does not only deal with aging aircraft. It contains all fatigue inspections, retirement lives and stress-corrosion inspections throughout the operational life of the aircraft. Even if

STRUCTURAL INTEGRITY PROGRAM (S.I.P)

Special Mandatory Inspection Document

. Containing:

- Fatigue inspection thresholds
- Fatigue inspection intervals
- Retirement lives
- Stress corrosion inspections

. Units used:

- Flight cycles for fatigue
- Calendar time for stress corrosion

. Reasons for a separate document:

- In principle safety is directly involved
- As shown by testing, analysis and service experience cracks are likely to occur
- Lives and crack location are fairly well predictable
- In view of the above, this category of inspections should be independent of routine maintenance upgradings
- Optimum probability of timely detection due to specific, directed, hard time inspections.

Figure 5

TYPICAL FEATURES OF THE FOKKER SIP DOCUMENT FOR COMPARISON WITH THE U.S. SID.

. The SIP covers not only aging aircraft, but all aircraft of the type.

. The typical contents of the SIP was issued shortly after the F28 entered operational service.

. The SIP contains all fatigue and stress corrosion inspections as well as retirement lives, resulting from testing, analysis and service experience.

. No items are ever removed from the SIP, since certainty that all aircraft in the fleet have undergone a terminating action can never be obtained.

. SIP items are given clear effectivities as in the sense of "pre- or post mod configuration"

Note: The fore runner of the SIP was a set of mandatory service bulletins

Figure 6

there is a terminating action, the associated inspection item will not be deleted from the SIP, since it is virtually impossible to obtain certainty that all aircraft are modified. All anticipated defects, resulting from testing, analysis and service experience, are covered.

1.4. F28 FATIGUE HISTORY - HIGHLIGHTS

Highlights of the F28 history related to fatigue are summarized in figure 7. These include the policy of carrying out extensive fatigue testing on almost the complete aircraft. In this context, however, the main feature is that a complete fatigue reevaluation was carried out in the early eighties. The reasons were that a new set of fatigue load spectra was issued, modifications and later developments had to be covered and the fatigue inspection program had to be extended beyond 60,000 flights, the life which hightime aircraft were approaching. This reevaluation was based on extrapolation of original test results, service experience and extensive analyses.

<u>F28 FATIGUE HISTORY (HIGHLIGHTS)</u>

. Extensive full scale fatigue testing carried out on all major components

. Extensive test evaluation and fatigue analysis carried out

. Inspections requirements determined up to 60,000 flights

. Inspection requirements laid down in mandatory Service Bulletins, later the Structural Integrity Program

. Re-evaluations carried out in the early 1980's due to:

- New, up-to-date fatigue load spectra were issued
- Several modifications to be covered
- Later developed Marks to be covered
- Inspection data to be extended to 90,000 flights.

Figure 7

1.5. CONCLUSION

It is attributed to the above described aspects, viz. the bonded structure, the very specific material treatment, the way the fatigue problem was tackled, the mandatory fatigue document and the reevaluation, that we have encountered few problems when carrying out the aging aircraft program. As will be shown presently, only one new potential problem area was found.

2. INTERNATIONAL MAINTENANCE TASK FORCE ACTIVITY

During the Federal Aviation Administration (FAA) Conference in June, 1988 on aging aircraft, the Air Transport Association (ATA) and the Aerospace Industries Association (AIA)

recommended to review maintenance practices which apply to hightime aircraft. This resulted in the so-called ATA/AIA Airworthiness Assurance Task Force. A steering committee directs and coordinates of 3 working groups, each covering one or more airplane manufacturers or airplane types (Boeing ; Douglas and Lockheed/Airbus/Convair/Fokker/BAe).
The working groups were to address a number of short-term goals selected to reflect the current condition of the fleet, and agree on recommendations, e.g. for mandatory actions, to ensure the airworthiness of hightime aircraft.
These objectives include:

* Review of Service Bulletins to determine the need for repair, modification or replacement action, rather than continued inspection.
* Corrosion prevention/control. Is there a common approach to tackle the problem?
* Structural fatigue: effectiveness of inspection programs
* Maintenance communications
* Review of repairs
* Review of inspections

Long-term objectives are:

* Human factors
* A research program on aircraft design, maintenance and inspection with special reference to hightime aircraft
* Training and qualification standards for aircraft mechanics and inspectors
* Improved aircraft maintenance communication systems

The next chapter describes the organization of the F28 working group, the way the above short-term objectives were pursued and the results of these activities.

3. THE PIEDMONT/USAir/FOKKER-F28 TASK GROUP ACTIVITY

At the end of 1988 the Piedmont/Fokker Task Group was formed to evaluate the Fokker F28 Selection of the group was based on the following criteria:

* The participants should have a thorough engineering knowledge of the F28
* The operators in the group should have an engineering staff which could make a valuable contribution to the evaluation
* The operators in the group should operate a representative part of the F28 aging fleet
* The operators should be representative of worldwide operation of the F28

The following operators were willing to participate in the task group:

* Alinord (Italy)
* Ansett Airlines (Australia)
* East West Airlines (Australia)
* Garuda (Indonesia)
* Linjeflyg (Sweden)
* Piedmont-USAir the (United States)
* Tourraine Air Transport (France)

Furthermore the Airworthiness Authorities of the following countries joined the task group as observers:

* Australia (CAA)
* The Netherlands (RLD)
* Italy (RAI)
* Indonesia (DCA)
* United States (FAA)

Of course Fokker, as the manufacturer also participated.
This group represents 72 % of the total F28 fleet with more than 30,000 flight cycles and all F28 aircraft with more than 50,000 flight cycles.
Although there is no African operator in the working group, all climates are represented. By this composition of the working group the aforementioned selection criteria are well met. The first meeting of the task group was in March 1989. During this meeting the task group started the evaluation of the F28 by establishing the working procedures.
The intention of the working procedures was to review the aircraft zone by zone, thereby considering the following items and assess possible interactions:

* Airframe-related Service Bulletins
* Bond quality of hightime aircraft
* Corrosion-prone areas and corrosion inspections
* Structural Integrity Program (SIP)
* Inspection methods
* Structural repairs

The next three meetings were used to review the complete airframe with respect to the above mentioned items.

In the fifth and last meeting of the short-term task group activity in May, 1990 also general subjects as life extension and service experience will be discussed.
The main item of the last meeting will be to present and approve the final document and determine long-term actions.

4. R E S U L T S

4.1. AIRFRAME RELATED MODIFICATIONS

The review of all airframe related modifications gave the following results:

* 7 existing Service Bulletins will be recommended for mandatory incorporation. These are Service Bulletins concerning:
  - vane rails
  - elevator torque tube and driving lever
  - flap mechanical actuators
  - wing lower skin reinforcing strap
* 8 new Service Bulletins will be issued. These are Service Bulletins concerning:
  - nose wheel bay
  - rear pressure bulkhead
  - front pressure bulkhead
  - window post
  - lap joints
  - wing to fuselage connection angle
  - door profiles in emergency exits and service emergency door
  - cargo door hinges

Incorporation of the above mentioned Service Bulletins will terminate SIP inspections, thereby ensuring structural integrity without relying on inspections only.

* 18 Service Bulletins dealing with improved corrosion protection and draining will be incorporated in the Corrosion Control Program.

The number of new Service Bulletins and existing ones to become mandatory is very limited. This shows that Fokker's approach concerning modifications and Service Bulletins has been successful, requiring only minor adjustments.

4.2. BOND QUALITY

The review of the bond quality in the fuselage revealed some problem areas. As a direct action those cold-bonded longitudinal lap joints, that might become fatigue-sensitive in case of debonding, were inspected on four hightime aircraft.

In addition, ten other aircraft were partially inspected. The result was that debonds did occur locally. These debonds, if large enough, could affect fatigue lives of the fuselage lap joints. Although no cracks were found, not even in two aircraft with more than 85,000 flight cycles, actions have been taken and modification Service Bulletins issued. For the time being the lap joints of the most critical configuration (ten early aircraft) are covered by mandatory inspection and actions will follow soon for the other aircraft.

4.3. CORROSION REVIEW

The review of corrosion in all zones of the F28 showed that some areas are prone to corrosion. In this case "prone" means that, if corrosion is found in an area whereafter the corrosion is removed and the area treated, the corrosion reappears. This means that the corrosion repair and protection of the area need improvement and even the inspection needs to be changed. A program is being compiled, containing improved repair and protection of corrosion-prone areas and some changes to inspection intervals and methods.
This program is called the Corrosion Control Program and the working group recommends to make the program mandatory.

4.4. STRUCTURAL INTEGRITY PROGRAM

During the review of the Structural Integrity Program 110 items were reviewed. It was recommended to terminate 12 inspection items by mandating existing Service Bulletins. These were: the rear pressure bulkhead, the wing-to-fuselage connection angle, the door frames and the nosewheel bay top cover, the vane rails, the elevator drive, a window post, the lap joint at str 73, the cargo door hinges, the door profiles, the wing-to-fuselage connection angle and the front pressure bulkhead.

With the exception of the rear pressure bulkhead all of these were already in the process of being issued. The rear pressure bulkhead modification had already been studied, but was turned down in discussions with operators.
As far as the Structural Integrity Program is concerned the conclusion is that things were well under control and only minor adjustments were required.

4.5. INSPECTION METHODS

The inspection methods were reviewed together with the applicable inspection items and although no major problems arose it was decided to arrange a separate meeting between NDT specialists of both the operators and Fokker.

4.6. REPAIR REVIEW

The repair review has looked at the quality of individual repairs and areas of multiple repairs.
For the individual repairs as described in the Structural Repair Manual, in stress reports of major repairs or other non-SRM but Fokker approved repairs, Fokker will make an analysis with respect to damage tolerance and fatigue behavior.
Additional requirements such as inspections or retirement lives on repairs will be established and published.
For non-Fokker approved repairs, no action has been determined yet.
How these new repair requirements will be implemented on the fleet has not been determined yet.
Guidance is awaited from the steering committee.
A complete review of each individual aircraft to document all repairs and take corrective actions for all non-approved repairs and fulfil all additional requirements, seems inevitable.

4.7. OPEN DISCUSSION ITEMS

A number of items are considered to deserve further discussion. These are:

* Should the design of repairs based on the use of the SRM be restricted to licensed personnel (engineering department)?
* All repairs should be documented properly (also retrospectively). Is it acceptable to make a distinction between "major" and "minor" and only document the repairs classified "major"?
* No postponement of mandatory inspections for planning or scheduling reasons
* No postponement of major maintenance for planning reasons
* The use of all necessary manuals must be ensured. No reliance on memory!
* Adherence to aging aircraft actions must be documented, especially for secondhand aircraft
* Maintenance programs, e.g. block type versus equalized
* Life extension
* Should flyable crack lengths be determined on the basis of limit load or ultimate load, even for very short periods? E.g. for the pressure cabin the ultimate load requirement for flyable crack lengths seems unduly severe.

5. SUMMARY

As far as the F28 in concerned it was Fokker's opinion that the original design with extensive application of bonded structure, excellent environmental protection and full-scale testing of all major components, together with a separate mandatory fatigue inspection document have produced an aircraft with a structural integrity that allow it to be operated for many years to come.
The working group activities over the last year have only strengthened this opinion, because no new major problem areas have been identified.

Admittedly, a few areas were found where stronger actions were judged necessary but these are rather improvements than drastic changes.
Seven existing Service Bulletins and eight new ones may have to be incorporated by the operators.
The eight new ones have already been written or have previously been turned down by operators, so Fokker has reacted adequately.

It also became evident that although the F28 is not a corrosion-prone aircraft, more guidance was needed from Fokker to assist operators in successfully controlling corrosion on their aircraft.
To this aim a corrosion control program is being prepared.
In that program attention if focused on treatment of corrosion sensitive critical areas.

With these actions finished and future actions with regard to repairs scheduled, Fokker as the manufacturer has fulfilled its role and responsibility to produce and guide operators in maintaining an aircraft which can be operated for a long time to come.

Furthermore the task group strongly feels that general practise of aircraft maintenance requires continuous attention and review. Firm recommendations will have to be made on open discussion items such as the use of manuals, record keeping, maintenance practices and adherence to task group recommendations.

After completion and publication of the task group recommendations, firm support of all operators and all authorities all over the world will be required to ensure and enforce compliance with the task group recommendations.
This is considered the most important recommendation, resulting from the F28 task group activities. It is felt to be an indispensable condition to improve structural integrity as a contribution to general air safety all over the world.

22. Test Facility for Evaluation of Structural Integrity of Stiffened & Jointed Aircraft Curved Panels

G. SAMAVEDAM, D. HOADLEY
Foster-Miller, Inc.

J. DAVIN
Transportation Systems Center, DOT

Summary

A new test facility for evaluating the fatigue and fracture strength of stiffened and jointed aircraft fuselage panels has been built at Foster-Miller. The test fixture can accommodate 40-deg curved sector panels of 120 in. in length and 58 in. in width. Internal pressure loading up to 20 psi and corresponding longitudinal and hoop restraining loads can be applied. Cyclic pressurization loads can be used for fatigue studies. The stress distribution in the full-size round fuselage structure can be accurately simulated in the central region of the test panel. Internal pressurization fluid can be air or water. The test facility is initially intended for (i) identifying causative factors contributing to multiple site damage in curved structures, (ii) generating fatigue data for comparing with the data from flat coupons, (iii) evaluating the residual strength of old aircraft panels, and (iv) validating the finite element based theoretical model currently being developed by Foster-Miller.

Introduction

Damage tolerant design is an important consideration in avoiding catastrophic consequences of fatigue cracks which can cause fracture instability during the service life of aircraft structures. The U.S. Air Force prepared a handbook on the subject in 1984 to support their MIL-A-83444 requirements. A similar document dealing with commercial aircraft requirements, apparently, is not available. With the exception of isolated works, (Swift, [1]), the authors have not found *detailed studies* of damage tolerant design published in the open literature.

Springer Series in Computational Mechanics
Atluri, Sampath, Tong (Eds.)
Structural Integrity of Aging Airplanes

Damage tolerant design of pressurized aircraft fuselage structures has been based on a single lead crack in the skin. Under ideal conditions, longitudinal fracture cracks (which are more severe than circumferential cracks due to larger hoop stress) will either be arrested at a frame or strap, or turn in the circumferential direction. The recent Aloha Airlines incident shows that other than ideal conditions can exist which can defeat fail-safe features in a lap-spliced fuselage design and allow a longitudinal fracture to propagate over several frame spacings. Aging, for example, is known to make fuselage skin panels susceptible to Multiple Site Damage (MSD), a phenomenon characterized by small cracks emanating from successive rivet holes along a longitudinal row in a lap-splice. MSD is especially troubling because it may not be readily detectable. Under certain conditions these cracks can link up to form one long fracture crack. The problem of MSD may be exascerbated by cracks in structural elements together with varying degrees of lap-splice disbond and corrosion. This combination can lead to cracks of critical length. With longer service life expected of aircraft, damage tolerance assessment may require consideration of new factors in some combination with others known to contribute to a common fracture path.

In view of the number of structural variables involved and their potential for interaction, testing will play a significant role. A dedicated test facility using full-size panels has been designed and built by Foster-Miller under a contract from the Transportation Systems Center of the U.S. DOT. The design of this facility and its performance are described in this paper.

Test Facility Design

The facility has been designed to test full-size curved panel sectors under internal pressure and biaxial stress conditions. The use of sectors of full-scale curved panels was preferred over other candidate concepts which were considered. Alternate concepts include: subscale models, flat panels, and full-size round fuselage structures.

Scale models, although attractive in view of the dimensional analysis and scaling laws that can be derived, have practical difficulties. The fracture toughness, K_C, an important parameter in the nondimensional parameters, depends on skin thickness. Scaling down the joints, skin thickness, rivet size, etc., is not practical.

The flat panel approach, although permitting easy simulation of biaxial stresses, has at least two problems. First, the lack of curvature would change the stress singularity levels at the crack tip and, hence, the stress intensity factor. This would introduce significant errors in crack propagation rates and fracture strength for reasonably large crack lengths. Although an empirical correction as developed by Swift [1] is available, its validity may be limited to the particular panel designed and tested by Swift. Second, the effect of nonlinear bulging at the crack tip due to internal fuselage shell pressure is important and difficult to simulate in the flat plate.

Full-size round fuselage structures have been used for testing by Boeing [4]. Half rounds with semicircular end diaphragms have also been used [5]. Using air as the pressurization medium in full-size structures raises significant safety issues. Use of water as the medium poses gravity head problems unless the structure is totally immersed. Finally, the number of tests that are required to generate the necessary data would be both expensive and time-consuming due to the size of the full-scale facility.

Curved aircraft panels have been tested using vacuum [1] rather than internal pressure as proposed here. The vacuum method limits the maximum pressure differential achievable in the test fixture. Also, synchronization between pressure and axial loading is difficult to achieve. As with systems powered exclusively with air, the vacuum method requires large, expensive power units to achieve even modest pressurization cycle rates.

Test Fixture Description

The general approach for panel testing is to anchor the two long sides and one end of the horizontal panel to the fixture and to load the panel by pulling on one curved end and pressurizing the panel on its concave side with water. The pressure load is contained by the fixture, an inflatable seal between the fixture and the panel around the panel periphery, and the panel itself.

The panel loading schematic is shown in Figure 1. The pressure is generated by pulling a hydraulic cylinder with an Instron® Universal Test Machine. Hydraulic oil under pressure is routed to four commonly plumbed hydraulic

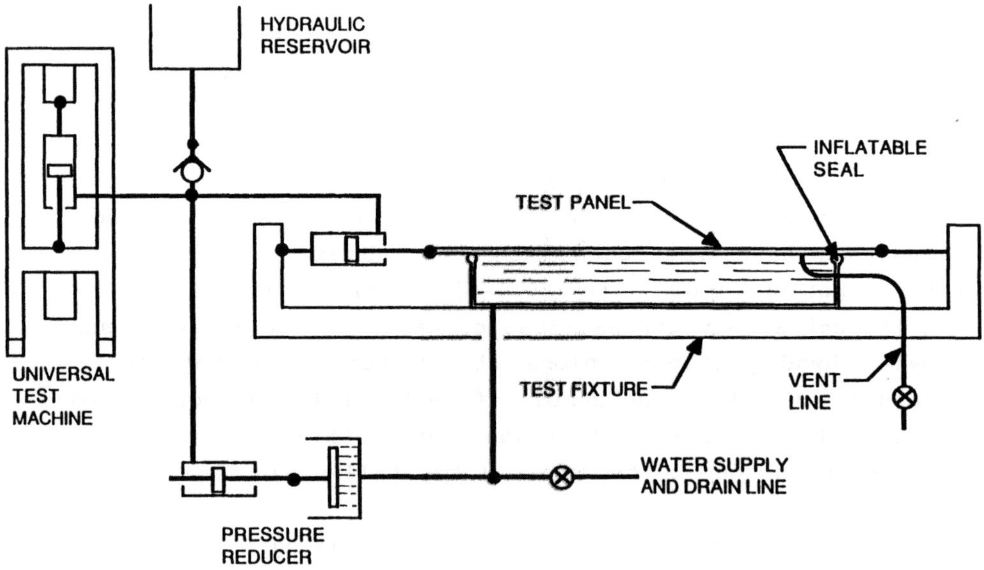

Fig. 1. Panel loading schematic

cylinders which pull the panel. The pressure is also routed from that cylinder to a special pressure reducer. Within this unit, the pressure line is connected to a small diameter hydraulic cylinder. This cylinder connects physically to a large diameter cylinder containing water. The pressure reduction from hydraulic oil to water is about 80:1. The low pressure side of the pressure reducer connects directly to the pressure vessel portion of the test fixture. Controlled loading of the universal test machine provides coordinated hoop and longitudinal loading of the test panel.

The test fixture structure is assembled from four major component parts: the base structure, inner seal frame, outer seal frame and upper structure. Figure 2 shows the general orientation of these items. The inner seal frame bolts to the base structure to form the pressure container. The outer seal frame bolts to the upper structure to form the upper frame assembly. This assembly bolts to the base structure; it is removed during panel installation. Inflatable seals are mounted to the outer and inner seal frames to sandwich the panel during pressure testing. The inner seal is the pressure seal. The outer seal balances the seal load on the panel. The seal arrangement is shown in Figure 3. An overview of the test fixture, the pressure reducer and the Instron machine is shown in Figure 4.

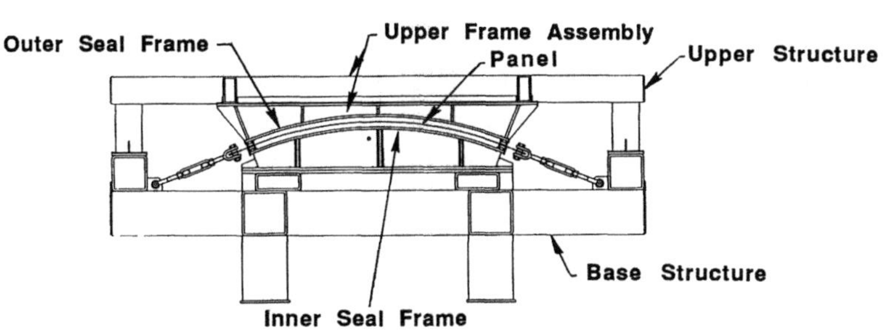

Fig. 2. Panel installation in test fixture

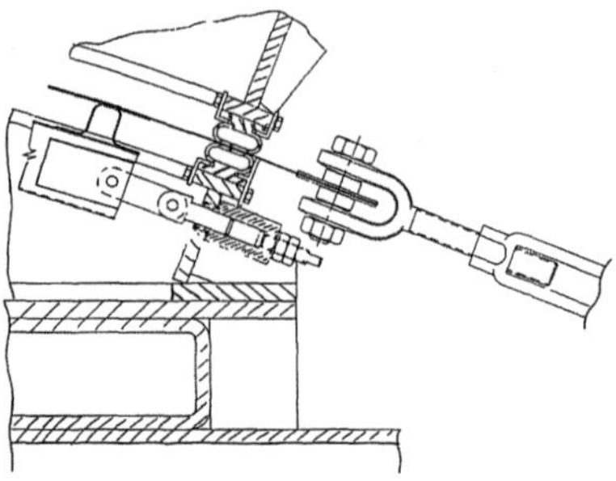

Fig. 3. Panel seal restraint

Fig. 4. Fixture and panel assembly

Panel Description

A number of curved panels are being fabricated for testing on the fixture described earlier. A typical test panel is shown in Figure 5. Major panel features are presented in Table 1.

Fig. 5. Typical test panel

Table 1. Typical panel features

Panel radius	75 in.
Panel length	120 in.
Panel width (developed)	68 in.
Number of frames	6
Number of tear straps	6 or 11
Number of stringers	7 or 6
Frame spacing	20 in.
Tear strap spacing	20 or 10 in.
Stringer spacing	8 or 9.6 in.

The initial test panels were fabricated from 0.040-in. thick clad sheet aluminum throughout. The panels include one central longitudinal splice positioned centrally over the middle stringer. The splice is joined by three longitudinal rivet rows. The center rivet row also attaches the stringer to the skin. One-inch wide tear straps are positioned in the hoop direction on 20-in. centers at each frame. Longitudinal filler strips, also 1 in. wide, are positioned between the tear straps at the seven stringer locations to provide a waffle-type configuration. Faying surfaces between the skins at the lap joint and between the skins, tear straps, and filler strips are adhesively bonded.

The stringers are of fabricated hat-section form and the frames are of fabricated Z-section form. Stringers are rivet-attached to the skin through the hat crown. The base of the stringer is rivet-attached to the frame flange by two universal-head rivets at each intersection. The frames are directly attached to the skin in each stringer bay by a fabricated angle shear-clip. All through-skin rivets are 5/32-in. diam, low profile, shear head, 100 deg countersunk rivets. An attempt has been made to fabricate panel configurations that have generic similarities to some of the older commercial airframes currently operating in the United States.

The four edges of the test panel are specially prepared for connection to the test fixture. Twenty-eight fingers are located along each of the two longitudinal panel sides and 16 fingers are located along each of the two curved ends. Each finger is reinforced by two aluminum reinforcing sheets bonded to each side of the skin. Each finger assembly includes a centrally positioned anchor hole. The panel periphery configuration has been selected to minimize attachment effects to the test fixture. Individual fingers provide a means of in-plane connection of the skin to the fixture. The longitudinal fingers transmit panel hoop loads to the test rig. These fingers have been designed to provide a minimum of longitudinal stiffness to the panel. Similarly, the fingers on the two curved ends transmit only longitudinal loads to the rig. Their contribution to panel hoop stiffness is minimal. The panel includes a short section of plane skin immediately inboard of all load fingers. This section is clear of frames, tear straps, fillers and stringers and is provided as a clear bearing surface for the inflatable peripheral seal.

The general panel configuration is such that a wide range of configurational changes can be accommodated within the bounds of the external size and the radius. Panel parameters that can be changed are shown in Table 2. It is currently planned that a number of these options will be included in some test panels.

Table 2. Variable panel parameter

Frame spacing
Stringer spacing
Tear strap spacing
Longitudinal lap joint location
Skin thickness
Stringer geometry
Frame geometry
Tear strap geometry
Shear clip geometry
Rivet spacing
Rivet size
Rivet style
Lap joint disbonds
Tear strap disbonds

Fixture Performance

A series of shakedown tests were conducted to verify the performance of the test fixture. Tests included:

- Component Pressure Testing. The hydraulic system and the pressure reducer were tested up to 20 psig.

- Axial and Pressure Load Coordination. Hydraulic and water (panel) pressures were measured and hydraulic pressures were converted to panel axial loads. Test points were compared to required loads to verify correct proportionality as shown in Figure 6. Frequency sensitivity was tested at frequencies up to 0.25 Hz.

A typical plot at 0.1 Hz (Figure 7) shows good synchronization during the pressure rise and hold phase of the cycle with some phase shift during pressure decrease. Test data show that a maximum cycle rate of 0.2 Hz (720 cycles/hr) can be achieved without significant phase shift between the pressure and axial loads on the panel.

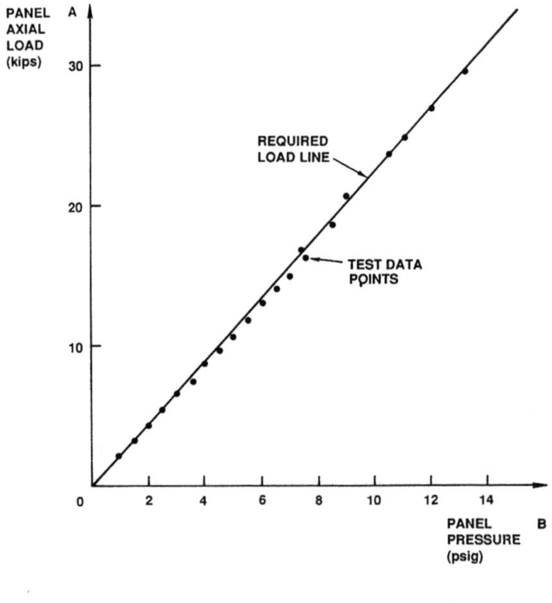

Fig. 6. Axial panel load versus pressure

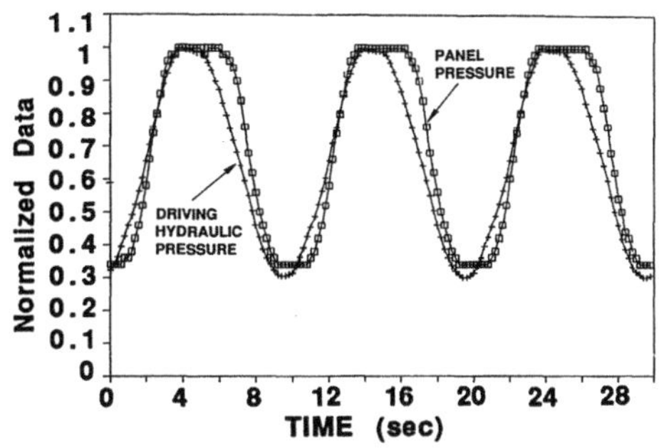

Fig. 7. Panel and hydraulic pressures at frequency = 0.1 Hz

Panel Test Methodology

Static, fatigue and fracture strength tests have been carried out to establish the panel test methodology.

Static Tests

Strain gauges were applied to the panel to determine the hoop and longitudinal skin stresses, the bending stresses in the splice joint region, and the loads in the frames and straps. Strain and pressure data have been recorded using the ACRO 900 system, with sampling rate of 5/sec.

Figures 8 and 9 show typical results for top (outer surface) skin stress (bending and membrane) in the circumferential and axial directions at different pressure levels. These figures show results for the case when a 16-in. crack was saw cut in the skin across a frame and tearstrap. The test data have been correlated with theoretical predictions obtained through a nonlinear finite element model and the NISA 360 code. The nonlinearity is associated with large radial displacements.

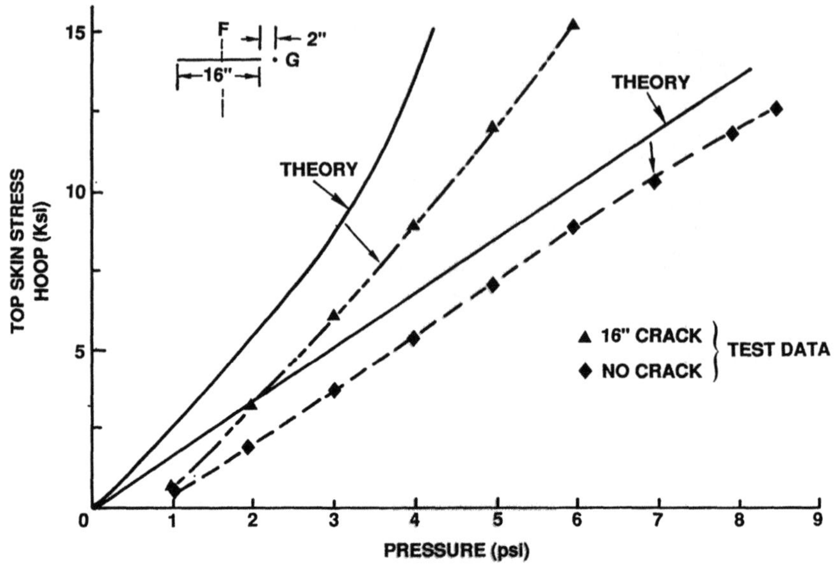

Fig. 8. Top skin stress in hoop direction

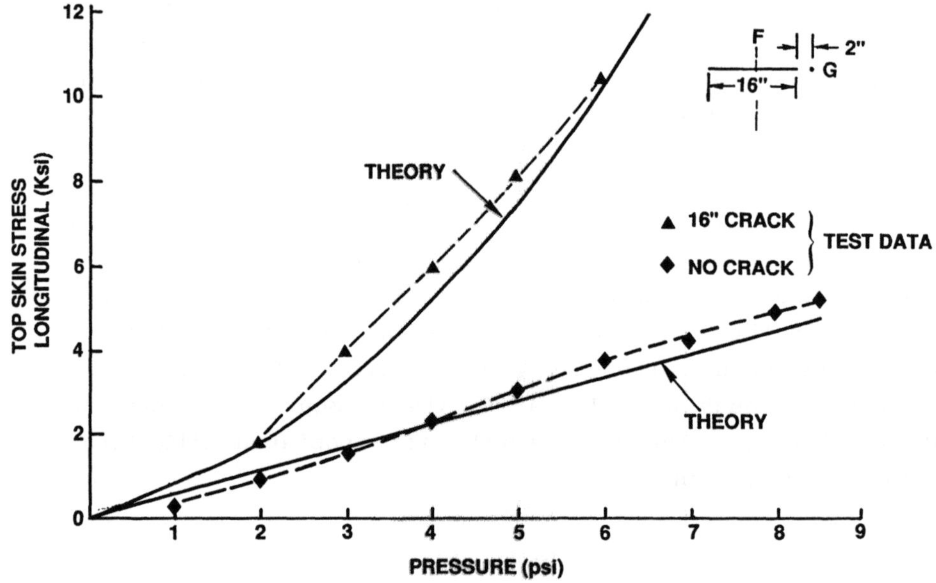

Fig. 9. Longitudinal top skin stress

Fatigue Tests

Fatigue tests on cracked panels were conducted at a rate of 720 cycles/hr. Crack growth was monitored using a traveling microscope. Typical results are shown in Figure 10 for a crack over a frame and a crack between two adjacent frames. The stress intensity factor when the crack tip is near a frame is much higher than for a mid-bay crack. Hence, the significant difference is the crack growth rates for the two cases.

Fracture Tests

Fracture tests were performed to evaluate the residual strength of cracked panels. Typical results for a 16-in. initial crack in the skin are shown in Figures 11 and 12. The top skin hoop stress nearest to the crack tip and the radial displacement at the tip were monitored.

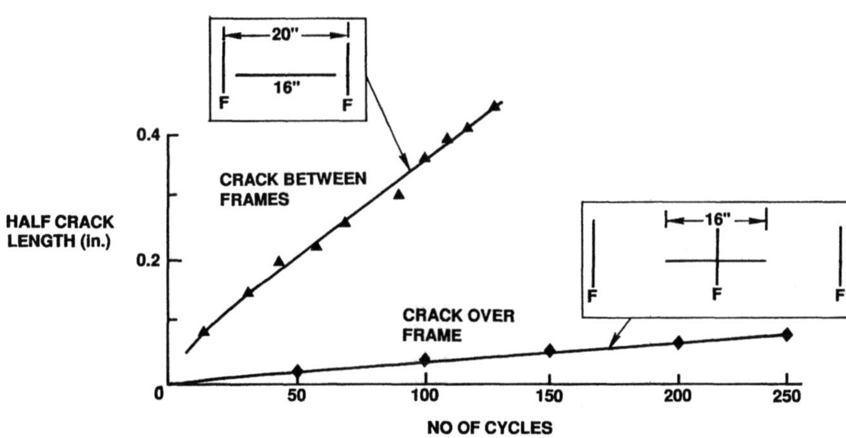

Fig. 10. Fatigue test - crack growth rate

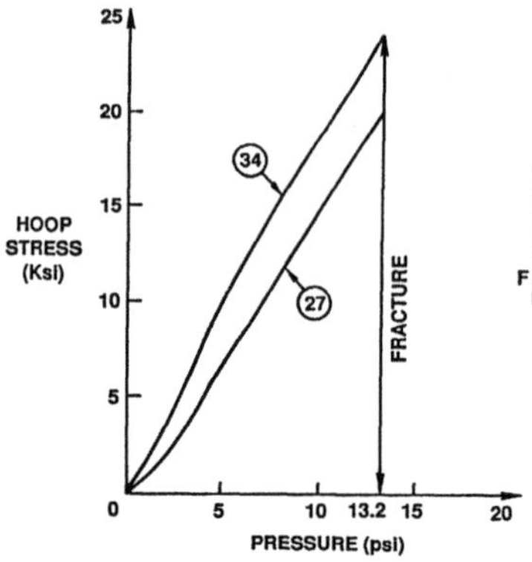

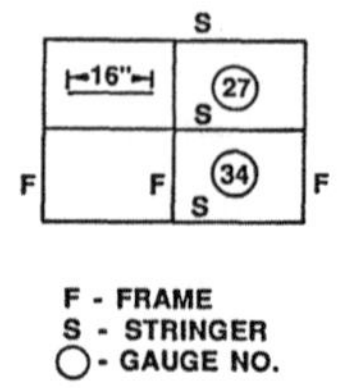

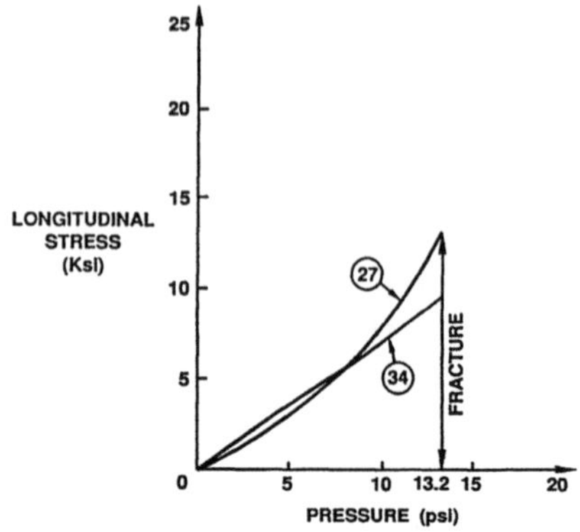

Fig. 11. Fracture test - top skin hoop stress

Fig. 12. Fracture test - top skin longitudinal stress

The crack experienced stable growth until it reached the rivet hole at the frame/strap location. At a pressure of 13.2 psig, the crack negotiated the rivet and grew suddenly by about 3.5 in. into the adjacent bay as shown by Figure 13. The fracture pressure of 13.2 psig compares well with the theoretical value predicted for the panel.

Bulging at the crack center is shown in Figure 14. At the time of fracture, the bulging deflection was about 1 in., quite close to the theoretical value. The observed nonlinearity between the pressure and the bulging deflection illustrates the importance of large deflection theory in the analysis.

Concluding Remarks

A curved panel pressure test facility has been designed, built and commissioned at Foster-Miller. This facility can test typical new or old stiffened fuselage panels safely and economically. Work capable of being performed with the use of the test facility includes: fracture and fatigue testing of multiple site damage; damage tolerance evaluation; repair

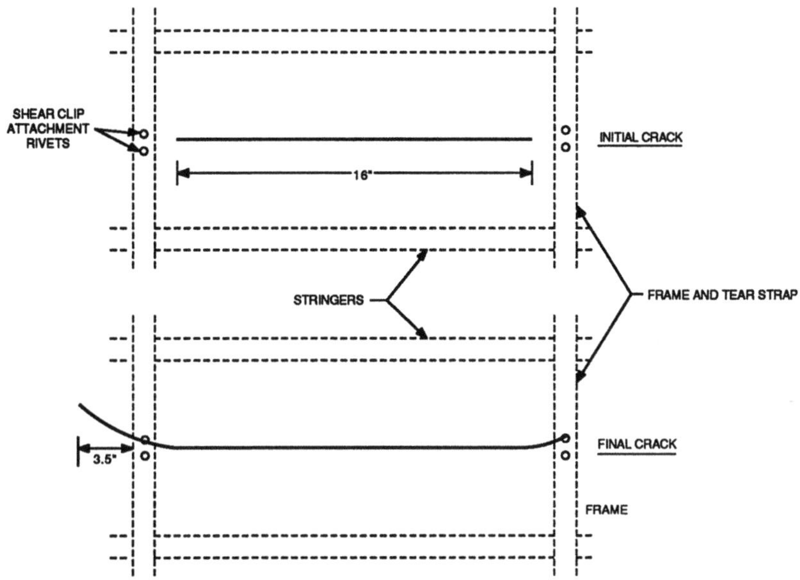

Fig. 13. Panel fracture

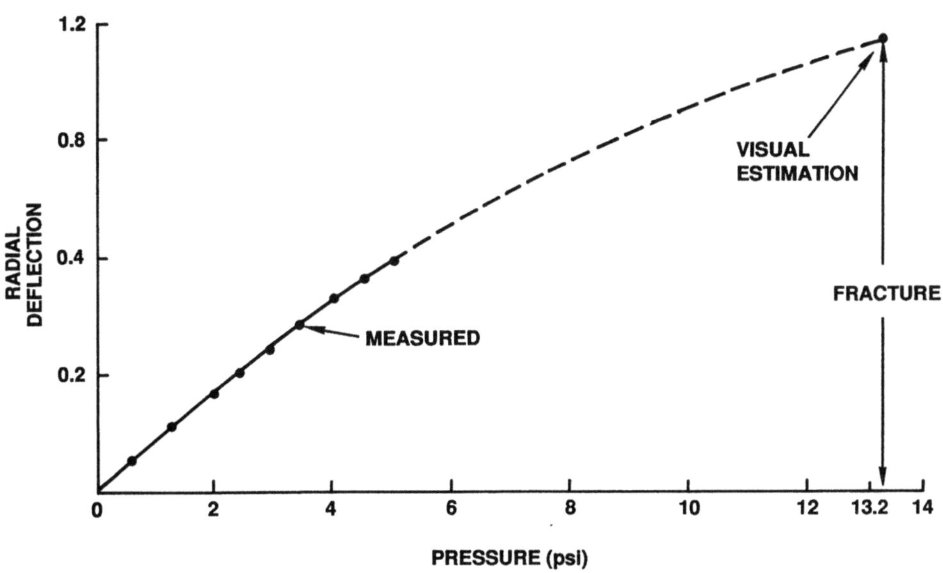

Fig. 14. Fracture test - bulging at crack center

patch performance evaluation; inspection technique development; and proof pressure testing.

This panel test facility and the engineering capability to design and fabricate panels is available to support future aging aircraft activities. The laboratory is equipped with an ACRO 900 system for data acquisition and NISA 386 software for computer simulation. Elasto plastic and large deflection analyses of stiffened cracked fuselage panels can be carried out for test data correlation.

Acknowledgements
The work on which this paper is based was sponsored by the Federal Aviation Authority as a part of its research program to investigate the effect of aging on structural integrity of aircraft in service. The authors wish to thank Dr. Sam Sampath and Dr. Oscar Orringer of TSC for their technical advice.

References

1. Swift, T., "The Application of Fracture Mechanics in the Development of the DC-10 Fuselage," Fracture Mechanics of Aircraft Structures, AGARD-AG-176.

2. Swift, T., "Important Considerations in Commercial Aircraft Damage Tolerance," Int. Journal of Vehicle Design, Vol. 7, Nos. 3/4, 1986.

3. Sampath, S., Broek, D. and Patil, S., "Estimation of Requirements for Inspection Intervals Associated with Multiple Site Damage Panels," International Symposium of Structural Integrity of Aging Airplanes," Atlanta, GA, March 1990.

4. Goranson, U. and Miller, M., "Boeing Aging Aircraft Programs," Int. Symposium of Structural Integrity of Aging Airplanes," Atlanta, GA, March 1990.

5. Ansell, H., "Bulging of Cracked Pressurized Aircraft Structures," Ph.D. Thesis, Dept. of Mech. Eng., Institute of Technology, Linköping University, Sweden, 1988.

23. Estimation of Requirements of Inspection Intervals for Panels Susceptible to Multiple Site Damage

S. Sampath
Transportation System Center

D. Broek
FractuREsearch Inc.

Summary

The objectives of the work reported in this paper were to assess the probability of detection of Multi-Site Damage (MSD) in fuselage lap-splices in aging airplanes. The results indicate that a reduction in the mandated period between inspections should be considered. Perforce, the model had to rely on NDI inspection reliability measured on samples devoid of MSD. A recommendation is made for a program to acquire such data for MSD.

Introduction

This paper derives from a larger study being conducted by the Transportation Systems Center on behalf of the Federal Aviation Administration. On account of concern about MSD in aging airplanes, alternative strategies for inspection of fuselages susceptible to MSD are concurrently pursued. MSD is caused by fatigue and is manifested in fuselage lap-joints in older commercial airplanes. The phenomenon is characterized by small, longitudinal cracks emanating from successive rivet holes. Their probabilities of detection associated with conventional Non-Destructive Inspection (NDI) methods is relatively low; yet, they can potentially link-up in domino fashion to cause an unstable longitudinal fracture which may override the built in fail-safe design.

The viability of pressure proof testing as a means for destructive inspection of fuselages to indentify MSD, was examined in a previous analytical study [1]. Although proof testing can be conducted reliably, the interval between proof tests would have to be too small (on the order of 500 flights) and the proof pressure too high, to be economically feasible.

Attention was then turned towards non-destructive inspection, assuming that both retirement and immediate action to terminate the risk of MSD are impractical options. A

Springer Series in Computational Mechanics
Atluri, Sampath, Tong (Eds.)
Structural Integrity of Aging Airplanes

cursory analysis indicated that the interval for eddy current inspection would also have to be rather short to provide the same reliability as proof testing. However, this result was based on pessimistic values for the detection probabilities for eddy current and visual inspection, derived through inference from experimental data [2], because the human factor of increased vigilance upon finding a crack was not accounted for. Neither was MSD modelled as cracks of assorted sizes. The present work has attempted to improve on the shortcomings of the previous effort.

MSD cracks will have unequal lengths because of the variation in hoop stress across each bay (between successive frames) and/or tear straps. Since the smaller cracks will have a lower probability of detection, the distribution of MSD crack sizes must be accounted for as well. It was the latter criterion that made the present analysis a rather complicated one.

The significance of the human factor is due to the following. If the inspector has not detected any cracks for a while, the probability of detection is lowered, because boredom will set in; on the other hand if a crack is detected, the inspector is alerted, so that there is a better chance that adjacent cracks will be detected (higher probability). An account for this effect was made in the present work by developing a procedure to assess cumulative probability of detection. It should be noted that the present analysis is still conservative, because the assumption of completely disbonded lap-joints is unavoidable.

A probabilistic model has been developed in which the initiation and growth of MSD cracks is dependent upon the stress level and the scatter in fatigue crack initiation in crack growth data. The stress distribution over a bay, as affected by frames, tear straps and stringers, was obtained from finite element analysis. The statistical distribution functions representing scatter in crack initiation and propagation were obtained from test data on lap-joint specimens. An array of holes has been considered, and initiation properties and crack growth properties randomly assigned in accordance with the distribution functions. Initiation and growth were assessed on the basis of the local stress level, where the effects of membrane stress, local bending, as well as fastener load transfer are included. A computer program was developed to perform the required calculations.

Fatigue damage and crack growth were incremented every 100 cycles, and simulated inspections performed at regular intervals. Probability of detection curves were again based upon the earlier cited inspection data; however, the probability of detection was adjusted after each 'detect' or 'miss'. In this manner, cumulative probabilities of detection during a series of inspections from crack initiation to link up could be determined.

Modelling Details

Stresses and Locations

The analysis considered 100 fastener holes in the critical row of a lap joint. (5 frame spacings/10 strap spacings). Stresses in the joint were calculated [3] using a finite element analysis (FEA) model and a commercially available code (NISA). The FEA model is shown in Figure 1. It consisted of 1 strap-bay in a curved panel subjected to (fuselage) pressure. The 3-row lap-joint was considered fully disbonded, so that the fasteners transferred all the loads. Fastener and fastener hole flexibility were accounted for using flexibility curves obtained from tests on lap-joint specimens [4]. Membrane stress, local bending stress and fastener forces were calculated at each fastener hole. These were used as input to the analysis.

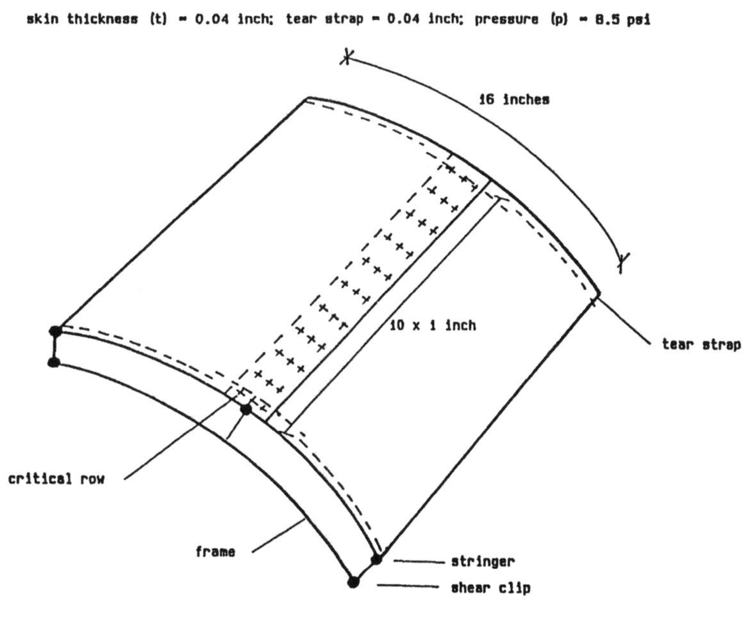

FIGURE 1. FINITE ELEMENT MODEL [1]

The stress analysis was performed assuming elastic action to prevail. The model parameters were: skin thickness = 0.04 inch; fastener shank diameter = 0.18 inch; fastener head diameter = 0.22 inch; fastener spacing = 1.00 inch; fuselage radius = 75 inch; fuselage pressure = 8.5 psi.

Membrane stress, bending stress, and fastener load are known for each of the holes included in the calculation. Due to the assumption of elastic deformations, the superposition principle dictates that a change in fastener spacing would scale the strap spacing and frame spacing as well. Hence scaling permits only nominal changes in fastener spacing, and diameter. Periodicity (frame spacing) was assumed in the assignment of values for membrane stress, bending stress and fastener force, calculated for one bay (20 holes).

Fatigue Damage and Crack Initiation

It was assumed that at the time of crack initiation a crack of size a_0 would be present, where

$$a_0 = 0.5(D_{HEAD} - D_{SHANK})$$

as illustrated in Figure 2. The argument for choosing this initiation crack size is obvious -- smaller cracks would not be detectable, so their history until they reach the size of a_0 is immaterial.

Fatigue tests on lap-joints subjected to constant amplitude [4] and variable amplitude loading [5] provided information on the scatter and distribution of $\Sigma\left(\frac{n}{N}\right)$ in Miner's rule.

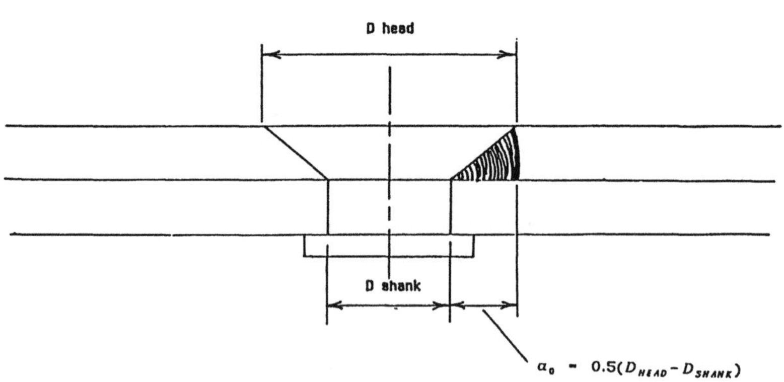

FIGURE 2. DEFENITION OF INITIATION CRACK SIZE

Analysis of these data showed that a Weibul distribution applies fairly well:

$$P_{fail} = 1 - e^{\left[-(DAMAGE-DAM_{min})-(DAM_{\lambda}-DAM_{min})^{DAM_{\alpha}}\right]} \qquad (1)$$

This is illustrated in Figure 3. In the above equation, DAMAGE is Miner's $\Sigma\left(\frac{n}{N}\right)$, where n is the number of cycles at a certain stress level and N the constant amplitude life at that same stress level. Miner's rule predicts occurrence of a_0 when $\Sigma\left(\frac{n}{N}\right) = 1$. Test data (Figure 3) show however, that this occurs for $\Sigma\left(\frac{n}{N}\right)$ values between 0.5 and 2, depending upon the type of loading. Figure 3 shows that the parameters in Eq (1) are [6,7]:

$$\Sigma\left(\frac{n}{N}\right)_{min} = 0.5(DAM_{min}); \quad \Sigma\left(\frac{n}{N}\right)_{\lambda} = 1.0(DAM_{\lambda}); \quad \alpha = 2.0(DAM_{\alpha})$$

The significance of Eq(1) and Figure 3 can be explained as follows: the first crack (of size a_0) occurs when the damage is $\Sigma\left(\frac{n}{N}\right) = 0.5$. When the damage is $\Sigma\left(\frac{n}{N}\right) = (DAM_{\lambda})$, both Eq(1) and Figure 3 show that the probability of failure would be 0.63, meaning that in 63 out of 100 cases a crack will occur prior to the damage reaching 1.0.

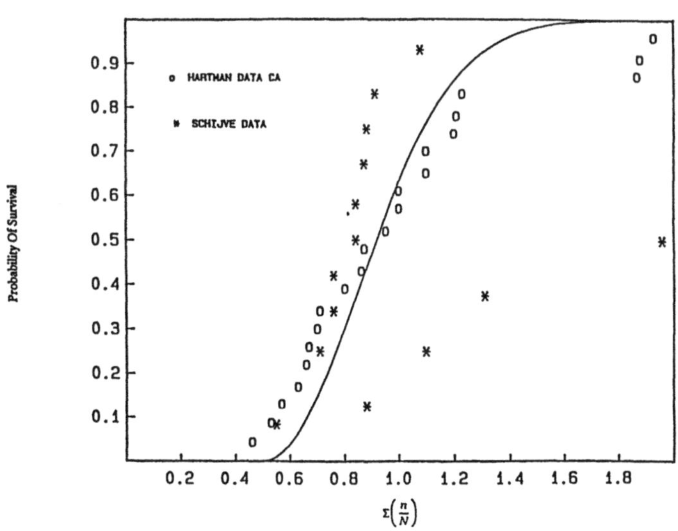

FIGURE 3. WEIBUL CURVE FOR DAMAGE: $\Sigma\left(\frac{n}{N}\right)$
SOLID CURVE: $DAM_{min} = 0.5$; $DAM_{\lambda} = 1.0$; $DAM_{\alpha} = 2.0$

In order to calculate the damage, the constant amplitude life, N, at the given stress level must be known. This life follows from constant amplitude tests. The data from Ref [4] are eminently suitable for this case, because they were obtained on lap-joint specimens of the configuration shown in the insert of Figure 4, which is almost exactly the same as the one considered here. The resulting fatigue life data are shown in Figure 4. It should be noted that the data in Figure 4 are based on applied (remote) stress. Therefore, all damage calculations need to consider the membrane stress (σ_θ) only.

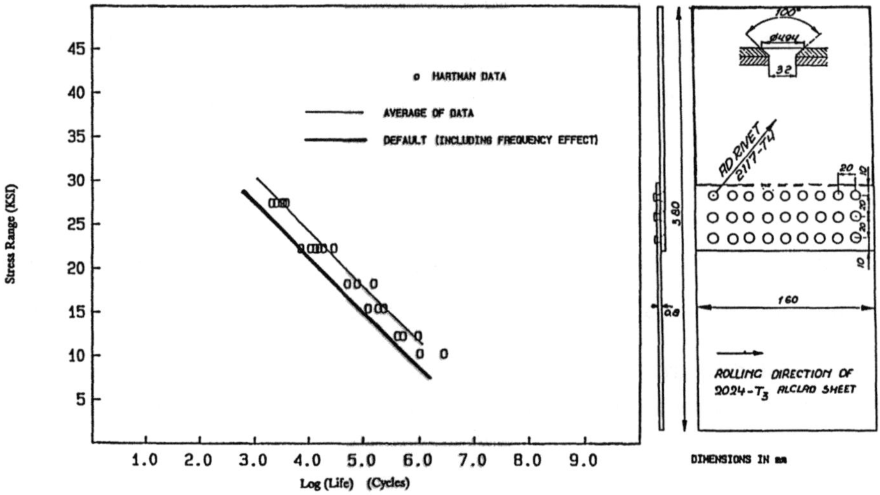

FIGURE 4. HARTMAN DATA [4] (SOLID LINE: CURVE FIT FOR RELEVANT STRESS RANGES FOR PARAMETERS SEE TEXT)

For σ_θ values of interest here, the data of Figure 4 can be well represented by the linear equation:

$$\sigma = 46.2 - 2.7\ln(N) = 0.92F_{ty} - 2.74\ln(N) \quad (2)$$

or

$$LIFE = N = e^{[(0.92 * F_{ty} - \sigma) / 2.74]} \quad (3)$$

The parameter values in the above equations provide the solid line in Figure 4, which accounts for a factor of 3 to allow for the effect of frequency and environment, and an R-ratio of 0. Hartman's data [4] are for $R = 0.1$, and for high frequency and laboratory air, whereas cabin pressure cycling rate is low, and the fuselage skin will be likely be subjected to a corrosive environment (cabin air pollution, atmospheric condition).

As mentioned above, only crack growth and crack detection are of interest here. Therefore, the time to initiation of the first crack is not accounted for, and the calculation begins with the (forced) appearance of the first crack according to the following scheme.

A hole is picked at random until one is found with a stress higher than average. To that particular hole a crack of size a_0 is assigned and the calculation proceeds. The time of occurrence of this crack is when $\Sigma\left(\frac{n}{N}\right) = 0.5$ or at $n = 0.5\ (Average)\ N$ for constant amplitude, so that the value of n can be obtained as shown in Figure 5(a), and calculated as $n = 0.5N$, where N follows from Eq(3).

All other holes have therefore experienced n cycles. Depending on their stress level and accumulated damage, some of the other holes might be cracked as well. This possibility is integrated in the calculations; should other holes be cracked, they are immediately assigned a_0 as well. The logic of the interrogation is as follows.

The damage distribution is known per Eq(1). On this basis a damage is assigned to each hole, before the calculation starts. The probability of any one being cracked is determined by its $\Sigma\left(\frac{n}{N}\right)$ and Eq(1). The probability, P, is increased from 0 to 1 in 100 steps of 0.01. By inverting Eq(1) one obtains:

$$DAMAGE = DAM_{min} + (DAM_{\lambda} - DAM_{min}) * [-\log(1.0 - P)]^{(1.0/DAM_{\alpha})} \tag{4}$$

for each increment of P. Random numbers are now picked between 1 and 100. Each pick represents a hole. The number (hole) that is picked is assigned the failure damage calculated above. The damage assignment is made for both the left side and the right side of the hole.

As pointed out above and in Figure 5(a,b) all holes initially have experienced n cycles (the actual value of n is immaterial). With knowledge of n (for every hole), the damage $\Sigma\left(\frac{n}{N}\right)$ for all other holes can be found. Given the stress level at a particular hole, and the

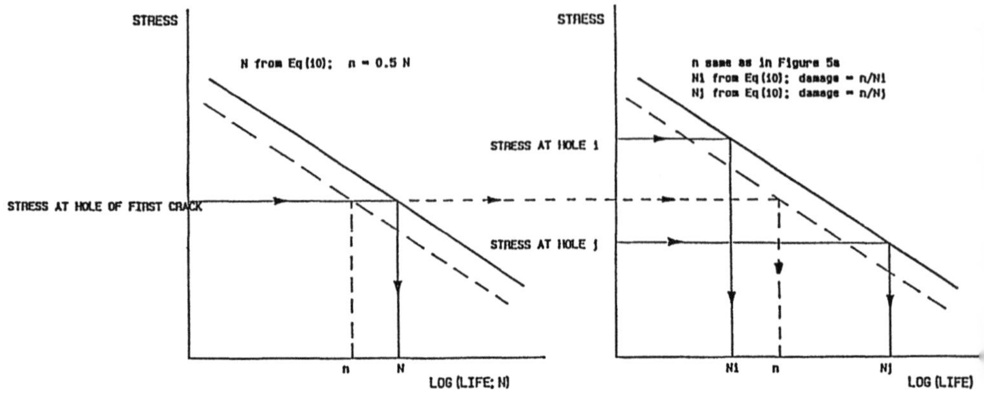

(a) Cycles n at first hole cracked at n/N = 0.5

(b) Damage at other holes at n cycles

FIGURE 5. DAMAGE AT ALL HOLES AT START OF CACULATION

number of cycles experienced, its life (See Figure 5(a,b)) is N, and its damage $\frac{n}{N}$. Should this damage be larger than the above assigned damage, the hole is cracked, otherwise it is not. During subsequent cycling, the increased damage is assessed every 100 cycles, taking due account of the increase of stress Due to cracks at adjacent holes.

Fatigue Crack Propagation

Growth of all cracks is accumulated every 100 cycles: $a = a + \Delta A$, $N = N + \Delta N$ (where $\Delta N = 100$ cycles). The crack growth rate $\frac{da}{dN}$ is assumed to follow the Paris equation for stress ratio R = 0 (fuselage pressure varies between 0 and p) as

$$\frac{da}{dN} = (C_p)(\Delta K^3) \text{ , where } \Delta A = \Delta N * \frac{da}{dN} \qquad (5)$$

Analysis of crack growth data of older 2024-T3 alclad aluminum shows that an exponent of 3 in the equation is a good fit for all cases [1].

A Weibul distribution for C_p was taken as (Figure 6):

$$P = 1 - e^{\left[-(C_p - C_{pmin})/(C_p - C_{p\lambda})^{C_{p\alpha}}\right]}$$

or (6)

$$C_p = C_{pmin} + (C_{p\lambda} - C_{pmin}) * [-\log(1.0 - P)]^{(1.0/C_{p\alpha})}$$

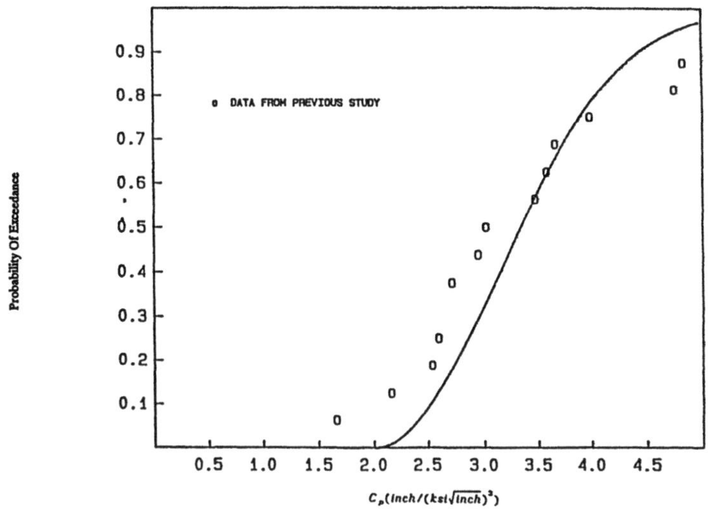

FIGURE 6. WEIBUL CURVE FOR C_p IN EQUATION (5); m = 3 (SOLID LINE: $C_{p_{min}}$ = 2; C_{p_λ} = 3.6; α = 3)

Default values for the parameters per Figure 6 were assigned. Before the analysis starts, each hole is assigned a C_p value in the same manner as initiation damage is assigned. The probability, P, in Eq(6) is increased from zero to 1 in steps of 0.01, whereupon C_p is calculated from the second of Eqs(6). A hole is than picked at random and assigned this C_p value and the incrementation continued until all holes have a value assigned. If a hole is picked to which a value had been assigned already, another hole is picked at random.

The stress intensity factor, ΔK, is defined as:

$$K = SIF = \bar{\beta}\sqrt{(\pi * A_c)} \tag{7}$$

where $\bar{\beta}$ is the geometry factor obtained from compounding and superposition [8], and A_c is the crack size. The loading situation is shown in Figure 7. It is a superposition of a membrane stress, a fastener load with bypass, and an out of plane bending stress. Thus,

$$K_A = K_B * 0.5(K_D + K_E) \tag{8}$$

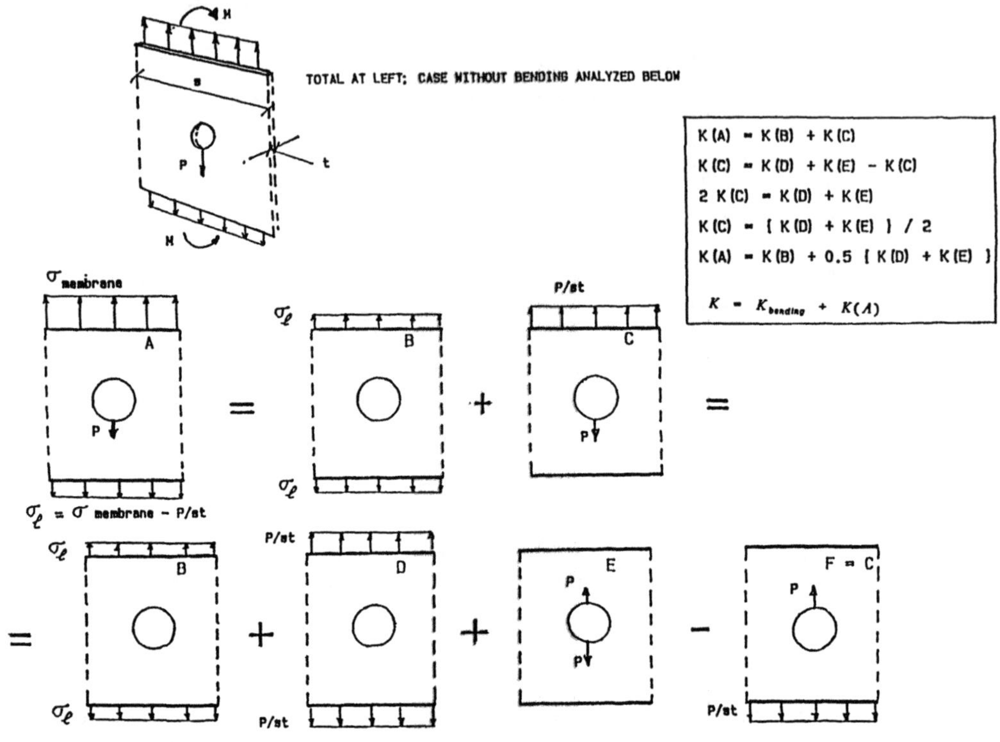

FIGURE 7. SUPERPOSITION OF STRESSES, EQUATIONS (8), (9)

and by further superposing the effect due to the bending stress, the total stress intensity factor becomes

$$K = K_{bending} + K_B + 0.5(K_D + K_E) \tag{9}$$

For cracks at holes of diameter D, the crack will behave [8] as if the hole is part of the crack as soon as $a/D > 0.1$. Note that $a_0/D = 0.02/0.18 > 0.1$. This leads to the stress intensity factors shown in Table 1. These equations are more easily written in terms of the effective crack size (including the hole), while, if K is based on the crack size, the geometry factor is equal to 1 (Figure 8, and Table 1). The geometry factors must be adjusted for the effect of adjacent cracks. As shown in Figure 8 and Table 1, use is made of an effective width, which leads to the correct expression for the interaction of co-linear cracks [9,10].

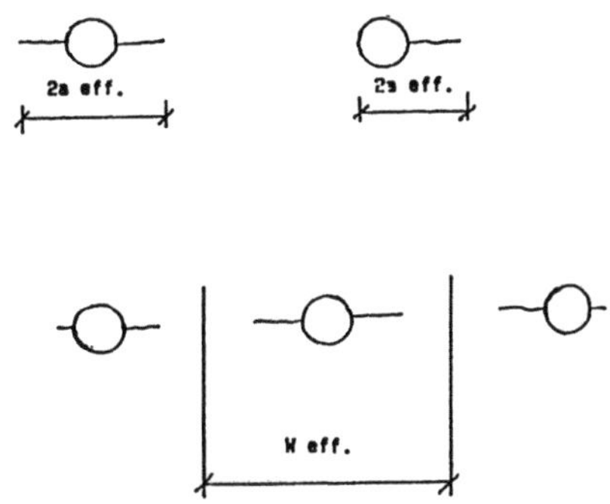

FIGURE 8. DEFINITIONS OF EFFECTIVE CRACK SIZE AND WIDTH

The geometry factor for the bending case was obtained through a curve-fit of Figure 9 of Ref [9], which is shown in Table 1. Note that this geometry factor is expressed in terms of effective crack size and effective width as well.

It is necessary to write the stress intensity for the fastener load, K_E, as a function of the effective crack size, and stress. As the membrane stress was chosen as the reference, one obtains the expression in Table 1. It should further be noted that K_B and $0.5K_D$ have the same geometry factor, so that these can be combined as shown in Table 1.

The final superposition of Eq(9) can then be made to provide the total geometry factor ($\bar{\beta}$) in Eq(7), while the final stress intensity factor is now based on the membrane stress as in reference stress [8].

Link-up and Failure

The link-up criterion was taken as the exceedance of net section yield, with some accounting for stress redistribution. The latter was effected by calculating the net membrane section stress at the fastener stations averaged over 3 fastener spacings. The calculation was continued until a total of 10 links had occurred (final failure).

The Probability of Crack Detection

The experimental data [2] that define the probability of detection curve has the general form as shown in Figure 9, which can be described by the equation:

$$p = 1 - e^{\{[-(A_{detect} - a_0)/(\lambda - a_0)]^{\alpha}\}} \tag{10}$$

TABLE 1. THE GEOMETRY FACTOR FROM SUPERPOSITION AND COMPOUNDING [8]

$$\left.\begin{aligned}
K &= K_{ben} + K^{B} + 0.5(K^{D} + K^{E}) \\
K_{ben} &= \beta_{ben}\,\sigma_{ben}\sqrt{\pi a} \\
K^{B} &= \beta_{ten}\left(\sigma_{mem} - P/ts\right)\sqrt{\pi a} = \beta_{ten}\,\sigma_{I}\sqrt{\pi a} \\
K^{D} &= \beta_{ten}\,P/ts\sqrt{\pi a} \\
K^{E} &= \frac{P}{t\sqrt{\pi a + D/2}}
\end{aligned}\right\} \quad (a)$$

So that:

$$K = \beta_{ben}\,\sigma_{ben}\sqrt{\pi a} + \beta_{ten}\left(\sigma_{mem} - 0.5\,P/ts\right)\sqrt{\pi a} + 0.5\,\frac{P}{t\sqrt{\pi a + D/2}} \quad (b)$$

For crack at hole [8] :

$$\left.\begin{aligned}
K_{hole} = \beta_{ten}\,\sigma\sqrt{\pi a}\,, \quad \text{with} \quad & \beta_{ten} = \sqrt{0.5 + 0.5\,D/a}\,, \quad \text{1 crack} \\
& \beta_{ten} = \sqrt{1.0 + 0.5\,D/a}\,, \quad \text{2 cracks}
\end{aligned}\right\} \quad (c)$$

Which can be written in effective crack size:

$$\left.\begin{aligned}
K_{ten} &= \sigma\sqrt{\tfrac{\pi}{2}(a + D)} = \sigma\sqrt{\pi a_{eff}}\,, \quad \text{1 crack} \\
K_{ten} &= \sigma\sqrt{\tfrac{\pi}{2}(2a + D)} = \sigma\sqrt{\pi a_{eff}}\,, \quad \text{2 cracks}
\end{aligned}\right\} \quad (d)$$

Continued on next page

TABLE 1. CONTINUED

With width effect:

$$K = \beta_w \sigma \sqrt{\pi a_{eff}} \quad ; \qquad \beta_w = \sqrt{\sec(\pi a_{eff}/W_{eff})} \qquad \text{(e)}$$

For effective crack size and width see Figure 8.

$$\beta_{ben} = 0.39\,\{1.0 + 0.16(a_{eff}/W_{eff})^2\} \;, \quad \text{with } \nu = 0.3 \quad [9] \qquad \text{(f)}$$

$$\left.\begin{aligned} K^E &= \frac{P}{t\sqrt{\pi a_{eff}}} = \frac{sP\sqrt{\pi a_{eff}}}{st \quad \pi a_{eff}} = \frac{s}{\pi a_{eff}}\sigma_p\sqrt{\pi a_{eff}} = \beta_p \sigma_p \sqrt{\pi a_{eff}} \\ \beta_p &= s/\pi a_{eff} \quad ; \quad \sigma_p = P/ts \end{aligned}\right\} \text{(g)}$$

Then total K becomes:

$$K = \left\{\beta_{ben}\frac{\sigma_{ben}}{\sigma_{mem}} + \beta_w\left(1.0 - 0.5\frac{\sigma_p}{\sigma_{mem}}\right) + 0.5\beta_p\frac{\sigma_p}{\sigma_{mem}}\right\}\sigma_{mem}\sqrt{\pi a_{eff}}$$

$$K = \bar{\beta}\,\sigma_{mem}\sqrt{\pi a_{eff}}$$

$$\bar{\beta} = 0.39\,\{1.0 + 0.16(a_{eff}/W_{eff})^2\}\frac{\sigma_{ben}}{\sigma_{mem}} + \sqrt{\sec(\pi a_{eff}/W_{eff})}\left(1.0 - 0.5\frac{\sigma_p}{\sigma_{mem}}\right) + \qquad \text{(h)}$$

$$+ 0.5\left(\frac{s}{\pi a_{eff}}\right)\frac{\sigma_p}{\sigma_{mem}}$$

Note that:

$$1.0 - 0.5\frac{\sigma_p}{\sigma_{mem}} = \frac{\sigma_{mem} - 0.5\sigma_p}{\sigma_{mem}} = \frac{\sigma_l + 0.5\sigma_p}{\sigma_{mem}} \quad \text{(Statement 4530)}$$

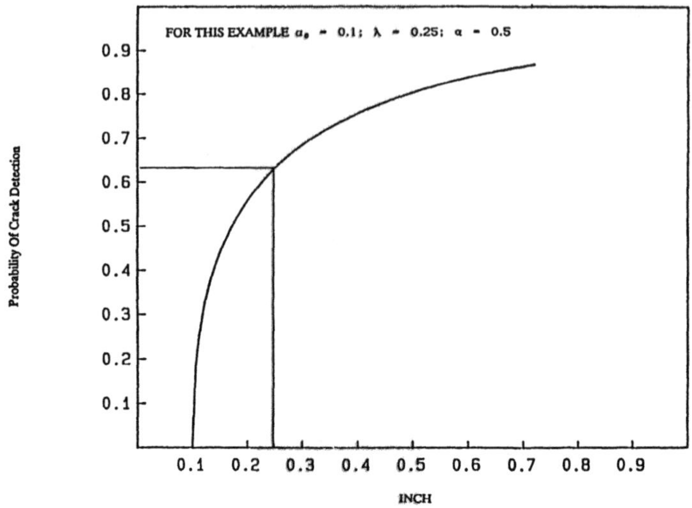

FIGURE 9. TYPICAL SHAPE OF CURVE FOR PROBABILITY OF CRACK DETECTION IN ONE INSPECTION

where a_0 is the crack size for which detection is near zero, λ and α are parameters determining the shape of the curve. The probability of non-detection is 1 - p. Note that Eq(10) is a curve-fitting equation; it does not make any statement about the statistics involved, despite the fact that it is the same as the Weibul equation. Any other equation providing a similar curve shape would be equally suitable.

A crack is likely to be subject to several inspections before link up. During each inspection it may be missed. Consider 100 cracks growing at equal rates, (same population), all in the same stage of growth (same size). Let the probability of detection at a certain inspection be $p = 0.2$. The probability that a crack will be missed is then $q = 1 - p = 0.8$. That means that 80 cracks will go undetected. The other 20 are detected and therefore must be repaired, so that these are eliminated. At the next inspection the remaining cracks are longer; let the probability of detection then be $p = 0.6$, so that $q = 0.4$. Thus of the remaining 80 cracks .4 x 80 = 32 cracks will go undetected, etc., and the cumulative probability that a crack will be missed in successive inspections is $Q = q_1 * q_2 * \ldots * q_n$. In the above example $Q = 0.8 \times 0.4 = 0.32$: of the 100 cracks 32 cracks remain undetected after two inspections. The cumulative probability of detection is $P = 1 - Q$. In the above example $P = 0.68$: of the 100 cracks 68 were detected, repaired and eliminated after two inspections, but 32 were missed. The cumulative probability of detection is then:

$$P = 1 - \pi(1 - p) \tag{11}$$

where p follows from a curve such as in Figure 9 or from Eq(10). The principle is discussed in more detail in the literature [8,11,12].

Consider a situation where 100 inspectors look at the same crack for which the probability of detection is known to be p. The results of these 100 inspectors are known beforehand for any probability p, because the distribution is binomial, as if these 100 inspectors were picking marbles out of a vase with white (detect) and red (miss) marbles. They either pick a white one or a red one. For a certain probability, p, the band in which a certain percentage of the observations of 100 inspectors must fall can be constructed a priori. For example, for $p = 0.8$ and $q = 0.2$, the standard deviation of this binomial distribution is:

$$s = \sqrt{(p \times q \times n)} = \sqrt{(0.8 \times 0.2 \times 100)} = 4 \tag{12}$$

where p is the probability of detection, $q = (1 - p)$ is the probability to miss, and n is the number of inspectors. Since the 90 per cent band is given by $1.64s = 1.64 \times 4 = 6.4$, the results are pre-determined. The expectation is for 80 inspectors to detect and for 20 to miss. There is 5 percent chance that only 80 - 6.4 = 73 inspectors would detect, and a 5 percent chance that 80 + 6.4 = 86 inspectors would detect the crack. Conversely, if 73 inspectors out of the 100 detect a certain crack, the probability of detection could still be as high as 0.8; if 86 inspectors would detect, the probability of detection could be as low as 0.8. The 90 per cent band for the detection ratio of 100 inspectors is between 0.73 and 0.86 if the actual probability of detection is $p = 0.8$. It follows that if 100 inspectors 'look' at two different cracks of the same size, and report 75 detects (25 misses) for one crack, and 40 detects (60 misses) for the other, then the probability of detection of these two cracks is different: although they may have the same size, they belong to different categories having different probability-of-detection curves.

For any probability curve, the 90 per cent band for 100 inspectors is established by Eq(12) as shown in Figure 10(a). If some data obtained by 100 inspectors fall outside this band, the cracks involved belong to different categories (populations). For example, the data [2]

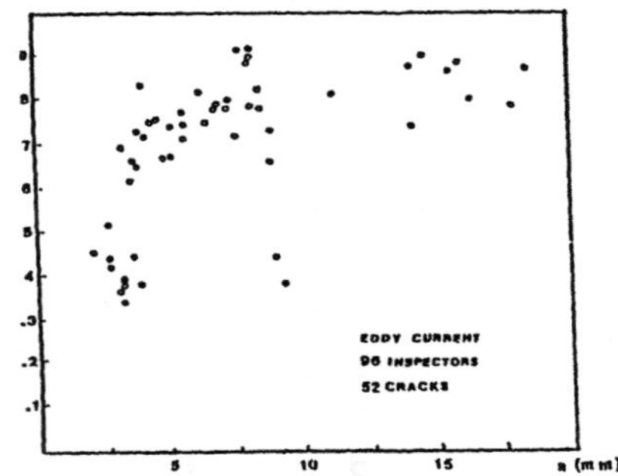

(a) Data from Lewis et al [2]

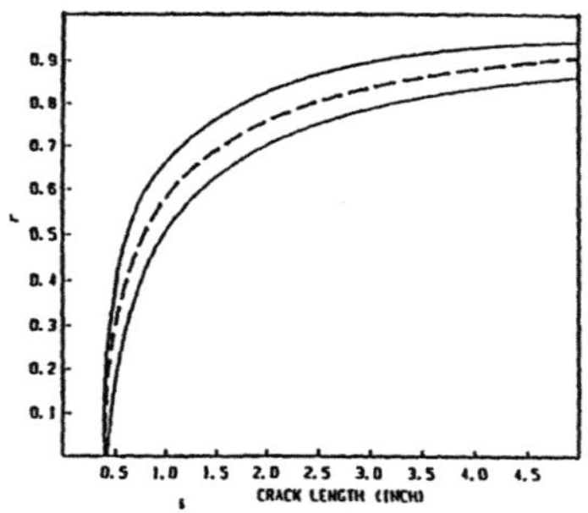

(b) Band for data of 100 inspectors

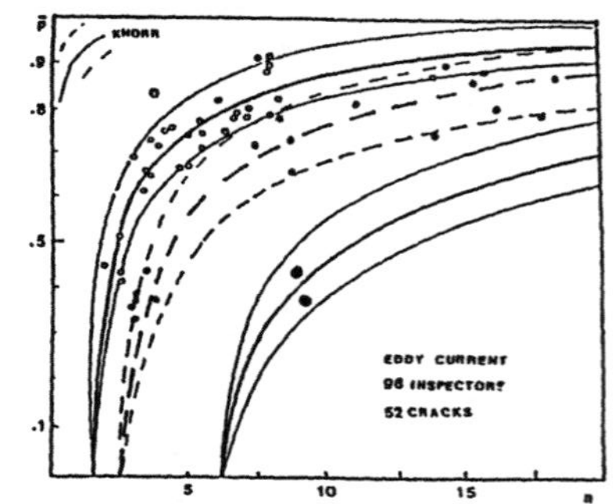

(c) Populations in data from Lewis et al

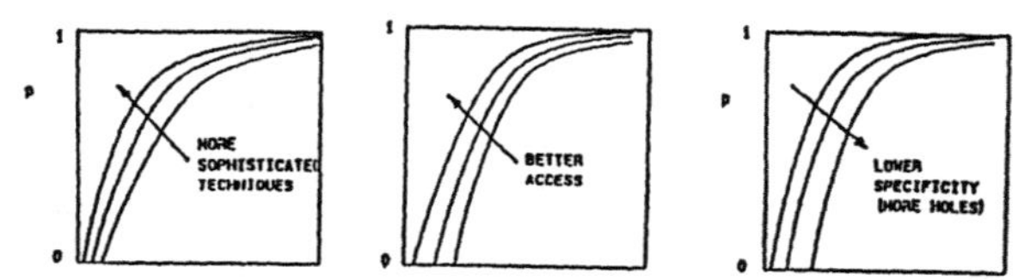

(d) Effect of circumstances on probability curves

FIGURE 10. PROBABILITY OF CRACK DETECTION

in Figure 10(b) show that certain cracks of approximately 9 mm size were detected with a ratio of 0.4, while others of the same size were detected with a ratio of 0.9. The above shows that the spread for 96 inspectors cannot be that large, so that these cracks of approximately 9 mm did not belong to the same segment of the population. Ergo, the data set reported in [2] contains more than one population and more than one probability curve.

The 'probability-of-detection' curve for cracks of any type in any structure is low and follows the lower bound of the data in Figure 10(b). Indeed, this low curve should be used if inspectors are assigned to look for cracks in a large structure somewhere. That is not the way inspections are specified in practice. The data of Figure 10(b) are repeated in Figure 10(c) to show categories of difficulty of inspection. Figure 10(c) also contains data [13] for a case where inspectors were assigned to examine one particular location in different aircraft: they knew where to inspect and what type of crack to look for (high specificity), as is normally the case; not surprisingly, their results are much better. Had all inspectors represented in Figure 10(c) received specific assignments their results would have been better as well. The 90 percent bands, known a priori as discussed on the basis of Eq(12) and Figure 10(a) for the 96 inspectors involved, can be drawn for certain probability curves. Three such bands are shown in Figure 10(c). They indicate that the data cover at least three different populations, determined by specificity (some components were small, others very large) and accessibility (some cracks were easily accessible, others were not).

Specificity and accessibility determine the applicable probability-of-detection curve (Figure 10(d)). The length of the inspection interval should be established such as to provide a consistent safety level (cumulative probability of detection), independent of the shape of the crack growth curve, the accessibility, and the specificity of the inspection. The aimed for cumulative probability of detection could be specified in damage tolerance requirements. Given the calculated crack growth curve and permissible crack size, and the probability-of-detection for the relevant specificity and accessibility, the cumulative probability of detection can be calculated for different lengths of the inspection interval by means of Eq(11).

When the results are plotted, the interval for the desirable probability of detection can obtained from the curve [8,11,12]. The interval will be different for different inspection methods, for different crack growth curves, accessibility, and specificity. A commercially available computer program [14] performs the calculation for different interval lengths. For a certain inspection interval it finds the crack sizes at which the inspection will take place, from the crack propagation curve. At each inspection (crack size) the probability of detection follows from the probability curve with the parameters appropriate for the inspection method and category. Eq(11) is then applied to obtain the cumulative probability. At least one newly developed aircraft has been certified by FAA on the basis of this procedure.

Inspection and Crack Detection Data Used in Present work

The inspection for MSD is one of very low specificity: a great many locations must be inspected. This provides for a low probability curve (Figure 10(d)). On the other hand the accessibility is good, which improves the situation somewhat. Based on a re-interpretation of the data by Lewis et al [2], the parameters for Eq(10) were determined [11,12]. These are the default values used in the present analysis for both eddy current and visual inspection for a non-specific inspection (very large area) with easy access. The associated probability of detection curves are shown in Figures 11 and 12 . The curve for visual detection is based on actual data [15], but it is unreliable because the effects of specificity and accessibility; had to be estimated; this is of no great consequence, because eddy current inspection is the principal inspection procedure for lap joints in aging aircraft.

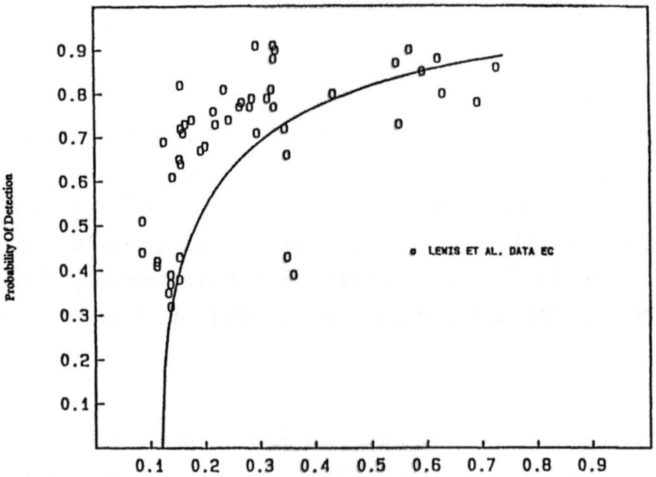

FIGURE 11.

PROBABILITY OF DETECTION IN ONE INSPECTION FOR EDDY CURRENT SOLID LINE REPRESENTS: a_0 = 0.12, λ = 0.25, α = 0.5 (easy access, low specificity)

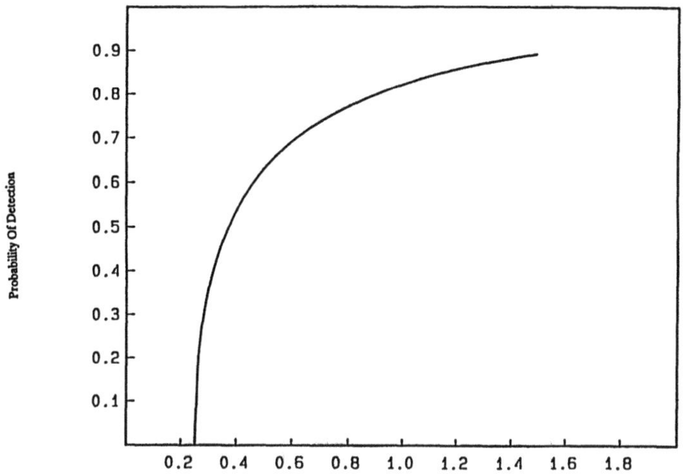

FIGURE 12. PROBABILITY OF DETECTION IN ONE VISUAL INSPECTION
$a_0 = 0.25$, $\lambda = 0.5$, $\alpha = 0.5$

One further, and essential, refinement was made. The inspection of MSD is of low specificity; the detection probability is low, because boredom soon sets in when thousands of fasteners must be inspected. On the other hand, if a crack is detected, the inspector will be back on 'alert', and have a better chance of detecting the next crack (higher probability), and so on. After one or more non-detects, the probability of detection will decrease again. Thus, the parameters describing the probability of detection should be adjusted after each 'detect' and each 'miss'. A quantitive basis for such adjustments does not exist. Thus the assumptions that had to be made are the weakest point in the present analysis.

The assumptions were: Upon detection of a crack, the parameters a_0 and λ in Eq(10) were decreased by 5 per cent. For the non-detection case, the parameters are increased again by 5 per cent. The above corrections were made until the increases did not exceed 50 per cent, and the decreases were less than -25 per cent. Resulting changes in the probability of detection curves are shown in Figure 13. Note that α in Eq(9) is not changed, because [11,12] $\alpha = 0.5$ fits all measured probability of detection curves.

The crack size at any time permits calculation of the probability of detection with Eq(10). In order to determine the cumulative probability of detection as per Eq(11) the probability to miss the crack must be determined. The product of successive probabilities to miss is as per Eq(11). The latter requires some more explanation.

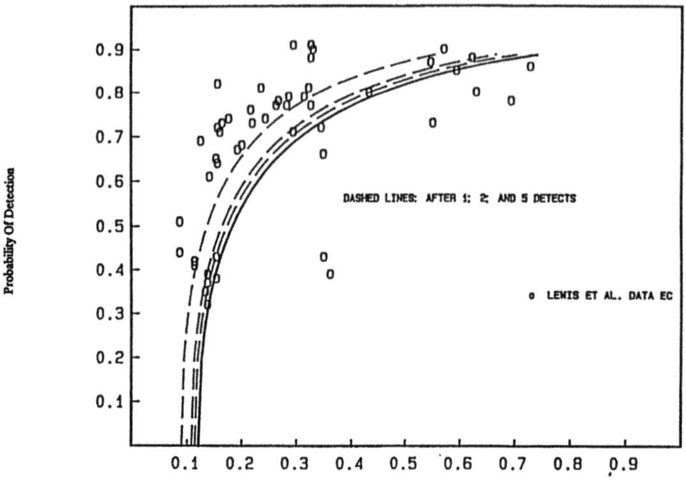

FIGURE 13.

EFFECT OF DETECTS AND NON-DETECTS ON PROBABILITY CURVE FOR EDDY CURRENT
DASHED LINES: STEP-WISE INCREASE AFTER DETECT; DECREASE AFTER MISS

The probability that all cracks are missed is the product of all the probabilities to miss for all cracks in all inspections. Although this is interesting information, more relevant is the probability that any particular crack is missed; even more important is whether the crack(s) eventually leading to failure are missed. Therefore, the probability to miss any individual crack must be determined as well.

The probability parameters depend upon whether or not a crack was detected at the previous hole. This does not follow directly from the above calculation of the probabilities for 'detects' and 'misses'. Suppose the probability of detection of a crack at a certain hole is 0.7 . A random number between 0 and 1 is taken. If this random number is less than 0.7 the crack is found, if it is larger than 0.7 it is missed. In this manner the crack is indeed found in 70 out of 100 cases as it should be for the above example of a probability of detection of 0.7. If a 'detect' is recorded the multiplier to the probability parameters discussed above is decreased; in the case of a 'miss', the multiplier is increased, with a minimum of -0.05 and a maximum of 0.10, in accordance with the discussion.

Repairs

The modelling feature provides the option to stimulate repairs made after the detection of a crack. The repair is assumed to cover 5 fastener spacings to the left and 5 fastener spacings to the right of a detected crack. If more than 1 crack is detected the repair will extend 5 spacings to the left of the leftmost detected crack and 5 spacings to the right of the rightmost detected crack. Although such cases are interesting, they are not relevant for the determination of cumulative probabilities of detection, because the latter must be based on the worst case (no repairs). Since the probability of detection is never equal to 1, one must count on any crack being missed.

Results

The following shows the result of a typical case in which all parameters were set at the default values. Normally, the analysis covers 6 inspection intervals (500, 1000, 2000, 3000, 4000, and 5000 flight cycles). In the following, graphical results are shown in most cases for all intervals, tabular results only for an interval of 3000 flight cycles.

Tables 2(a) and 2(b) show the initial assignments of stresses and fastener loads, and the random assignment of crack growth properties (Cp) and failure damage; these assignments were made in the manner described earlier. Note that holes 1, 21, 41 etc. are at a frame, while holes 1, 11, 21 etc. are at a strap.

In this particular case, the first crack was introduced at hole 43 on the left side. This can be seen in Table 3(b), where Tables 3(a, b, c) show the begin-situation. The first column in Table 3(a, b, c) gives the hole location, the second the membrane stress as in Table 2(a,b). In the third column appears the number of cycles expired, which is zero here because the table is for the begin-situation. The present damage , as calculated on the basis of Figure 5, is shown for the left and right side of every hole in columns 4 and 5. Should any other hole have an initial damage larger than its assigned failure damage, it would be immediately cracked as well.

As random failure damage assignments are made in every case, while the first crack appears at random at a hole with higher than average stress, the begin-situation of Table 3(a,b,c) varies from case to case. Typically however, the number of initial cracks is small (from 1 to 5). Columns 6-9 provide crack size and the number of cycles during which the crack existed, for both the left and the right side of each hole. Since this is the begin-situation, most holes are as yet uncracked (but see hole 43 in Table 3(b)). The number of cycles for crack growth is zero throughout. The last column provides the crack-case (CCRASE) extending over 3 holes, each hole represented by 0. If there is no crack at any of these holes, the case is represented as 0 .0. 0. If there is a crack, a '-' is included. Thus the situation at hole 43 in Table 3(b) is denoted as 0 -0. 0, indicating that

TABLE 2a . ASSIGNMENTS STRESS, CP, AND FAILURE DAMAGE; PAGE 1

MSDDET

LOCATION	MEMBRANE STRESS	BENDING STRESS	FASTENER LOAD	Cp(*E9)	DAM. LEFT AT FAILURE	DAM. RIGHT
1	11.200	4.000	244.000	2.546	0.869	1.276
2	12.700	5.100	207.000	3.390	0.887	0.687
3	13.900	6.700	207.000	4.166	1.005	0.828
4	14.500	7.300	211.000	3.088	0.972	0.702
5	14.800	7.500	215.000	3.639	0.953	0.571
6	14.900	7.400	216.000	3.732	1.125	1.259
7	14.800	7.400	216.000	2.691	0.654	0.613
8	14.500	7.100	214.000	2.160	1.259	0.805
9	13.800	6.700	211.000	3.218	1.072	0.881
10	12.600	5.200	214.000	5.165	1.019	0.892
11	11.100	3.900	253.000	4.684	1.106	1.214
12	12.600	5.200	214.000	4.130	1.089	0.500
13	13.800	6.700	211.000	4.377	1.397	0.749
14	14.500	7.100	214.000	3.371	0.780	1.089
15	14.800	7.400	216.000	3.313	0.959	0.857
16	14.900	7.400	216.000	3.831	0.799	0.910
17	14.800	7.500	215.000	4.871	0.881	1.295
18	14.500	7.300	211.000	2.279	1.064	0.928
19	13.900	6.700	207.000	2.798	0.743	0.723
20	12.700	5.100	207.000	3.884	1.097	0.922
21	11.200	4.000	244.000	4.243	1.012	0.768
22	12.700	5.100	207.000	4.428	0.979	1.243
23	13.900	6.700	207.000	3.430	1.189	0.736
24	14.500	7.300	211.000	2.491	1.365	0.587
25	14.800	7.500	215.000	2.227	0.892	0.716
26	14.900	7.400	216.000	2.838	0.840	0.762
27	14.800	7.400	216.000	2.956	1.228	0.959
28	14.500	7.100	214.000	4.204	0.679	1.056
29	13.800	6.700	211.000	2.818	0.966	0.966
30	12.600	5.200	214.000	3.553	1.276	0.840
31	11.100	3.900	253.000	3.351	0.898	0.662
32	12.600	5.200	214.000	3.940	1.243	1.072
33	13.800	6.700	211.000	3.999	1.144	1.041
34	14.500	7.100	214.000	3.969	0.904	0.947
35	14.800	7.400	216.000	4.483	1.056	1.026
36	14.900	7.400	216.000	2.621	0.749	1.365
37	14.800	7.500	215.000	2.645	0.644	0.624
38	14.500	7.300	211.000	3.685	0.787	1.436
39	13.900	6.700	207.000	2.878	0.947	1.189
40	12.700	5.100	207.000	3.332	1.049	1.080
41	11.200	4.000	244.000	4.330	0.768	0.756
42	12.700	5.100	207.000	3.574	0.916	1.125
43	13.900	6.700	207.000	4.543	1.214	0.916
44	14.500	7.300	211.000	3.031	0.863	1.019
45	14.800	7.500	215.000	3.662	0.736	1.315
46	14.900	7.400	216.000	3.490	0.624	0.985
47	14.800	7.400	216.000	2.734	0.587	0.953
48	14.500	7.100	214.000	2.917	1.080	1.049
49	13.800	6.700	211.000	3.181	1.134	0.834
50	12.600	5.200	214.000	2.323	0.500	0.780

MSDDET

MSDDET

TABLE 2b . ASSIGNMENTS STRESS, CP, AND FAILURE DAMAGE; PAGE 2

MSDDET

LOCATION	MEMBRANE STRESS	BENDING STRESS	FASTENER LOAD	Cp(*E9)	DAM. LEFT AT FAILURE	DAM. RIGHT
51	11.100	3.900	253.000	5.434	0.828	0.654
52	12.600	5.200	214.000	3.617	1.315	0.869
53	13.800	6.700	211.000	2.777	0.709	1.034
54	14.500	7.100	214.000	4.609	0.928	0.972
55	14.800	7.400	216.000	3.780	1.026	0.934
56	14.900	7.400	216.000	3.756	0.687	1.012
57	14.800	7.500	215.000	4.030	0.999	0.730
58	14.500	7.300	211.000	2.936	1.436	1.106
59	13.900	6.700	207.000	3.256	0.992	0.999
60	12.700	5.100	207.000	4.062	0.756	1.064
61	11.200	4.000	244.000	3.162	0.716	0.635
62	12.700	5.100	207.000	2.519	0.941	0.799
63	13.900	6.700	207.000	3.237	0.934	0.743
64	14.500	7.300	211.000	2.431	0.846	1.097
65	14.800	7.500	215.000	3.450	0.662	0.601
66	14.900	7.400	216.000	3.470	0.793	1.134
67	14.800	7.400	216.000	2.597	1.166	1.177
68	14.500	7.100	214.000	2.975	0.635	1.228
69	13.800	6.700	211.000	3.532	1.201	0.816
70	12.600	5.200	214.000	3.294	0.550	1.573
71	11.100	3.900	253.000	2.756	1.041	1.397
72	12.600	5.200	214.000	2.572	0.702	0.679
73	13.800	6.700	211.000	4.095	1.489	1.489
74	14.500	7.100	214.000	3.050	0.723	0.904
75	14.800	7.400	216.000	3.200	1.573	0.992
76	14.900	7.400	216.000	3.125	1.115	1.166
77	14.800	7.500	215.000	2.994	0.822	0.822
78	14.500	7.300	211.000	2.713	0.922	0.852
79	13.900	6.700	207.000	3.275	0.875	1.115
80	12.700	5.100	207.000	2.668	0.601	1.005
81	11.200	4.000	244.000	3.013	0.671	0.644
82	12.700	5.100	207.000	4.769	0.762	0.793
83	13.900	6.700	207.000	2.362	1.295	1.144
84	14.500	7.300	211.000	4.285	0.694	0.774
85	14.800	7.500	215.000	3.511	1.339	0.671
86	14.900	7.400	216.000	3.805	0.985	0.898
87	14.800	7.400	216.000	3.595	0.834	0.709
88	14.500	7.100	214.000	4.996	1.034	0.694
89	13.800	6.700	211.000	3.857	0.571	1.155
90	12.600	5.200	214.000	3.708	0.811	0.863
91	11.100	3.900	253.000	2.898	0.500	0.787
92	12.600	5.200	214.000	2.398	1.177	0.979
93	13.800	6.700	211.000	3.106	0.857	0.887
94	14.500	7.100	214.000	2.000	0.730	1.201
95	14.800	7.400	216.000	3.069	0.852	0.875
96	14.900	7.400	216.000	2.462	1.155	1.339
97	14.800	7.500	215.000	3.410	0.910	0.846
98	14.500	7.300	211.000	3.144	0.613	0.811
99	13.900	6.700	207.000	2.858	0.805	0.941
100	12.700	5.100	207.000	3.911	0.774	0.550

MSDDET

MSDDET

TABLE 3a . BEGIN SITUATION; TYPICAL RESULTS FOR DEFAULT CASE; PAGE 1
INTERVAL 3000

MSDDET

SKIN THICKNESS	=	0.0400
DIA FASTENER SHANK	=	0.1800
DIA FASTENER HEAD	=	0.2200
FASTENER DISTANCE	=	1.0000
RADIUS FUSELAGE	=	75.0000
FUSELAGE PRESSURE	=	8.5000
YIELD STRENGTH	=	50.0000
MIN. FAILURE DAMAGE	=	0.5000
LAMBDA DAMAGE DISTR	=	1.0000
ALPHA DAMAGE DISTR	=	2.0000
POWER S-N CURVE	=	2.7400
Cp MINIMUM	=	2.0000
LAMBDA Cp DISTR	=	3.6000
ALPHA Cp DISTR	=	2.0000
Ao EDDY CURRENT	=	0.1200
LAMBDA EDDY CURRENT	=	0.2500
ALPHA EDDY CURRENT	=	0.5000
Ao VISUAL	=	0.2500
LAMBDA VISUAL	=	0.5000
ALPHA VISUAL	=	0.5000
INSP.PROB. INCR	=	0.0500
CRACK GROWTH INCR.	=	100.0000

MSDDET

LOCATN	STRESS	NTOTAL	DAM LE	DAM RI	a-LEFT	NCR LE	a-RGHT	NCR RI	CP	EQN	CRCASE
1	11.200	0	0.187	0.187	0.000	0	0.000	0	2.546	0	.0. 0
2	12.700	0	0.323	0.323	0.000	0	0.000	0	3.390	0	.0. 0
3	13.900	0	0.500	0.500	0.000	0	0.000	0	4.166	0	.0. 0
4	14.500	0	0.622	0.622	0.000	0	0.000	0	3.088	0	.0. 0
5	14.800	0	0.694	0.694	0.000	0	0.000	0	3.639	0	.0. 0
6	14.900	0	0.720	0.720	0.000	0	0.000	0	3.732	0	.0. 0
7	14.800	0	0.694	0.694	0.000	0	0.000	0	2.691	0	.0. 0
8	14.500	0	0.622	0.622	0.000	0	0.000	0	2.160	0	.0. 0
9	13.800	0	0.482	0.482	0.000	0	0.000	0	3.218	0	.0. 0
10	12.600	0	0.311	0.311	0.000	0	0.000	0	5.165	0	.0. 0
11	11.100	0	0.180	0.180	0.000	0	0.000	0	4.684	0	.0. 0
12	12.600	0	0.311	0.311	0.000	0	0.000	0	4.130	0	.0. 0
13	13.800	0	0.482	0.482	0.000	0	0.000	0	4.377	0	.0. 0
14	14.500	0	0.622	0.622	0.000	0	0.000	0	3.371	0	.0. 0
15	14.800	0	0.694	0.694	0.000	0	0.000	0	3.313	0	.0. 0
16	14.900	0	0.720	0.720	0.000	0	0.000	0	3.831	0	.0. 0
17	14.800	0	0.694	0.694	0.000	0	0.000	0	4.871	0	.0. 0
18	14.500	0	0.622	0.622	0.000	0	0.000	0	2.279	0	.0. 0
19	13.900	0	0.500	0.500	0.000	0	0.000	0	2.798	0	.0. 0
20	12.700	0	0.323	0.323	0.000	0	0.000	0	3.884	0	.0. 0

MSDDET

TABLE 3b . BEGIN SITUATION; TYPICAL RESULTS FOR DEFAULT CASE; CONTINUED; PAGE 2

INTERVAL 3000

MSDDET

LOCATN	STRESS	NTOTAL	DAM LE	DAM RI	a-LEFT	NCR LE	a-RGHT	NCR RI	CP	EQN	CRCASE
21	11.200	0	0.187	0.187	0.000	0	0.000	0	4.243	0	.0. 0
22	12.700	0	0.323	0.323	0.000	0	0.000	0	4.428	0	.0. 0
23	13.900	0	0.500	0.500	0.000	0	0.000	0	3.430	0	.0. 0
24	14.500	0	0.622	0.622	0.000	0	0.000	0	2.491	0	.0. 0
25	14.800	0	0.694	0.694	0.000	0	0.000	0	2.227	0	.0. 0
26	14.900	0	0.720	0.720	0.000	0	0.000	0	2.838	0	.0. 0
27	14.800	0	0.694	0.694	0.000	0	0.000	0	2.956	0	.0. 0
28	14.500	0	0.622	0.622	0.000	0	0.000	0	4.204	0	.0. 0
29	13.800	0	0.482	0.482	0.000	0	0.000	0	2.818	0	.0. 0
30	12.600	0	0.311	0.311	0.000	0	0.000	0	3.553	0	.0. 0
31	11.100	0	0.180	0.180	0.000	0	0.000	0	3.351	0	.0. 0
32	12.600	0	0.311	0.311	0.000	0	0.000	0	3.940	0	.0. 0
33	13.800	0	0.482	0.482	0.000	0	0.000	0	3.999	0	.0. 0
34	14.500	0	0.622	0.622	0.000	0	0.000	0	3.969	0	.0. 0
35	14.800	0	0.694	0.694	0.000	0	0.000	0	4.483	0	.0. 0
36	14.900	0	0.720	0.720	0.000	0	0.000	0	2.621	0	.0. 0
37	14.800	0	0.694	0.694	0.000	0	0.000	0	2.645	0	.0. 0
38	14.500	0	0.622	0.622	0.000	0	0.000	0	3.685	0	.0. 0
39	13.900	0	0.500	0.500	0.000	0	0.000	0	2.878	0	.0. 0
40	12.700	0	0.323	0.323	0.000	0	0.000	0	3.332	0	.0. 0
41	11.200	0	0.187	0.187	0.000	0	0.000	0	4.330	0	.0. 0
42	12.700	0	0.323	0.323	0.000	0	0.000	0	3.574	0	.0.-0
43	13.900	0	0.500	0.500	0.020	0	0.000	0	4.543	0	-0. 0
44	14.500	0	0.622	0.622	0.000	0	0.000	0	3.031	0	.0. 0
45	14.800	0	0.694	0.694	0.000	0	0.000	0	3.662	0	.0. 0
46	14.900	0	0.720	0.720	0.000	0	0.000	0	3.490	0	.0. 0
47	14.800	0	0.694	0.694	0.000	0	0.000	0	2.734	0	.0. 0
48	14.500	0	0.622	0.622	0.000	0	0.000	0	2.917	0	.0. 0
49	13.800	0	0.482	0.482	0.000	0	0.000	0	3.181	0	.0. 0
50	12.600	0	0.311	0.311	0.000	0	0.000	0	2.323	0	.0. 0
51	11.100	0	0.180	0.180	0.000	0	0.000	0	5.434	0	.0. 0
52	12.600	0	0.311	0.311	0.000	0	0.000	0	3.617	0	.0. 0
53	13.800	0	0.482	0.482	0.000	0	0.000	0	2.777	0	.0. 0
54	14.500	0	0.622	0.622	0.000	0	0.000	0	4.609	0	.0. 0
55	14.800	0	0.694	0.694	0.000	0	0.000	0	3.780	0	.0. 0
56	14.900	0	0.720	0.720	0.000	0	0.000	0	3.756	0	.0. 0
57	14.800	0	0.694	0.694	0.000	0	0.000	0	4.030	0	.0. 0
58	14.500	0	0.622	0.622	0.000	0	0.000	0	2.936	0	.0. 0
59	13.900	0	0.500	0.500	0.000	0	0.000	0	3.256	0	.0. 0
60	12.700	0	0.323	0.323	0.000	0	0.000	0	4.062	0	.0. 0

MSDDET

MSDDET

TABLE 3c . BEGIN SITUATION; TYPICAL RESULTS FOR DEFAULT CASE; CONTINUED; PAGE 3

INTERVAL 3000

MSDDET

LOCATN	STRESS	NTOTAL	DAM LE	DAM RI	a-LEFT	NCR LE	a-RGHT	NCR RI	CP	EQN CRCASE
61	11.200	0	0.187	0.187	0.000	0	0.000	0	3.162	0 .0. 0
62	12.700	0	0.323	0.323	0.000	0	0.000	0	2.519	0 .0. 0
63	13.900	0	0.500	0.500	0.000	0	0.000	0	3.237	0 .0. 0
64	14.500	0	0.622	0.622	0.000	0	0.000	0	2.431	0 .0. 0
65	14.800	0	0.694	0.694	0.000	0	0.000	0	3.450	0 .0. 0
66	14.900	0	0.720	0.720	0.000	0	0.000	0	3.470	0 .0. 0
67	14.800	0	0.694	0.694	0.000	0	0.000	0	2.597	0 .0. 0
68	14.500	0	0.622	0.622	0.000	0	0.000	0	2.975	0 .0. 0
69	13.800	0	0.482	0.482	0.000	0	0.000	0	3.532	0 .0. 0
70	12.600	0	0.311	0.311	0.000	0	0.000	0	3.294	0 .0. 0
71	11.100	0	0.180	0.180	0.000	0	0.000	0	2.756	0 .0. 0
72	12.600	0	0.311	0.311	0.000	0	0.000	0	2.572	0 .0. 0
73	13.800	0	0.482	0.482	0.000	0	0.000	0	4.095	0 .0. 0
74	14.500	0	0.622	0.622	0.000	0	0.000	0	3.050	0 .0. 0
75	14.800	0	0.694	0.694	0.000	0	0.000	0	3.200	0 .0. 0
76	14.900	0	0.720	0.720	0.000	0	0.000	0	3.125	0 .0. 0
77	14.800	0	0.694	0.694	0.000	0	0.000	0	2.994	0 .0. 0
78	14.500	0	0.622	0.622	0.000	0	0.000	0	2.713	0 .0. 0
79	13.900	0	0.500	0.500	0.000	0	0.000	0	3.275	0 .0. 0
80	12.700	0	0.323	0.323	0.000	0	0.000	0	2.668	0 .0. 0
81	11.200	0	0.187	0.187	0.000	0	0.000	0	3.013	0 .0. 0
82	12.700	0	0.323	0.323	0.000	0	0.000	0	4.769	0 .0. 0
83	13.900	0	0.500	0.500	0.000	0	0.000	0	2.362	0 .0. 0
84	14.500	0	0.622	0.622	0.000	0	0.000	0	4.285	0 .0. 0
85	14.800	0	0.694	0.694	0.000	0	0.000	0	3.511	0 .0. 0
86	14.900	0	0.720	0.720	0.000	0	0.000	0	3.805	0 .0. 0
87	14.800	0	0.694	0.694	0.000	0	0.000	0	3.595	0 .0. 0
88	14.500	0	0.622	0.622	0.000	0	0.000	0	4.996	0 .0. 0
89	13.800	0	0.482	0.482	0.000	0	0.000	0	3.857	0 .0. 0
90	12.600	0	0.311	0.311	0.000	0	0.000	0	3.708	0 .0. 0
91	11.100	0	0.180	0.180	0.000	0	0.000	0	2.898	0 .0. 0
92	12.600	0	0.311	0.311	0.000	0	0.000	0	2.398	0 .0. 0
93	13.800	0	0.482	0.482	0.000	0	0.000	0	3.106	0 .0. 0
94	14.500	0	0.622	0.622	0.000	0	0.000	0	2.000	0 .0. 0
95	14.800	0	0.694	0.694	0.000	0	0.000	0	3.069	0 .0. 0
96	14.900	0	0.720	0.720	0.000	0	0.000	0	2.462	0 .0. 0
97	14.800	0	0.694	0.694	0.000	0	0.000	0	3.410	0 .0. 0
98	14.500	0	0.622	0.622	0.000	0	0.000	0	3.144	0 .0. 0
99	13.900	0	0.500	0.500	0.000	0	0.000	0	2.858	0 .0. 0
100	12.700	0	0.323	0.323	0.000	0	0.000	0	3.911	0 .0. 0

MSDDET

NUMBER OF CRACKS 1 ; NUMBER OF LINKS 0

TABLE 4a . SITUATION AT INSPECTION 8 ; TYPICAL RESULTS FOR DEFAULT CASE; PAGE 1

INTERVAL 3000

MSDDET

SKIN THICKNESS	=	0.0400
DIA FASTENER SHANK	=	0.1800
DIA FASTENER HEAD	=	0.2200
FASTENER DISTANCE	=	1.0000
RADIUS FUSELAGE	=	75.0000
FUSELAGE PRESSURE	=	8.5000
YIELD STRENGTH	=	50.0000
MIN. FAILURE DAMAGE	=	0.5000
LAMBDA DAMAGE DISTR	=	1.0000
ALPHA DAMAGE DISTR	=	2.0000
POWER S-N CURVE	=	2.7400
Cp MINIMUM	=	2.0000
LAMBDA Cp DISTR	=	3.6000
ALPHA Cp DISTR	=	2.0000
Ao EDDY CURRENT	=	0.1200
LAMBDA EDDY CURRENT	=	0.2500
ALPHA EDDY CURRENT	=	0.5000
Ao VISUAL	=	0.2500
LAMBDA VISUAL	=	0.5000
ALPHA VISUAL	=	0.5000
INSP.PROB. INCR	=	0.0500
CRACK GROWTH INCR.	=	100.0000

MSDDET

LOCATN	STRESS	NTOTAL	DAM LE	DAM RI	a-LEFT	NCR LE	a-RGHT	NCR RI	CP	EQN CRCASE
4	14.500	24000	0.961	0.702	0.000	0	0.106	16200	3.088	0 -0--0
5	14.800	24000	0.955	0.696	0.144	11200	0.229	23900	3.639	0--0--0
6	14.900	24000	1.128	1.263	0.061	5900	0.037	2300	3.732	0--0--0
7	14.800	24000	0.696	0.696	0.185	23900	0.186	23900	2.691	0--0--0
8	14.500	24000	1.207	0.805	0.000	0	0.041	6000	2.160	0--0- 0
14	14.500	24000	0.781	0.910	0.066	8400	0.000	0	3.371	0 -0.-0
15	14.800	24000	0.961	0.859	0.043	3400	0.089	11400	3.313	0 -0--0
16	14.900	24000	0.799	0.912	0.169	17300	0.122	10300	3.831	0--0--0
17	14.800	24000	0.882	1.011	0.090	8000	0.000	0	4.871	0--0. 0
24	14.500	24000	1.037	0.623	0.000	0	0.124	23900	2.491	0 .0--0
25	14.800	24000	0.893	0.716	0.088	13600	0.121	22100	2.227	0--0--0
26	14.900	24000	0.840	0.763	0.134	16400	0.154	20400	2.838	0--0--0
27	14.800	24000	1.232	0.960	0.022	400	0.056	7000	2.956	0--0--0
28	14.500	24000	0.679	1.054	0.159	18400	0.000	0	4.204	0--0- 0
35	14.800	24000	0.981	1.027	0.000	0	0.054	4400	4.483	0 -0--0
36	14.900	24000	0.750	1.369	0.130	21500	0.051	5100	2.621	0--0--0
37	14.800	24000	0.696	0.696	0.186	23900	0.187	23900	2.645	0--0--0
38	14.500	24000	0.787	0.934	0.099	12500	0.000	0	3.685	0--0. 0
43	13.900	24000	0.500	0.917	0.200	23900	0.031	1100	4.543	0 -0--0
44	14.500	24000	0.864	1.021	0.023	600	0.029	1900	3.031	0--0--0

MSDDET

TABLE 4b . SITUATION AT INSPECTION 8 ; TYPICAL RESULTS FOR DEFAULT CASE; CONTINUED; PAGE 2
INTERVAL 3000

MSDDET

LOCATN	STRESS	NTOTAL	DAM LE	DAM RI	a-LEFT	NCR LE	a-RGHT	NCR RI	CP EQN	CRCASE
45	14.800	24000	0.736	1.322	0.177	20300	0.086	6700	3.662	0--0--0
46	14.900	24000	0.721	0.988	0.234	23900	0.168	13300	3.490	0--0--0
47	14.800	24000	0.696	0.955	0.152	23900	0.078	8800	2.734	0--0--0
55	14.800	24000	1.028	0.935	0.021	100	0.094	10900	3.780	0 -0--0
56	14.900	24000	0.721	1.014	0.233	23900	0.132	9400	3.756	0--0--0
57	14.800	24000	1.000	0.730	0.106	7500	0.206	20900	4.030	0--0--0
64	14.500	24000	0.846	1.103	0.048	2300	0.410	3200	2.431	0--0--0
65	14.800	24000	0.696	0.696	0.410	23000	0.410	23000	3.450	0--0--0
66	14.900	24000	0.794	1.135	0.410	18200	0.069	2200	3.470	0--0--0
67	14.800	24000	0.974	1.180	0.000	0	0.020	100	2.597	0--0--0
68	14.500	24000	0.636	1.055	0.138	22700	0.000	0	2.975	0--0- 0
74	14.500	24000	0.723	0.906	0.092	14100	0.029	1600	3.050	0 -0- 0
77	14.800	24000	0.823	0.823	0.099	12700	0.099	12700	2.994	0--0--0
78	14.500	24000	0.924	0.852	0.023	600	0.026	1400	2.713	0--0- 0
84	14.500	24000	0.695	0.776	0.172	16900	0.131	10900	4.285	0--0--0
85	14.800	24000	1.342	0.696	0.052	3200	0.194	23900	3.511	0--0--0
86	14.900	24000	0.987	0.900	0.104	10000	0.105	10100	3.805	0--0--0
87	14.800	24000	0.834	0.709	0.179	14700	0.230	22700	3.595	0--0--0
88	14.500	24000	1.036	0.695	0.119	7400	0.200	16900	4.996	0--0--0
89	13.800	24000	0.571	0.723	0.096	13500	0.000	0	3.857	0--0- 0
94	14.500	24000	0.730	0.977	0.063	13400	0.000	0	2.000	0 -0.-0
95	14.800	24000	0.852	0.875	0.079	10100	0.071	8400	3.069	0 -0- 0
97	14.800	24000	0.911	0.847	0.074	7200	0.116	14200	3.410	0 -0--0
98	14.500	24000	0.623	0.812	0.171	23900	0.105	11200	3.144	0--0--0

MSDDET

NUMBER OF CRACKS 93 ; NUMBER OF LINKS 21

the left side of hole 43 is cracked, and hole 42 is denoted as 0 .0.-0, indicating the left side of hole 43 is cracked.

The life to failure (10 links) in this case was 26000 cycles. Therefore, the last inspection took place at 24000 cycles. The situation at this last inspection is printed out in the same manner as in Table 3(a,b,c), but only cracked holes are represented. It is shown in Tables 4(a,b). At this time there were 93 cracks as indicated at the bottom of Table 4(b).

Tables 5(a,b) show the development over 24000 flight cycles. Note that after 100 cycles there were already 14 cracks in this case. The cumulative probability of detection in this table is for the detection of (any) one crack. It is emphasized that this is not a useful indication of the probability of detection. The last column in Table 5(a,b) provides the size of the largest crack not yet linked.

TABLE 5a . STATISTICS FOR TYPICAL RESULTS FOR DEFAULT CASE; INTERVAL 3000 ; PAGE 1
NO REPAIRS

MSDDET

CYCLES	#CRACKS	#LINKS	ACTUAL DETECTS #DET E	#DET V	CUM.PROB. EC AT LEAST ONE	VI	LAR.CR
100.	14.	0.	0.	0.	0.0000	0.0000	0.0200
500.	14.	0.	0.	0.	0.0000	0.0000	0.0225
1000.	14.	0.	0.	0.	0.0000	0.0000	0.0257
1500.	16.	0.	0.	0.	0.0000	0.0000	0.0289
2000.	17.	0.	0.	0.	0.0000	0.0000	0.0321
2500.	18.	0.	0.	0.	0.0000	0.0000	0.0354
3000.	18.	0.	0.	0.	0.0000	0.0000	0.0387
3500.	19.	0.	0.	0.	0.0000	0.0000	0.0419
4000.	21.	0.	0.	0.	0.0000	0.0000	0.0452
4500.	21.	0.	0.	0.	0.0000	0.0000	0.0486
5000.	22.	0.	0.	0.	0.0000	0.0000	0.0519
5500.	22.	0.	0.	0.	0.0000	0.0000	0.0553
6000.	23.	0.	0.	0.	0.0000	0.0000	0.0587
6500.	23.	0.	0.	0.	0.0000	0.0000	0.0621
7000.	24.	0.	0.	0.	0.0000	0.0000	0.0657
7500.	26.	0.	0.	0.	0.0000	0.0000	0.0694
8000.	28.	0.	0.	0.	0.0000	0.0000	0.0733
8500.	28.	0.	0.	0.	0.0000	0.0000	0.0772
9000.	28.	0.	0.	0.	0.0000	0.0000	0.0812
9500.	29.	0.	0.	0.	0.0000	0.0000	0.0853
10000.	31.	0.	0.	0.	0.0000	0.0000	0.0896
10500.	33.	0.	0.	0.	0.0000	0.0000	0.0939
11000.	35.	0.	0.	0.	0.0000	0.0000	0.0983
11500.	38.	0.	0.	0.	0.0000	0.0000	0.1029
12000.	38.	0.	0.	0.	0.0000	0.0000	0.1075
12500.	38.	0.	0.	0.	0.3241	0.0000	0.1124
13000.	41.	0.	0.	0.	0.3241	0.0000	0.1174
13500.	43.	0.	0.	0.	0.3241	0.0000	0.1225
14000.	47.	0.	0.	0.	0.3241	0.0000	0.1278
14500.	47.	0.	0.	0.	0.3241	0.0000	0.1333
15000.	48.	0.	0.	0.	0.3241	0.0000	0.1391
15500.	49.	0.	0.	0.	0.8291	0.0000	0.1450
16000.	52.	0.	0.	0.	0.8291	0.0000	0.1512
16500.	53.	0.	0.	0.	0.8291	0.0000	0.1578
17000.	56.	0.	0.	0.	0.8291	0.0000	0.1646
17500.	57.	0.	0.	0.	0.8291	0.0000	0.1718
18000.	58.	0.	0.	0.	0.8291	0.0000	0.1794
18500.	59.	0.	6.	0.	0.9971	0.1286	0.1876
19000.	60.	0.	6.	0.	0.9971	0.1286	0.1963
19500.	60.	0.	6.	0.	0.9971	0.1286	0.2057
20000.	62.	0.	6.	0.	0.9971	0.1286	0.2160
20500.	62.	0.	6.	0.	0.9971	0.1286	0.2277
21000.	64.	0.	6.	0.	0.9971	0.1286	0.2410

MSDDET

MSDDET

TABLE 5b . STATISTICS FOR TYPICAL RESULTS FOR DEFAULT CASE; INTERVAL 3000 ; PAGE 2

NO REPAIRS

MSDDET

CYCLES	#CRACKS	#LINKS	ACTUAL DETECTS #DET E	#DET V	CUM.PROB. EC AT LEAST ONE	VI	LAR.CR
21500.	64.	0.	7.	1.	1.0000	0.5411	0.2568
22000.	67.	0.	7.	1.	1.0000	0.5411	0.2766
22500.	69.	0.	7.	1.	1.0000	0.5411	0.3058
23000.	71.	0.	7.	1.	1.0000	0.5411	0.4756
23500.	73.	2.	7.	1.	1.0000	0.5411	0.2245
24000.	76.	2.	7.	1.	1.0000	0.5411	0.2340
24500.	78.	2.	10.	3.	1.0000	0.9940	0.2454
25000.	79.	2.	10.	3.	1.0000	0.9940	0.2585
25500.	79.	3.	10.	3.	1.0000	0.9940	0.2744
26000.	79.	3.	10.	3.	1.0000	0.9940	0.2997
26500.	80.	6.	10.	3.	1.0000	0.9940	0.3326

MSDDET

MSDDET

Table 6 shows a summary for all inspections (at 3000 cycles intervals). It should be noted again that the probabilities of detection of (any) one crack shown in Table 6 are of little relevance. Finally, Tables 7(a,b) shows the cumulative probability of detection of individual cracks. This is the most relevant information, especially for those cracks that linked-up. For the assessment of safety, one needs to know the probability of detection of those cracks that, when missed, would cause a final failure of the fuselage. Note that Table 7(a,b) covers a different case than the previously cited tables.

Graphical results of a typical case are shown in Figure 14 (a,b,c) are discussed below. Since each run is different because of the random assignments of damage, crack growth properties and appearance of first crack, the displays differ from those in Tables 2-7.

A typical situation at the last inspection before final failure is exhibited in Figure 15. The associated cumulative probability of detection of the individual cracks is shown in Figure 16. Indeed, the probabilities shown in Figure 16, are relevant to the problem. Most cracks occur in the bay between the straps as can be ascertained from Figure 15. This is in accordance with observations of actual cracks [16]. The cracks that linked up first (between holes 10 and 20 in Figure 15), have the highest cumulative probability of detection in Figure 16, as might be expected (larger cracks are more easily detectable).

TABLE 6 . LAST INSPECTION; INTERVAL: 3000 ;PAGE 1
TYPICAL RESULTS FOR DEFAULT CASE; NO REPAIRS

MSDDET

SKIN THICKNESS	=	0.0400
DIA FASTENER SHANK	=	0.1800
DIA FASTENER HEAD	=	0.2200
FASTENER DISTANCE	=	1.0000
RADIUS FUSELAGE	=	75.0000
FUSELAGE PRESSURE	=	8.5000
YIELD STRENGTH	=	50.0000
MIN. FAILURE DAMAGE	=	0.5000
LAMBDA DAMAGE DISTR	=	1.0000
ALPHA DAMAGE DISTR	=	2.0000
POWER S-N CURVE	=	2.7400
Cp MINIMUM	=	2.0000
LAMBDA Cp DISTR	=	3.6000
ALPHA Cp DISTR	=	2.0000
Ao EDDY CURRENT	=	0.1200
LAMBDA EDDY CURRENT	=	0.2500
ALPHA EDDY CURRENT	=	0.5000
Ao VISUAL	=	0.2500
LAMBDA VISUAL	=	0.5000
ALPHA VISUAL	=	0.5000

MSDDET

CYCLES	CUMIS E	CUMDETE	# DET E	CUMIS V	CUMDETV	# DET V	# CRACKS	# LINKS
	ALL	ONE		ALL	ONE			
3000	1.0000	0.0000	0	1.0000	0.0000	0	18	0
6000	1.0000	0.0000	0	1.0000	0.0000	0	23	0
9000	1.0000	0.0000	0	1.0000	0.0000	0	28	0
12000	0.6759	0.3241	0	1.0000	0.0000	0	38	0
15000	0.1709	0.8291	0	1.0000	0.0000	0	48	0
18000	0.0029	0.9971	6	0.8714	0.1286	0	58	0
21000	0.0000	1.0000	7	0.4589	0.5411	1	64	0
24000	0.0000	1.0000	10	0.0060	0.9940	3	76	2

MSDDET

TOTAL DETECTED EC : 46 ; TOTAL DETECTED VISUAL : 8

Other interesting information is plotted in Figure 17, namely the number of cracks and the number of link-ups (left scale), and the size of the largest crack (right scale). The number of cracks increases in discrete steps of at least one crack. As accounting is done every 100 cycles, the number stays constant over at least one such interval. Occasionally, more than one new crack initiates in such an interval, especially towards the end.

TABLE 7a . DETECTION RECORD; DEFAULT CASE ; PAGE 1
INSPECTION # 10 ; INTERVAL: 3000 ; CYCLES 30000

MSDDET

LOCATN	PER CRACK Pdet-E	Pdet-V	#det-E	#det-V	CYCL-E	CYCL-V	REPR-E	REPR-V	CRCASE
1	0.000	0.000	0	0	30000	30000	0	0	O .O. O
2	0.000	0.000	0	0	30000	30000	0	0	O .O. O
3	0.000	0.000	0	0	30000	30000	0	0	O .O--O
4	0.000	0.000	0	0	30000	30000	0	0	O--O- O
5	0.000	0.000	0	0	30000	30000	0	0	O-.O. O
6	0.000	0.000	0	0	30000	30000	0	0	O .O--O
7	0.000	0.000	0	0	30000	30000	0	0	O--O--O
8	0.344	0.000	0	0	30000	30000	0	0	O--O- O
9	0.000	0.000	0	0	30000	30000	0	0	O-.O. O
10	0.000	0.000	0	0	30000	30000	0	0	O .O. O
11	0.000	0.000	0	0	30000	30000	0	0	O .O. O
12	0.000	0.000	0	0	30000	30000	0	0	O .O. O
13	0.000	0.000	0	0	30000	30000	0	0	O .O. O
14	0.000	0.000	0	0	30000	30000	0	0	O .O.-O
15	0.191	0.000	1	0	30000	30000	0	0	O -O--O
16	0.495	0.000	0	0	30000	30000	0	0	O--O- O
17	0.000	0.000	0	0	30000	30000	0	0	O-.O--O
18	0.000	0.000	0	0	30000	30000	0	0	O--O- O
19	0.000	0.000	0	0	30000	30000	0	0	O-.O. O
20	0.000	0.000	0	0	30000	30000	0	0	O .O. O
21	0.000	0.000	0	0	30000	30000	0	0	O .O. O
22	0.000	0.000	0	0	30000	30000	0	0	O .O.-O
23	0.000	0.000	0	0	30000	30000	0	0	O -O- O
24	0.000	0.000	0	0	30000	30000	0	0	O-.O--O
25	0.544	0.000	1	0	30000	30000	0	0	O--O--O
26	0.000	0.000	0	0	30000	30000	0	0	O--O. O
27	0.000	0.000	0	0	30000	30000	0	0	O .O. O
28	0.000	0.000	0	0	30000	30000	0	0	O .O. O
29	0.000	0.000	0	0	30000	30000	0	0	O .O. O
30	0.000	0.000	0	0	30000	30000	0	0	O .O. O
31	0.000	0.000	0	0	30000	30000	0	0	O .O. O
32	0.000	0.000	0	0	30000	30000	0	0	O .O.-O
33	0.000	0.000	0	0	30000	30000	0	0	O -O--O
34	0.608	0.000	0	0	30000	30000	0	0	O--O--O
35	0.000	0.000	0	0	30000	30000	0	0	O--O--O
36	0.937	0.446	1	1	30000	30000	0	0	O--O--O
37	0.739	0.477	1	1	30000	30000	0	0	O--O--O
38	0.990	0.838	1	1	24000	30000	0	0	O--O--O
39	0.742	0.482	1	0	30000	30000	0	0	O--O--O
40	0.000	0.000	0	0	30000	30000	0	0	O--O. O
41	0.000	0.000	0	0	30000	30000	0	0	O .O. O
42	0.000	0.000	0	0	30000	30000	0	0	O .O. O
43	0.000	0.000	0	0	30000	30000	0	0	O .O--O
44	0.000	0.000	0	0	30000	30000	0	0	O--O--O
45	0.000	0.000	0	0	30000	30000	0	0	O--O--O
46	0.000	0.000	0	0	30000	30000	0	0	O--O--O
47	0.000	0.000	0	0	30000	30000	0	0	O--O. O
48	0.000	0.000	0	0	30000	30000	0	0	O .O. O
49	0.000	0.000	0	0	30000	30000	0	0	O .O. O
50	0.000	0.000	0	0	30000	30000	0	0	O .O. O

MSDDET

MSDDET

TABLE 7b . DETECTION RECORD; DEFAULT CASE ; PAGE 2
INSPECTION # 10 ; INTERVAL: 3000 ; CYCLES 30000

MSDDET

LOCATN	PER CRACK Pdet-E	Pdet-V	#det-E	#det-V	CYCL-E	CYCL-V	REPR-E	REPR-V	CRCASE
51	0.000	0.000	0	0	30000	30000	0	0	O .O. O
52	0.000	0.000	0	0	30000	30000	0	0	O .O. O
53	0.000	0.000	0	0	30000	30000	0	0	O .O. O
54	0.000	0.000	0	0	30000	30000	0	0	O .O--O
55	0.000	0.000	0	0	30000	30000	0	0	O--O--O
56	0.000	0.000	0	0	30000	30000	0	0	O--O- O
57	0.000	0.000	0	0	30000	30000	0	0	O-.O. O
58	0.000	0.000	0	0	30000	30000	0	0	O .O. O
59	0.000	0.000	0	0	30000	30000	0	0	O .O. O
60	0.000	0.000	0	0	30000	30000	0	0	O .O. O
61	0.000	0.000	0	0	30000	30000	0	0	O .O. O
62	0.000	0.000	0	0	30000	30000	0	0	O .O. O
63	0.000	0.000	0	0	30000	30000	0	0	O .O.-O
64	0.000	0.000	0	0	30000	30000	0	0	O -O--O
65	0.619	0.000	0	0	30000	30000	0	0	O--O--O
66	0.980	0.644	1	1	24000	30000	0	0	O--O--O
67	0.000	0.000	0	0	30000	30000	0	0	O--O--O
68	0.000	0.000	0	0	30000	30000	0	0	O--O- O
69	0.000	0.000	0	0	30000	30000	0	0	O-.O. O
70	0.000	0.000	0	0	30000	30000	0	0	O .O. O
71	0.000	0.000	0	0	30000	30000	0	0	O .O. O
72	0.000	0.000	0	0	30000	30000	0	0	O .O.-O
73	0.000	0.000	0	0	30000	30000	0	0	O -O--O
74	0.841	0.672	1	1	30000	30000	0	0	O--O--O
75	0.992	0.850	1	0	27000	30000	0	0	O--O--O
76	0.766	0.488	1	0	30000	30000	0	0	O--O--O
77	0.000	0.000	0	0	30000	30000	0	0	O--O--O
78	0.000	0.000	0	0	30000	30000	0	0	O--O- O
79	0.000	0.000	0	0	30000	30000	0	0	O-.O. O
80	0.000	0.000	0	0	30000	30000	0	0	O .O. O
81	0.000	0.000	0	0	30000	30000	0	0	O .O. O
82	0.000	0.000	0	0	30000	30000	0	0	O .O.-O
83	0.000	0.000	0	0	30000	30000	0	0	O -O. O
84	0.000	0.000	0	0	30000	30000	0	0	O .O.-O
85	0.710	0.000	1	0	27000	30000	0	0	O -O--O
86	0.000	0.000	0	0	30000	30000	0	0	O--O--O
87	0.000	0.000	0	0	30000	30000	0	0	O--O- O
88	0.000	0.000	0	0	30000	30000	0	0	O-.O. O
89	0.000	0.000	0	0	30000	30000	0	0	O .O. O
90	0.000	0.000	0	0	30000	30000	0	0	O .O. O
91	0.000	0.000	0	0	30000	30000	0	0	O .O. O
92	0.000	0.000	0	0	30000	30000	0	0	O .O. O
93	0.000	0.000	0	0	30000	30000	0	0	O .O.-O
94	0.000	0.000	0	0	30000	30000	0	0	O -O--O
95	0.414	0.000	0	0	30000	30000	0	0	O--O--O
96	0.727	0.450	1	1	30000	30000	0	0	O--O--O
97	0.995	0.836	1	1	18000	30000	0	0	O--O--O
98	0.729	0.454	1	0	30000	30000	0	0	O--O- O
99	0.000	0.000	0	0	30000	30000	0	0	O-.O. O
100	0.000	0.000	0	0	30000	30000	0	0	O .O. O

MSDDET

NUMBER OF CRACKS 86 ; NUMBER OF LINKS 21

```
FIRST CRACK AT LOC.  78 ; STRESS =  14.5 ; AVERAGE STRESS =  13.62872
PRESENT SITUATION AT CYCLE NUMBER      100
# CRACKS          3;          LINK UPS       0
LARGEST CRACK     0.02000; AT LOCATION   78

INSPECTION INTERVAL NOW IS   3000 CYCLES
```

F = FRAME; S = STRAP; O = FASTENER; - = CRACK; # = LOCATION OF LARGEST CRACK

```
..F....O....O....O....O....O....O....O....O....O....S....O....O....O....O....O..
..O....O....O....O....F....O....O....O....O....O....O....O....O....O....S....O..
..O....O....O....O....O....O....O....O....F....O....O....O....O....O....O....O..
..O....O....S....O....O....O....O....O....O....O....O....O....F....O....O....O..
..O....O....O....O....O....O....S....O....O....O....O....O....O...-#....O....O..
..F....O....O....O....O....O....O....O....O....O....S....O....O....O....O....O..
..O....O....O....O..
```

```
FIRST CRACK AT LOC.  78 ; STRESS =  14.5 ; AVERAGE STRESS =  13.62872
PRESENT SITUATION AT CYCLE NUMBER   15000
# CRACKS          20;          LINK UPS       0
LARGEST CRACK     0.11709; AT LOCATION   25

INSPECTION INTERVAL NOW IS   3000 CYCLES
LAST INSPECTION WAS  #   5. AT  15000 CYCLES
CUM.PROB.DET. EC/VIS :  0.0959  0.0000
# DETECTS      EC/VIS :   0   0
# REPAIRED     EC/VIS :   0   0

..F....O....O....O...-O....O....O-...O....O....O....S....O....O....O....O....O..
..O....O....O....O....F....O....O....O...-#-..-O....O....O....O....O....S....O..
..O....O-..-O....O....O....O....O....O....F....O....O....O....O....O....O-...O..
..O....O....S....O....O....O...-O....O....O-..-O....O....O....F....O....O....O..
..O...-O-...O....O-...O....O....S....O....O....O....O....O....O...-O....O....O..
..F....O....O....O....O...-O-..-O...-O....O....O....S....O....O....O....O...-O..
..O....O....O....O..
```

FIGURE 14a. SCREEN DISPLAYS OF SUCCESSIVE CRACK DEVELOPMENT
TOP: BEGIN SITUATION; BOTTOM: AFTER 15000 CYCLES

```
FIRST CRACK AT LOC.  78 ; STRESS =  14.5 ; AVERAGE STRESS =  13.62872
PRESENT SITUATION AT CYCLE NUMBER   25000
# CRACKS          45;          LINK UPS       0
LARGEST CRACK     0.28425; AT LOCATION   86

INSPECTION INTERVAL NOW IS   3000 CYCLES
LAST INSPECTION WAS  #   8 AT  24000 CYCLES
CUM.PROB.DET. EC/VIS :  0.9998  0.6014
# DETECTS      EC/VIS :   8   2
# REPAIRED     EC/VIS :   0   0

..F....O....O....O...-O-...O...-O-...O....O....O....S....O....O....O-...O....O..
..O....O....O....O....F....O....O-...O-..-O-..-O-..-O....O....O....O....S....O..
..O....O-..-O....O-...O....O....O....O....F....O....O....O-...O-..-O....O-..-O..
..O....O....S....O....O....O-..-O....O..-O-..-O-...O....O....F....O....O....O..
..O-..-O-...O....O-...O....O....S....O....O....O....O-..-O...-O...-O-...O....O..
..F....O....O....O....O...-#-..-O-..-O-...O....O....S....O....O....O-...O...-O-.
.-O....O....O....O..
```

```
FIRST CRACK AT LOC.  78 ; STRESS =  14.5 ; AVERAGE STRESS =  13.62872
PRESENT SITUATION AT CYCLE NUMBER   27300
# CRACKS          53;          LINK UPS       3
LARGEST CRACK     0.31314; AT LOCATION   25

INSPECTION INTERVAL NOW IS   3000 CYCLES
LAST INSPECTION WAS  #   9 AT  27000 CYCLES
CUM.PROB.DET. EC/VIS :  1.0000  0.9979
# DETECTS      EC/VIS :  14   6
# REPAIRED     EC/VIS :   0   0

..F....O....O....O...-O-...O...-O-...O....O....O....S....O....O....O-...O....O..
..O....O....O....O....F....O....O-...O-..-#-..-O-..-O....O....O....O....S....O..
..O...-O-..-O....O-...O....O....O....O....F....O....O....O-...O-..-O...-O-..-O..
..O....O....S....O....O....O-..-O....O...-O-..-O-...O....O....F....O....O....O..
..O-..-O-...O....O-...O....O....S....O....O....O....O-..-O...-O...-O-...O-...O..
..F....O....O....O...-O----O----O----O-...O....O....S....O....O....O-..-O-..-O-.
.-O....O....O....O..
```

FIGURE 14b.

SCREEN DISPLAYS OF CRACK DEVELOPMENT (CONTINUED)
TOP: AFTER 24000 CYCLES; BOTTOM: AFTER 27000 CYCLES (FIRST LINK-UPS)

```
FIRST CRACK AT LOC.  78 ; STRESS =  14.5 ; AVERAGE STRESS =  13.62872
PRESENT SITUATION AT CYCLE NUMBER   30000
# CRACKS          67;        LINK UPS       9
LARGEST CRACK    0.34251; AT LOCATION   96

INSPECTION INTERVAL NOW IS   3000 CYCLES
LAST INSPECTION WAS  #  10 AT  30000 CYCLES
CUM.PROB.DET. EC/VIS :  1.0000  1.0000
# DETECTS      EC/VIS :  25  15
# REPAIRED     EC/VIS :   0   0

..F....O...-O....O...-O-..-O-..-O-...O....O....O....S....O....O....O-...O-...O..
..O-...O....O....O....F....O....O----O----O----O----O-...O....O....O....S....O..
..O...-O-..-O-...O-...O....O....O....O....F....O....O....O-...O-..-O...-O-..-O..
..O....O....S....O....O....O-..-O-...O-..-O-..-O-...O....O....F....O....O....O..
..O-..-O-..-O....O-...O....O....S....O....O....O....O-..-O-..-O-..-O-...O-...O..
..F....O....O...-O----O----O----O----O----O-...O....S....O....O....O-..-O-..-#-.
.-O....O....O....O..

FIRST CRACK AT LOC.  78 ; STRESS =  14.5 ; AVERAGE STRESS =  13.62872
PRESENT SITUATION AT CYCLE NUMBER   34100
# CRACKS          85;        LINK UPS      21
LARGEST CRACK    0.29398; AT LOCATION   57

INSPECTION INTERVAL NOW IS   3000 CYCLES
LAST INSPECTION WAS  #  10 AT  30000 CYCLES
CUM.PROB.DET. EC/VIS :  1.0000  1.0000
# DETECTS      EC/VIS :  25  15
# REPAIRED     EC/VIS :   0   0

--0----0----0----0----0----0----0----0----0----0----0----0----0----0----0--
--0----0----0----0----0----0----0----0----0----0----0----0----0----0----0--
--0----0----0----0----0----0----0----0----0----0----0----0----0----0----0--
--0----0----0----0----0----0----0----0----0----0----0----0----0----0----0--
--0----0----0----0----0----0----0----0----0----0----0----0----0----0----0--
--0----0----0----0----0----0----0----0----0----0----0----0----0----0----0--
--0----0----0----0--
```

FIGURE 14c.

**SCREEN DISPLAYS OF CRACK DEVELOPMENT (CONTINUED)
TOP: AFTER 30000 CYCLES (CONTINUED LINK-UP; BOTTOM: FINAL FAILURE**

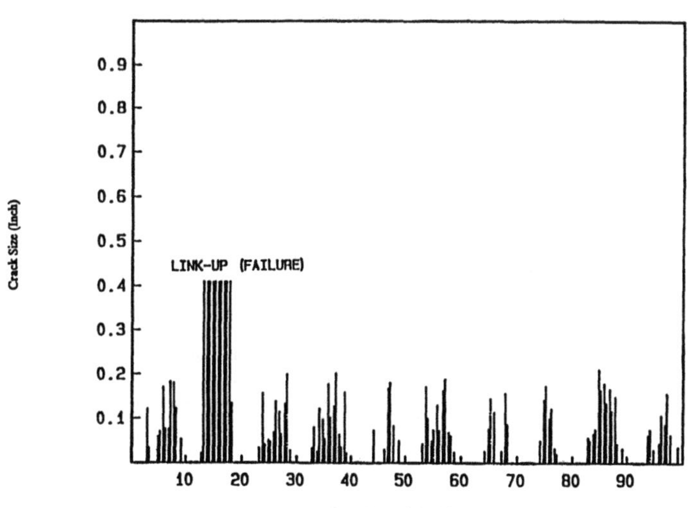

FIGURE 15.

**FINAL CRACK SIZES AT ALL LOCATIONS; SAME AS IN FOLLOWING FIGURE
SOLID LINE: LEFT CRACK; DASH-DOT : RIGHT CRACK**

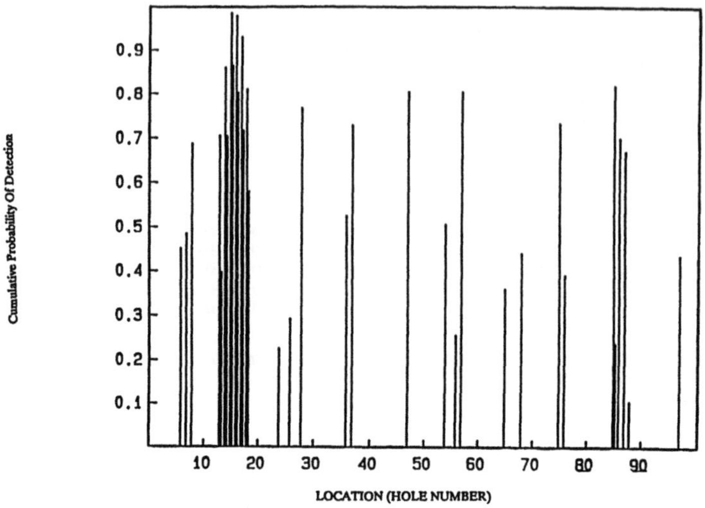

FIGURE 16. CUMULATIVE PROBABILITY OD DETECTION PER LOCATION
SOLID LINE: EDDY CURRENT; DASH-DOT LINE: VISUAL INSPECTION
INTERVAL 3000 CYCLES; TYPICAL CASE

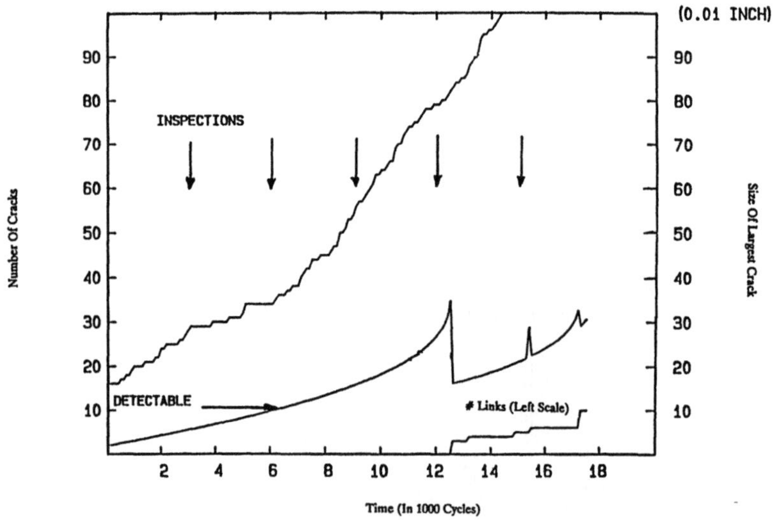

FIGURE 17. NUMBER OF CRACKS; NUMBER OF LINK-UPS AND SIZE OF LARGEST CRACK NOT LINKED; TYPICAL CASE:
INSPECTION INTERVAL 3000 CYCLES

The size of the largest crack drops after link-ups (Figure 17), because, when the largest crack links up, it is no longer accounted for, and the curve is continued with the next largest crack not yet linked. The largest crack is not always at the same location. If there is only one (initial) crack, that crack is naturally the largest, but another crack occurring later may grow faster and then become the largest. This can be seen in Figures 14 (a,b,c), where the largest crack is at location 78 (first crack in Figure 14(a), top), but later at locations 25, 86, 25, 96, and 57 (Figures 14 (a,b,c)).

Figure 18 shows the cumulative probability of detection. The top curve is for the probability that any (one) crack, no matter where, is detected. This probability is very high, as might be expected. Most relevant is the curve for the probability of detection of the cracks that linked up and caused final failure. The average curve for all cracks is interesting, but not of great value. The 'average' curves for all cracks and those that linked-up were obtained by literally taking averages for those cracks concerned.

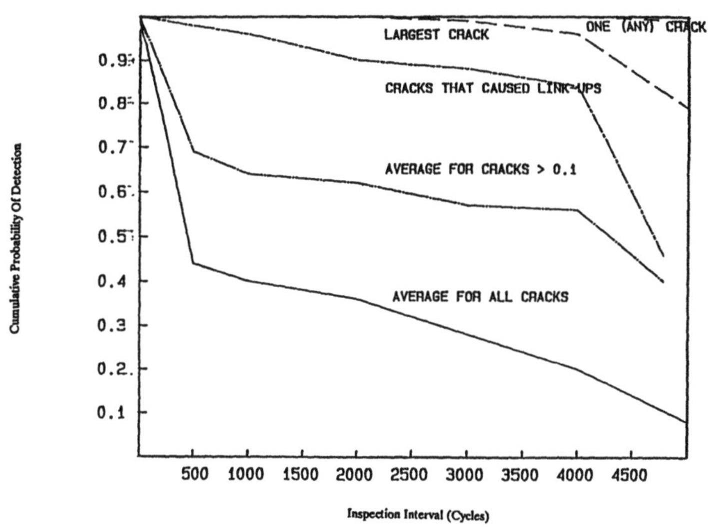

FIGURE 18. CUMULATIVE PROBABILITY OF DETECTION WITH EDDY CURRENT AS A FUNCTION OF LENGTH INSPECTION INTERVAL TYPICAL CASE

Sensitivity Analysis

Since many parameters are assigned randomly, they vary statistically over a wide range. Therefore, and also because many assumptions are involved, a sensitivity study was necessary to investigate whether the analysis results are meaningful, or whether they depend so strongly on the statistical distributions and assumptions that no definite conclusions can be drawn.

Fortunately, the sensitivity analysis showed that the results are very consistent, showing little or no effect of 'random picks', the statistical parameters, or even the assumptions. This seems a little surprising at first but can be logically explained.

All sensitivity analysis was done using as a reference a situation with a cabin pressure of 7.8 psi, and a skin thickness of 0.036 inch. Because of random assignments the results of each run will be different. By making a series of runs with identical parameters the variability of the results can be assessed.

Complete runs, each with six inspection intervals were made. The most telling information is the numbers of cracks in each case. These are shown in the following table.

Results of 5 runs

	# of cracks in case of inspection interval (cycles)						
Run #	500	1000	2000	3000	4000	5000	average
1	107	89	103	97	82	82	92
2	97	85	83	73	90	88	86
3	88	87	98	102	93	105	96
4	85	92	95	86	87	85	88
5	76	80	85	96	103	88	88
Average	91	87	93	91	91	90	90 (overall)

The number of cracks in some of these runs is plotted in Figure 19 (in order to avoid clutter not all results are shown). Figure 20 shows the growth of the largest crack in a number of runs. The discontinuities in these curves were explained already on the basis of Figure 17.

Despite the variability in Figures 19 and 20, the essential results are not much affected. As this work attempted to determine the most suitable inspection interval for the most dangerous cracks, the results shown on Figure 21 are more interesting. As was the case for Figure 18, the curves in Figure 21 were obtained by the average of the cumulative

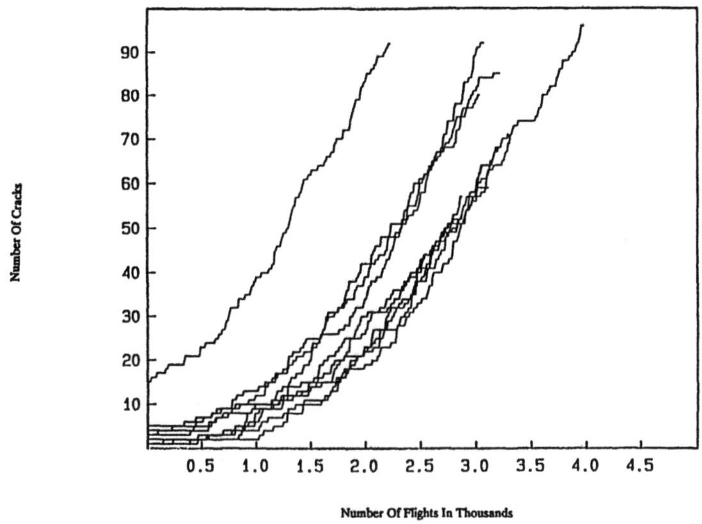

FIGURE 19. NUMBER OF CRACKS IN 10 RUNS

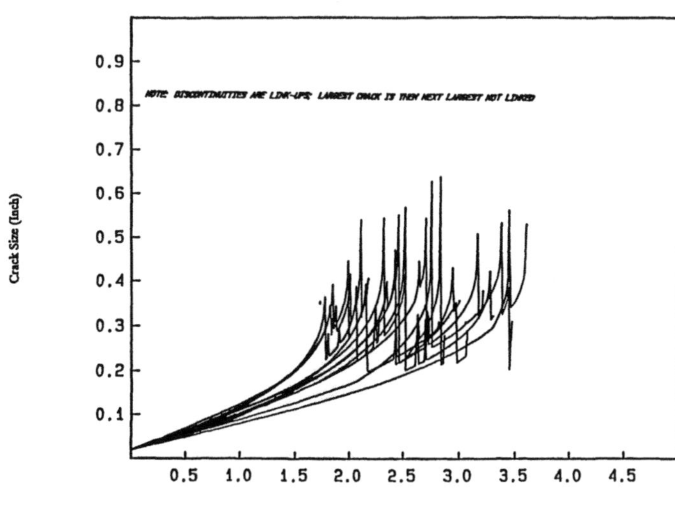

FIGURE 20. GROWTH OF LARGEST CRACK IN 10 RUNS

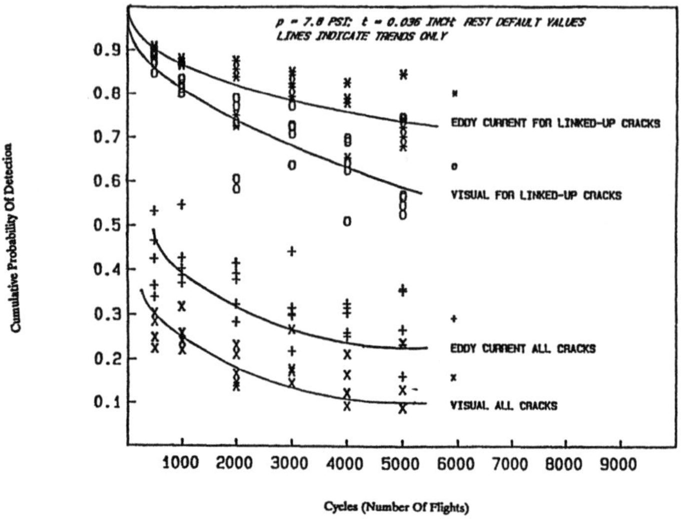

FIGURE 21. STATISTICAL VARIABILITY OF RESULTS IN FIVE IDENTICAL RUNS VARIABILITY DUE TO STATISTICAL VARIATION OF PARAMETERS

probabilities for all cracks. It can be taken as an indicator for comparisons, but taking averages is not a valid procedure for the determination of inspection intervals. Despite this averaging, there is considerable variability. Nevertheless, a definite trend emerges. Longer inspection intervals lead to lower probability of detection, but this is still trivial. Most interesting is the comparison with Figure 22.

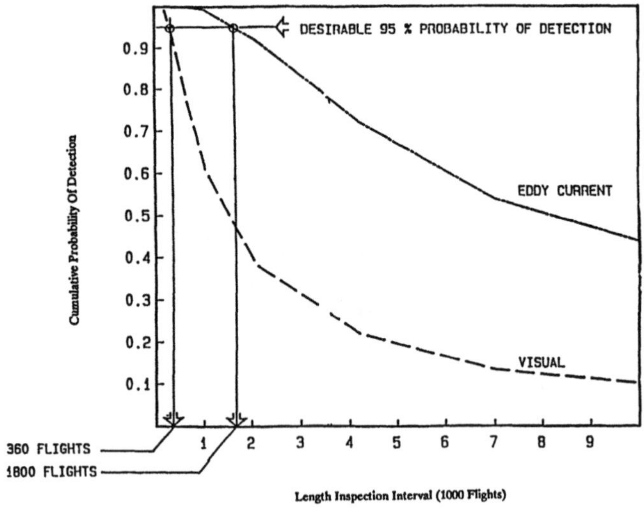

FIGURE 22.
CUMULATIVE PROBABILITY OF DETECTION. EASY ACCESS; LARGE AREA; MANY HOLES [1]

For the longer inspection intervals, Figures 21 and 22 give essentially the same results for eddy current inspection. This can be explained tentatively. Although Figure 13 is based on an assumption, it is obvious that the probability of detection increases somewhat after a 'detect', due to greater alertness of the inspector, and vice versa. But because MSD cracks are clustered (Figure 15), these effect apply only very briefly. Besides, most of the cracks are below the detection limit during most of their 'life', so that the above effect hardly comes into play, except for very short inspection intervals. For practicable inspection intervals, Figures 21 based on averages for critical cracks, cannot differ very much from Figure 22. It is at least an indication of the consistency of the analysis.

Further confirmation of this logic is obtained from Figures 23-26. These figures show the results for an inspection interval of 3000 cycles. In order to 'dramatize' the differences these data were plotted for the cases of the smallest number of cracks (73) and the largest number of cracks (102). The only cracks that really count for safety are the ones that link up and cause a failure. No matter what the statistical distribution, these cracks behave in essentially the same manner, since they always occur at the worst locations. The statistical distribution only affects the location where they occur, hardly their actual growth. It is rather immaterial whether the failure occurs at hole 1 or at hole 100; the result is the same.

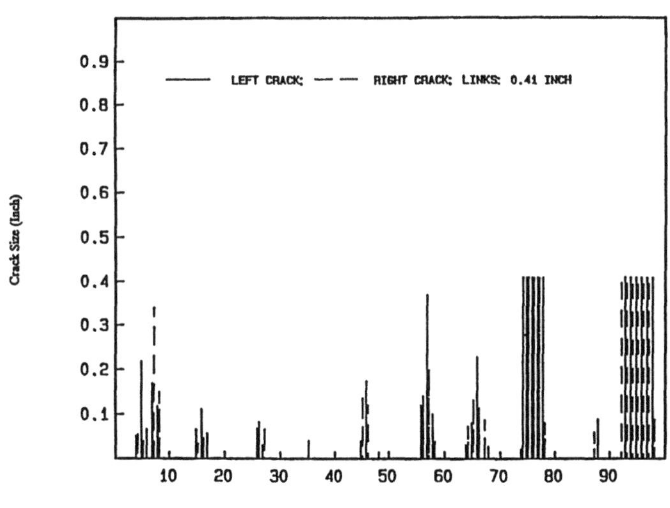

FIGURE 23. CRACK SIZES AT LAST INSPECTION IN CASE WITH LEAST CRACKS
p = 7.8 PSI; t = 0,036 INCH; INTERVAL = 3000 CYCLES
NUMBER OF CRACKS: 73

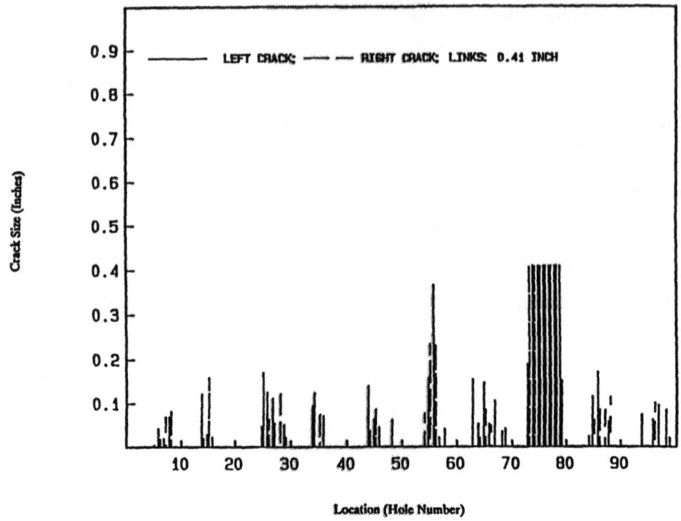

FIGURE 24. CRACK SIZES AT LAST INSPECTION IN RUN WITH THE MOST CRACKS
p = 7.8 PSI; t = 0.036 INCH; NUMBER OF CRACKS: 102

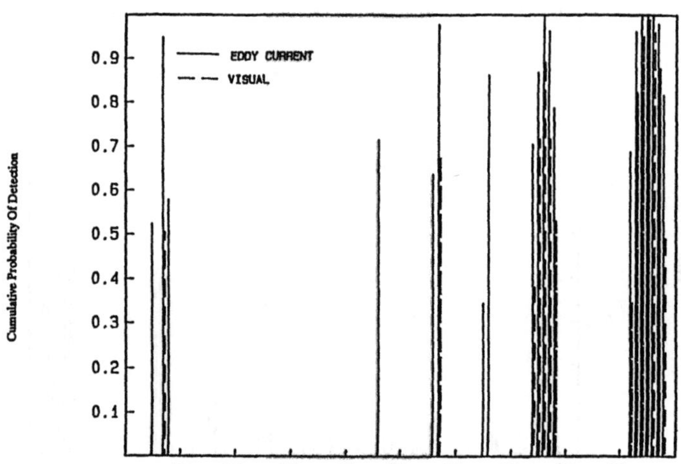

FIGURE 25. CUMULATIVE PROBABILITY OF DETECTION INDIVIDUAL CRACKS
CASE WITH LEAST CRACKS = (73)
p = 7.8 PSI; t = 0.036; INTERVAL: 3000 FLIGHTS

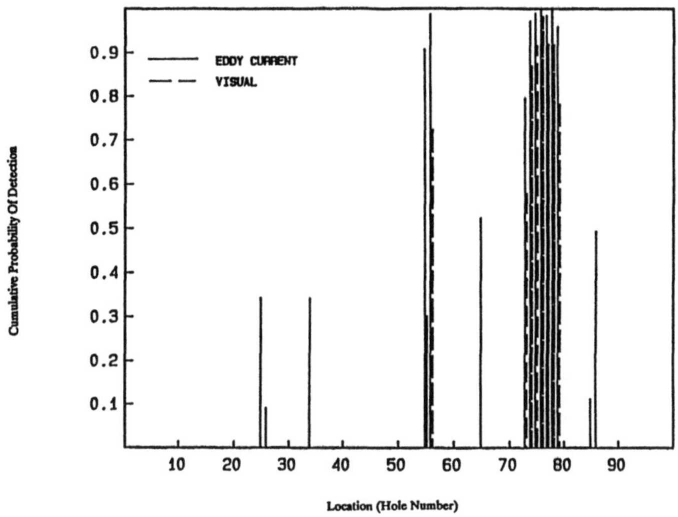

FIGURE 26. CUMULATIVE PROBABILITY OF DETECTION OF INDIVIDUAL CRACKS CASE WITH MOST CRACKS = (102)
p = 7.8 PSI; t = 0.036 INCH; INTERVAL: 3000 FLIGHTS

Effect of initiation damage and crack growth properties

The distribution functions for these two variables were shown in Figures 3 and 6, respectively. The effect of different fits was investigated, as shown in Figure 27 for the damage, and in Figure 28 for crack growth. Results are shown in Figures 29 and 30. These figures again are averages for the cracks that linked up as well as for all cracks; while the average is not a good measure of the probability of detection, it does provide a basis for comparison in a sensitivity analysis. The variability in Figure 29 is primarily due to the random assignments of crack initiation damage.

Figure 30 shows that even crack growth properties have little effect, unless they become extreme (right end of Figure 30). This is understandable, if it is realized that cracks do not become detectable until they reach a certain size (Figure 17). At that point the remaining time for detection is short anyway, regardless of the growth rate, because only 3 to 4 inspection opportunities occur after the cracks become detectable (Figure 17).

Effect of Stress Level

The stress is affected by many parameters. But to assess its effect, it is not necessary to vary all of them, as long as one is chosen that causes stress variations over a sufficiently

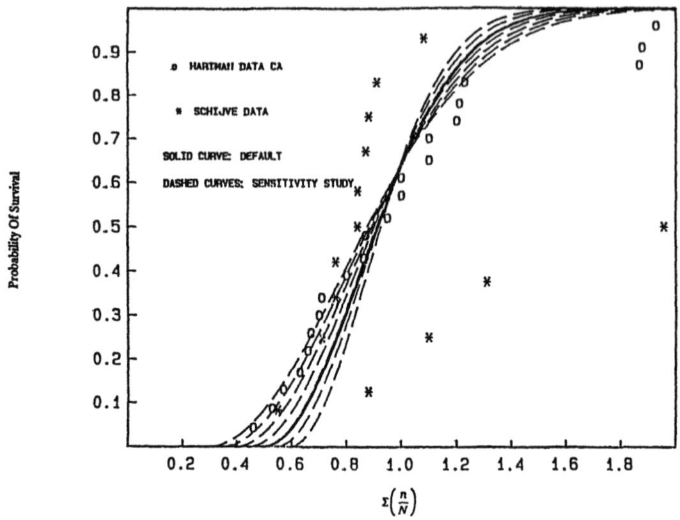

FIGURE 27. WEIBUL CURVE FOR DAMAGE $\Sigma\left(\frac{n}{N}\right)$
SOLID CURVE IS FOR: $DAM_{min} = 0.5$; $DAM_{\lambda} = 1.0$; $DAM_{\alpha} = 2.0$

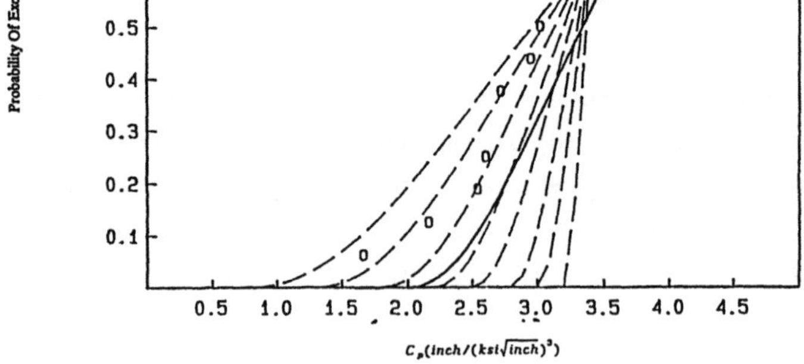

FIGURE 28. WEIBUL CURVE OF C_p VALUES IN PARIS EQUATION
DEFAULTS: $C_{p_{min}} = 2$; $C_{p_{\lambda}} = 3.6$; $\alpha = 3$ SOLID LINE: DEFAULTS;
m = 3 THROUGHOUT

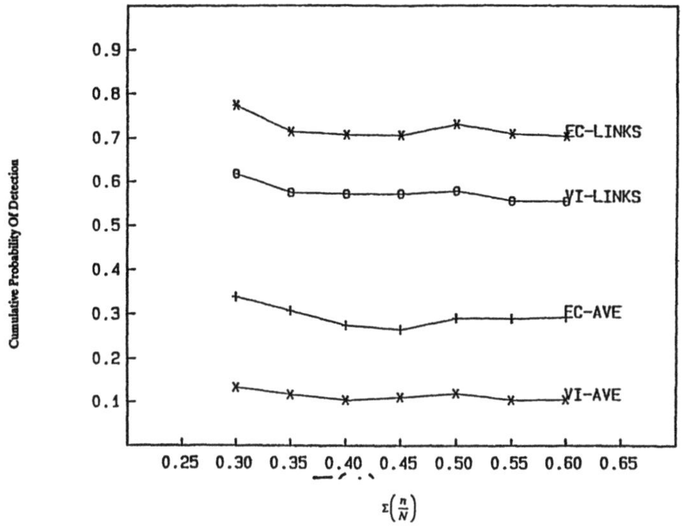

FIGURE 29. $\Sigma\left(\frac{n}{N}\right)$ OF DAMAGE DISTRIBUTION
p = 7.8 PSI; t = 0.036; OTHER PARAMETERS HAVE DEFAULT VALUES
INSPECTION INTERVAL = 3000 CYCLES (FLIGHTS)

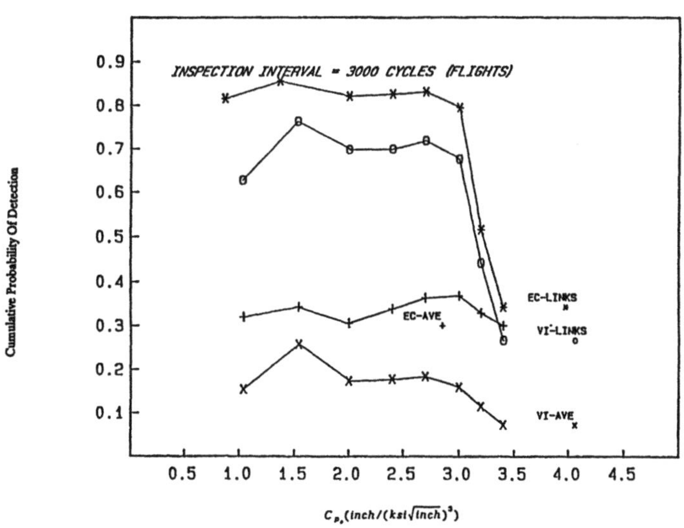

FIGURE 30. EFFECT OF C_{p_0} IN DISTRIBUTION OF C_p VALUES IN PARIS EQUATION
p = 7.8 PSI; t = 0.036 INCH;

large range. Such a parameter is the cabin pressure. The effect of cabin pressure, varied from 7.4 psi to 8.6 psi, is shown in Figure 31 for a 3000 flights inspection interval. The variability is largely due to assignments of crack growth properties. The figure shows little effect of stress level either, except for extreme cases.

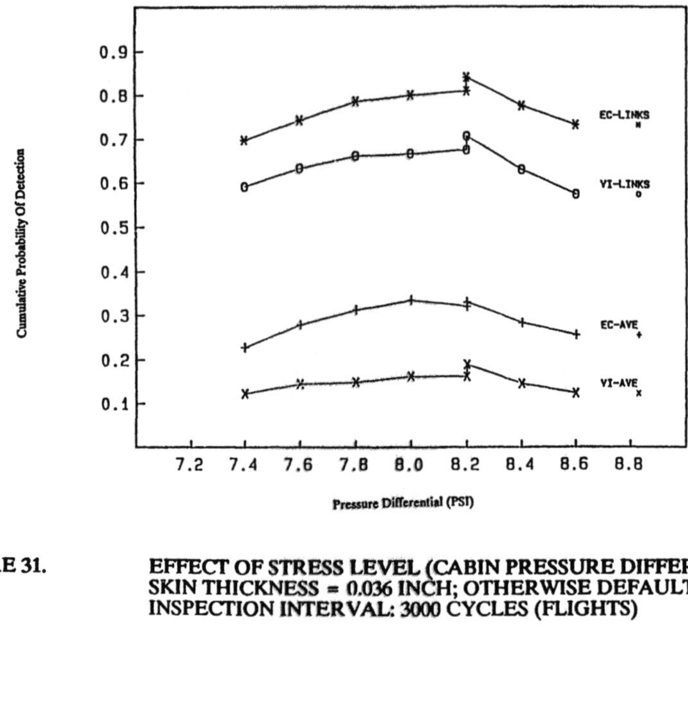

FIGURE 31. EFFECT OF STRESS LEVEL (CABIN PRESSURE DIFFERENTIAL)
SKIN THICKNESS = 0.036 INCH; OTHERWISE DEFAULT VALUES
INSPECTION INTERVAL: 3000 CYCLES (FLIGHTS)

Other Effects

Other parameters for which a sensitivity analysis might be performed are repairs, link-up conditions, and the effect of the assumption regarding increased/decreased probability of detection after a 'detect'/'miss'. Such sensitivity runs were not made, because the effect of the above parameters can be obtained in a logical fashion.

Finally, what was called earlier the 'most crucial assumption' could have been considered. However, the general effect of probability-of-detection parameters already showed the results to be insensitive. Therefore, it can be concluded without further analysis, that a refinement of these parameters by adjusting them, is really a secondary effect. If first order effects do not change the results, second order effects certainly will not.

Significance of Results and Conclusion

The analysis appears insensitive to statistical parameters and assumptions. It means that the results can be used to draw definitive conclusions.

We will now focus attention to the cumulative probability of detection of the second largest crack. In this respect Figure 32 is important, because it shows that the above probability is not significantly affected by any of the parameters involved, at least not for an inspection interval of 3000 cycles.

There is a slight, but discernable, effect of the total crack growth life on the cumulative probability of detection of the second largest crack, as can be appreciated from Figure 33. This effect is not of great relevance either, because the actual crack growth life is a statistical variable even for constant parameter values.

The objectives of this work were (1) to obtain insight in the cumulative probability of detection of the cracks leading to failure (link-up), and (2) to conclude whether the presently used inspection interval provides adequate probability of detection, or if not, to arrive at suggestions for a 'safe' inspection interval. The first objective was accomplished,

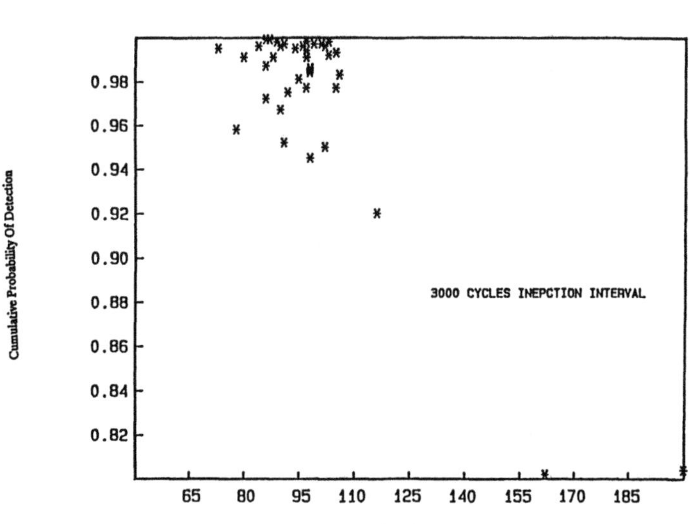

FIGURE 32. PROBABILITY OF DETECTION OF SECOND LARGEST CRACK VERSUS NUMBER OF CRACKS (INSPECTION INTERVAL = 3000 CYCLES)

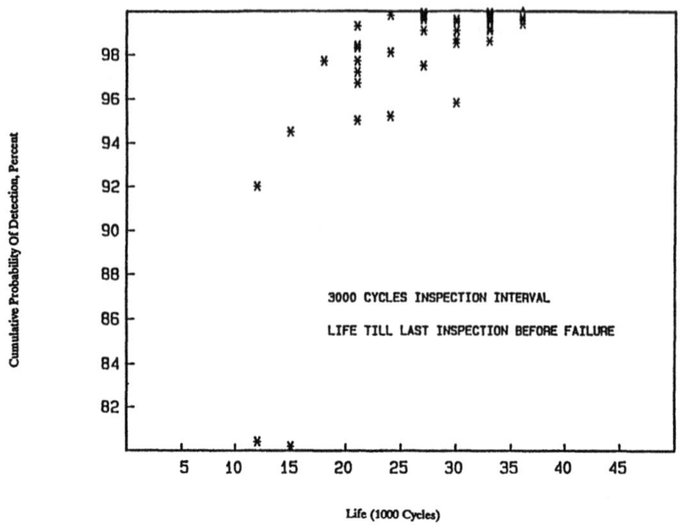

FIGURE 33. PROBABILITY OF DETECTION OF SECOND LARGEST CRACK VERSUS CRACK GROWTH LIFE FROM FIRST CRACK TO LINK UP

and especially such plots as Figures 15, 16 and 23 through 26 provide the information desired. The second objective requires an interpretation of the data with regard to the safety implied by a certain probability of detection. Such an interpretation presents several difficulties.

For purposes of comparison, use has been made of the average probability of detection for all cracks and for the cracks that linked up. Although this is certainly permissible for a comparison of the relative effects of the various parameters involved, a simple averaging is not in accord with a probability analysis as used here. Using more advanced probability theory it might be possible to arrive at a more useful 'total probability of failure', but it should be noted that this would require many new assumptions, the most important being the amount of disbonding.

From Figure 18 it can be concluded that there is rather a precipitous drop in probability of detection if the inspection interval is longer than 3500 flights. Although this effect becomes more elusive in Figure 21, the latter still shows that intervals in excess of 3000 flights introduce more uncertainty (scatter). This means that it may be prudent to reduce the present inspection interval of 4500 flights to 3000 flights or less. On the basis of Figure 22, one might conclude that for worst cases an inspection interval of about 1800 flights would provide a cumulative probability of only 95 per cent, so that covering the worst case would require a further reduction of the interval.

Figures 16, 25, and 26 show that the cumulative probability of detection of the very largest crack is practically equal to 1 in most cases, at least for an inspection interval of 3000 cycles. However, detection of the second largest crack is almost always less than 1. This second largest crack may be far away, so that it still could lead to failure. It does not make any difference whether a critical crack at hole 100 is detected and repaired, while another at hole 1 subsequently causes a failure. It should be realized as well that in an actual aircraft not just 100 holes are involved but many thousands. Therefore, the probability of detection of the second largest crack was investigated. Some results for this case were discussed already. More important, are the data in Figures 34 and 35.

In accordance with previous figures, Figure 34 again shows a precipitous drop in the probability of detection when the inspection interval exceeds 3000 flights. Figure 35 shows that for an interval of 3000 flights the cumulative probability of detection of the second largest crack exceeds 95 per cent in about 90 per cent of the cases investigated. When loosely interpreted, this figure would indicate that there is about 90 per cent confidence that the second largest crack would be detected with 95 per cent probability. Then the line drawn in Figure 34 could be loosely interpreted as well, as providing about 90 per cent confidence.

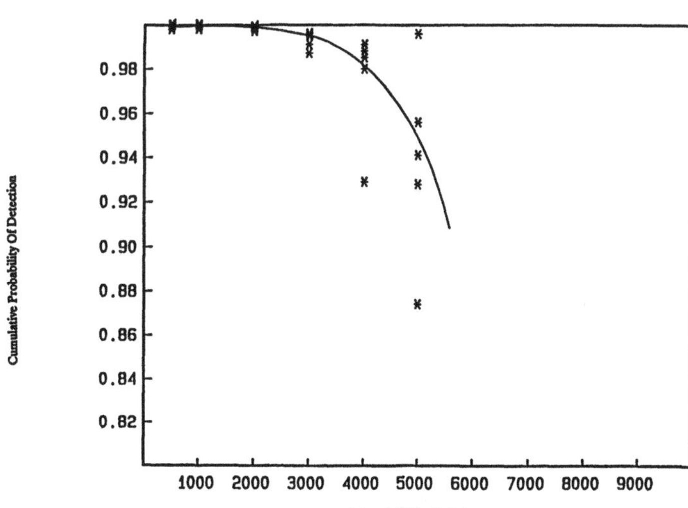

FIGURE 34. CUMULATIVE PROBABILITY OF DETECTION OF SECOND LARGEST CRACK IN FIVE RUNS

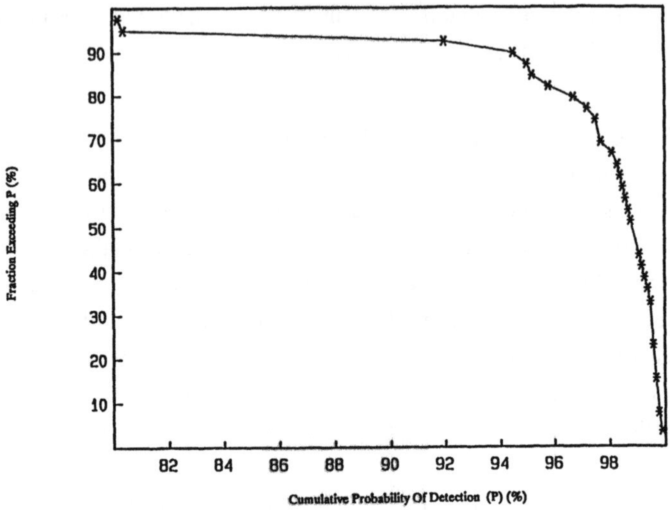

FIGURE 35. FRACTION OF CASES IN WHICH CUMULATIVE PROBABILITY OF DETECTION OF SECOND LARGEST CRACK EXCEEDS VALUE (INSPECTION INTERVAL = 3000 CYCLES)

While it would be difficult to pinpoint which inspection interval would be acceptable, it can be concluded that a reduction of the interval from the present 4500 flights to 3000 flights would result in considerable higher safety, while the effect of reductions to below 3000 flights cause only marginal improvements. The above can be stated as a general conclusion only.

References

1. D. Broek, "The feasibility of proof testing", FractuREsearch Inc., Report to the Transportation Systems Center (1989)

2. W.H. Lewis et al, "Reliability of non-destructive inspection", San Antonio Air Logistics Center, Report SA-ALC/MME 76-6-38-1 (1978)

3. S. Patil, Finite element analysis of lap joint. Foster-Miller Inc. (1990)

4. A. Hartman, "Fatigue tests on 3-row lap-joints in clad 2024-T3 manufactured by riveting and adhesive bonding", Nat. Aerospace Lab., NLR, Amsterdam, Report TN M-2170 (1967)

5. J. Schijve, "The endurance under program fatigue testing", in Full-Scale Fatigue Testing of Aircraft Structures, pp 41-59, Pergamon (1961)

6. L. Jarfall "Review of some Swedish work on fatigue of aircraft structures during 1975-1977", Institute for Aviation Research, FFA, Stockholm, Report TN HE-1918

7. T. Haas, "Spectrum fatigue tests on typical wing joints", Material Pruefung 2 (1960), 1, pp 117

8. D. Broek, The Practical Use of Fracture Mechanics, Kluwer Ac. Publ., (1988)

9. D.P. Rooke and D.J. Cartwright, Compendium of Stress Intensity Factors, HM Stationery Office, London (1976)

10. D. Broek, Elementary Engineering Fracture Mechanics, Kluwer Ac. Publ. 4th edition (1986)

11. D. Broek, "Fracture control by periodic inspection with fixed cumulative probability of crack detection, in Structural Failure, Product Liability and Technical Insurance", Rossmanith Ed. pp 238-258, Interscience Ent. (1987)

12. D. Broek, USAF ASIP Conference, Dayton (1986)

13. E. Knor, "Reliability of the detection of flaws and of the determination of flaw size", AGARDograph 176, pp 396-412 (1974)

14. Anon, Software for damage tolerance analysis, FractuREsearch Inc. (1989)

15. U. Gorenson, Paper presented at ICAF meeting Toulouse (1983)

16. R. Mayville, "A Laboratory Study of Fracture in the Presence of Lap-Splice Multiple Site Damage", Presented at the International Symposium on Stuctural Integrity of Aging Airplanes, Atlanta, Georgia (1990)

24. Evaluation Methodologies Applied for Pressurized Fuselages of Airbus A/C

H.-J. SCHMIDT

Airbus Industrie / Deutsche Airbus
'Fatigue and Fracture Mechanics' Department

A B S T R A C T

All AIRBUS A/C have been designed with special consideration to the fatigue behaviour and damage tolerance aspects from the beginning of the seventies.
The AIRBUS philosophy and the essential evaluation methodologies, used for type certification, which were used for AIRBUS A/C designed prior to or after introduction of the damage tolerance requirements as specified by FAR 25 Amendment 45 and JAR 25.571 are presented. The calculation methods and the extensive full scale fatigue tests, used for certification, are described, as well as special investigations to assess the effect of possible corrosion on fatigue and crack propagation.

In addition details and the status of activities related to ageing A/C, such as the establishment of the supplemental structural inspection program and FAA initiated ageing A/C tasks, are explained. This includes initial results from research programs to study the effect of periodical overloads on the fatigue and crack propagation behaviour of pressurized fuselages as well as activities to optimize crack propagation calculation methods for fuselages using stress-time-histories.

1. Introduction

From the beginning, in the seventies, all AIRBUS INDUSTRIE A/C have been designed with special consideration to the fatigue behaviour and damage tolerance aspects. For verification of the design goals, extensive analyses and testing have been performed for all A/C types. The first A/C type, the AIRBUS A300, was certified in 1974 and a structural inspection program has been established using the MSG2 (EMSG) logic and with consideration to the fatigue and damage tolerance results as well as an estimated effect of corrosion.

Springer Series in Computational Mechanics
Atluri, Sampath, Tong (Eds.)
Structural Integrity of Aging Airplanes

Since the corrosion effect on fatigue and damage tolerance is a function of the design features, post-certification studies have been performed to assess the fuselages of ageing A/C. The results of pre- and post-certification investigations, including the tear down results from full scale testing, have been used for the establishment of the A300 SSIP in the second half of the eighties.

These activities were followed by the FAA initiated ageing A/C task force unit, which is under progress. Furthermore internal research activities have been initiated to study the possibility of applying periodical overpressures for increasing the inspection intervals of the fuselage. In addition optimized crack propagation calculation methods for typical stress-time-histories for upper fuselage areas are under development.

The technical aspects of the a.m. procedures with regard to the A300 fuselage are presented in the following.

2. Fatigue and Damage Tolerance Justification for Certification

2.1 General

The A300B2 has been certified according to FAR Part 25 Amendment 19 in 1974. At that time the regulations required the consideration of lower repeated loads and their effect on the initiation of cracks within the structure based on fatigue analyses and tests. Furthermore the structure had to withstand limit load in case of a pre-defined crack.

In addition to the required fatigue and fail safe justification, crack propagation calculations had been performed as required later by FAR Part 25 Para. 25.571 including Amendment 45. Therefore the basic A300 MRB had taken into account the individual fatigue and damage tolerance characteristics of the concerned SSI.
The above mentioned justification had been performed by analyses and tests which are explained in the following.

2.2 Fatigue and Damage Tolerance Calculations for Certification

In the fuselage (with the exception of the cockpit and center area) more than 100 SSI's have been analyzed regarding their fatigue and damage tolerance behaviour, i.e. the time to detectable crack length, the crack propagation period between detectable and critical crack length and the critical crack length, with the methods briefly described in the following:

The evaluation of the structure under fatigue strength calculation methods, is intended to ensure that catastrophic failure is extremely improbable (safe life design) and that economic targets, such as low costs of maintenance and repair (damage tolerance design), are met. All A300 fatigue life evaluations are based on the so-called hypothesis of the Miner Rule with a subsequent evaluation of the occurence probability of a detectable flaw at economic repair life. The a.m. fatigue analysis, including the subsequent probability calculation, takes into account the following aspects to obtain the correct life to detectable crack length:

I. Appropriate S-N data extrapolated with a certain slope for consideration of fatigue damage for low amplitudes

II. Influence of finite number of test specimens and number of fatigue critical locations

III. Influence of occurence probability and standard deviation for both, S-N data and load spectra

IV. Influence of spectrum type on allowable fatigue damage calculated by Miner's Rule

V. Definition of allowable probability of damage taking into account economic aspects for damage tolerance structure ($P_A = 10^{-1}-10^{-2}$ at economic repair life (ERL)) or safety aspects for safe life structure ($P_A = 10^{-3}-10^{-5}$ at ERL)

VI. Consideration of full ground-air-ground cycle resulting from rain-flow counting

The damage tolerance analyses have been performed by using deterministic methods for the majority of the SSI's, which are inspectable for less than complete load path failure. The remaining SSI's, with multiple load path and not inspectable prior to complete load path failure, have been analyzed by applying probability methods.

The first category of SSI's have been analyzed using linear elastic fracture mechanics to assess the effect of cracks present in the structure, i.e. to determine rate of crack growth and maximum tolerable crack length, for establishment of intervals for the structural inspection program. The second category of SSI's have been evaluated by fatigue analysis assuming load redistribution after failure of the critical neighbouring element and pre-damage for one life time.

2.3 Fatigue and Damage Tolerance Tests for Certification

The aim of the A300 fatigue testing was to justify the design goal of 48 000 flights ERL (A300B2) and to verify fatigue and damage tolerance calculations.

Four fatigue tests were performed by multi-section testing with the following separate specimens, see Fig. 1.

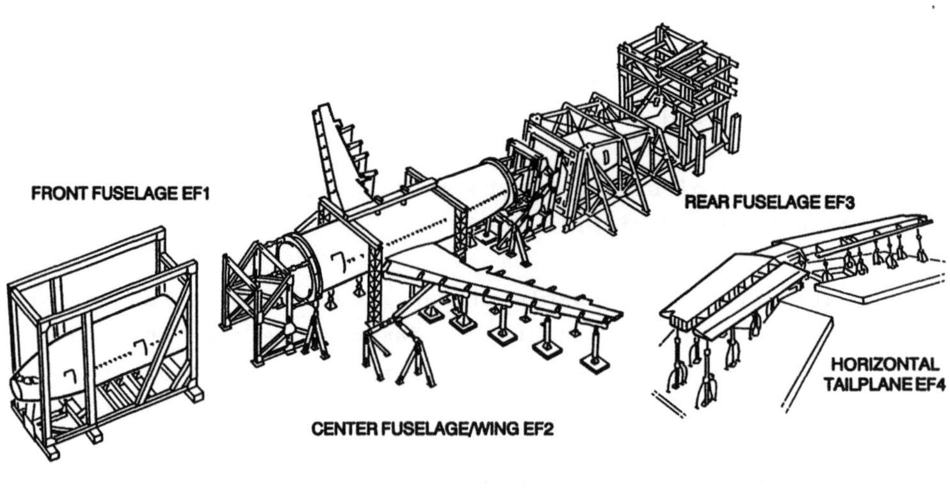

Fig. 1: Airbus A300 Full Scale Fatigue Test Specimens

According to the AIRBUS philosophy a minimum of two life times have been simulated, which was neither required nor common practice at that time. With this long-term testing all locations having limited fatigue life have been detected, where damages may occur before or after exceeding ERL. For all major findings, regardless of the time of detection, relevant solutions, on production and in-service A/C, have been launched. In general each major full scale test finding at the pressurized fuselage led to the establishment of a retrofit modification included in a recommended service bulletin (SB) and the establishment of an alternative inspection program distributed by a mandatory inspection service bulletin (ISB). Consequently the number of additional actions, defined by the present FAA initiated ageing A/C task unit, is very limited.

Furthermore the simulation of at least two life times for A300 provides a high safety standard, since scatter in loads and material data is covered sufficiently.
The following number of simulated flights are accomplished, see Fig. 2.

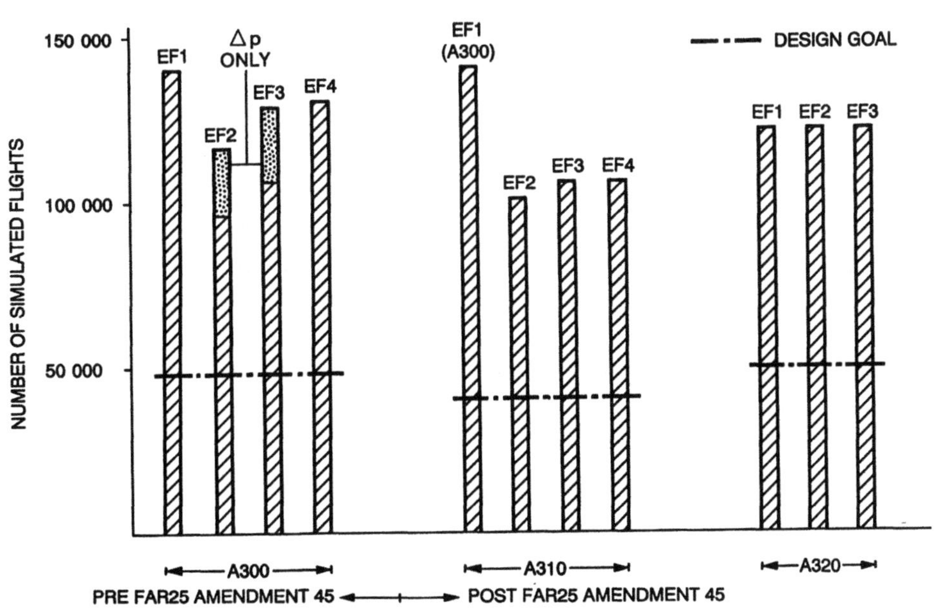

Fig. 2: Full Scale Fatigue Testing of Metallic Structure
Number of Simulated Flights

The importance of the long-term fatigue testing can be seen in Fig. 3 which shows the time of detection of natural damages for center fuselage/wing specimen and rear fuselage specimen. Up to simulation of one life time (ERL) approximately 30 % of all detected damages have been occured only.

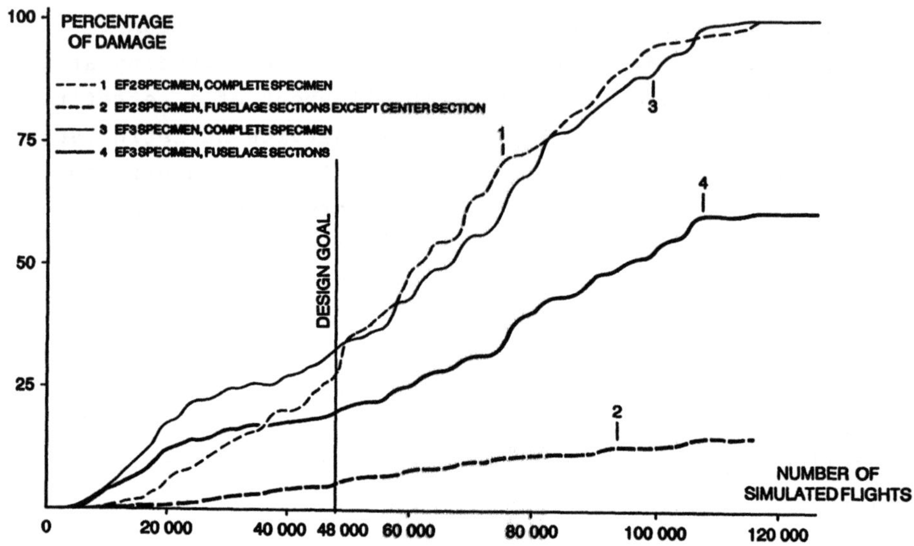

Fig. 3: Time of Detection of Natural Damages

In the second phase of all full scale fatigue tests, damage tolerance investigations have been performed to support the damage tolerance analysis not usual at that time. More than 40 artificial damages have been installed in the described fuselage sections and the crack propagation has been monitored up to the end of test or repair.

The a.m. full scale fatigue tests revealed only a few cases of multiple site damage. In the described fuselage area only three cases of multiple site damage have been detected between 1.6 and 2.2 times ERL. For those areas modifications and inspection programs have been defined beginning at 1/3 of time-to-detection with repeat inspections according to the crack propagation behaviour and the capability of the defined NDI method.

3. Ageing A/C Activities and Status

3.1 Effect of Corrosion on Fatigue Life and Crack Propagation

Prior to present FAA ageing A/C activities several investigations, including tests, have been launched to study environmental effects on main load carrying parts. For example the effect of debonding due to corrosion on the fatigue life of the longitudinal joints has been investigated and is presented in the following.

The longitudinal lap joints are reinforced by bonded doublers and riveted with titanium rivets, see Fig. 4. Different types of possible defects have been investigated, e.g. debonding of inner, outer or both doublers. For evaluation of these effects a full scale barrel test with a length of 14 frame bays has been accomplished besides complementary coupon testing. The barrel test revealed a fatigue life of more than three times the ERL for an intact longitudinal joint and deteriorations depending on the location and length of defect as shown in Fig. 4. These deteriorations are the result of increased bearing stress, reduced net section and changed eccentricity leading to a change of crack origin and crack shape up to a certain length.

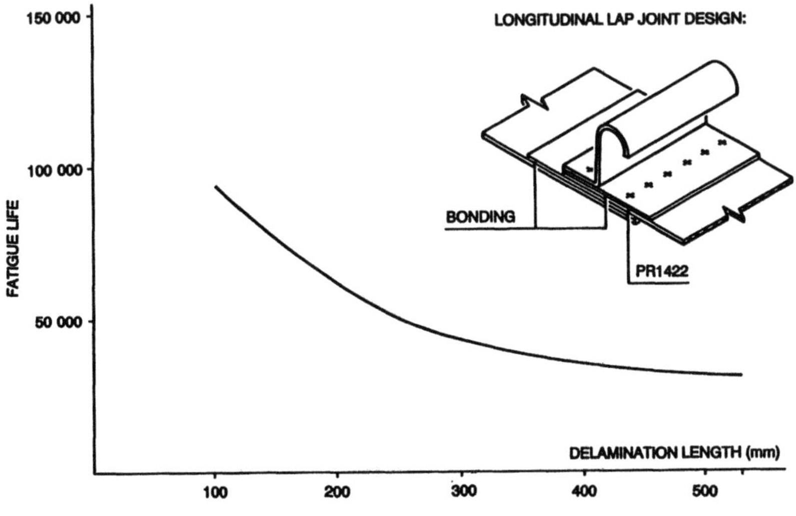

Fig. 4: Fatigue Life of Longitudinal Joints

The effect of debonding on the crack propagation life was investigated by analysis and test which indicated a reduction of the crack propagation life up to factor 2 depending on the length of debonding.

The results of the a.m. study and similar investigations for other parts have been considered for complementary structural inspection programs for high-time A/C.

3.2 <u>Establishment of Supplemental Structural Inspection Program</u>
The publications in 1978 and 1981 of the CAA airworthiness notice No. 89 and FAA advisory circular 91-56 required the establishment of a Supplement Structural Inspection Program (SSIP) for transport A/C certified prior to issue of FAR 25 Amendment 45. This implies a complete fatigue and damage tolerance evaluation for the A/C structure.

For AIRBUS A300 the SSIP was completed in 1989 after re-evaluation of the entire structure with special consideration of the results from full scale fatigue and damage tolerance testing. At that time the lead A/C had accumulated approx. 28 000 F which is ca. 60 % of ERL. The SSIP activities contain the following six steps for determination of the inspection program (threshold, interval and method) for each selected 'Priniciple Structural Element' (PSE):

I. detailed analysis of full scale fatigue test results and in-service findings
II. complete fatigue and damage tolerance evaluation for 3 basic A/C types
III. modification analysis
IV. mission profile parametric study
V. calculation of the inspection task requirements for the inspected aircraft
VI. lead fleet aircraft selection

The major activities are steps I, II and IV. Steps I and II contain the evaluation of fatigue and damage tolerance for representative A300B4 conditions regarding structure and mission profile after detailed analysis of the structural behaviour of fatigue test findings, see Fig. 5.

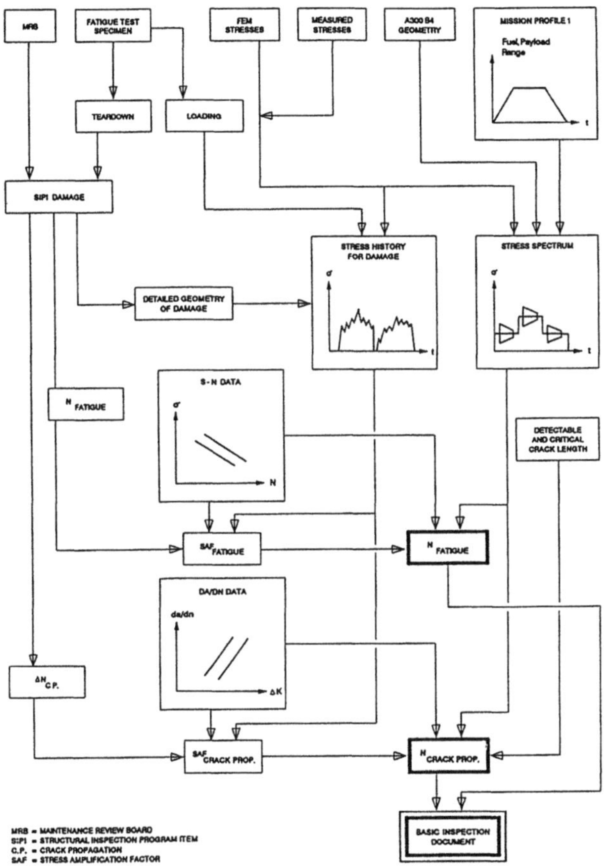

Fig. 5: Analysis of Findings and Fatigue and Damage Tolerance Evaluation

The influence of different mission data as range, payload and fuel at landing has been analyzed for all PSE's. The investigations, carried out for seven different mission profiles, show, that the range variation is the most important parameter for both, fatigue and crack propagation. This is valid for fuselage items, where the maximum once per flight stress is a result of internal pressure and external loading. An example of these investigations is given in Fig. 6.

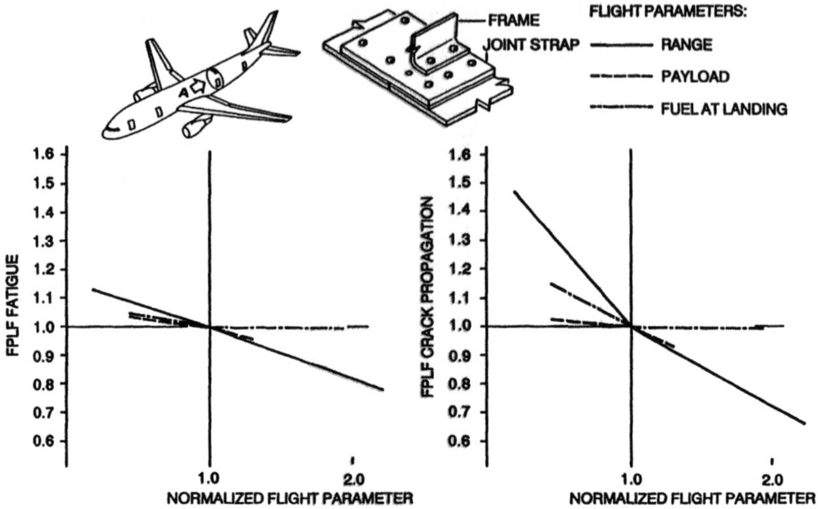

Fig. 6: SSIP - Flight Parameter Life Factors (FPLF)
Splice Plate at Circumferential Joint

As explained in chapter 2.3 corrective actions were launched immediately after fatigue test findings for most of the concerned areas. Therefore the SSIP covers exceptional cases and tear down findings only.

The items to be covered by the SSIP are split into two categories regarding fatigue life and criticality, i.e. 100 %-items covered by inspection service bulletins and fleet leader program items included in the 'Supplemental Structural Inspection Document'.

3.3 FAA initiated Activities for Ageing A/C and Status

Following several incidents with ageing transport aircraft the FAA has initiated a review of former actions launched by the manufacturers to maintain the safety of the A/C:

- Review of service bulletins and inspection service bulletins
- Establishment of a baseline program for corrosion inspection and corrosion prevention
- Repair quality assessment
- Review of SSIP and maintenance program

For performing of the a.m. activities, three working groups have been created with participants from airworthiness authorities, airlines and manufacturers. The AIRBUS A300 task group started the activities in the second half of 1988 with the following status to date:

- Review of SB's and ISB's:
  - completion of activities in the first half of 1989
  - review of 64 SB's and 65 ISB's (30 SB's / 33 ISB's for the described fuselage area)
  - task unit group rendered 5 recommended SB's and 7 recommended ISB's to mandatory (4 SB's and 1 ISB for the described fuselage area)

- Establishment of a baseline program for corrosion inspection and corrosion prevention:
  - completion of activities in February 1990
  - review of airline's and manufacturer's experience regarding corrosion for the entire A300 A/C
  - review of existing maintenance programs regarding their effectiveness to detect and prevent corrosion
  - establishment of a corrosion inspection and corrosion prevention program for all zones of the aircraft
  - establishment of implementation rules considering safety and economic aspects

- Repair quality assessment and review of SSIP and maintenance program:
  - harmonization of philosophy to be applied with authorities, airlines and other manufacturers is in progress

4. Present Research Activities for Ageing A/C

For the assessment of the structural integrity of the fuselage two in-house research programs are of special interest, i.e. determination of the effect of overloads on the fatigue behaviour of complex joints and crack propagation behaviour of fuselage structure under simplified and complex stress-time-histories.

4.1 Effect of Overloads on the Fatigue Behaviour of Complex Joints

During the 2nd FAA initiated Ageing Aircraft Conference in Baltimore in 1989 several contributions dealt with the well-known proof pressure concept. The idea of this concept is to periodically apply overpressures to the fuselage, to verify that no crack exceeds critical crack length and to retard crack growth of existing cracks for a certain time. The overall goal of this proof pressure concept is to minimize or replace complicated NDT-inspections for fatigue sensitive areas such as longitudinal panel joints. This is of special interest, because of the great extent of these panel joints on large transport A/C.

The risk of the original proof pressure test producing large defects in the fuselage is quite high, since the condition of the structure is not known in detail. Therefore the a.m. concept has been developed further by Deutsche Airbus.

The new idea is either to commence with periodic overloads on relatively young A/C or after accomplishment of NDT inspections of all fatigue sensitive areas. Before applying the first proof pressure, all areas with findings have to be repaired, i.e. it can be assumed, that at time of proof pressure testing undetected cracks are less than the detectable crack length by NDT means. The proof pressure test has to be repeated in such intervals, that a repetition of the NDT inspections can be minimized or deleted.

For development of this method representative coupon tests have been carried out. The test specimens represent two rivet row longitudinal lap joints reinforced by thin bonded doublers and jointed with titanium rivets. The constant amplitude test has been performed on three stress levels for the following three specimen conditions:

- both doublers bonded
- inner doubler completely debonded
- both doublers completely debonded

The a.m. tests have been carried out with and without overload at a R ratio of 0.1. The applied overload amounted to 1.33 times the maximum stress of the constant amplitude test. The overload has been applied first at one third of the life of the specimens tested without overload with a repetition in intervals of 1 000 to 6 000 cycles. Fig. 7 shows the first test results.

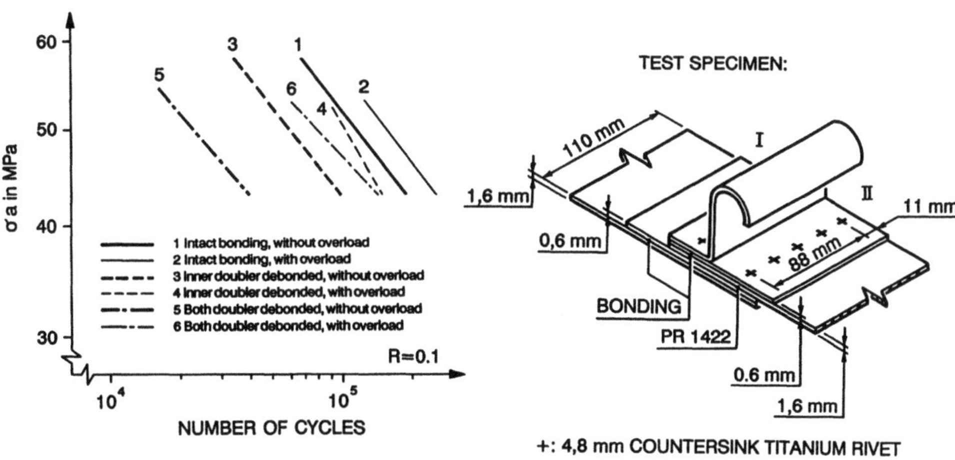

Fig. 7: Coupon Tests with and without Overload
Specimen Design and Test Results

For specimens without debonding the comparison of the S-N data shows an improvement of fatigue life of 50 % due to overload. The results reveal no significant differences for the above described range of interval overload application. Furthermore an earlier start of periodical overloads at one sixth of the specimen life gained, without overload, leads to approximately the same results.

For specimens with inner doubler debonded the benefit resulting from overload is up to 100 %. The overload leads to a fatigue life of debonded joints which is slightly lower than the fatigue life of the joints without debonding and no overload. No significant differences in life have been observed for the different overload intervals.

The specimens with both doublers debonded show an improvement of the fatigue life of greater than factor 2 due to overload. Significant differences, due to variation of the intervals of overloading, have not been observed.

From the a.m. results it can be concluded, that the application of overload leads to an improvement of the fatigue life which is significantly higher for joints with debonded doublers. This effect is the result of the higher net stress level for debonded joints, which leads to early fatigue crack initiation. The subsequent crack growth is retarded by the overloads. The achieved results, i.e. improvement up to factor 2, are valid for the titanium riveting which leads, at each rivet hole, to an expansion of approximately 0.3 to 0.5 mm and therefore to high permanent stresses.

The achieved results have to be verified by further tests with larger panels before drawing any conclusions for application on in-service A/C. Additionally detailed studies have to be carried out for the entire fuselage regarding the structural behaviour, i.e. static strength, fatigue life and crack propagation, due to overpressure. Furthermore crack propagtion tests with complex structure are to be performed to evaluate the effect of overloads for multiple cracks slightly less than the detectable crack length by NDT means; such tests are under preparation.

4.2 Crack Propagation Behaviour of Fuselage Structure under Simplified and Complex Stress-Time-Histories

The A300 crack propagation calculations have been carried out using linear elastic fracture mechanics method (Forman equation) without consideration of retardation effects due to non-constant amplitude loading.

Most of the retardation models such as PREFFAS and CORPUS are especially developed and verified of standardized spectra for bottom wing skin of transport and fighter A/C, e.g. TWIST and FALSTAFF spectrum. Since the spectrum shapes of the most fatigue sensitive fuselage areas (e.g. upper shell) are

completely different, special investigations in conjunction with the university of Hamburg-Harburg, West-Germany, have been launched to assess the retardation effect for typical fuselage spectra.
For basic research the crack propagation of the aluminium alloy 2024 T3 and the new aluminium-lithium alloy 2091 T8 (material composition from 1985) from Pechiney has been compared using CCT-specimens. The main topic was the crack propagation for constant amplitude and simplified and complex stress-time-histories. The test results have been compared with predictions calculated with Forman equation as well as σ_{open}= const. according to the proposal of SCHIJVE.

Fig. 8 shows the stress-time-histories. Type I represents a realistic flight type for the A310 full scale fatigue test. The simplifications of the stress-time-histories are performed in such a way, that the influence of simplified incremental spectra (types B, F, G and H) can be evaluated separately and compared with crack propagation of the ground-air-ground spectrum (type A), peak-to-peak spectrum (type C) and realistic flight type (I).

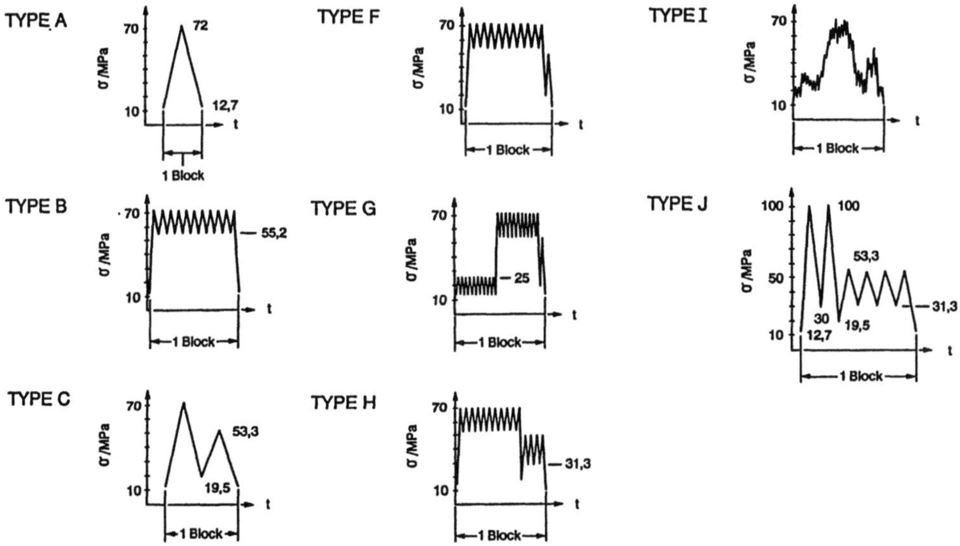

Fig. 8: Stress-time-histories of Crack Propagation Tests with CCT-Specimens

Fig. 9 shows the crack growth for 2024 T3 as a function of the maximum stress intensity revealing, that the crack growth of the peak-to-peak spectrum is greater than that of the ground-air-ground spectrum due to the additional peak. The crack growth of the real spectrum (type I) is sligthly greater compared to the result of type C. The crack growth data of types B, F, G and H are similar, i.e. landing impact, taxi and gust loads at lower mean stress have no significant influence on the crack growth behaviour. The effect of periodic overloads has been evaluated by simulation of a mixture of flight type I (249 times) and flight type J (1 time) showing significant reduction of crack growth compared with flight type I only.

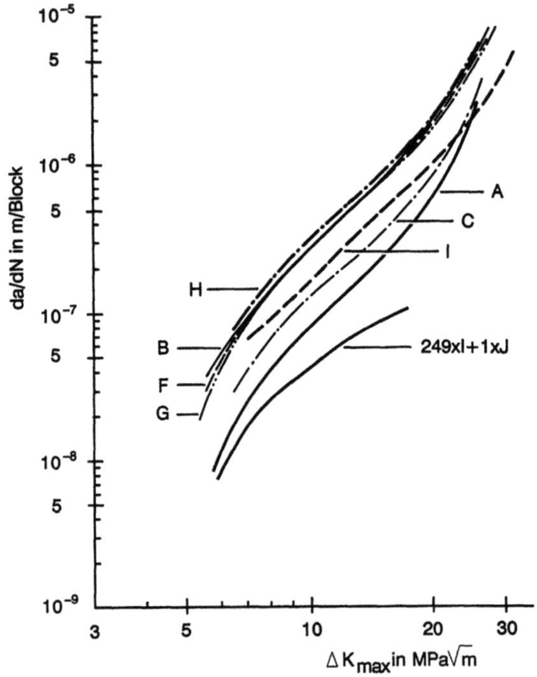

Fig. 9: Crack Propagation Behaviour of 2024 T3 for Different Stress-Time-Histories (CCT-Specimens, Width 160 mm)

The comparison between 2024 T3 and 2091 T8 alloy is presented in Fig. 10. The comparison shows only slight differences in crack growth of both alloys for the real spectrum (type I) and the simplified spectrum (type B).

Major differences occured for constant amplitude loading (type A) due the change of the fracture mechanism in the range of about da/dn = 10^{-7}m/cycle.

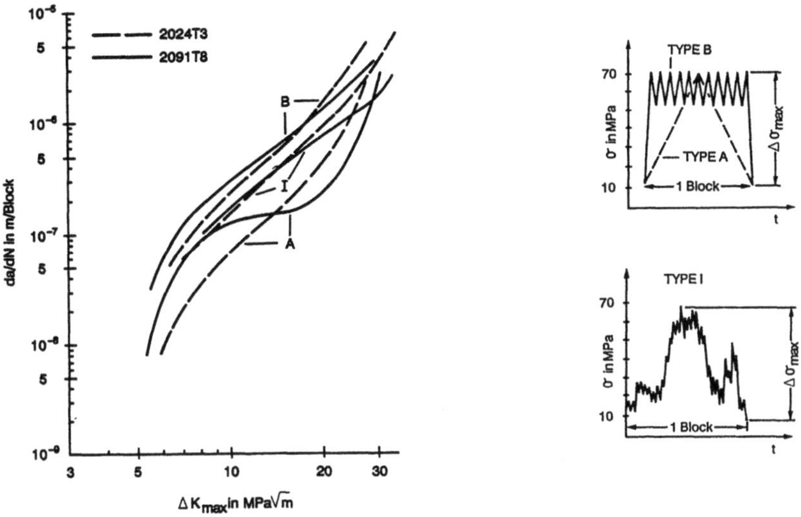

Fig. 10: Comparison of Crack Propagation Behaviour of 2024 T3 and 2091 T8

5. Conclusion

The applied evaluation methodologies for all AIRBUS A/C types included extensive analyses and testing to assess the structural behaviour regarding fatigue and damage tolerance. The early application of the conclusions of these pre- and post-certification investigations led to a safe operation without surprises for the fuselages of the ageing A300 A/C, which have reached up to 30 000 flights to date.
The presented research program relating to the effect of over-pressure shows the possibility to increase the inspection intervals for fuselage joints which are difficult to inspect. However, before drawing any conclusion about the application to in-service A/C, representative verification tests have to be accomplished in addition to assessing the implications of the overpressure on the static strength, fatigue life and crack propagation for the remaining fuselage areas.

25. Experience in Specifying/Prolonging the Airframe Time Limits

A. F. SELIKHOV, V. L. RAIKHER, V. G. LEIBOV, C. I. NESTERENKO

Central Aerodynamic Institute (TsAGI)
Zhukovsky, Moscow region, USSR

Abstract
Problems of providing the safety of transport airplanes with respect to the fatigue strength have been considered by Soviet researchers since the mid 1950's. The adopted approaches basically correspond to those used all over the world, however some aspects of both technical and managerial efforts have different features. The main one is that the airframe is considered as an aging one from the beginning of the aircraft fleet service. At the present time this is a complete system, which has shown a high efficiency. The methodological and managerial base of the system and some scientific and technical aspects are discussed in this report.

Basic Features of the System

The system is based on a set of fundamental principles. The central one is the principle of permanent monitoring, i.e. the safety level planned at the design stage and manufacturing of the airframe, must be kept throughout the airplane life. The extremely high level of safety that is required can be achieved only through systematic supervision over a changing operational situation. In other words, the principle of feedback is used. Service experience is considered to be a key aspect, i.e. the information about actual operating conditions and about the actual state of an airframe.

In order to follow the principle of monitoring, the concept of specified service life limits is used. This is the time interval after which operations must be stopped, independent of the actual state of the airframe. Thus, we impose a restriction, but this restriction is a current one and should be revised timely. Based on this, a procedure of establishing/prolonging the service life limits is organized and placed in bounds of a certain technical policy. The procedure is regulated by the Civil Aircraft Airworthiness Requirements [1]. This procedure reduces the

Springer Series in Computational Mechanics
Atluri, Sampath, Tong (Eds.)
Structural Integrity of Aging Airplanes

principle of increasing stage-by-stage the specified service life limits. The practice of prolonging service life has shown that the duration of every stage must be as great as 5,000-10,000 flight hours. This time interval is short enough to assure the required validity of the prediction, and is long enough for us to have the opportunity to thoroughly consider incoming information or to perform additional investigations (for example, laboratory tests). Work between these intervals should be conducted according to a special schedule. How to establish the initial service life limits and specify the next value are stated in the special document - Regulations for establishing the service life and permissible operational conditions. Here, one point must be mentioned: The regulations rely upon the principle of the simultaneous responsibility of both the Aircraft Manufacturers and the Aircraft Operators for the safety of the airframes. The first must design appropriate structures and demonstrate the required characteristics, where the second must provide the specified service conditions and must estimate the actual state of the fleet. Therefore, the development of the regulations is coupled with a procedure of obtaining, exchanging and analyzing the information about the performance of structures and the forthcoming operation conditions.

The results of this collaboration are contained in the document - Substantiations to the Specified Service Life Limits. This document represents a scientific and technical basis of the regulations. It is the intent of the regulations to establish the next service life limits and to specify the conditions under which the level of safety required by the Civil Aircraft Airworthiness Requirements will be provided. These conditions are contained in three parts of the regulations: (A) the list of structures subject to replacement; (B) the list of modifications to the structure; and (C) the list of inspections in operation or at scheduled repairs. For each item of these lists, a range of operating time is stated within the limits of which the required action must be performed. For inspections the so-called rejection condition is stated, i.e. the characteristic of the condition of the structure at which further operations must not be permitted and a repair is required.

Some Main Technical Aspects of the Problem

Superduty Elements, Critical Areas and Critical States

The superduty elements are those which in case of single failure would cause an accident or catastrophic situation. Such elements can be present as components of a control system, landing gear, or some other functional system of the airplane. As for an airframe, some development is performed to the concept of superduty elements. The concept of a critical area in a structure is introduced: Briefly, a structural zone where occurrence of dangerous fatigue damage in operation is a possibility. An a priory information makes it possible to draw the first version of the list of the superduty elements to be used for designing the aircraft and at working out the effective programs of full-scale fatigue tests. However, the full-scale fatigue tests, including the thorough tear-down inspection after completion of the tests, can introduce essential corrections to the list of critical areas. Thus, they form the second, main approximation to the desired completeness of the list. The list is completed in the course of prolonging the service life limits through the use of operational experience. Critical states are defined as states of a damage structure which should be prevented by the system. An allowable extent of damage is defined by the necessary residual strength and by the residual stiffness which must be great enough to exclude the occurrence of dangerous aeroelastic phenomena, for example, flutter.

In connection with unpredictability of a pattern of multiple-site damages, critical dimensions may be specified in this case, conservatively, assuming well synchronous development of the cracks at all sites. The schematic example in Fig. 1 represents the numerical measure of the effect of this assumption in application to a wing lower surface (D16AT material).

The Fatigue Loading

Data about stress history at critical areas is a fundamental characteristic necessary for determining the allowable operating time. The critical area stress history investigation method starts from the flight parameters in advance, at the present time. This is related to the possibility of using the data from different flight accident recorders, which helped to obtain extensive statistics on repetition rates of load factors at the airplane center of gravity (Fig. 2). The same records may be used for an evaluation of individual load levels.

The On-Board Fatigue Loading Meter

The design concept consists of mounting the strain gauges in the vicinity of the critical areas. An on-board computer processes signals by a corresponding algorithm and gives a calculated value of the fatigue. Now, use is made of a simplified algorithm (Fig. 3): (1) outlining the complete sysles; (2) transforming them equivalently to a set of zero-to-maximum cycles; (3) applying S-N curves of known power type; (4) summing the fatigue rates. Rational variations in some "free" parameters of the algorithm as well as a correct choice of both area and directions for mounting the strain-gauges allow us to "monitor" an actual fatigue accummulation at these critical areas.

Some Important Aspects of the Full-Scale Tests

-The remaining difference between a real service loading and a test loading necessitates an introduction of special equivalents for critical areas. It would be best to make the full-scale test program have the uniform equivalent for every critical area, but if it is not achieveable the real equivalents must be known.

-The system of establishing/extending stage-by-stage the service life limits envisages the testing of the structures sampled out of operation. Effects of other degradation processes may be revealed in this case.

-In many cases, only the fractography gives information about a crack growth rate. Special markers are proposed to be introduced into the test program to code the place of the time axis; an example is given in Fig. 4.

-As mentioned earlier, the full-scale specimen must be disassembled after the test completion for a special tear-down detection procedure.

The maximum undetectable fatigue crack size (MUS) is a very important service life characteristic. It is defined as a crack size that may be missed at inspection with an allowable probability (of approximately 5%). The MUS determined by the full-scale test results is adopted as a basic value. The corresponding value for the service stage should be obtained by means of multiplying the basic one by an empirical correction factor of 5-10 or even more. This factor takes into account the so-called "service disturbing environment", connected with service inspectability level, the operator's qualification, etc.

The System of Reliability Coefficients

These coefficients must compensate for the scatter in loading and fatigue endurance and other likely errors in a realization of the structural safety system. The mean values obtained from full-scale tests must be divided by reliability coefficients to establish the values to be specified for the service. Ranges of these coefficients are shown in Table 1.

TABLE 1

η_1	Error of predicting the equivalents	1/1.5
η_2	Possible additional errors(for uninspected zones)	1/1.2
η_3	Scatter of individual loading	1/1.5
η_4	Scatter of fatigue endurance	3/5

The reliability coefficient, η , for the first inspection and for the inspection time intervals is established as

$$\eta = \eta_1 \cdot \eta_2 \cdot \eta_3 \cdot \eta_4$$

It is seen, for example, that in case of using the on-board fatigue loading there is no necessity to divide its readings by the coefficient η_3 , hence, the specified service life or inspection intervals will be increased by a factor of 1.5 on the average.

Here, the maximum value is the reliability coefficient η_4, which takes into account the fatigue endurance scatter. Its values for different numbers of identical tested specimens (based on the adopted standard deviation of the endurance logarithm equal to 0.15) are given in Table 2.

TABLE 2

Number of specimens η	1	2	3	4	5	≥ 6
η_4^{stand}	5	4	3,5	3,2	3,1	3
$\frac{N_{max}}{N_{min}} = \delta_{stand}$		1.4	1.7	1.9	2.1	2.3

The Civil Aircraft Airworthiness Requirements adopt the use of these standard coefficients only when the endurance ratio - N_{max} to N_{min} - does not exceed the standard limits shown in the table. Otherwise the value of the coefficients must be increased according to the formula.

$$\eta_4^{real} = \left(\eta_4^{stand}\right)^{\frac{\delta_{real}}{\delta_{stand}}}$$

In case of structures samples from service this procedure may compensate implicitly for possible effects of other degradation processes.

In order to prevent the structure from the initiation of multi-site damage the specified service life limits must be established even if the service complies with the damage tolerance principle. In this case the standard reliability coefficient may be decreased, but the factor to divide must not be greated than 1.5 (a time interval from the first failure to the second one among independent critical points give a ratio approximately constant for a very wide sample size range, see Fig. 5). The minimum value of the reliability coefficient for total service life specified according to the damage tolerance principle must be $\eta = 2$.

The analysis of service experience is the most important item in the system. It must be used for two purposes: for correction of the "Critical Areas" list and for confirmation (or necessary correction) of the values of full-scale test equivalents. The first problem is clear. The second problem should be solved on the basis of special statistical analysis by comparing the structure state data from both tests and service experience. The form of this information is very specific (Fig. 6) since the experimental points may have different specific senses. For example, if the damage is not revealed this means only that its size is equal to or less than the maximum undetectable size and therefor the corresponding point on the picture must have an arrow down. There is a scientific basis for some methods, and the necessary mathematical "tools" to develop an effective algorithm for those "undetermined" samples. A range of equivalent deflections from the predicted values is shown in Fig. 7.

Conclusions

The information about the Fatigue Safety System given in this report is only generic. The authors hope that a possible interest in the system may produce additional contacts in the future.

References

1. Edinyye Normy lyotnoy godnosti grazhdanskikh transportnykh samolyotov stran-chlenov SEV, glava 4, razdel 4.9, Moscow, 1985.

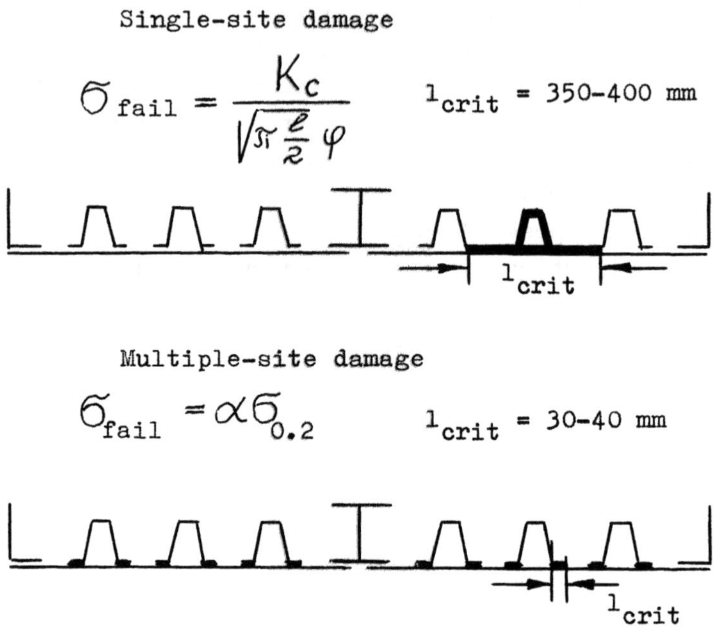

Fig.1 Effect of multiple-site damaging (conservative assumption)

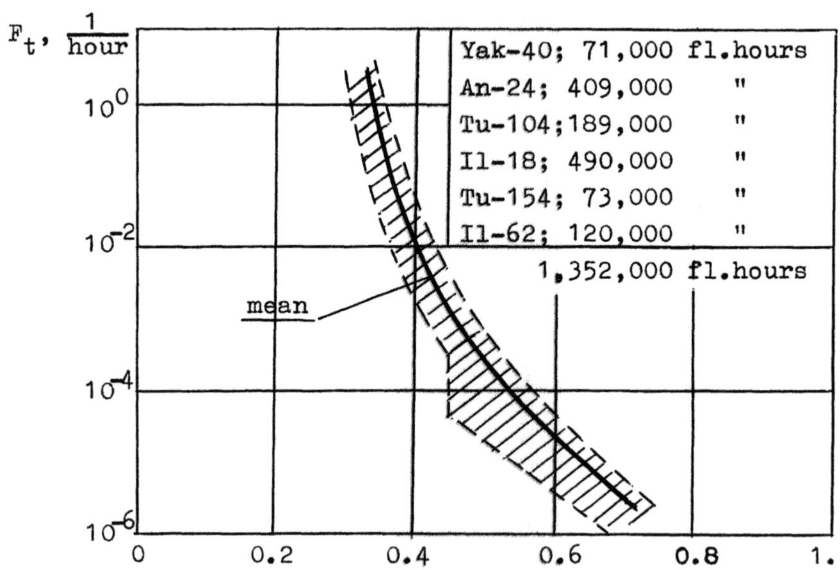

Fig.2 The loading factor occurence cumulative frequency (in-service statistics for several types of aircrafts)

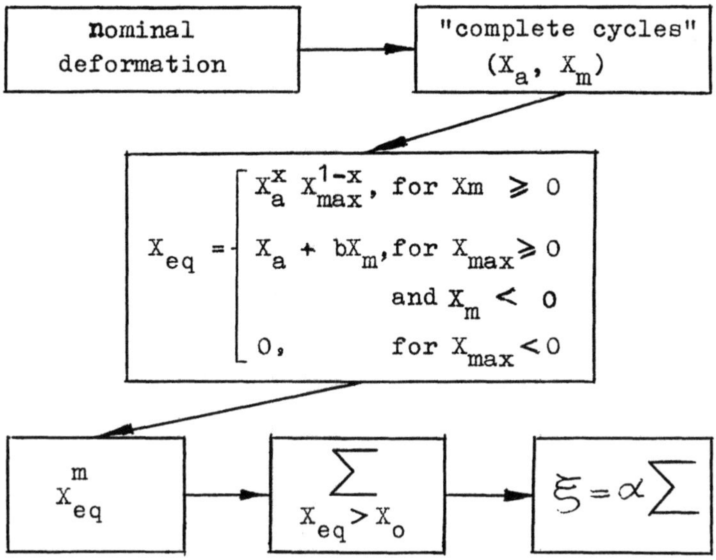

Fig.3 A simple on-board fatigue loading meter algorithm (x, b, m, X_o, α - var)

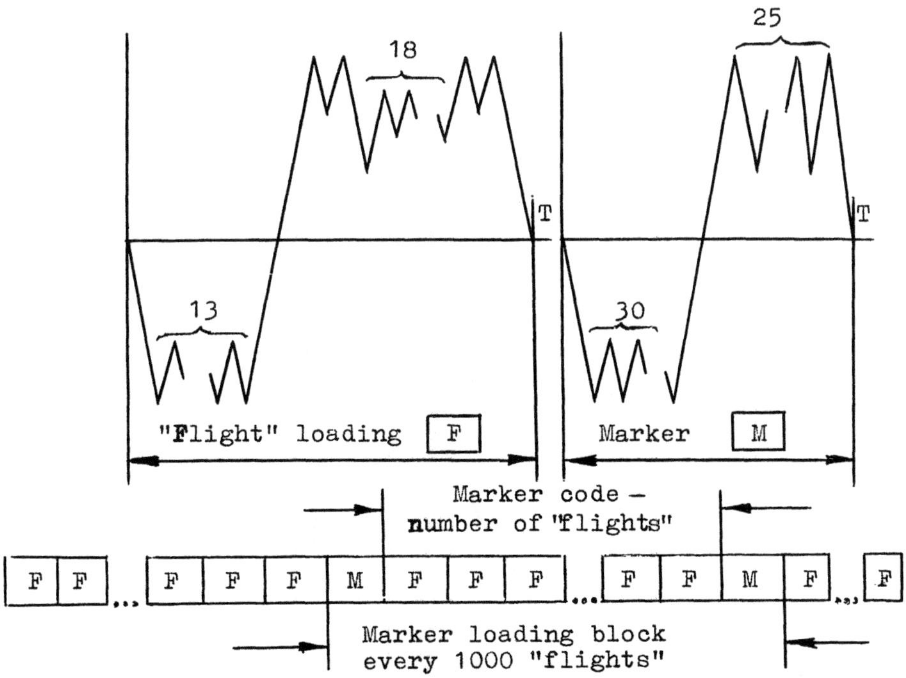

Fig.4 An example of marking by loading variations

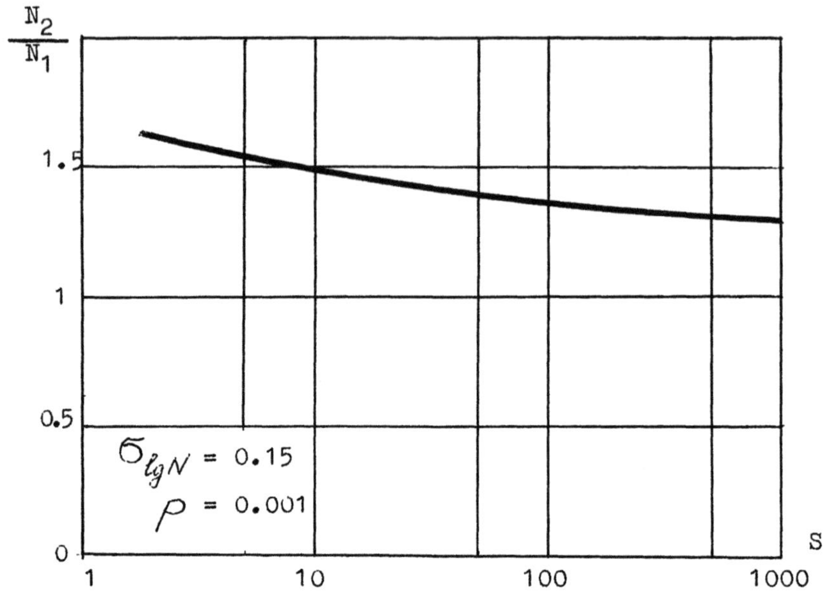

Fig.5 Ratio of the second fatigue endurance to the first one in the sample with S independent sites

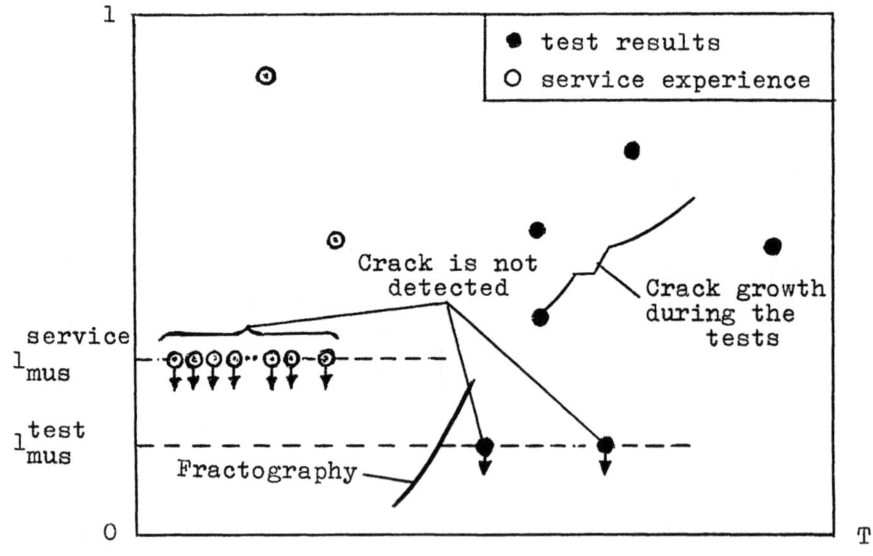

Fig.6 A typical test/service data interpretation

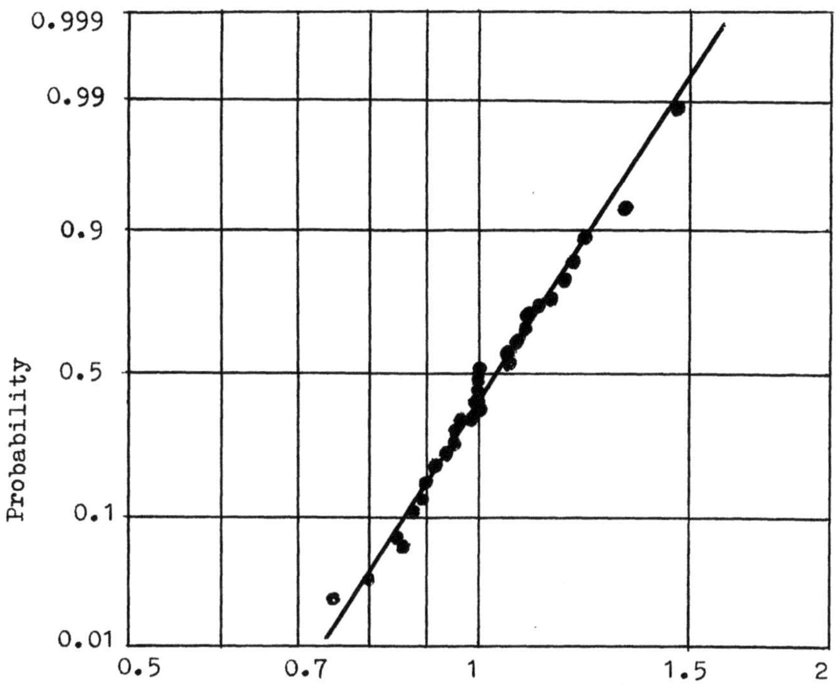

Fig.7 Ratio of service experience equivalents to predicted values

26. Optical Methods of Stress Analysis Applied to Cracked Components

C. W. Smith

Department of Engineering Science and Mechanics
Virginia Polytechnic Institute and State University
Blacksburg, VA 24061-0219 U.S.A.

SUMMARY

After briefly describing the principles of frozen stress photoelastic and moire interferometric analyses, and the corresponding algorithms for converting optical data from each method into stress intensity factors (SIF), the methods are applied to the determination of crack shapes, SIF determination, crack closure displacement fields, and pre-crack damage mechanisms in typical aircraft component configurations.

INTRODUCTION

Perhaps the most common scenario for the fracture of aircraft structural components begins with material damage at the surface (or interface) at a stress raiser, followed by crack initiation, and, after a period of stable fatigue crack growth, the unstable extension of the crack to fracture. Initial damage is often associated with various surface mechanisms (corrosion, fretting, etc.) and the stable growth regime often involves stress gradients due to complex geometry producing curved crack fronts. Moreover, fatigue loading is usually accompanied by crack closure.

We shall begin by briefly describing two optical methods of analysis, frozen stress and moire interferometry, and appropriate algorithms for extracting SIF values from the optical data. However, the focus of this paper will be directed towards the results obtained by applying these methods to quantify SIF values during stable growth, crack closure displacement fields, and pre-crack damage conditions in typical aircraft structural geometries.

OPTICAL METHODS OF ANALYSIS

The methods of analysis used as noted above were frozen stress photoelasticity and moire interferometry:

Springer Series in Computational Mechanics
Atluri, Sampath, Tong (Eds.)
Structural Integrity of Aging Airplanes

i) Frozen Stress Photoelasticity - In principle, photoelasticity [1] is the result of the proportionality of applied load to temporary double refraction which results when an isotropic transparent plate of material is loaded and observed in a field of polarized light which traverses the plate thickness. Dark and bright regions are observed in the stressed material and, for a crossed circular polariscope the stress "fringes" are dark and the half fringes bright. The magnitude of the maximum shearing stress (τ^{nz}_{max}) in the plane normal to the light propagation direction has a constant value along each fringe and the fringes arrange themselves like contours. The frozen stress method was introduced by Oppel [2] in 1937. It utilizes a transparent epoxy material which exhibits, in simplest concept, diphase mechanical and optical properties. That is, at room temperature, its mechanical response is viscoelastic. However, above it's "critical" temperature, it's viscous coefficient vanishes, and it's behavior becomes purely elastic, exhibiting a Modulus of Elasticity of about 0.2% of it's room temperature value and some twenty times more sensitivity to load. Thus, by loading the photoelastic models above critical temperature, cooling under load, and then removing the load, only negligible elastic recovery occurs at room temperature and the stress fringes and deformations produced above critical temperature are retained.

In order to determine useful optical data from frozen stress analysis, one needs to suppress deformations near the crack tip in the photoelastic material in its rubbery state above critical temperature and to be able to produce the same crack shape and size produced in the fatigue test. In order to accomplish the first objective, applied loads are kept very small, and a polariscope modified to accomodate the tandem application of Post partial mirror fringe multiplication [3] and Tardy compensation [4] is employed to accurately measure partial fringe orders. Normally, fifth multiples of fringe patterns are read to a tenth of a fringe thus providing adequate data within about 1 or 2 mm of the crack tip.

Surface crack shapes are obtained by introducing a starter crack at the desired location in the photoelastic model of the structure before stress freezing by striking a sharp blade held normal to the crack surface with a hammer. The starter crack will emanate from the blade tip and propagate dynamically a short distance into the model and then arrest. Further growth to the desired size is produced when loaded monotonically above critical temperature. The shape of the crack is controlled by the body geometry and loads. By comparing crack shapes grown in photoelastic models by this process to those grown under tension-tension fatigue loads in steel, excellent correlation has been obtained [5] even when some crack closure was present on the free surface of the latter. It appears that the cracked body geometry and loads control the crack shape in thick bodies and that the stress ratio $R = \sigma_{min}/\sigma_{max}$ (as long as it is positive) and plasticity or closure effects are of secondary importance.

By removing thin slices of material which are oriented mutually orthogonal to the crack front and the crack plane locally, analysis of these slices will yield the distribution of the maximum shear stress in the nz plane of Fig. 1. Then, algorithms appropriate to the local mode of deformation are used to extract the SIF from the optical data. These algorithms are given in the Appendix (See also Ref. [6]).

ii) Moire Interferometry - Use of this method of analysis for measuring displacement was introduced by Weller et al. [7] in 1948 and involves the generation of fringes by interference between two gratings, one of which is attached to the specimen surface and deforms, and the other which is the undeformed or "master" grating. The fringes are then proportional to the displacements (and/or rotations) at the surface of the deformed body. By splitting a laser, and reflecting one part from a mirror back into the other part, walls of constructive and destructive interference are formed and constitute a high density "virtual" master grating of some 1200 ℓ/m. For the following applications, the fringes are proportional to the displacement field in the vertical direction and were obtained using horizontal gratings.

APPLICATIONS OF THE METHODS

The results of three applications of the foregoing methods will now be briefly described.

i) SIF Values for a Pin-Loaded Cracked Hole in a Plate by the Frozen Stress Method - The specimen geometry and slice locations for these tests are pictured in Fig. 2. Geometries covered the following ranges:

a/c: 1.08 to 2.08

a/T: 0.19 to 0.88

$c/\bar{r}$: 0.34 to 1.68

$2\bar{r}/T$: Two values, approximately 0.5 and 1.0.

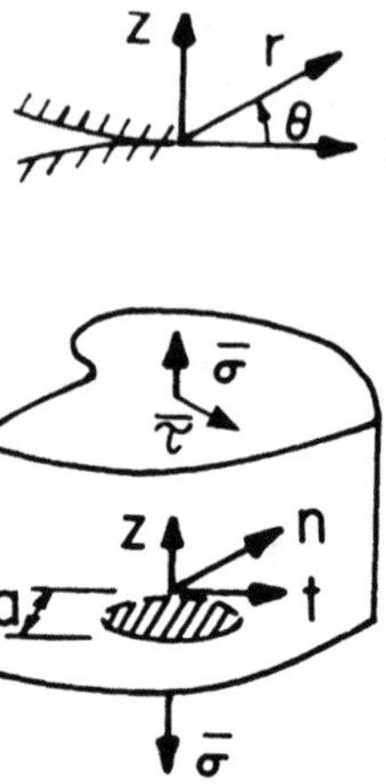

Fig. 1. General Problem Geometry and Notation

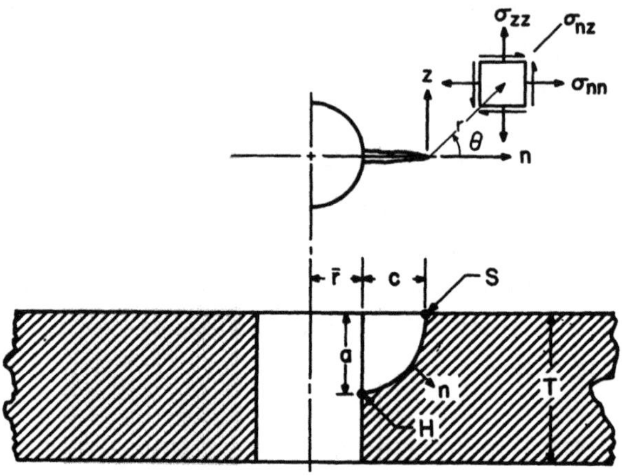

Fig. 2. Pin-Loaded Cracked Hole Geometry [7]

Typical stress fringe signatures (unmultipled) at S and H are pictured in Fig. 3. Slice H (Fig. 3a) revealed a Mode I signature (Fig. A-1). However, slice S shows a Mixed Mode Signature and this was true for all cases. Fig. 4 provides a Mode I pattern before and after partial mirror fringe multiplication but before application of Tardy compensation. Fig. 5 presents typical graphs from S & H slices for the linear zones described in the Appendix for Mode I and the Mixed Mode Case. Here, $K_H^* = (K_I)_H$. SIF values for the pin loaded holes were found to be two to four times higher than for holes loaded only by remote loads [8]. However, the experimental results were lower than those obtained numerically [9]. Details of this study are found in Ref. [10].

ii) <u>Displacement Fields Associated with Crack Closure by Moire Interferometry</u> - When a crack is grown under fatigue loading, a plastic wake is generated above and below the crack surface. Thus, upon unloading after fatiguing, the crack often closes especially on or near where it intersects a free surface. If the crack is reloaded in tension, a certain stress level is required to just open the crack. Many engineers measure

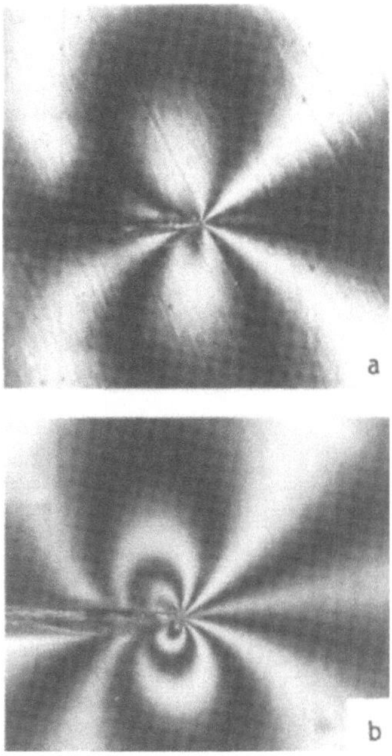

Fig. 3. Typical Fringe Patterns (a) Hole Slice (dark field) (b) Surface Slice (bright field) [7]

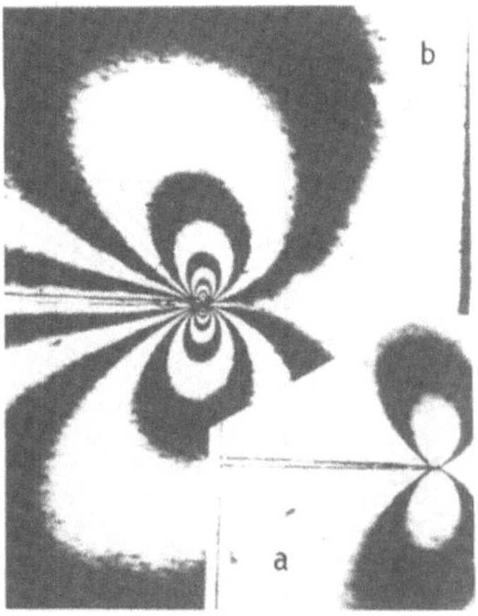

Fig. 4. Near Tip Mode I Fringes (a) Before and (b) After Partial Mirror Fringe Multiplication showing (a) unmultiplied and (b) fifth multiple

the ΔK associated with fatigue crack growth from this opening K level and refer to it as the effective ΔK or $(\Delta K)_{eff}$. Fig. 6 gives the geometry of an edge notched 7075-T6 Aluminum specimen which had an equivalent horizontal grating of 2400 ℓ/mm applied ahead of the notch. Fig. 7a shows a photograph of the moire pattern for u_z after the crack was extended in fatigue at $R = \sigma_{min}/\sigma_{max} = 0.1$ for 6 mm and then unloaded. Fig. 7b shows the u_z pattern when the crack is reloaded until it just opens.

$$\therefore (u_z)_{eff} = u_z - (u_z)_{op} \tag{1}$$

Now, assuming a state of plane stress (alternatively a plane strain state may be assumed) in the singularity dominated zone at the free surface, we have:

$$u_z = \frac{2(1+\nu)}{E} \left(\frac{r}{2\pi}\right)^{\frac{1}{2}} \left[\frac{(7-\nu)}{1+\nu} \sin\frac{\theta}{2} - \sin\frac{3\theta}{2}\right]$$

Along $\theta = \pi/2$

$$u_z = \frac{(3-\nu)}{E} K_1 \left(\frac{r}{2\pi}\right)^{\frac{1}{2}}$$

$$\therefore (\Delta K)_{eff} = \frac{E}{3-\nu} \left(\frac{2\pi}{r}\right)^{\frac{1}{2}} [u_z - (u_z)_{op}] \tag{2}$$

where E is Young's Modulus, r is measured along $\theta = \pi/2$, u_z is the displacement under load and $(u_z)_{op}$ are displacements at crack opening.

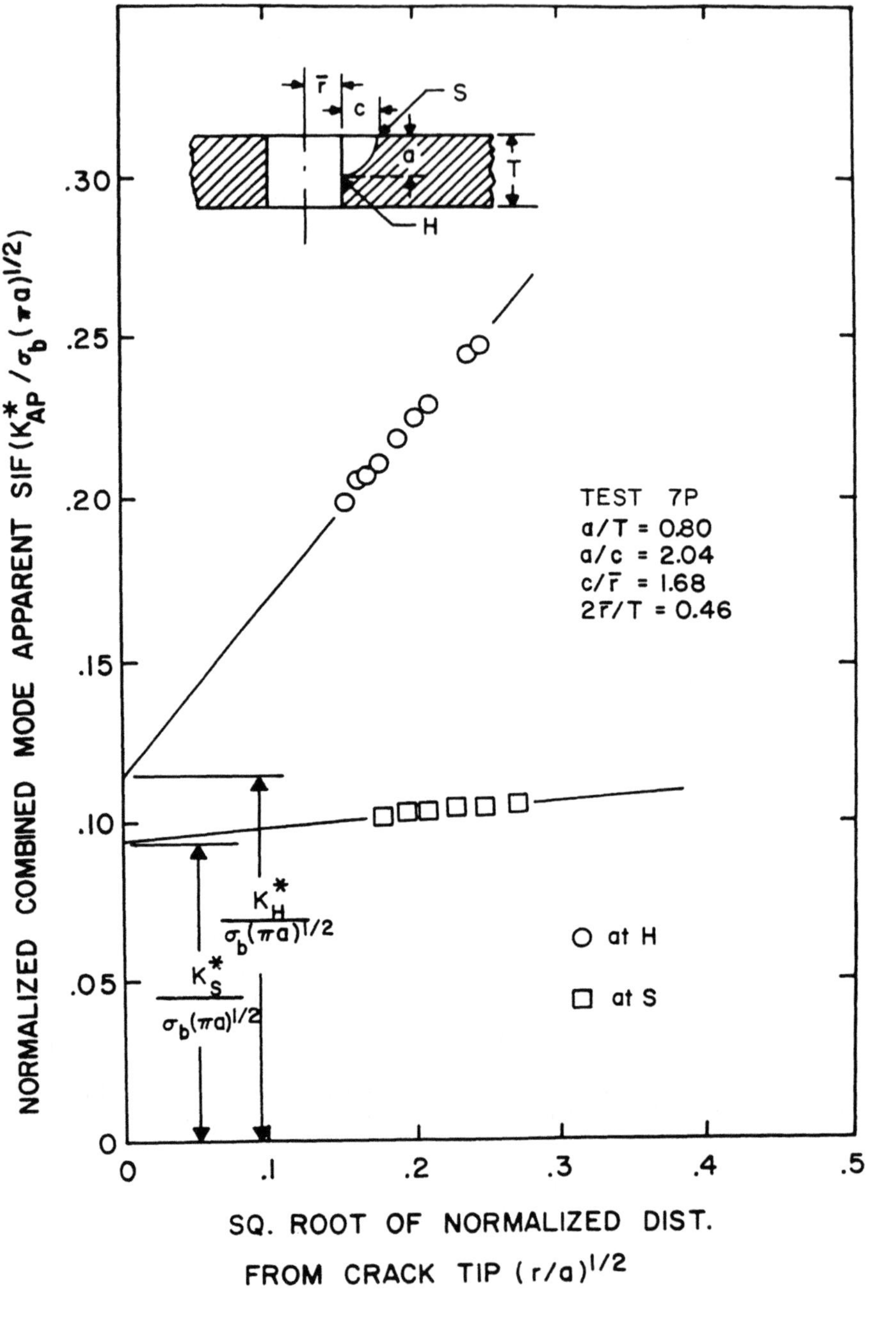

Fig. 5. Typical Sets of Data in Linear Zone for K* Determination [7]

While measurements are made only on the free surface, they do record accurately the instant of full crack opening. Details of this study are found in Ref. [11].

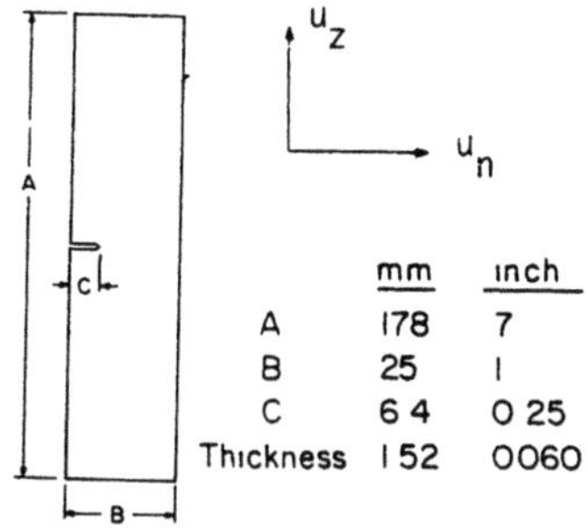

Fig. 6. Test Configuration for Fatigue Cracked Aluminum Specimen [9]

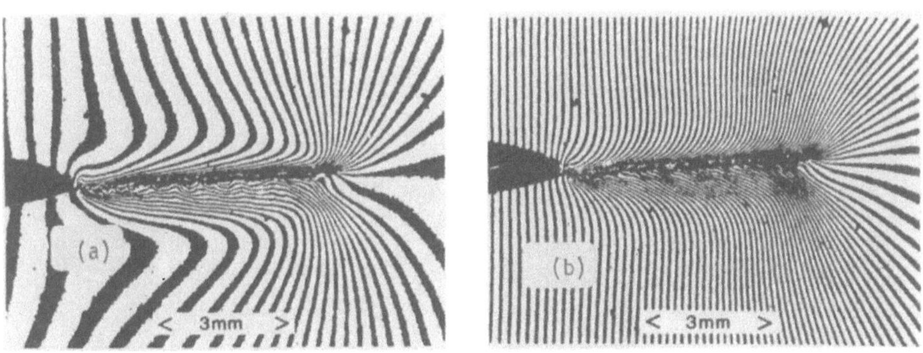

Fig. 7. a) Crack closure and b) Crack Opening Moire Displacement Component Normal to Crack Plane (u_z) [9]

iii) <u>Pre-crack Damage in a Dovetail Joint by Moire-Interferometry</u> - The test setup for this study is pictured in Fig. 8. The test specimen was a 3 mm thick steel sheet with two blades of the same thickness dovetailed into it. The dovetail was form ground, leaving all edges sharp. A cross-grating equivalent to 2400 ℓ/mm was applied across the dovetail after fatigue cycling and test piece and specimens were then loaded in tension. It was desired to measure the displacement components tangent and normal to the dovetail testpiece interface. The moire patterns for the vertical displacement component obtained from the horizontal grating is shown in Fig. 9. The vertical displacement component is given by:

$$u_z = \frac{n}{2F} \qquad (3)$$

where n is the moire fringe order and F is the grating density (1200 ℓ/m).

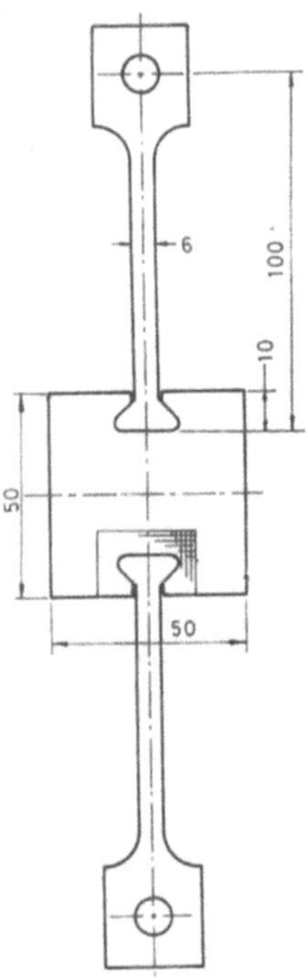

Fig. 8. Test Setup for Dovetail Loading (Courtesy D. Post) [11]

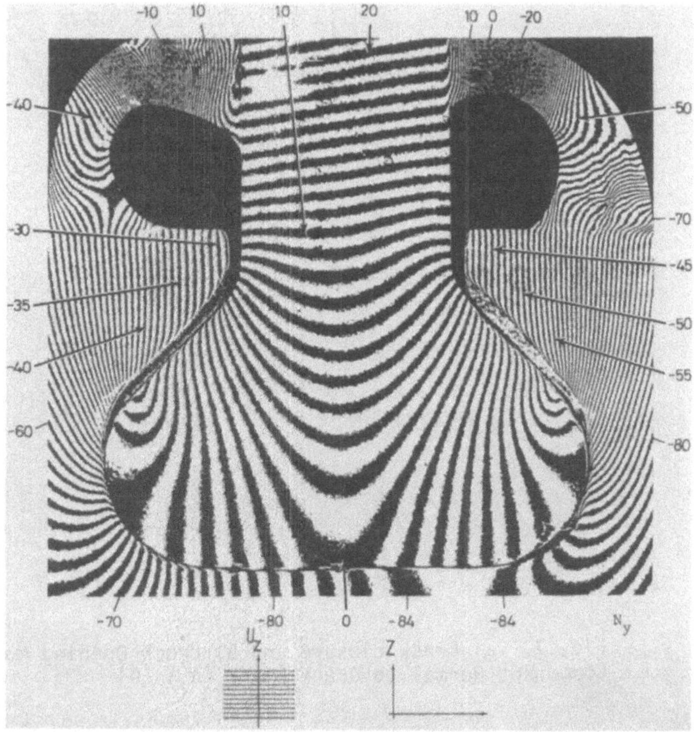

Fig. 9. Moire Pattern for (U_z) for Loaded Dovetail Joint (Courtesy D. Post [11])

It can be seen from Fig. 9 that the grating deteriorated (delaminated) along the 45° flanks of the contact region of the dovetail but fringes could still be extrapolated to the interface. When compared to a finite element solution which modelled both contact-separation and stick-slip phenomena, but with constant coefficient of friction [12], the experimental results revealed global asymmetry as shown in Fig. 7, and a non-linear friction effect especially during the changeover from stick to slip. This study is described in detail in Ref [13].

CONCLUSIONS

Two optical methods of analysis have been briefly described and used to study the pre-crack damage, SIF values during stable growth in a stress gradient region and crack closure phenomena. There are many potential sources of error in the experiments, even with the indicated refinements. For Mode I photoelastic analysis, errors of ±2 to ±5% are not uncommon. For mixed mode, the error may be twice as great as for Mode I. The moire results should be within ±2%.

It is important to emphasize that these methods are used to best advantage when coupled with numerical methods, especially in three dimensional or contact problems such as those described here. It should also be pointed out that, although not emphasized here, service environment may significantly influence structural component behavior. Despite the above limitations, continued refinement of optical methods can be expected to provide improved quantitative estimates of many practical aircraft structural problems.

ACKNOWLEDGEMENTS

The author gratefully acknowledges the contributions of former students J. J. McGowan, W. H. Peters and M. Jolles and colleagues C. Ruiz, D. Post and R. Czarnek and the support of NASA Langely Research Center for parts of this work.

REFERENCES

[1] Maxwell, J. C. "On the Equilibrium of Elastic Solids," Trans. of Royal Society of Edinburgh, Vol. XX, Part I. (1853), 87-120.

[2] Oppel, G. "Polarisationsoptische Untersuchung Raumlicher Spannungs und Dehnungszustande," Forsch. Geb. Ingenieurw, Vol. 7 (1936) 240-248.

[3] Post, D. "Isochromatic Fringe Sharpening and Fringe Multiplication in Photoelasticity," Proc. of Society for Experimental Stress Analysis, Vol. XII, No. 2 (1955), 143-156.

[4] Tardy, M. H. L. "Methode Practique D'examen de Messure de la Birefringence des verres D'optique," Rev. Optics, Vol. 8 (1929), 59-69.

[5] Smith, C. W. "Stress Intensity and Flaw Shape Variations in Surface Flaws," J. of Experimental Mechanics, Vol. 20, No. 4, (1980), 126-133.

[6] Smith, C. W. and Kobayashi, A. S. "Experimental Fracture Mechanics," Chapter 20 of Handbook of Experimental Mechanics, A. S. Kobayashi, Ed., Prentice-Hall, New York (1987) 891-956.

[7] Weller, R. and Shepard, "Displacement Measurement by Mechanical Interferometry," Proc. of Society for Experimental Stress Analysis, Vol. VI, No. 1, 1948, 35-38.

[8] McGowan, J. J. and Smith, C. W. "Stress Intensity Factors for Deep Cracks Emanating from the Corner Formed by a Hole Intersecting a Plate Surface," Mechanics of Crack Growth, ASTM-STP 590, American Society for Testing and Materials, 1975, 460-476.

[9] Shah, R. C. "Stress Intensity Factors for Through and Past Through Cracks Emanating from Fastener Holes," Mechanics of Crack Growth, ASTM-STP 590, American Society for Testing and Materials (1975), 429-459.

[10] Smith, C. W., Jolles, M. and Peters, W. H., "Stress Intensities for Cracks Emanating from Pin-Loaded Holes," Flaw Growth and Fracture, ASTM-STP 631, American Society for Testing and Materials, 1977, 190-201.

[11] Post, D., Czarnek, R. and Smith, C. W. "Patterns of U and V Displacements Around Cracks in Aluminum by Moire Interferometry," Application of Fracture Mechanics to Materials and Structures, G. C. Sih, E. Sommer and R. H. Dahl, Eds., Martinus-Nijhoff (1984) 644-709.

[12] Boddington, P. H. B., Chen, K. C. and Ruiz, C. "The Numerical Analysis of Dovetail Joints," J. of Computers and Structures, Vol. 24 (1984), 208-217.

[13] Ruiz, C., Post, D., and Czarnek, R. "Moire Interferometric Study of Dovetail Joints," J. of Applied Mechanics, Vol. 52, March (1985), 109-114.

Appendix - Photoelastic Algorithms for the SIF

Figure A-1 shows the form of the photoelastic stress fringes for near tip Mode I and Mixed Mode (i.e., I and II) deformation. Appropriate algorithms for these cases have been developed elsewhere [6] so only the results of those developments are included here for the convenience of the reader.

i) Mode I - (Fig. A-1-a) We compute a truncated form of τ^{nz}_{max}(Fig. 1, $\bar{\tau} = 0$) along $\theta = \pi/2$ for a two parameter model. The result is:

$$\frac{K_{Ap}}{\bar{\sigma}\,(\pi a)^{\frac{1}{2}}} = \frac{K_I}{\bar{\sigma}(\pi a)^{\frac{1}{2}}} + \frac{B(8)^{\frac{1}{2}}}{\bar{\sigma}}\left(\frac{r}{a}\right)^{\frac{1}{2}} \qquad \text{(A-1)}$$

where $K_{Ap} = \tau^{nz}_{max}(8\pi r)^{\frac{1}{2}}$ is defined as an "apparent" SIF.

$\bar{\sigma}$ = remote stress
a = crack depth
K_1 = Mode I SIF
B = Non-Singular stress parameter
r = Distance from crack plane along $\theta = \pi/2$

Equation A-1, when plotted as

$$\frac{K_{Ap}}{\bar{\sigma}(\pi a)^{\frac{1}{2}}} \text{ versus } \left(\frac{r}{a}\right)^{\frac{1}{2}} \qquad \text{(A-2)}$$

yields a straight line which when extrapolated to the origin will yield $K_I/\ \bar{\sigma}(\pi a)^{\frac{1}{2}}$ the normalized SIF (See Fig. 5).

ii) Mixed Mode - (Fig. A-1-b) Here both $\bar{\sigma}$ and $\bar{\tau}$ (Fig. 1) are present. Proceeding as before, but not specifying θ, $K^{*}_{Ap} = \tau^{nz}_{max}(8\pi r)^{\frac{1}{2}}$ as before, and plotting, $\frac{K^{*}_{Ap}}{\bar{\sigma}(\pi a)^{\frac{1}{2}}}$ vs $\left(\frac{r}{a}\right)^{\frac{1}{2}}$ as before, we obtain $K^{*}/\bar{\sigma}(\pi a)^{\frac{1}{2}}$ (See Fig. 5). Here

$$K^{*}_{Ap} = \left[\{(K_I)_{Ap}\sin\theta_m + 2(K_{II})_{Ap}\cos\theta_m\}^2 + \{(K_{II})_{Ap}\sin\theta_m\}^2\right]^{\frac{1}{2}} \text{ and}$$

$$K^{*} = \left[(K_I\sin\theta^o_m + 2K_{II}\cos\theta^o_m)^2 + (K_{II}\sin\theta^o_m)^2\right]^{\frac{1}{2}} \qquad \text{(A-3)}$$

where θ^o_m is obtained from a plot of θ_m vs r/a (Fig. A-2) Moreover, from a near tip geometric fringe condition,

$$\frac{K_{II}}{K_I} - \frac{4}{3}\left(\frac{K_{II}}{K_I}\right)\cot 2\theta^o_m - \frac{1}{3} = 0 \qquad \text{(A-4)}$$

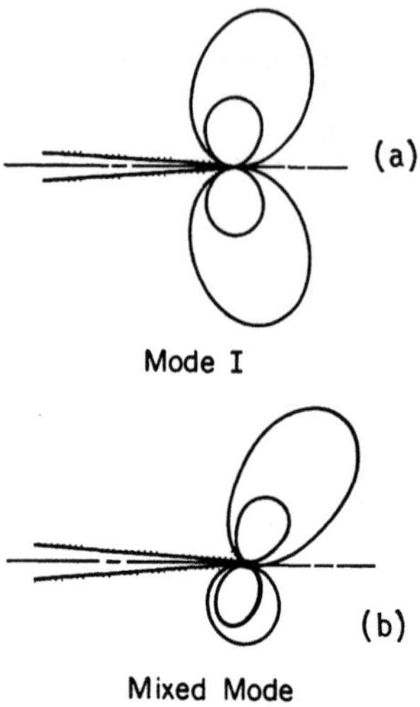

Fig. A-1 Stress Fringe Signatures for a) Mode I and b) Mixed Mode Loading

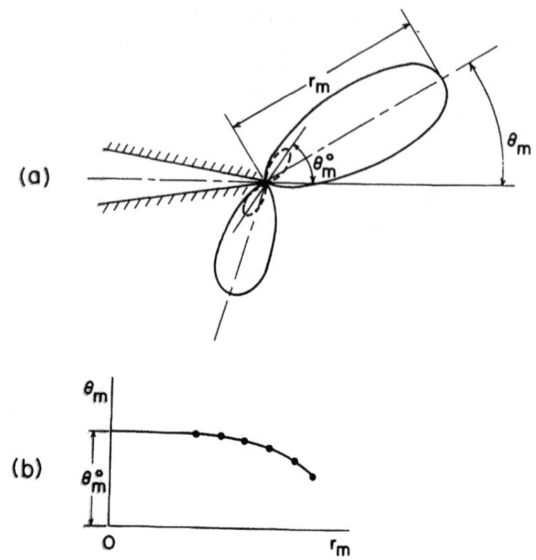

Fig. A-2 Determination of θ_m^o from θ_m (Note that only forward leaning loops are used).

Knowing θ_m^o, Eqs. (A-3) and (A-4) may be solved for K_I & K_{II} the Mode I and Mode II SIF values.

27. Repairs to Damage Tolerant Aircraft

T. SWIFT

Federal Aviation Administration
3229 E Spring Street
Long Beach, Calif. 90806-2425

Summary

The results of displacement compatibility analysis, representing a variety of repair doubler and lap splice configurations, are presented with a view to illustrating how structural repairs can degrade the fatigue initiation life and damage tolerance capability of primary transport aircraft structure. Examples show that fatigue initiation life is directly related to the peak loads induced in the first fastener rows at the edges of repair doublers. Design of repairs to an equal or better static strength capability and the associated static strength analysis will not normally highlight these peak loads which can result in considerable degradation of structural fatigue life. Critical fastener loads, based on displacement compatibility analysis accounting for fastener flexibility, are parametrically presented for a variety of skin and doubler thicknesses. Suggestions are made on how repair designs can be modified to improve fatigue initiation life and subsequent fatigue crack detectability particularly in the event of multiple-site damage (MSD). The importance of riveting quality during repairs, often not up to initial manufacturing standards, is discussed with respect to fatigue initiation life. A simplified but conservative method to generate crack growth curves is discussed with a view to easing the analytical burden for the small modifiers. It is hoped that this information, together with conservative fatigue Sn data, will help the many small repair and modification stations gain an appreciation of the fatigue and damage tolerance quality of structural repairs. It is pointed out in the paper that the FAA regulations were amended in December, 1978 for new transport category aircraft and in May, 1981 for aging transport aircraft to include a damage tolerance philosophy. This means that any repair to a transport category airframe, which may effect threshold, frequency and type of inspection of principal structural elements, must be evaluated for it's damage tolerance capability.

Introduction

In December, 1978 the Federal Aviation Administration amended their Fatigue Evaluation requirements for Transport Category Airplanes to include a damage tolerance philosophy. Prior to this time FAR 25.571, Fatigue Evaluation of Flight Structure, included an option to design to either "Fail-Safe" or "Safe-Life" principles. The manufacturers had generally adopted the Fail-Safe option with the exception of a few components such as

Springer Series in Computational Mechanics
Atluri, Sampath, Tong (Eds.)
Structural Integrity of Aging Airplanes

engine mounts and landing gear. In satisfying the regulation to the Fail-Safe option they had designed structure to be redundant so that catastrophic failure would not result after fatigue failure or obvious partial failure of a single principal structural element. However, the so-called Fail-Safe approach had not included a disciplined engineering evaluation of crack growth and residual strength characteristics of each principal structural element using fracture mechanics technology necessary to specify inspection methods, threshold and frequency which would detect damage prior to catastrophic failure. There had been a number of accidents in the mid seventies on aircraft designed to the Fail-Safe principle which may have been prevented had such inspections been imposed. Thus, in December, 1978 the FAA released amendment 45 to FAR 25.571 requiring that new structure be designed to "Damage Tolerant" principles unless it could be shown that this approach would be impractical whereupon a "Safe-Life" option could be used. In May, 1981 an advisory circular AC-91.56 was issued to provide guidance material for the issue of Supplemental Inspection Documents (SIDs) for existing Large Transport Category Airplanes. It was expected that advanced fracture mechanics principles would also be adopted to develop these SIDs. Thus, for both new designs and existing aircraft it is expected that engineering evaluation of the structure under typical load environmental spectra must show that catastrophic failure due to fatigue, accidental damage or corrosion will be avoided throughout the operational life of the aircraft. Over the last decade this process has been carried out. Inspection programs for both new designs and existing older aircraft have been based on a damage tolerance philosophy. However, up to the time of this writing, a system is not in place to require that all repairs or modifications to principal structural elements on these aircraft be evaluated to damage tolerance principles. Currently, the majority of these repairs are designed to an equal or better static strength requirement. Inspection procedures specified for the basic airplane are not being modified to reflect a change in design detail as a result of the repair or modification. Although the repairs or modifications may have equal or better static strength capability their effect on the fatigue quality and inspectability of the structure may be considerable. It is understood that a number of the major airframe manufacturers are currently working towards including damage tolerant evaluated standard repairs in their structural repair manuals. Others may follow this lead. This paper is intended to highlight this problem and perhaps outline some guidance to those occupied in aircraft modification and repair.

Degradation of Structure Due to Repair

In general, any repair to an airframe structure can substantially degrade fatigue life if extreme care in design detail is not taken. In the past, engineering evaluation of repairs has been based on equal or better static strength only without too much consideration for fatigue life. This philosophy can easily lead to static strength overdesign with consequent considerable loss in fatigue quality compared to the original structure. This phenomenon can easily be demonstrated analytically by considering a very simple example case. Figure

1a shows a typical fuselage skin element 0.04 inches thick riveted to a typical stringer. This element is subjected to cyclic hoop (circumferential) stress due to internal cabin pressure σ_h. Assume for our example this stress is 15 KSI with stress ratio R = 0. The fatigue life of this element will be based on the life at the row of rivet holes shown. These rivet holes are initially unloaded in the hoop direction. For our example case it will be assumed that the rivet holes have poor quality. This will be explained in detail later. The fatigue life of this row of rivet holes can be obtained from Figure 2 and as shown is about 160,000 cycles for 15 KSI with $\sigma_{br}/\sigma_{gr} = 0$. Suppose now we rivet a doubler to the skin as shown on Figure 1b. For our example we are not assuming any damage to the primary skin and we are not assuming the skin is cut. In other words we are not depending on all load being transferred out of the skin into the doubler. We are merely considering a doubler fastened to a piece of undamaged skin. Following the usual convention of increasing the doubler thickness one gauge, in the case of damaged skin, we will assume the doubler thickness is 0.05 inches. Whether we need it or not there will be load transfer out of. the skin into the doubler due to strain compatibility between the two pieces. The two pieces are fastened together and will be attempting to strain together. The highest rivet loads, induced from the skin into the doubler, will be at the first row as shown in Figure 1b. Figure 1c illustrates the induced fastener load F in the skin reacted by the load F applied to the doubler. The gross stress σ_h, along line AB in Figure 1c, will be unchanged in the presence of the doubler but now we have induced a bearing stress in the hole σ_{br} which will degrade the fatigue life. This is illustrated by the various Sn curves of Figure 2. It can be seen that for constant amplitude stress the life is reduced with increasing σ_{br}/σ_{gr} ratio. The stress in the sheet will reduce to the bypass stress σ_{bp} and the stress in the doubler will increase to σ_d. In most cases the critical location for fatigue will be at row 1 in the sheet where the gross stress is still σ_h and the bearing stress in the sheet is highest due to the highest fastener load. In order to determine the fatigue life at this location it is necessary to determine the fastener loads in the first row. This is done through a displacement compatibility analysis of the joint.

A typical strip, as indicated in Figure 1b, is idealized as shown in Figure 3. Each rivet is simulated as an elastic spring under shear load and each portion of the skin and doubler strip is idealized as a bar. Bar displacements are obtained simply from equation 1:

$$\text{Bar displacement} \quad \delta_{bar} = PL/AE \tag{1}$$

Where P = Bar load
L = Bar length
A = Bar area
E = Bar modulus

Rivet shear displacements are given by empirical equation 2 [1]:

$$\text{Rivet displacement} \quad \delta_{riv} = F[A+B(D/t_D + D/t_S)]/(ED) \tag{2}$$

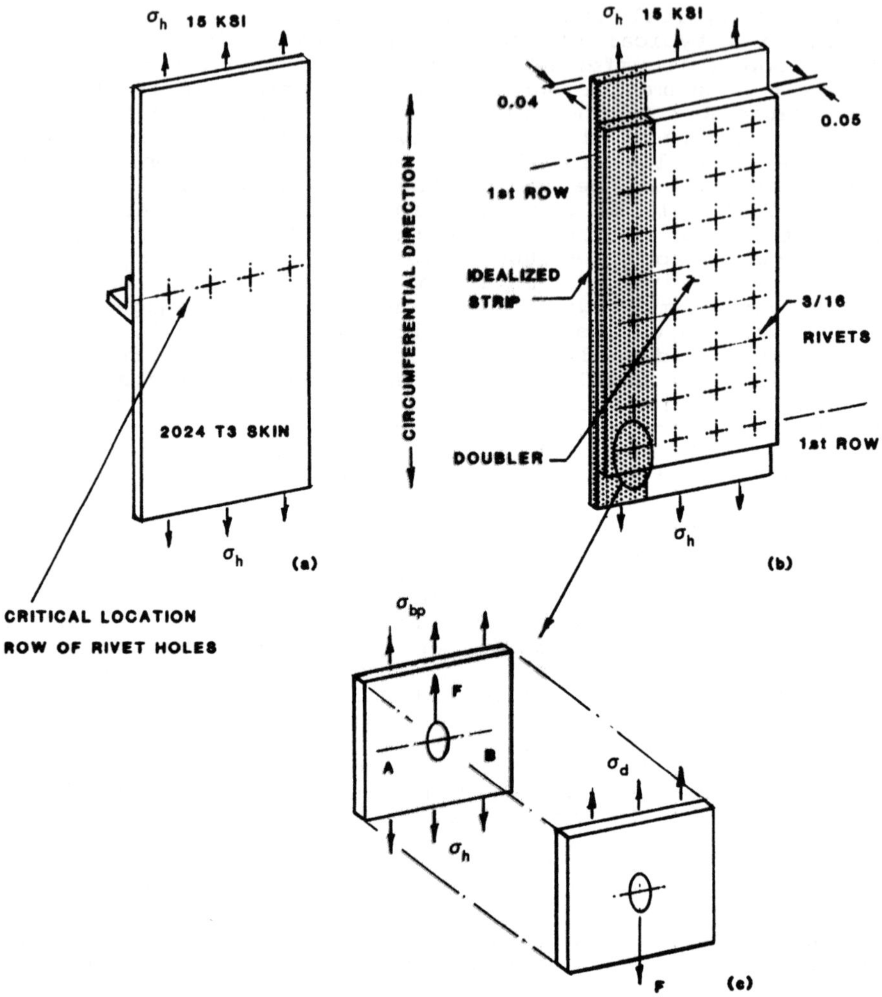

Fig. 1. Example of Doubled Element

Where F = Rivet shear load
D = Rivet diameter
E = Sheet modulus
t_d = Doubler Thickness
t_s = Skin thickness
A = 5.0 for aluminum rivets and 1.666 for steel fasteners
B = 0.8 for aluminum rivets and 0.86 for steel fasteners

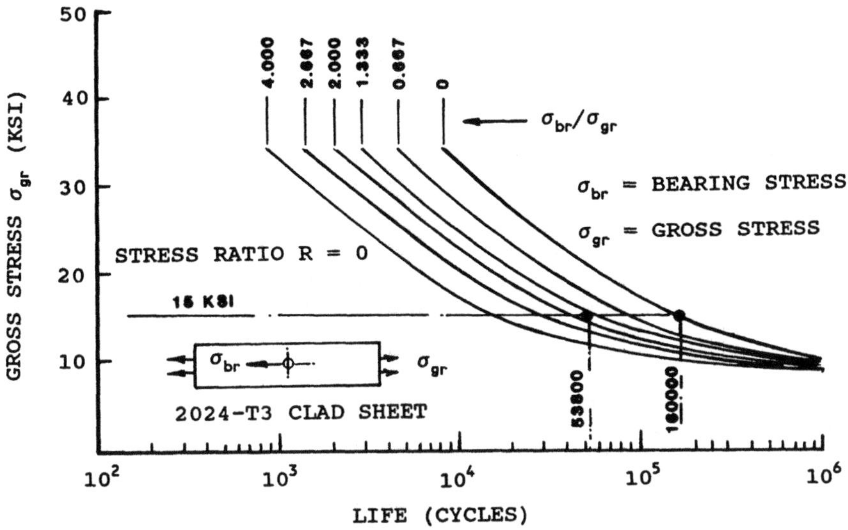

Fig. 2. Open Hole Sn Data 2024-T3

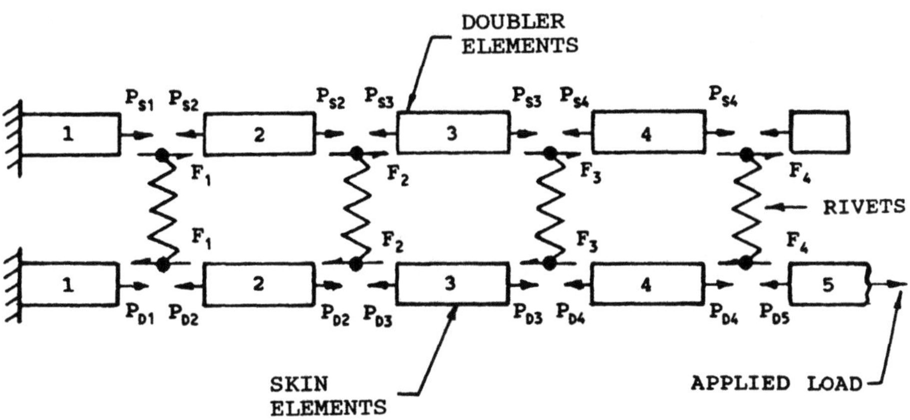

Fig. 3. Idealization of Strip

Displacements in the skin at each rivet are made compatible with those in the doubler after accounting for rivet displacement. Load compatibility is merely assumed, for example as:

$$P_{S1} - P_{S2} = F_1, \quad P_{S2} - P_{S3} = F_2 \quad \text{etc.}$$

$$\text{and} \quad P_{D1} - P_{D2} = F_1, \quad P_{D2} - P_{D3} = F_2 \quad \text{etc.}$$

Where S and D denote skin and doubler.

The resulting load distribution in our example case is shown in Figure 4. It can be seen the highest loads are at the first row and are 187.2 lbs per rivet. The skin bearing stress is therefore 187.2/(0.04 x .19) = 24632 psi. Bearing stress to gross stress ratio σ_{br}/σ_{gr} = 24362/15000 = 1.642. The calculated life is therefore 53800 cycles as indicated by Figure 2. Thus, the addition of the 0.05 inch thick doubler, even though for our example case is not transferring load across a damage area, has caused a reduction in fatigue life from 160,000 cycles to 53,800 cycles because of load transfer from the skin into the doubler due to displacement compatibility. Figure 5 shows decreasing skin life with increasing doubler thickness obtained by similar analysis. Further reductions in skin life are shown when steel fasteners are used due to higher first fastener load due to increased stiffness.

It is shown in Figure 5 that the fatigue life of the skin is extremely sensitive to doubler thickness and that the life is controlled by the first fastener load. In order to provide some guidance to repairers and modifiers of airframe structure a number of typical configurations have been analyzed. Displacement compatibility analysis was performed to give critical fastener loads for a variety of skin doubler and lap splice combinations. Figure 6 shows the idealization used for the skin/doubler combinations. Sufficient rivets (5) were considered so that rivet load away from the doubler edge was insignificant. The rivet load distribution is shown in Figure 6 for a typical configuration assuming a unit 1 KSI applied gross stress. Strip and rivet displacements were based on

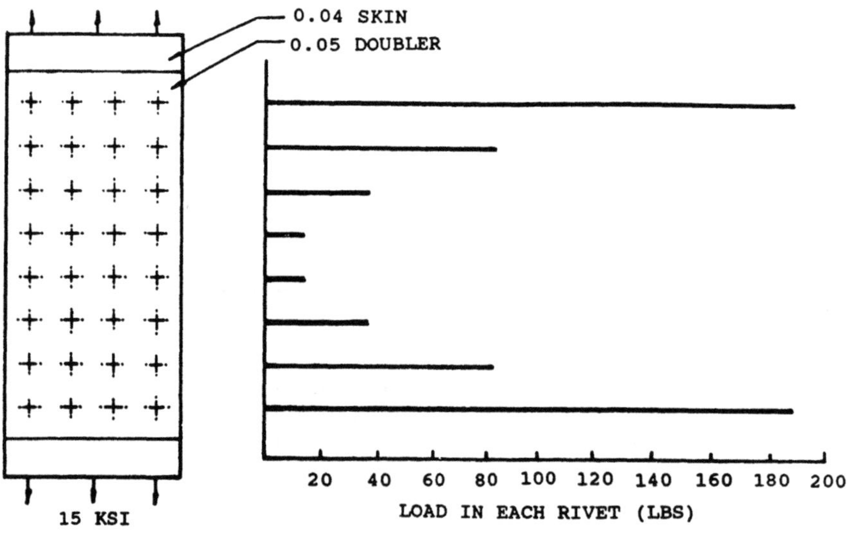

Fig. 4. Rivet Load Distribution For Example Case

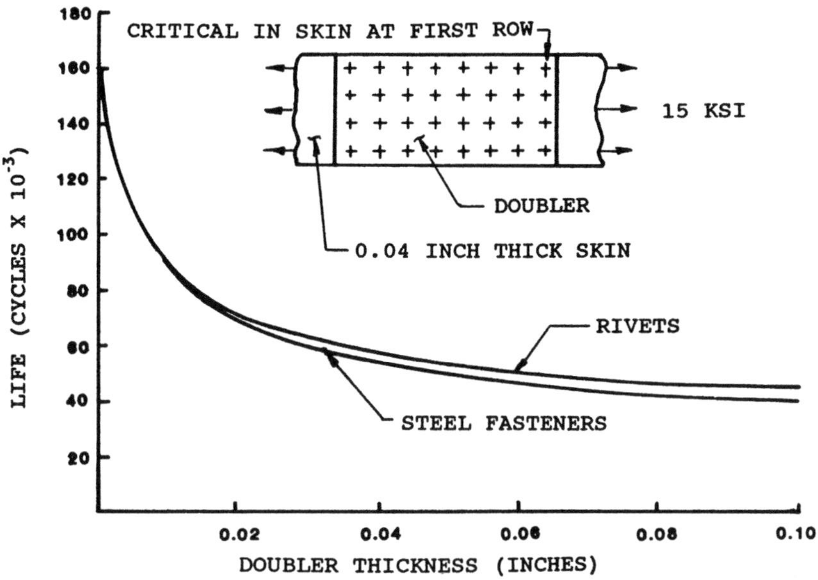

Fig. 5. Skin Fatigue Life At First Fastener

equations 1 and 2 respectively. Figure 7 shows first and critical fastener load as a function of basic skin and doubler thickness for 1 KSI unit applied gross stress when 3/16 inch diameter aluminum rivets were assumed for the skin/doubler configuration. Figure 8 shows similar data when 3/16 inch diameter steel fasteners are assumed. Figures 9, 10 and 11 show critical fastener loads for 3, 4 and 5 rivet lap splices. Aluminum rivets, 3/16 inch diameter, were assumed for the lap splice configurations.

In the case of the lap splice configurations represented by Figures 9, 10 and 11 the following rules will be helpful.

1) When t_2 is thinner than t_1 the first attachment load is the highest.

2) When t_2 is thicker than t_1 the last attachment load is the highest.

3) When t_1 and t_2 are equal the first and last attachment loads are equal.

Skin bearing stress σ_{br} can be calculated as a function of applied gross stress σ_{gr} from Figures 7 through 11. An estimate of conservative fatigue lives, not including local bending stresses due to eccentricity, can be made using this information along with open hole fatigue data shown in Figures 12, 13 and 14 for 2024-T3 material at stress ratios $R = \sigma_{min}/\sigma_{max}$ of -0.2, 0.0 and +0.2 respectively.

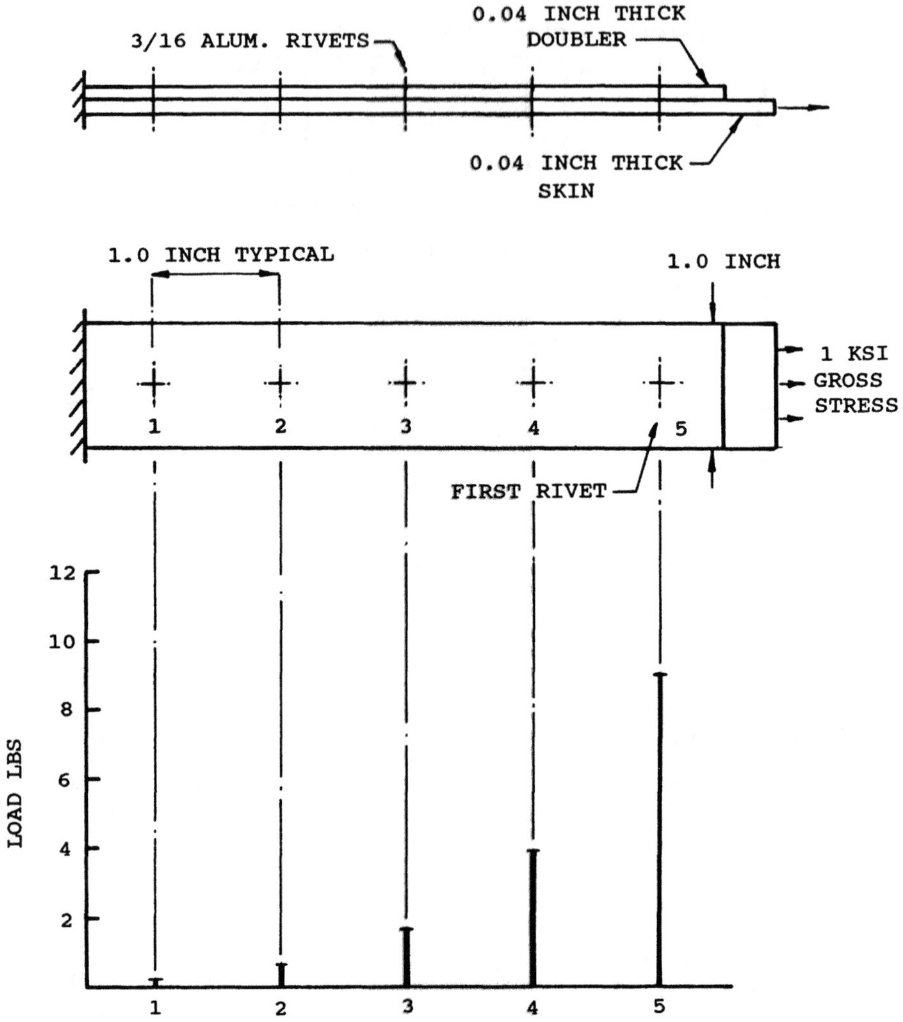

Fig. 6. Rivet Load Distribution in a Example Idealized Strip

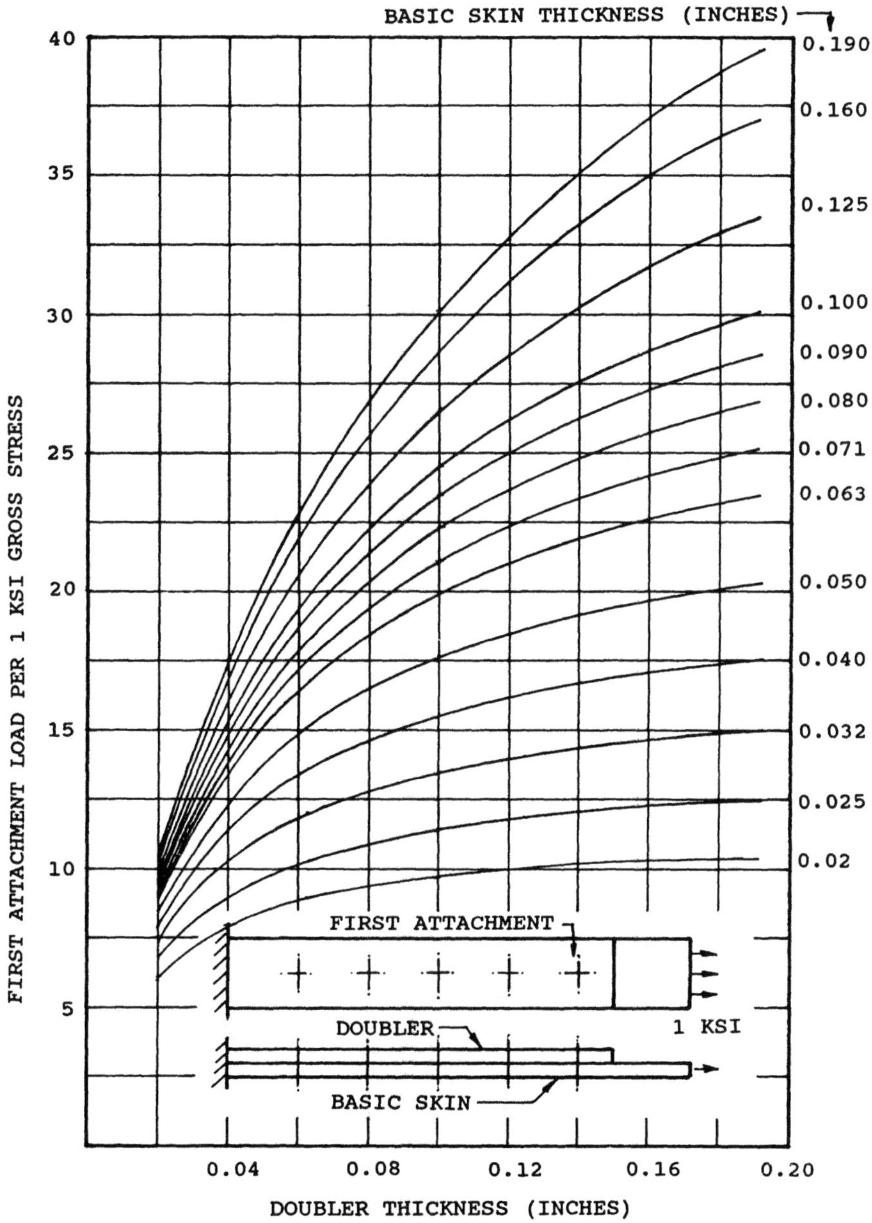

Fig. 7. First Fastener Load Per 1 KSI Gross Stress - 3/16 Inch Diameter Aluminum Rivets

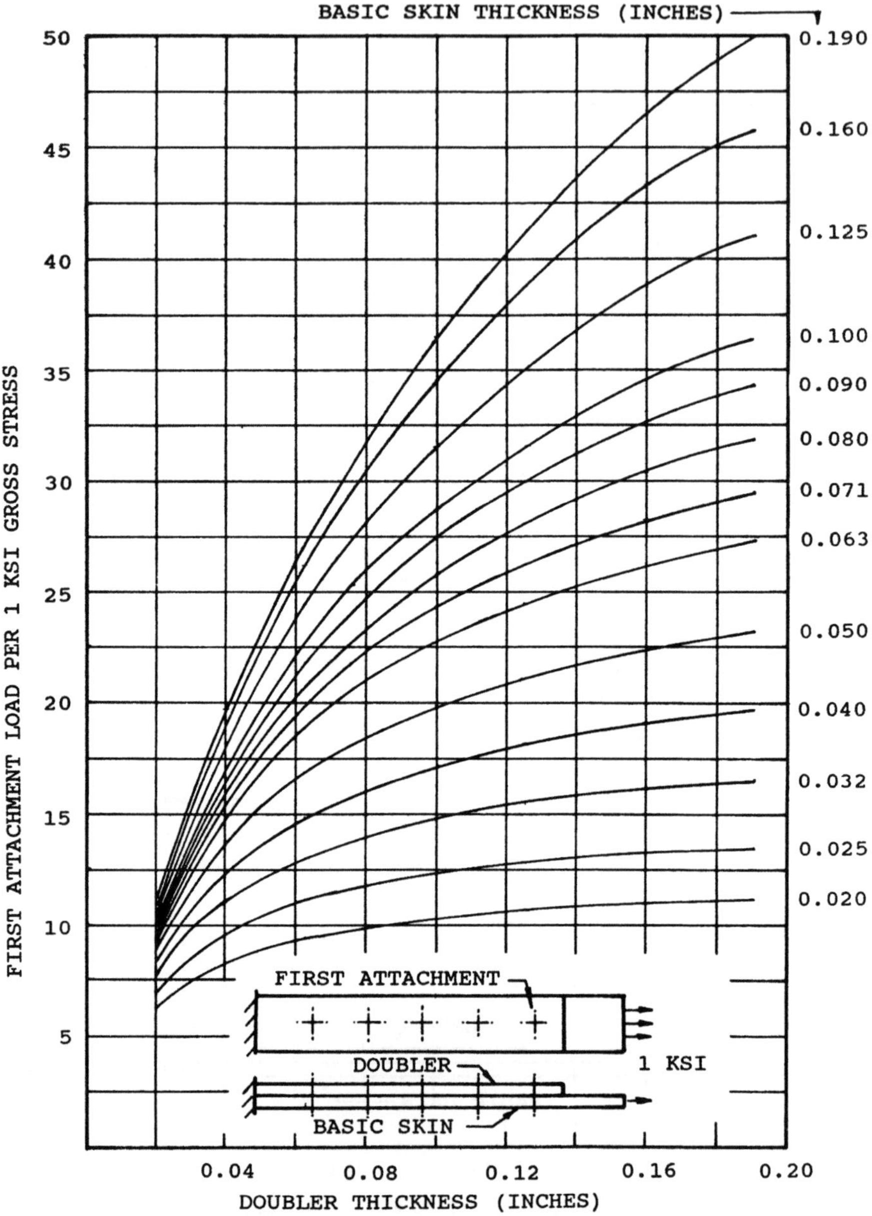

Fig. 8. First Fastener Load Per i KSI Gross Stress - 3/16 Inch Diameter Steel Fasteners

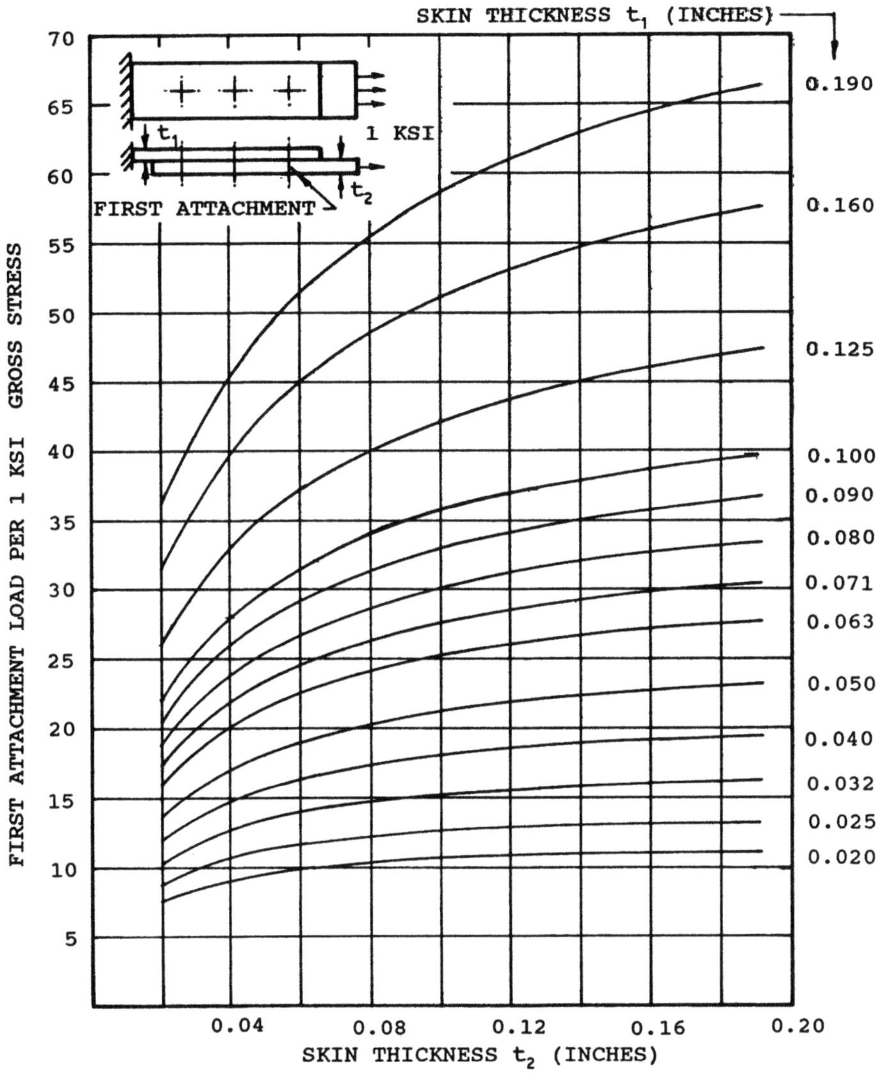

Fig. 9. First Fastener Load Per 1 KSI Gross Stress. 3 Attachment Lap Splice - 3/16 Inch Diameter Aluminum Rivets

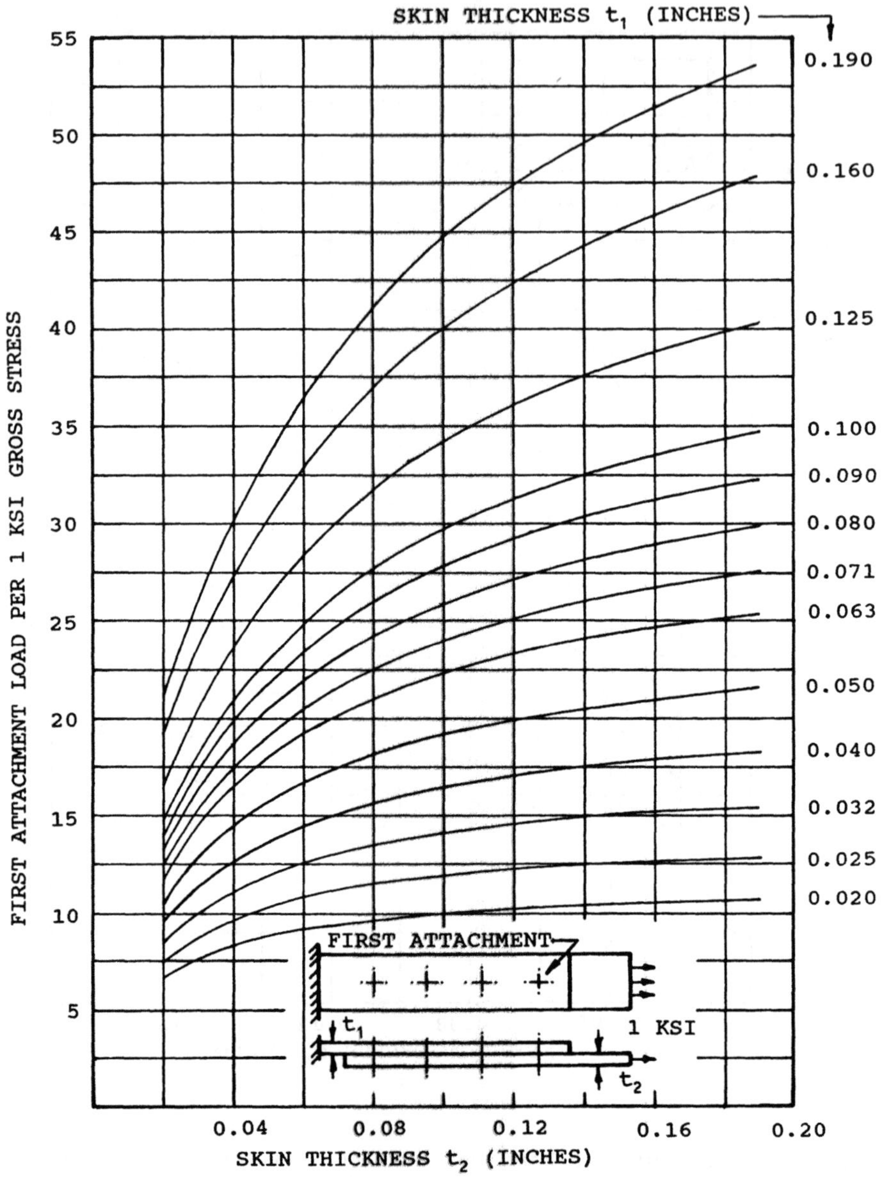

Fig. 10. First Fastener Load Per 1 KSI Gross Stress. 4 Attachment Lap Splice - 3/16 Inch Diameter Aluminum Rivets

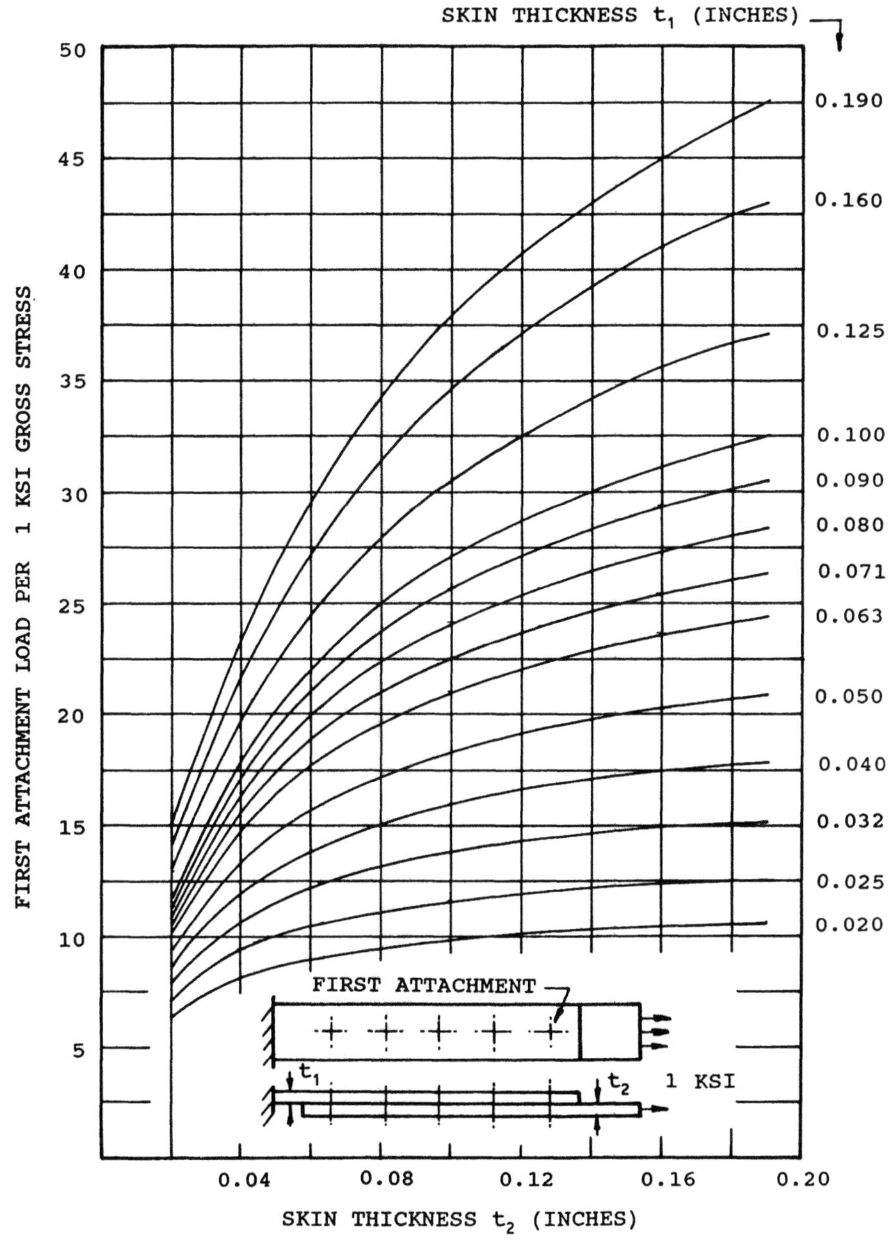

Fig. 11. First Fastener Load Per 1 KSI Gross Stress. 5 Attachment Lap Splice - 3/16 Inch Diameter Aluminum Rivets

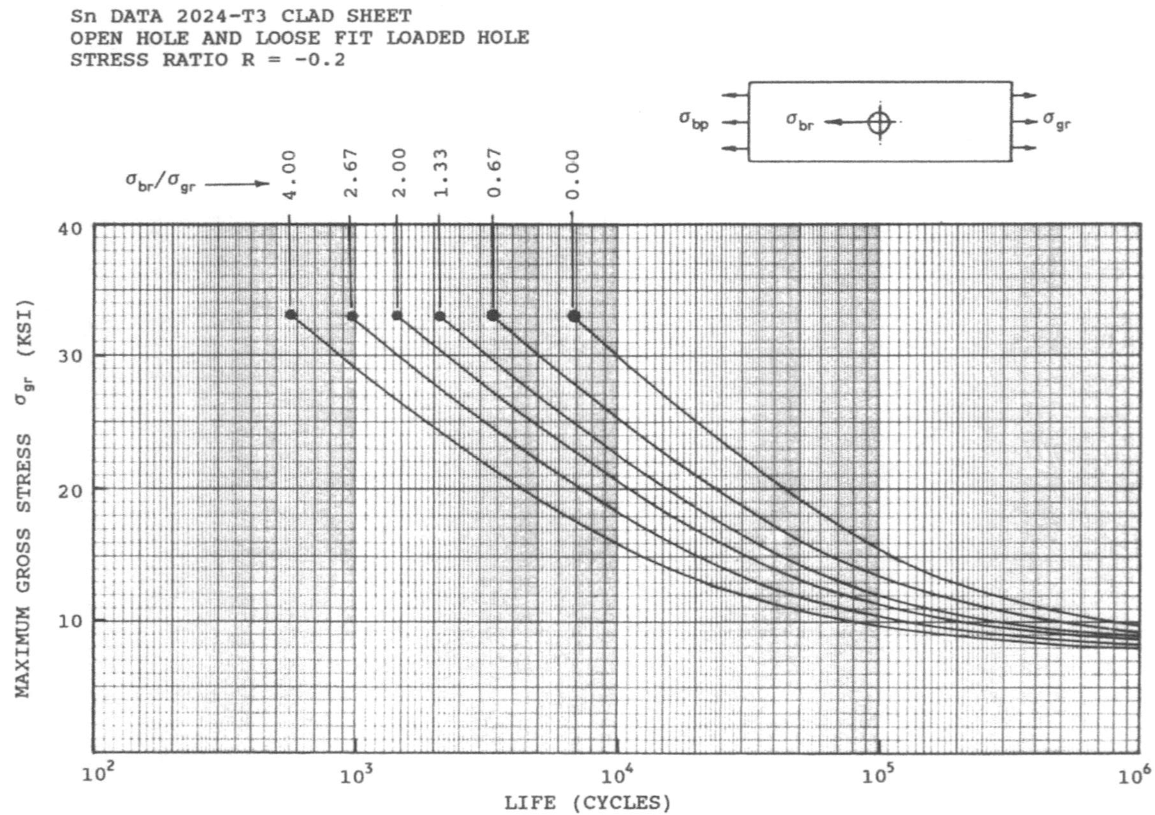

Fig. 12. Fatigue Sn Data 2024-T3, Stress Ratio R = -0.2

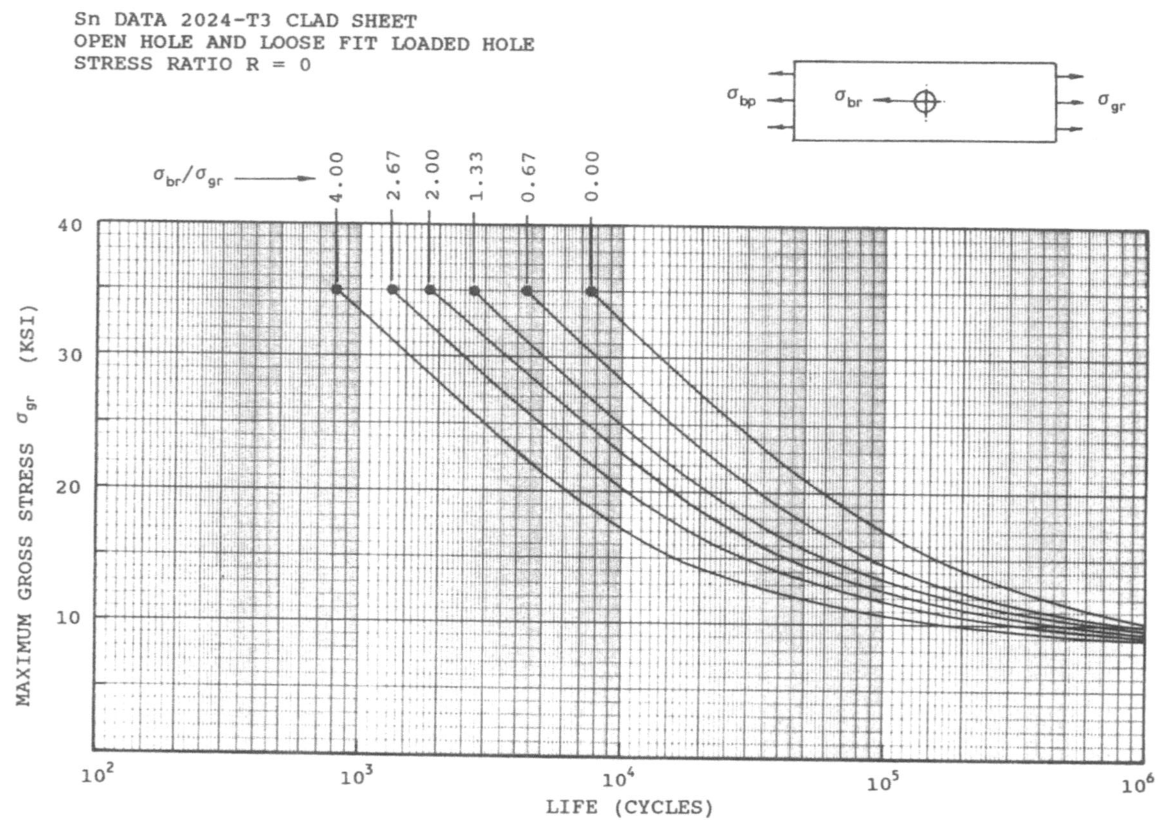

Fig. 13. Fatigue Sn Data 2024-T3, Stress Ratio R = 0.0

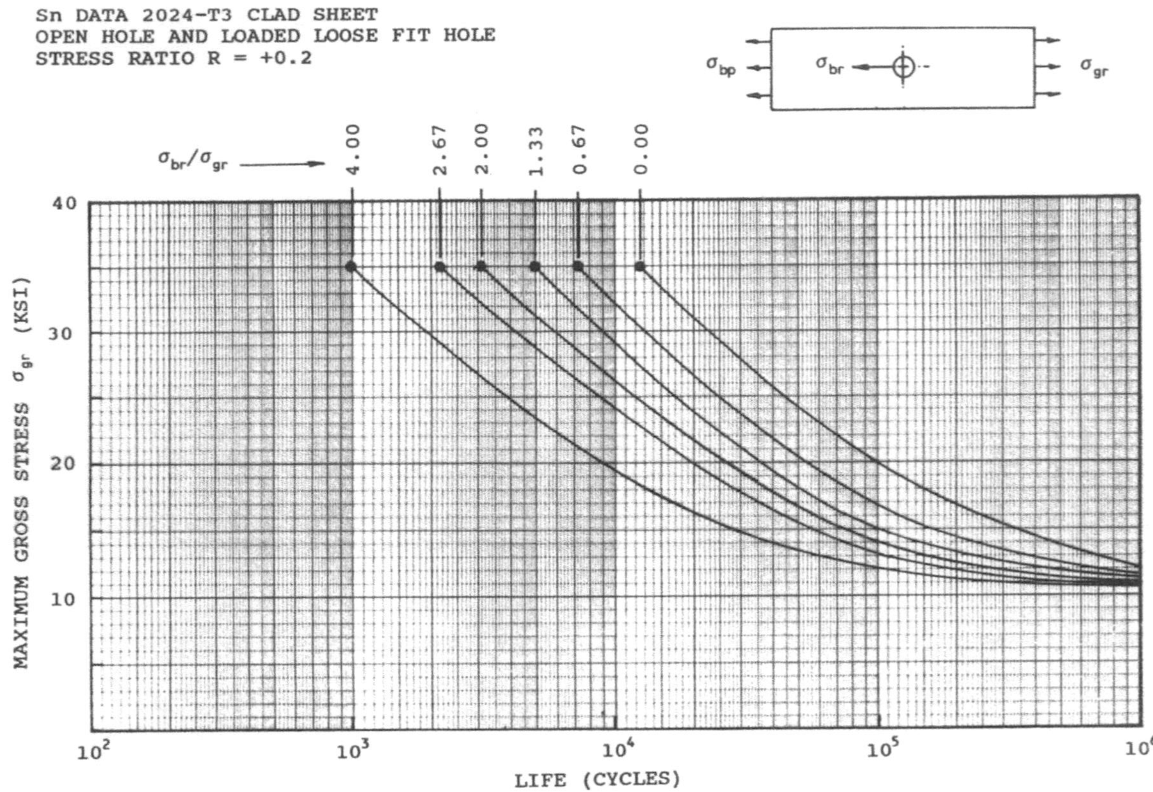

Fig. 14. Fatigue Sn Data 2024-T3, Stress Ratio R = +0.2

CYCLES REQUIRED TO GROW CRACK
FROM 0.02 INCHES TO 0.35 INCHES
AT 15.0 KSI R = 0

DOUBLER THICKNESS (INCHES)	INDUCED FASTENER LOAD (LBS)	CYCLES
0	0	27508
0.050	187.2	25524
0.063	203.4	25456
0.071	211.2	25423
0.100	232.2	25337

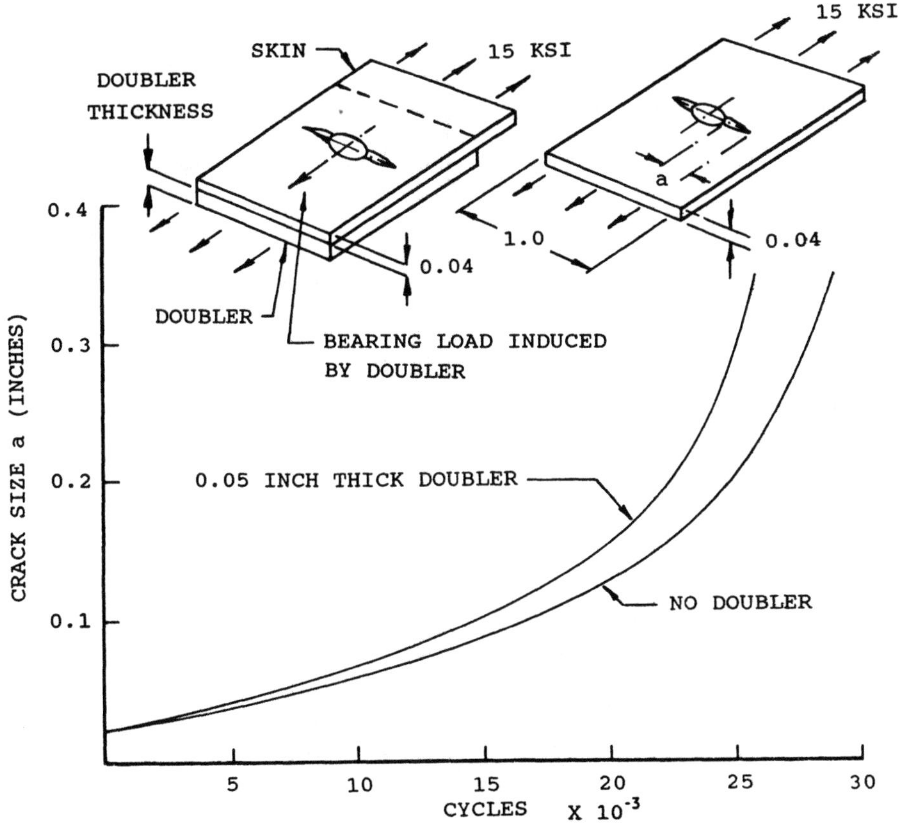

Fig. 15. Effect of Doubler Thickness on Crack Growth Life

Figure 5 shows how overall fatigue life can be drastically reduced with increasing doubler thickness. Crack propagation life is also influenced by fastener load induced by transfer of load out of the skin into the doubler. However, the effect on crack growth life is not so great as on overall fatigue life. The effect of fastener load is a very localized effect taking place near the hole boundary. As the crack forms and starts to propagate the fastener bearing load becomes less significant in it's influence on the crack tip stress intensity factor. The major difference in overall fatigue life is created in the nucleation portion of the life when the crack is extremely small. The effect of doubler thickness on crack growth life is illustrated by Figure 15. This figure shows the number of constant amplitude cycles, at a gross stress of 15.0 KSI with stress ratio R = 0, to propagate a crack in a 0.04 inch thick by 1.0 inch wide strip of 2024-T3 material. The initial crack size is 0.02 inches on both sides of a 3/16 inch diameter hole and the final crack size is 0.35 inches on both sides of the hole. The table on Figure 15 shows some difference in crack growth life up to a doubler thickness of 0.05 inches but beyond this doubler thickness there is little difference. Induced fastener loads shown in Figure 15 assumed the configuration illustrated by Figure 7. It should be noted, however, that skin bending due to eccentricity of the load reaction due to doubler thickness is not accounted for. The curves on Figure 15 show the difference in life between a configuration with no doubler and one with a 0.05 inch thick doubler. This analysis was performed using the computer programs GEOFAC and LICAFF [2]. Forman's equation was used to represent the crack growth rate as follows:

$$da/dN = [4.372 \times 10^{-7} (\Delta K)^{2.7229}]/[(1-R)80 - \Delta K] \qquad (3)$$

Repaired Structure Life Improvement

It has been illustrated by previous examples that repairs can drastically degrade fatigue life of basic structure. It was analytically demonstrated that this degradation is primarily due to first fastener induced load caused by displacement compatibility. This fastener load can be reduced and the fatigue life improved by tapering the doubler. The objective in tapering is to reduce the doubler thickness between the first and second fasteners. Merely tapering the doubler between the edge and the first fastener has no effect on reducing the first fastener load.

Figure 16 illustrates an example of how tapering can reduce the first fastener load. The fastener load distribution is compared to the original example case of 0.04 inch thick skin with a uniform 0.05 inch thick doubler. The reduced fastener load, resulting in a reduced bearing stress σ_{br} and therefore reduced σ_{br}/σ_{gr} ratio, results in a life of 59,000 cycles at a gross stress σ_{gr} of 15.0 KSI with R = 0. This is a life improvement of nearly 10% compared to the life of 53,800 cycles obtained for the uniform 0.05 inch thick doubler. It should be noted that these comparisons are being made with conservative open hole fatigue data. Greater improvements could be shown using fatigue Sn data representative of good quality riveting. However, this

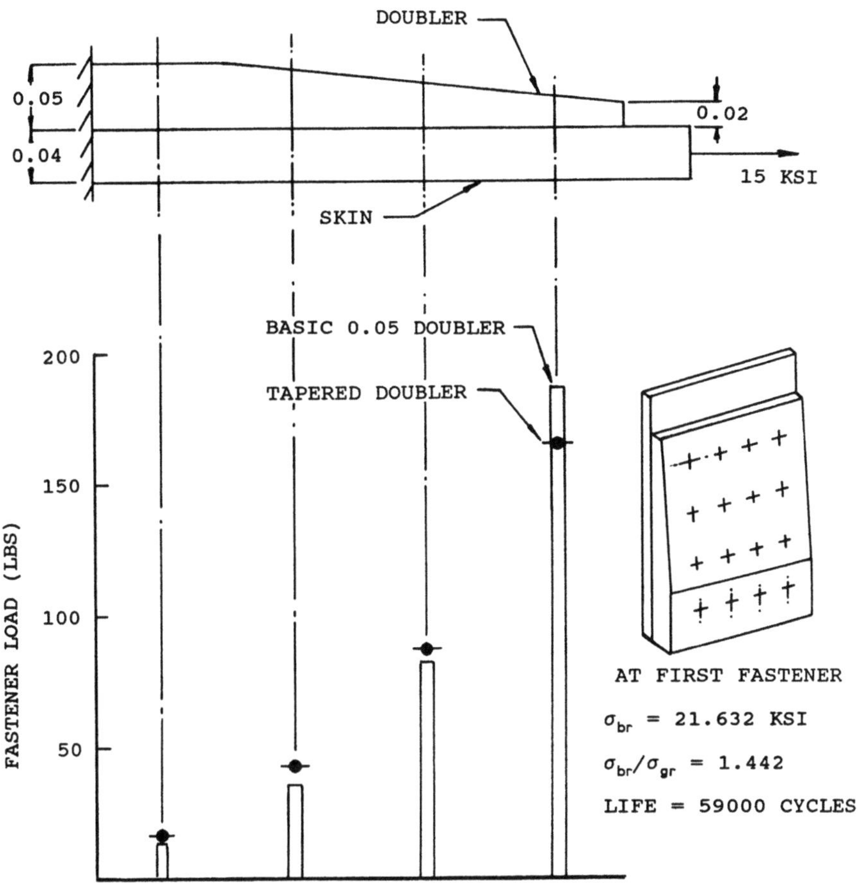

Fig. 16. Effect of Doubler Tapering on First Fastener Load And Fatigue Life

type of riveting is rare in field repaired structure usually due to space restrictions as will be discussed later.

Further reductions in first fastener load resulting in life improvement can be achieved if the typical 0.05 inch thick doubler is replaced by multiple doublers. This is usually easier to accomplish in field repairs anyway. As an example, a doubler system involving standard gauges of 0.025 inches and 0.032 inches were analyzed for comparison. The configuration is shown in Figure 17a. The 0.025 inch thick doubler is extended another fastener row. For this comparison strain compatibility analysis accounting for fastener flexibility results in first fastener load of 122.4 lbs resulting in skin bearing stress of 16.105 KSI and σ_{br}/σ_{gr} = 1.074. This results in a fatigue life of 70,500 cycles at a gross stress of 15.0 KSI. Thus, laminating the

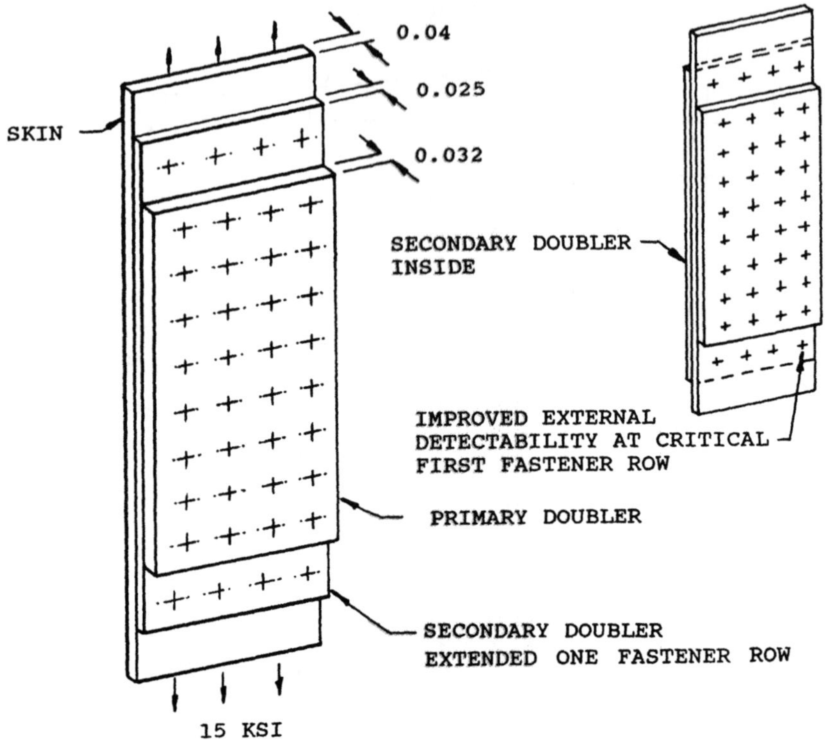

ADVANTAGES OF THIS REPAIR DESIGN

- FIRST FASTENER LOADS IN SKIN REDUCED FROM 187.2 TO 122.4 LBS
- SKIN BEARING STRESS σ_{br} REDUCED FROM 24.632 KSI TO 16.105 KSI
- BEARING STRESS TO GROSS STRESS RATIO σ_{br}/σ_{gr} REDUCED FROM 1.642 TO 1.074
- FATIGUE LIFE INCREASED FROM 53,800 CYCLES TO 70,500 CYCLES

Fig. 17. Improved Fatigue Life by Doubler Lamination

doubler increases the fatigue life by 31% over the life obtained for the example case using a single 0.05 inch thick doubler.

Improvements in Inspectability

Any repair doubler applied externally to the fuselage structure such as that shown in Figure 1(b) will degrade the external detectability of the skin at the critical fastener row. Skin cracking at the first fastener will be hidden externally by the repair doubler. Of course the skin will still be visually inspectable internally but then internal lining would have to be removed. Also external inspectability of this type of repair is still possible with more sophisticated low frequency eddy current NDI. The inspectability of the repair can be improved by placing the secondary doubler inside and the primary doubler outside as shown in Figure 17b.

Further improvements in external inspectability can be made by removing alternate fasteners at the first critical row to improve residual strength capability and increase multi-site-damage (MSD) critical crack sizes. For example, Figure 18 shows our laminated doubler case but with alternate fasteners in the critical first row removed. This allows the net cross sectional area in the skin at the critical first fastener row to be increased. In the case of MSD in 2024-T3 material the residual strength is governed approximately by net section yield criteria between crack tips. Assuming net section yield governs the residual strength capability then the critical MSD crack size can be expressed as:

$$a_{cr} = [\sigma_y(P-D) - \sigma P]/[2(\sigma_y)] \qquad (4)$$

Where σ_y = material yield strength
P = fastener spacing
D = fastener diameter
σ = applied gross stress

Assuming a B value of σ_y = 43.0 KSI for 2024-T3 unclad sheet material in the LT direction from MIL-HDBK-5 the residual strength diagram for the MSD case is shown on Figure 18 for P = 1.0 and 2.0 inches with rivet diameter 0.19 inches. It can be seen that the critical crack size is increased from 0.2306 with 1.0 inch fastener spacing to 0.5566 inches with 2.0 inch fastener spacing. This provides for increased detectability and longer crack growth life. However, the overall fatigue life will be decreased with this configuration. Displacement compatibility analysis indicates that when alternate fasteners are removed the remaining fasteners in the critical first row try to load up a wider strip of the secondary doubler thus increasing the first fastener load. Analysis for our typical example at an applied stress of 15 KSI indicates the first fastener load is increased from 122.4 lbs to 184.2 lbs. This increases the bearing stress to 24.237 KSI resulting in a bearing to gross stress ratio σ_{br}/σ_{gr} = 1.616. The resulting overall fatigue life becomes 54,500 cycles. Therefore removing alternate fasteners in the critical first row in the secondary doubler has degraded the overall fatigue life of the laminated configuration but considerably improved crack detectability.

The fatigue life can be improved by effectively tapering the inner doubler. This is done by fingering as illustrated by Figure 18. The fingers effectively reduce the inner doubler area between the first and second fastener rows thus relieving the first fastener load. This results in a lower skin bearing stress and improved crack initiation life. This configuration offers the highest fatigue life coupled with the best external detectability.

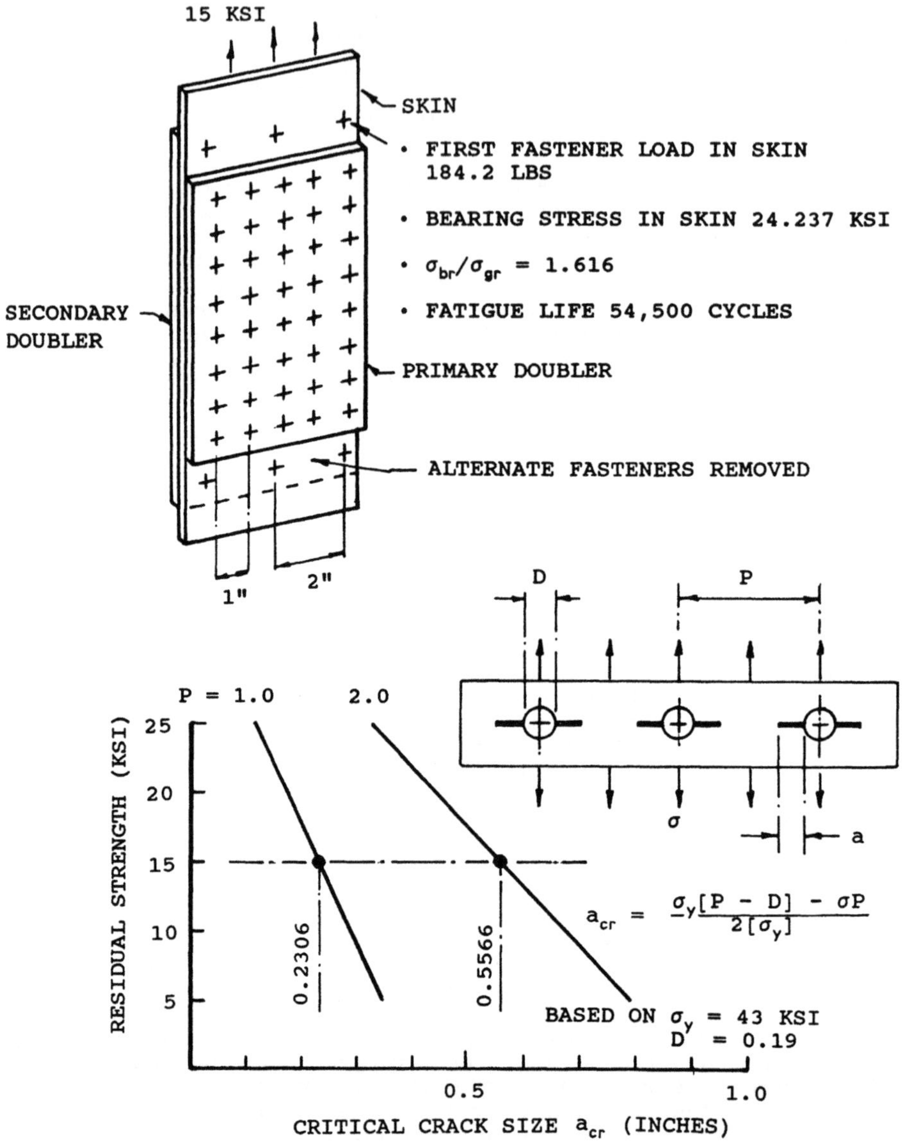

Fig. 18. Improved Crack Detectability By Removing Alternate Fasteners In Critical Row

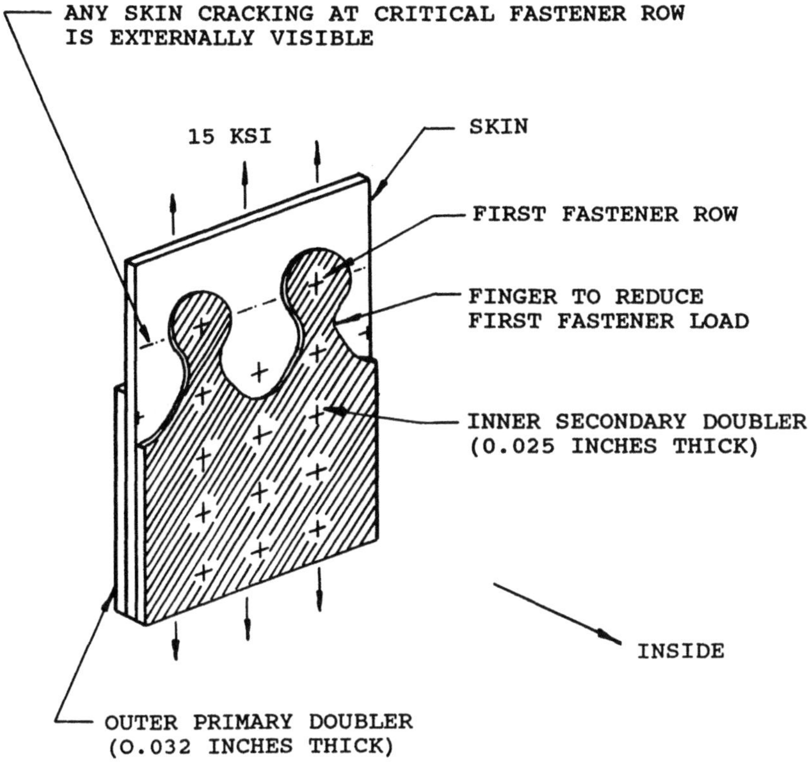

- FIRST FASTENER LOAD IN SKIN 174 LBS
- BEARING STRESS IN SKIN 22.895 KSI
- σ_{br}/σ_{gr} = 1.526
- FATIGUE LIFE 57,000 CYCLES

Fig. 19. Finger Doubler Configuration To Reduce First Fastener Load, Improve Fatigue Initiation Life And Skin Crack Detectability

Remember, again the fatigue lives quoted are for open hole quality.

Riveting Quality

The fatigue initiation life of the basic skin in the presence of a repair doubler can be related to the quality of the riveting operation. If the rivets do not properly fill the hole after "bucking" then poor fatigue quality can be expected as reflected in the open hole fatigue data of Figures 12 through 14. Open hole fatigue data reflects a stress concentration factor, K_t, of approximately 3.0 for wide thin sheet when $\sigma_{br}/\sigma_{gr} = 0$. This concentration factor results when the boundary of the rivet hole is allowed to deflect as illustrated in Figure 20(a). If the hole is filled with a pin capable of restraining the free boundary displacement of the hole then the stress concentration factor is reduced to approximately 2.0 as illustrated by Figure 20(b). This reduction in stress concentration factor at the hole boundary will result in improved fatigue life. Further reductions in effective stress concentration factor can be achieved if the rivet is bucked properly. Figure 20(d) illustrates that as the rivet is bucked it will swell in the hole and create the equivalent of an interference fit pin applying a radial pressure to the hole boundary which in turn creates local radial and tangential stresses in the sheet around the rivet. With little interference the material near the hole boundary becomes plastically deformed. It can easily be shown analytically how this effect can reduce the effective stress concentration factor at the hole boundary. Elastic-plastic stress analysis used in pressurized thick cylinders can be used to illustrate this effect. Using this theory it is possible to develop an equation relating the amount of interference to the radius of the plastic front as shown below:

$$2\mathrm{Ln}(\rho/r)+(1+\nu_s)/(1-\nu_r)(E_r/E_s)(\rho/r)^2 = I\surd 3E_r/[2r\sigma_y(1-\nu_r)] \qquad (5)$$

This equation is solved by iteration for ρ/r. The value of ρ is then used to calculate the resulting stress distribution as follows:

Stresses in the elastic region

$$\sigma_{RP} = \sigma_y/\surd 3[2\mathrm{Ln}(R/\rho) - 1] \qquad (6)$$

$$\sigma_{\phi P} = \sigma_y/\surd 3[2\mathrm{Ln}(R/\rho) + 1] \qquad (7)$$

Stresses in the elastic region

$$\sigma_{RE} = \sigma_y/\surd 3(\rho/R)^2 \qquad (8)$$

$$\sigma_{\phi E} = \sigma_y/\surd 3(\rho/R)^2 \qquad (9)$$

Where σ_{RP} = radial stress in plastic region

$\sigma_{\phi P}$ = tangential stress in plastic region

σ_{RE} = radial stress in elastic region

$\sigma_{\phi E}$ = tangential stress in elastic region

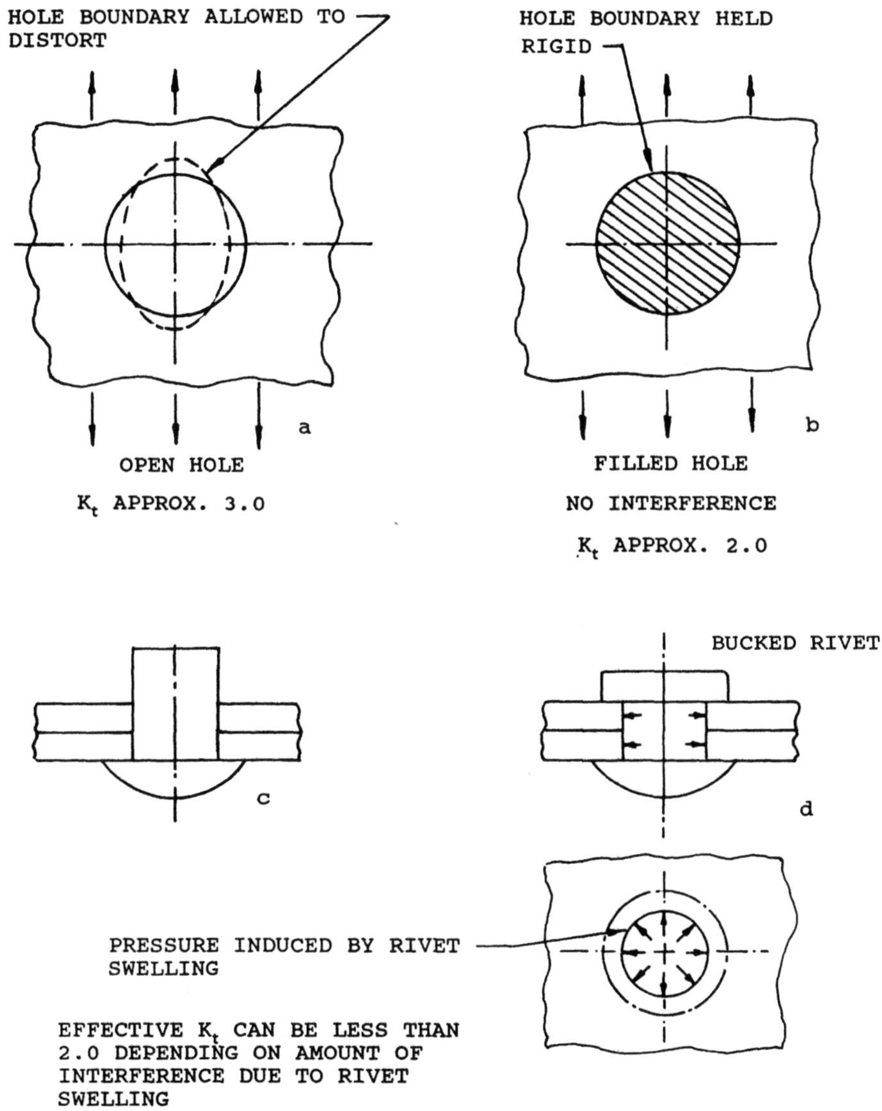

Fig. 20. Effect Of Riveting On Stress Concentration Factor

I = diametral interference

r = hole radius

ρ = radius of plastic front

E_r = modulus of rivet material

E_s = modulus of sheet material

ν_r = Poisson's ratio for rivet material

ν_s = Poisson's ratio for sheet material

R = radius at which stresses are required

σ_y = material yield stress

Equation 5 was solved for diametral interference values, I, of 0.003 and 0.004 inches respectively to illustrate how proper rivet swelling can provide beneficial residual compressive stresses at the hole boundary. Rivet diameter was assumed to be 0.19 inches in 2024-T3 sheet material with yield strength equal to 43 KSI. Poisson's ratio in the plastic range was assumed to be 0.5. Figure 21 shows the resulting stress distribution in both plastic and elastic regions. The compressive residual stresses would be subtractive from the applied tension stresses effectively reducing the stress concentration at the hole boundary and subsequently improving the fatigue life. However, to obtain substantial interference the rivet must be bucked squarely. Very often, especially in field repairs where space may be restrictive, the rivets are not bucked squarely resulting in a "clinched" installation. This situation often occurs when the working space is restrictive not allowing the bucking bar to be held squarely. When clinching occurs the hole is not properly filled and rivet swelling does not occur. Thus the beneficial residual compressive stresses are not present. In fact, severe clinching can cause a situation where there may be a gap between the hole boundary and the rivet wall thus allowing some free boundary displacement as shown in Figure 20(a). When this occurs, reflective of poor quality riveting, the fatigue life will be more representative of open hole Sn data shown in Figures 12 to 14. Figure 22 shows an example of rivet clinching taken from an actual failed field repair. The sketch on Figure 22 illustrates the gap which can exist between the rivet and the hole boundary. This is graphically illustrated by Figure 23 which shows a cross sectioned purposely clinched rivet. The gap between the hole boundary and the rivet wall is clearly visible. Figure 24 illustrates the variation in fatigue life which can be expected with variations in riveting quality in 2024-T3 sheet at R = 0.

Figure 25 shows fatigue Sn data which may be reasonably expected with good quality 3/16 inch diameter rivets in 2024-T3 material where the riveting is square and hole filling but with no load transfer. This data is a little difficult to use in the presence of load transfer but this author has found the methods of Lars

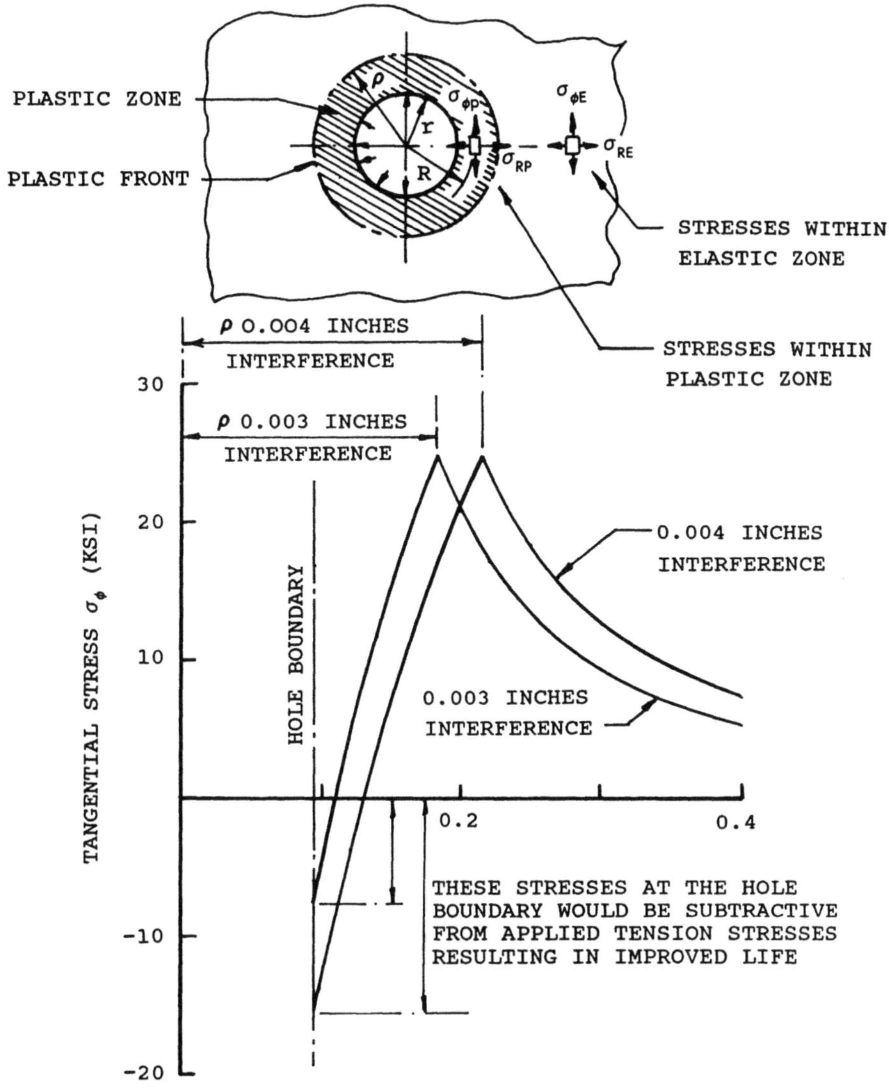

Fig. 21. Elastic-Plastic Stress Distribution Around a 0.19 Inch Diameter Rivet In 2024-T3 Sheet With Varying Degrees Of Equivalent Interference

EXAMPLE OF RIVET CLINCHING
FROM AN ACTUAL FIELD REPAIR

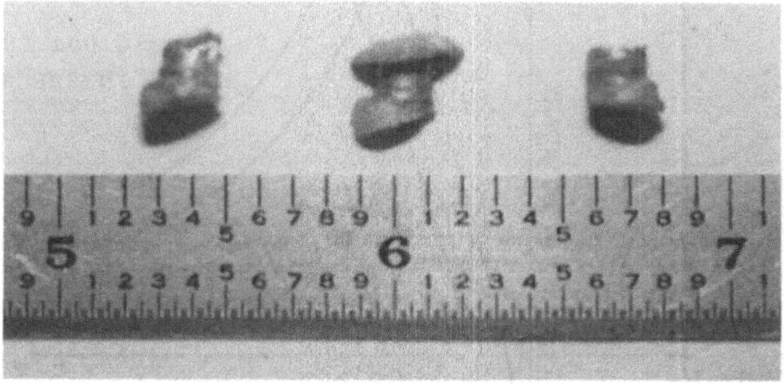

RIVETS REMOVED FROM THE
REPAIR

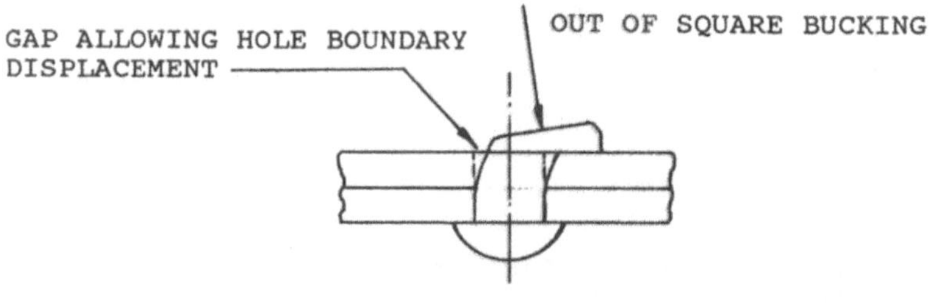

Fig. 22. Examples Of Rivet Clinching

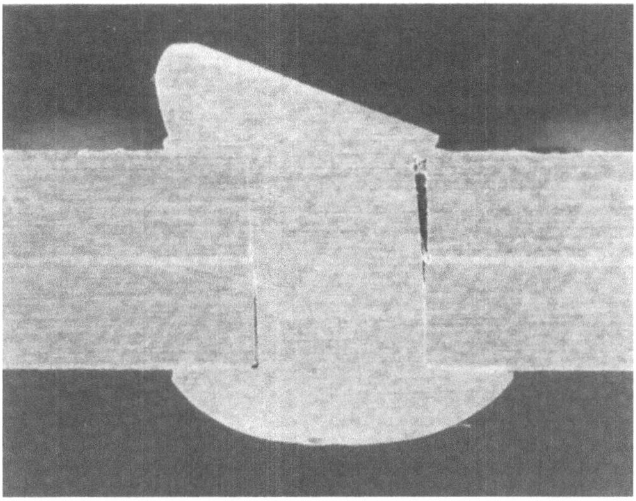

Fig. 23. Lack Of Rivet Hole Filling Due To Clinching

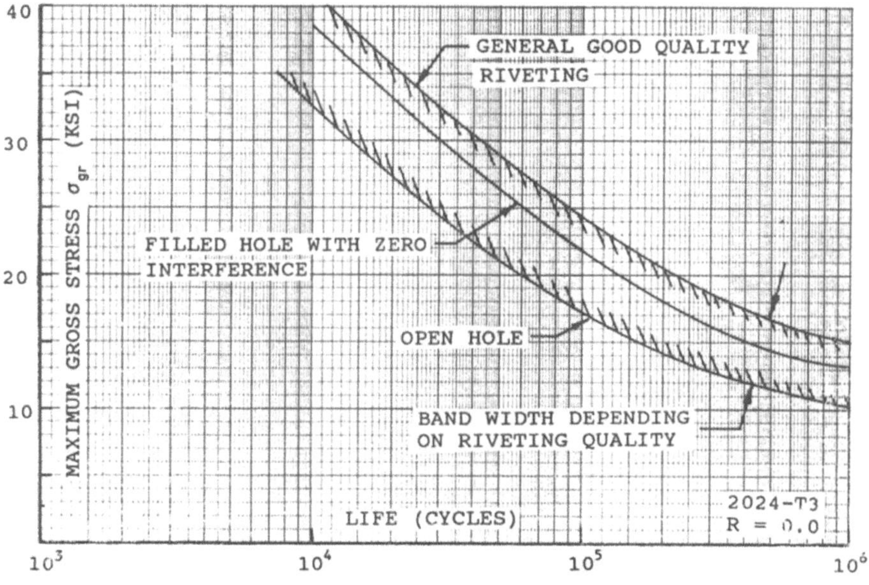

Fig. 24. Effect Of Riveting Quality On Fatigue Life

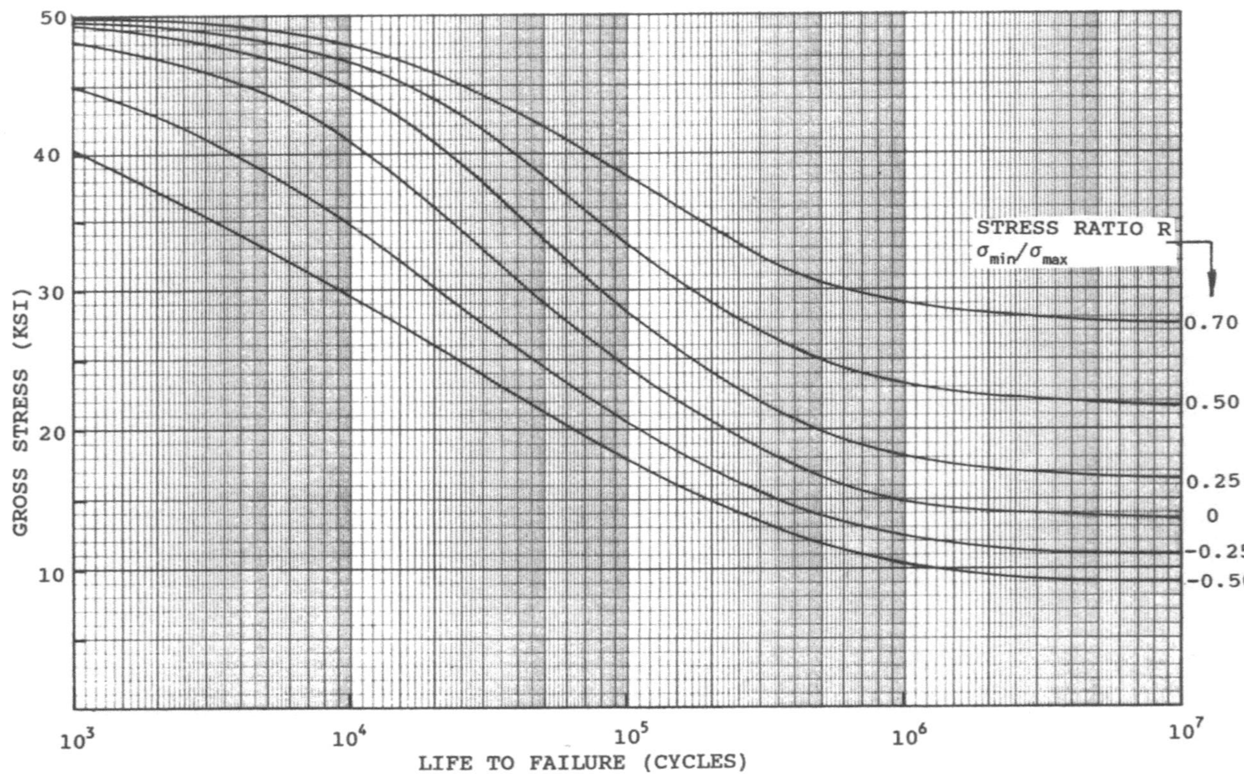

Fig. 25. Fatigue Sn Data For Good Quality 3/16 Inch Diameter Rivets In 2024-T3 Material

Jarfall [3,4] to be useful in obtaining an effective stress σ_{eff} which may be used with the Sn data of Figure 25 to approximate the fatigue failure life. By reviewing the work of Jarfall it is possible to determine an effective stress as follows:

$$\sigma_{eff} = [1.3K_{tbr}\ \sigma_{br} + K_{tbp}\ \sigma_{bp}]/K_{tbp} \tag{10}$$

Where σ_{eff} = effective stress

σ_{br} = bearing stress

σ_{bp} = bypass stress

K_{tbr} = stress concentration factor relating peak stress at edge of hole to σ_{br}

K_{tbp} = stress concentration factor relating peak stress at edge of hole to σ_{bp}

Of course the study of Jarfall's results should be used to obtain the proper parameters but this author has found the following reduction of equation 10 to give reasonable results with Sn data of Figure 25 for good quality riveting. Again this data does not reflect possible bending in the joint due to eccentricities.

$$K_{tbp} = 3.2, \quad K_{tbr} = 1.4$$

$$\text{Therefore} \quad \sigma_{eff} = 0.569\sigma_{br} + \sigma_{bp} \tag{11}$$

Stresses in equations 10 and 11 are defined in Figures 12, 13 and 14.

It should again be pointed out that field repairs are very often not of the fatigue quality represented by Figure 25. If advise could be given it would be not to consider fatigue quality beyond that depicted by Figures 12, 13 and 14 for field repairs. Even then, a conservative scatter factor should be considered to establish threshold for future detailed inspection of the repair. The reason for including Figure 25 was mainly to illustrate the differences in fatigue life between good quality riveting and what may be expected in field repairs where access may not be as good as in the initial assembly of the airplane.

<u>Degradation of Inspectability Due to Repairs</u>

It has been shown how repairs can degrade the overall fatigue quality of the basic structure. It has also been shown that the most critical locations for future fatigue damage are in the basic skin at the first attachment row in the doubler. In addition to degradation in overall fatigue life the repair itself can degrade the inspectability of the basic structure considerably. In order to illustrate this assume that a crack growing in a basic fuselage skin becomes visually detectable with high reliability at say a half crack length of 0.5 inches. Assuming a constant amplitude cyclic stress of 15 KSI with stress ratio R = 0 with crack growth rate represented by Forman's equation 3, the crack growth life would be 29,600 cycles as illustrated by Figure 26. Now consider a repair doubler 9.0 inches long has been riveted to the basic skin. The most critical location for future fatigue damage is in the skin at the first row of fasteners. Still assuming the visual half crack length of 0.5 inches the skin crack would be able to propagate to a half crack length of 4.5 + 0.5 = 5.0 inches before it could be detected visually externally. Thus the safe inspectable crack growth life has been reduced from 29600 cycles to 1200 cycles. It can be seen that a doubler length slightly longer than 9.0 inches would leave no visually detectable life from an external inspection standpoint. Of course, a less desirable internal visual inspection could be adopted but this would require internal lining removal. Also a more sophisticated low frequency eddy current inspection could be adopted. However, the more desirable external visual inspection capability could be completely degraded by this type of repair.

The external visual inspectability can be improved by including an internal doubler as indicated by Figure 27. This configuration was addressed in Figure 17 and shown to have improved fatigue life of 31% for our example case over the doubler type shown in Figure 26. The first fastener row in the internal doubler is critical for fatigue cracking in the skin. A crack propagating in the skin at this row will be externally detectable and so the crack growth life would be about 29600 cycles as shown by Figure 26 for our example case. The residual strength capability of this type of repair in the presence of multi-site damage can be improved by eliminating alternate fasteners in the critical row. This was indicated by Figure 18. Also the external visual detectability of this configuration would be improved like the straight edged internal doubler also shown on Figure 27. However there is a penalty on overall fatigue life by eliminating alternate fasteners. The purpose of the fingers, as previously explained, is to restore some of this fatigue life by reducing the first fastener loads through fingering or effectively tapering the internal doubler. The finger doubler repair is, in the opinion of the author, the best compromise for a permanent repair to basic fuselage structure.

Damage Tolerance Evaluation of Repairs

Current regulations mandate that repairs or modifications to commercial transport category airplanes which have been certified to the damage tolerance regulations FAR 25.571, or are operating under a Supplemental Inspection Document (SID), must be evaluated for damage tolerance. The reason for this is that the in-service structural safety of these aircraft is being managed through inspection programs based on a damage tolerance philosophy. Inspections for principal structural elements are developed through an engineering evaluation of crack growth and residual strength, both of which are extremely sensitive to geometry. Any change in the geometry of principal structural

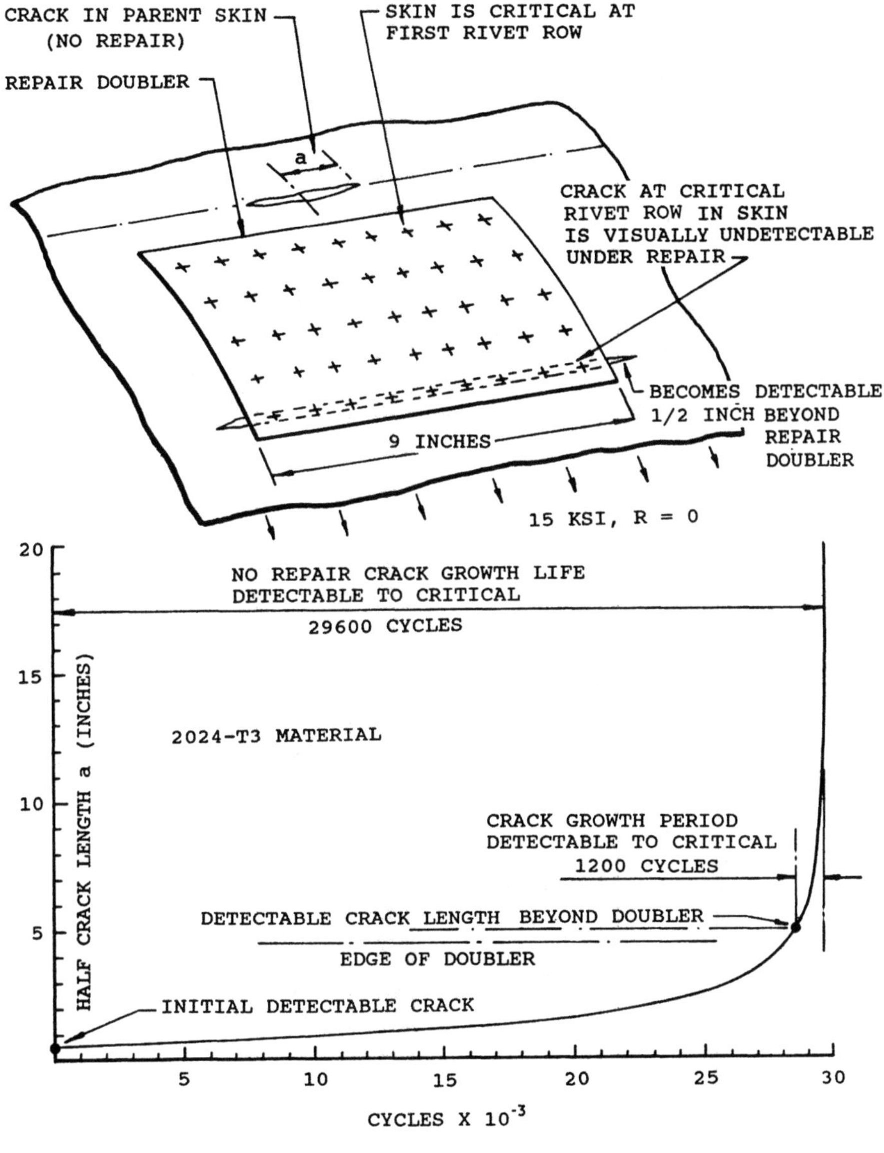

Fig. 26. Example Of Impaired Crack Growth Detectability With Repair Doubler

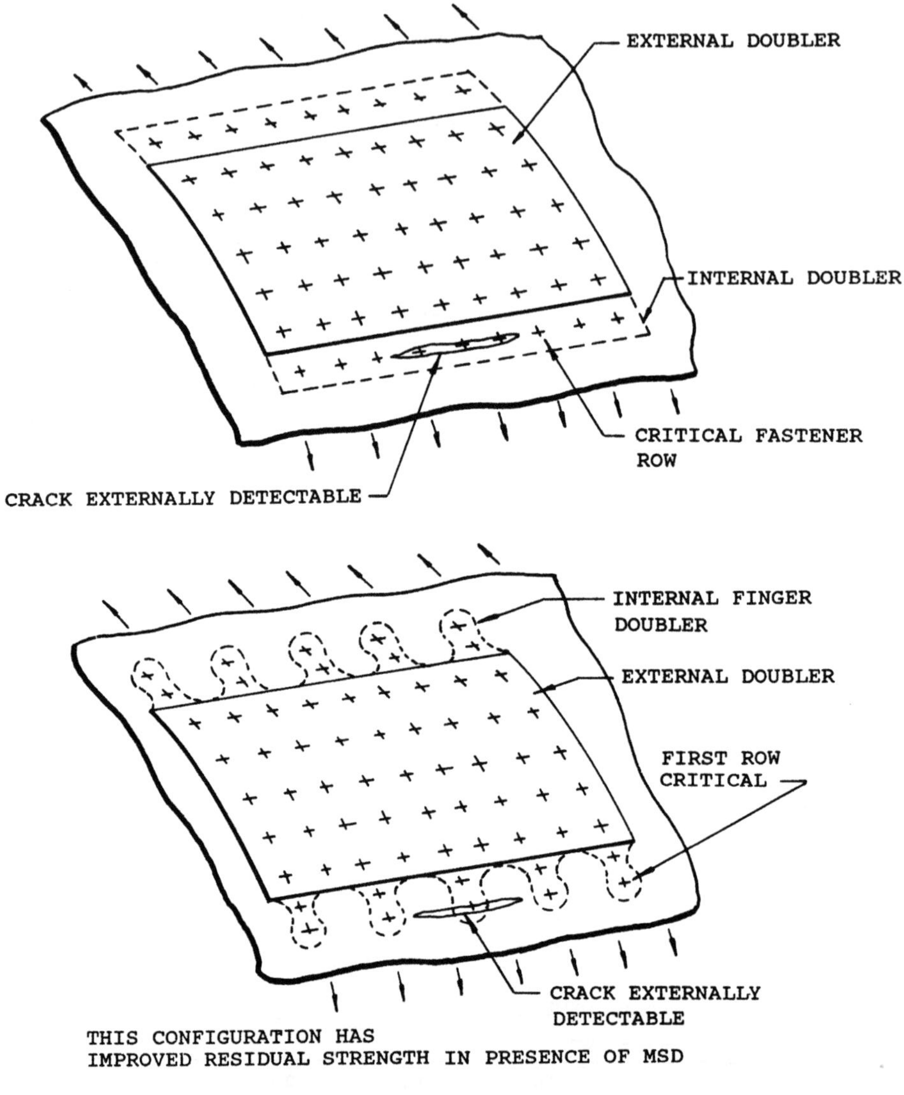

Fig. 27. Improved Doubler Design For External Detectability And Residual Strength Improvement

elements such as repairs or modifications can drastically change the threshold, frequency and method of inspection and can also effect the residual strength capability of the structure. Obviously the original manufacturers of the airplanes are in the best position to do this evaluation. They usually have the know-how and are in possession of all the structural loads and stress data necessary for such an evaluation. However, the current system is such that there are hundreds of repair stations, small modification companies, operators and private DERs who are involved in the repair and modification of airplane structures. In order to support the aging fleet over the next decade the number of these repair stations and small modification companies must increase. Information contained in a recent General Accounting Office (GAO) report on aging aircraft states that by the year 2000 there will be 4474 commercial transport airplanes, manufactured by Boeing, Douglas and Lockheed, world wide, which will be operating beyond the initial 20 year life goal. The manufacturers of these airplanes will need considerable assistance to keep them operating safely and repaired. The small repairers and modifiers, currently performing repair analysis based on equal or better static strength only, will need to develop capability to perform damage tolerance evaluations. Conservative, simplified methods need to be developed to assist these small modifiers and repairers in performing crack growth and residual strength analysis in order to ease the burden on the major manufacturers. These manufacturers need to concentrate on producing new aircraft to replace the aging fleet. These statements are obviously over simplified. Exact damage tolerance evaluation is an extremely sophisticated process requiring experts in fracture mechanics who must also be experts in basic airframe stress analysis. The number of these experts in the industry is small and the job over the next decade will be overwhelming unless we do something about it now. As a first step, then, simplified methods need to be developed that will enable hundreds of modifiers and repairers to perform conservative analysis to support inspection programs for repairs and modifications. It is the opinion of the author that all repairs receive 100% continual inspection beyond a conservative threshold. Because of operational schedules and the economic need to keep airplanes flying the repair system may have to be in two stages. Immediate temporary repairs good for a limited time followed by permanent repairs supported by a damage tolerance evaluation. The greatest problem to the small modifier in performing damage tolerance evaluations is crack growth analysis. This generally needs original manufacturers load and stress data. Residual strength evaluation is not necessarily a considerable problem to a good fracture mechanics analyst since this falls into the same category as equal or better static strength analysis. ie the residual strength can usually be assessed from a knowledge of the airframe geometry. The first step then is to develop simplified, conservative, crack growth analysis methods. This of course has been done to some extent by a number of researchers. The most appealing method, at least to this author, is the method developed by Boeing's Damage Tolerance Technology group. Sufficient information exists in the literature [5, 6, 7, 8, 9] to provide an outline of the method. Some of the necessary data needed to complete this method does not exist in the open literature but this data can be readily developed with some effort.

Development of a simplified crack growth method

In general a crack growth analysis requires the solution of the following equation:

$$N = \int_{a_i}^{a_f} da/[F(\sigma_{max},R,a,M,G,\sigma_y)]$$

Where σ_{max} = maximum applied stress
R = stress ratio $\sigma_{min}/\sigma_{max}$
a_i = initial crack length
a_f = final crack length
M = material influence
G = geometric influence
σ_y = material yield strength

The majority of large airplane manufacturers have developed sophisticated numerical integration crack growth computer programs to solve this equation for a variety of materials and geometries and load spectra variations. These programs are being used to design new aircraft to meet the damage tolerance regulations and in some cases to develop SIDs for aging aircraft. Some manufacturers are using these methods to assess standard repairs with a view to updating their structural repair manuals. The crack growth method developed by Boeing allows rapid development of crack growth curves. Boeing has added sophistication to this method by considering load sequence effects under spectrum loading. It is the opinion of this author that for a conservative analysis of repairs additional sophistication needs to be left to the large manufacturers at this stage. The following is an outline of the development of the Boeing method but with parameters altered to fit in with familiar fracture mechanics methodology. The development here neglects the sophistication of load sequence effects included in the Boeing approach and which generally produce more accurate and less conservative results. In other words crack growth retardation is not included in the following approach. The approach is made possible through the following basic assumptions:

- Walker type simulation of crack growth rate data
- Separation of geometrical effects from stress and material so that integration of geometrical effects can be performed separately
- Standardization of complex randomized load spectra into a single repeatable flight
- Simulation of this single flight by a single constant amplitude cycle with zero stress ratio
- Subsequent constant amplitude crack growth analysis to produce a crack growth versus flights curve.

The Walker crack growth equation is expressed as:

$$da/dN = C\{(1-R)^q K_{max}\}^P \quad (12)$$

The parameters for this equation are defined on Figure 28. The equation is basically a straight line equation when plotted on Log-Log paper. When da/dN is plotted against the effective stress intensity factor range, $K_{eff} = (1-R)^q K_{max}$, a single line is produced as shown to the left of Figure 28. Plotting da/dN as a function of the true stress intensity factor range, ΔK, allows several lines at different stress ratio, R, to be drawn as shown to the right of Figure 28. The width of the band of lines is controlled by the exponent q on the (1-R) term. This enables a better fit to be made of the test data points. The value of P is the slope of the lines as indicated.

Inverting equation 12 gives:

$$dN/da = 1/C\{(1-R)^q K_{max}\}^{-P}$$

$$N = \int_{a_i}^{a_f} 1/C\{(1-R)^q K_{max}\}^{-P} \, da$$

$$N = 1/C\{(1-R)^q\}^{-P} \int_{a_i}^{a_f} (K_{max})^{-P} \, da$$

Substituting for K_{max}:

$$N = \underbrace{1/C\{1/[\sigma_{max}(1-R)^q]\}^P}_{\text{load and material terms}} \underbrace{\int_{a_i}^{a_f} da/(\sqrt{\pi a}\,\beta)^P}_{\text{geometry terms}}$$

$$\text{let} \quad G^{-P} = \int_{a_i}^{a_f} da/[(\sqrt{\pi a}\,\beta)^P]$$

$$\text{Then } N = 1/C\{\sigma_{max}(1-R)^q\}^{-P} G^{-P}$$

$$\text{Or } N = 1/C\{1/[G\sigma_{max}(1-R)^q]\}^P \quad (13)$$

$$\text{Where} \quad G = \left\{\int_{a_i}^{a_f} da/[(\sqrt{\pi a}\,\beta)^P]\right\}^{-1/P} \quad (14)$$

Equation 13 gives the number of cycles N to propagate a crack from an initial size a_i to a final size a_f with the effects of geometry defined by equation 14.

The term G will depend on the geometrical configuration. The results of equation 14 for all possible geometries is not included in the literature given by references [5, 6, 7, 8, 9,].

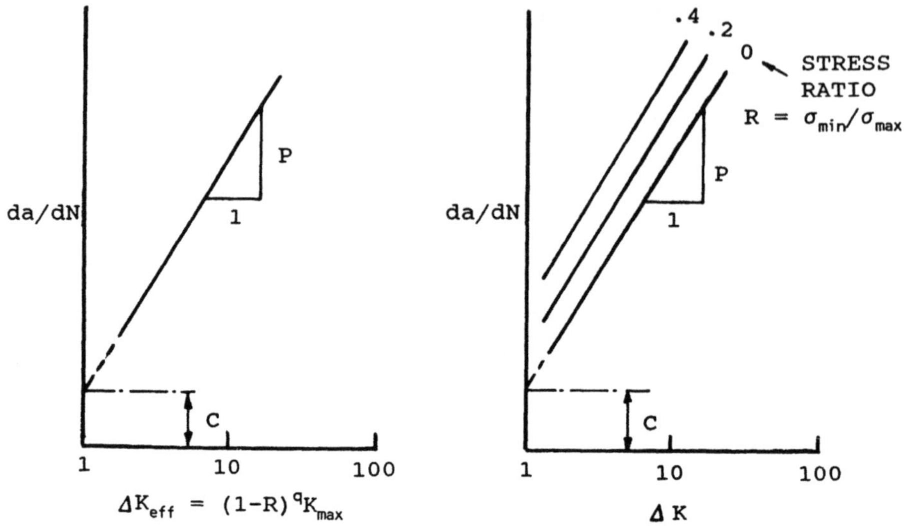

$$da/dN = C\{(1-R)^q\ K_{max}\}^P$$

where C = Intercept with K = 1.0 ordinate

R = Stress Ratio $\sigma_{min}/\sigma_{max}$

q = Exponent to control width of R ratio band of lines

K_{max} = Maximum stress intensity factor $\sigma_{max}\sqrt{\pi a}$

σ_{max} = Maximum applied gross stress

β = Geometry effect

P = Slope of da/dN line

Fig. 28. Walker Crack Growth Rate Equation

However, this term can easily be obtained by numerical integration for each specific geometry. There are many references in the literature providing the term β or the effect of geometry on the stress intensity factor K. For example, one of the most comprehensive of these is reference [10]. All other terms in equation 14 are defined.

Only in the simplest of cases can equation 14 be analytically integrated, for example, in the case of an infinitely wide unstiffened panel, where = 1.0.

$$G^{-P} = \int_{a_i}^{a_f} (\sqrt{\pi a})^{-P}\, da \qquad = \pi^{-P/2} \int_{a_i}^{a_f} (a)^{-P/2}\, da$$

$$\text{Therefore} \quad G^{-P} = (\pi)^{-P/2} \left[\frac{a^{1-P/2}}{1-P/2} \right]_{a_i}^{a_f}$$

$$G = \left\{ (\pi)^{-P/2} \left[\frac{(a_f)^{1-P/2} - (a_i)^{1-P/2}}{1-P/2} \right] \right\}^{-1/P}$$

Figure 29 shows a plot of this equation for two values of a_f. There is a need to develop figures such as Figure 29, based on equation 14, for a variety of typical geometries by numerical integration.

For the benefit of those not familiar with this approach consider the case of a through crack in a panel of finite width 20.0 inches. The value of β for this case is generally accepted to be $[\mathrm{Sec}(\pi a/W)]^{1/2}$ reference [11]. From equation 14

$$G^{-P} = \int_{a_i}^{a_f} (\sqrt{\pi a}\, \beta)^{-P}\, da$$

Therefore for our example case

$$G^{-P} = \int_{a_i}^{a_f} \{\sqrt{\pi a}[\mathrm{Sec}(\pi a/W)]^{1/2}\}^{-P}\, da \tag{15}$$

Now we need to plot the term $\{\sqrt{\pi a}[\mathrm{Sec}(\pi a/W)]^{1/2}\}^{-P}$ as a function of a. Figure 30 shows this plot for $a_i = 1.0$ to $a_f = 4.0$ for our 20 inch wide panel example. In this case the slope of the da/dN curve, P, for aluminum alloy has been assumed 3.7. This value will vary from alloy to alloy but a value of 3.7 is a reasonable average value to use for aluminum. We now need a plot of G versus initial crack size a_i. Assume for a start $a_i = 1.0$. The value of G^{-P} will be the area under the curve ABCDEFGH shown in Figure 30. This can be obtained either by the mid-ordinate or Simpson's rule. For our case this value is approximately 0.09345. Therefore $G = [0.09345]^{-1/3.7} = 1.89767$. Similarly for $a_i = 2.0$ the value of G^{-P} will be the area under the curve BCDEFG.

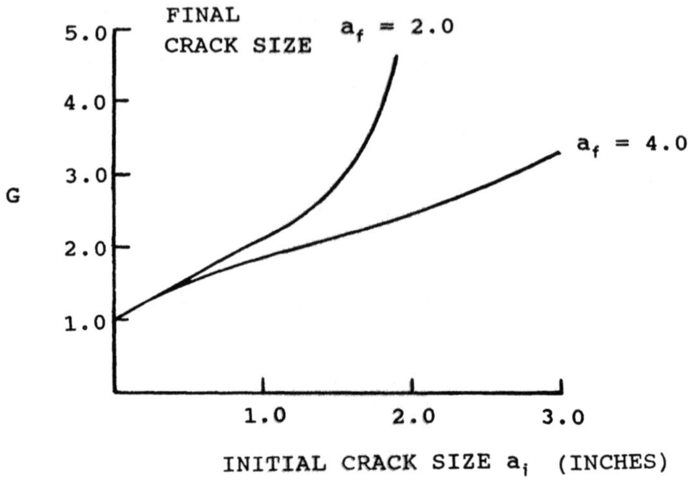

Fig. 29. Example Of The Geometric Term G For An Infinitely Wide Panel

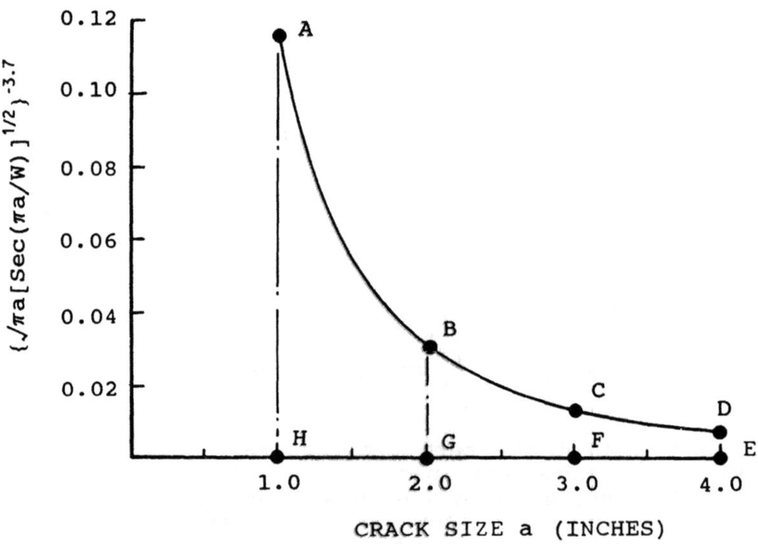

Fig. 30. Curve Of $(\sqrt{\pi a}\,\beta)^{-P}$ For Finite Width Panel

For our case this is approximately 0.02948. Therefore G = $[0.02948]^{-1/3.7}$ = 2.59204. Using this simple approach a curve of G versus a_i can be plotted. The result for our case of a 20 inch wide panel is shown on Figure 31. This procedure to obtain a plot of G can be used for any complex geometry provided the term β, as a function of crack size a, is known. As mentioned previously β can be found for many geometries in the literature.

A useful tip to consider is that the term $(\sqrt{\pi a}\,\beta)^{-P}$ can be integrated in pieces. Therefore, for example:

$$G^{-P} = \int_{a_i}^{a_f} (\sqrt{\pi a}\,\beta)^{-P}\, da = \int_{a_i}^{a_1} (\sqrt{\pi a}\,\beta)^{-P}\, da + \int_{a_1}^{a_2} (\sqrt{\pi a}\,\beta)^{-P}\, da + \int_{a_i}^{a_f} (\sqrt{\pi a}\,\beta)^{-P}\, da \quad \text{etc}$$

This is illustrated by Figure 32.

Therefore $G^{-P} = [G_1^{-P} + G_2^{-P} + G_3^{-P}]$ or $G = [G_1^{-P} + G_2^{-P} + G_3^{-P}]^{-1/P}$

In this case G_1^{-P} is the area under the curve ABGH, G_2^{-P} is the area under curve BCFG and G_3^{-P} is the area under curve CDEF. The usefulness of being able to integrate the geometrical term G^{-P} in pieces is best illustrated if one considers that the objective of a crack growth analysis is to produce a crack growth versus cycles curve. In this case it is an advantage to plot a G curve for the maximum value of a_f expected. This single

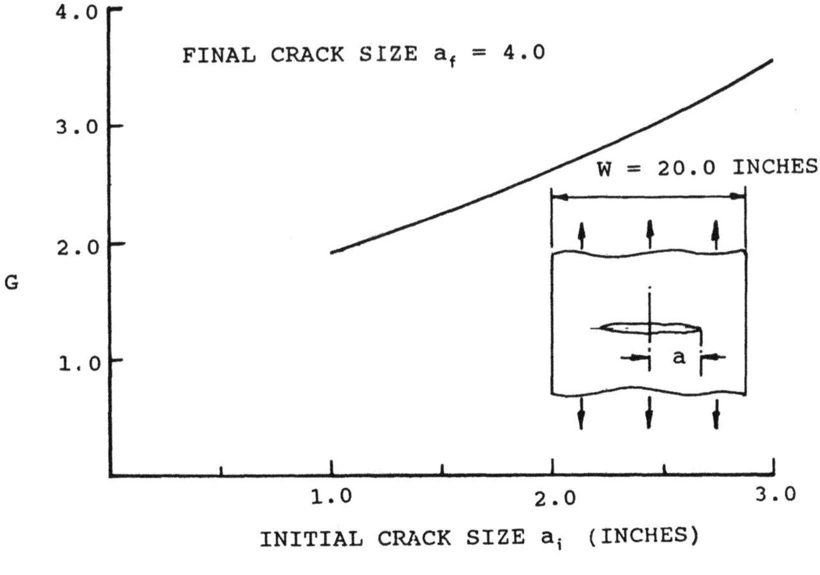

Fig. 31. Example Plot Of G For 20.0 Inch Wide Panel

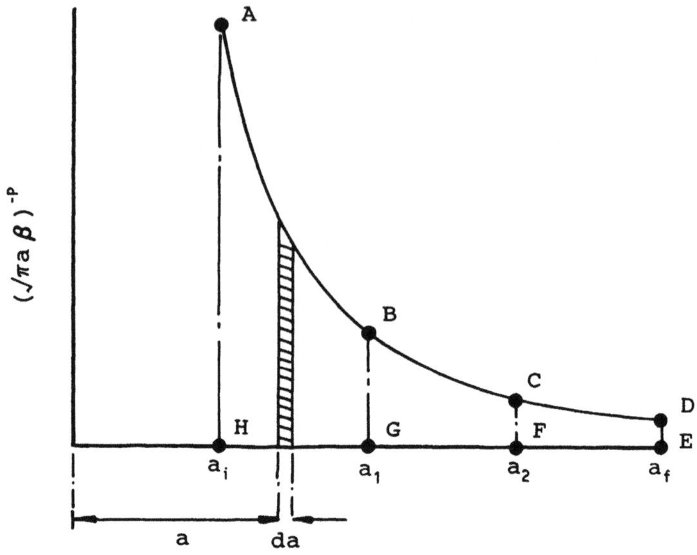

Fig. 32. Step Integration of $(\sqrt{\pi a}\,\beta)^{-P}$

curve can then be used to obtain G for lower values of a_f. This concept is illustrated by Figure 33. For example if we have a curve of G having been developed for $a_f = 8.0$ as shown to the left of Figure 33 but we want to consider crack growth from $a_i = 1.0$ to $a_f = 6.0$ we can say:

$$G[6.0 - 1.0] = \{(G[8.0 - 1.0])^{-P} - (G[8.0 - 6.0])^{-P}\}^{-1/P} \qquad (16)$$

Terms similar to G[6.0 - 1.0] mean G obtained between the limits 1.0 to 6.0. We can obtain the respective values of G[8.0 - 1.0] and G[8.0 - 6.0] from the curve to the left of Figure 33 and use these values in equation 16 to obtain G[6.0 - 1.0]. This is

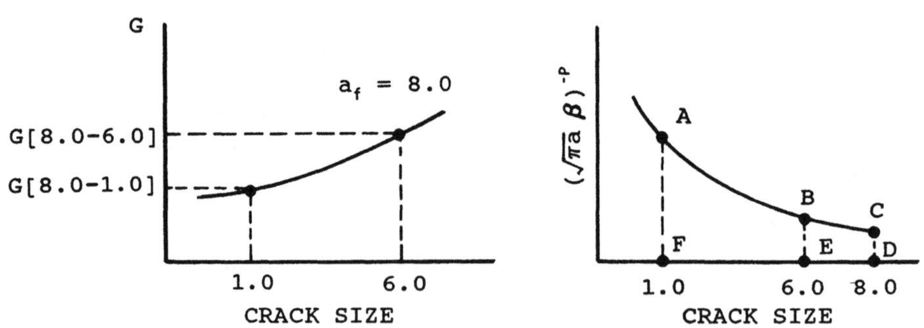

Fig. 33. Integration of G^{-P} In Pieces

possible since $\{G[8.0 - 1.0]\}^{-P}$ is the integral of $(\sqrt{\pi a}\,\beta)^{-P}$ between the limits 1.0 and 8.0 which is the area of the curve shown to the right of Figure 33 represented as ABCDEF. Similarly $\{G[6.0 - 1.0]\}^{-P}$ is the integral of $(\sqrt{\pi a}\,\beta)^{-P}$ between the limits 1.0 to 6.0. Since the curve to the left of Figure 33 is based on the upper limit of 8.0 only then we use this to obtain G[6.0 - 1.0} by subtracting area BCDE from area ABCDEF and this is equivalent to area ABEF representing $\{G[6.0 -1.0]\}^{-P}$. We then raise this area to the power -1/P and this represents G[6.0 - 1.0].

Expanding the concept illustrated by Figure 33 we can calculate more points on the crack growth curve if we only have one curve of G versus a_f by the same procedure. Figure 34 illustrates the procedure for 4 points on a typical crack growth curve. The term $\{G[8.0 -1.0]\}^{-P}$ represents the entire area under the curve of $(\sqrt{\pi a}\,\beta)^{-P}$ from 1.0 to 8.0 shown to the right of Figure 33. The number of cycles to propagate the crack from 1.0 to other crack lengths are obtained by subtracting the remaining area of the curve of $(\sqrt{\pi a}\,\beta)^{-P}$ beyond the crack length required as illustrated in Figure 34. The values of G again are obtained from the G versus a_i curve as illustrated by the curve to the left of Figure 33.

$$G[2.0-1.0] = \{[G(8.0-1.0)]^{-P} - [G(8.0-2.0)]^{-P}\}^{-1/P}$$

$$G[3.0-1.0] = \{[G(8.0-1.0)]^{-P} - [G(8.0-3.0)]^{-P}\}^{-1/P}$$

$$G[4.0-1.0] = \{[G(8.0-1.0)]^{-P} - [G(8.0-4.0)]^{-P}\}^{-1/P}$$

$$G[5.0-1.0] = \{[G(8.0-1.0)]^{-P} - [G(8.0-5.0)]^{-P}\}^{-1/P}$$

$$N_1 = N[2.0-1.0] = 1/C\{[G(2.0-1.0)]S\}^{-P}$$

$$N_2 = N[3.0-1.0] = 1/C\{[G(3.0-1.0)]S\}^{-P}$$

$$N_3 = N[4.0-1.0] = 1/C\{[G(4.0-1.0)]S\}^{-P}$$

$$N_4 = N[5.0-1.0] = 1/C\{[G(5.0-1.0)]S\}^{-P}$$

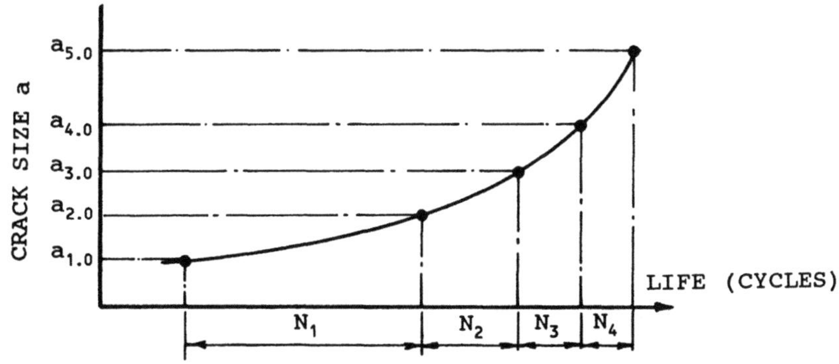

Fig. 34. G For Various Points On Crack Growth Curve

To simplify the crack growth analysis under spectrum loading the first assumption is to assume all flights are similar. That is the crack is propagating under a single repeatable flight. This assumption is more reasonable for commercial transport aircraft than it would be for say a military fighter. This single repeatable flight is then simulated to a single cycle having the same crack growth as the repeatable flight.

From equation 13 $N = 1/C\{1/[G\sigma_{max}(1-R)^q]\}^P$

Or $N = 1/(CG^P)\{1/[\sigma_{max}(1-R)^q]\}^P$

Crack growth per cycle

$$1/N = CG^P\{\sigma_{max}(1-R)^q\}^P \qquad (17)$$

Consider a single repeatable flight as shown in Figure 35. The total crack growth for this flight can be calculated using equation 17 by summing the growth for each cycle. It is usual practice to truncate the negative stresses so the cycles considered are those represented by the heavy lines between dots on Figure 35. However, some prefer to leave in the negative stresses, take the full range of stress from negative to positive, and then use the appropriate crack growth rate corresponding to the negative R ratio. There is usually very little difference between the two results. The total crack growth for the single flight is therefore:

$$T = CG^P \sum_{i=1}^{i=n} \{\sigma i_{max}(1-R_i)^q\}^P \qquad (18)$$

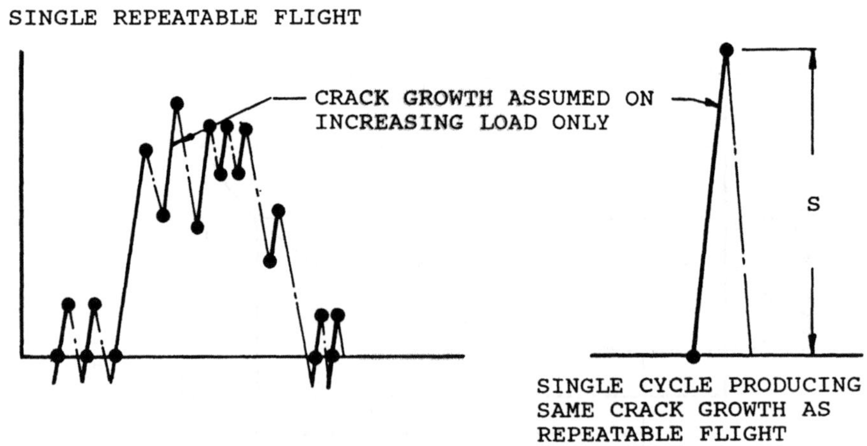

Fig. 35. Simulation Of Repeatable Flight To A Single Cycle

The stress level S is then obtained, at a stress ratio R = 0, which will give the same crack growth as the single flight. This can be obtained by substituting R = 0 and σ_{max} = S into equation 17. Therefore the growth for the single cycle is:

$$C(GS)^P \tag{19}$$

We can equate equations 18 and 19 to obtain the value of S:

$$CG^P \sum_{i=1}^{i=n} \{\sigma i_{max}(1-R_i)^q\}^P = CG^P(S)^P$$

$$\text{Therefore} \quad S = \left[\sum_{i=1}^{i=n} \{\sigma i_{max}(1-R_i)^q\}^P\right]^{1/P}$$

$$S = [\{\sigma 1_{max}(1-R_1)^q\}^P + \{\sigma 2_{max}(1-R_2)^q\}^P + \text{etc.}]^{1/P} \tag{20}$$

This value of S, at a stress ratio R = 0, is substituted into equation 13 to obtain the number of cycles to propagate the crack from a_i to a_f with G determined for a_i to a_f.

$$\text{Therefore } N = 1/C\{1/[GS]\}^P$$

$$\text{Or } N = 1/C(GS)^{-P} \tag{21}$$

Crack Growth Evaluation of Specific Repair

Simplified methods to allow small repairers and modifiers to develop crack growth curves have been discussed. However, even to perform this simplified analysis it is necessary to have a knowledge of stress levels in the basic structure. Without this knowledge it is impossible to perform even a simple damage tolerance evaluation. If, however, a crack growth curve for a particular element, prior to repair, were available then it may be possible to use this information to obtain a crack growth curve for the modified structure. Consider the very simple example of an original structural element as shown in Figure 36a which may be the vertical web of a wing spar cap. This element may have been repaired by adding a reinforcement strap as shown by Figure 36b perhaps to replace corroded material. If the original element had been classified as a principal structural element in the initial damage tolerance evaluation, performed to either certify the airplane or to develop the SID, then a crack growth curve must have been developed by the manufacturer. This curve may have been generated using a more sophisticated cycle-by-cycle crack growth analysis method including the effects of retardation and using a particular equation to simulate da/dN. What is needed is to determine the constant amplitude stress level S at a R ratio of zero which will match the crack growth curve for the original element. This would be done by calculating G for the original configuration using equation 14 and then varying S in equation 21 to match the overall crack growth life covered by the original crack growth curve. There are some minor problems associated with this approach. A very simple example will help illustrate these

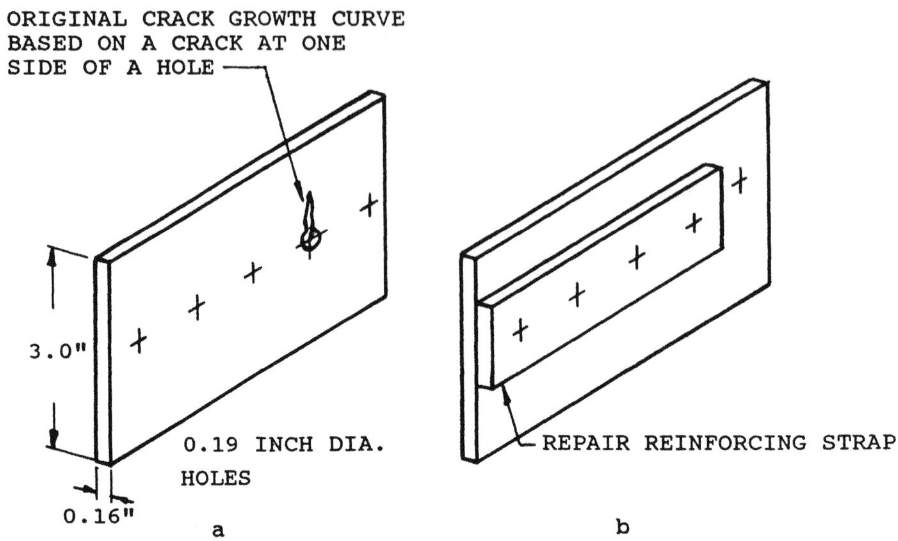

Fig. 36. Example Of Repaired Element

problems. Suppose the original stress spectrum for the element shown in Figure 36a was as illustrated in Figure 37. This has been simplified to illustrate a point. Suppose the existing crack growth curve were based on an initial 0.05 inch crack at one side of a hole as shown on Figure 36a. Suppose the original crack growth curve was based on Forman's equation given by equation 3. Crack growth retardation may have been accounted for using the Willenborg model [12]. The resulting crack growth curve, produced by RECYCL [2], would be as shown on Figure 38 curve A. Now if equation 21 were used to match the overall life for this curve, using equation 14 for G, and varying the stress level S in equation 21, then a life match would be established at a stress S of 19.45 KSI using a Walker equation with $C = 3.88 \times 10^{-10}$. The value of C is not important since the same value would be used to generate a crack growth curve for the modified structure. The resulting curve is illustrated by curve B on Figure 38. Even though the overall life has been exactly matched the shape of this curve is not the same. The primary reason for this is that retardation was considered to develop curve A and the peak stress of 17 KSI in the spectrum created a plastic zone at the crack tip which effected the growth of subsequent cycles and reduced the slope of the crack growth curve when the crack was small. Since retardation was not considered for curve B the slope at the start of the curve was greater. If the original curve had been generated neglecting retardation effects but using the same Forman's equation 3 then the result would be as shown by curve C, developed by LICAFF [2], on Figure 38. Again using the value of G developed from equation 14 and varying the stress level S in equation 21, the overall crack growth life would be matched at a stress of 22.43 KSI using the Walker constant $C = 3.88 \times 10^{-10}$. The resulting curve is shown as curve D on Figure 38. Now it can be seen that the shape of this curve is a closer fit to the curve developed under spectrum loading without the effects of retardation. The slight difference in shape is due to the differences in the crack growth equation used. The Forman equation, used to generate curve C, is not a straight line equation as is the Walker equation. The spectrum analysis without retardation was repeated using the Walker equation with $C = 3.88 \times 10^{-10}$, $q = 0.6$ and $P = 3.7$. The results are shown on Figure 39. Equation 21 was then used to match this curve and a perfect match was achieved at a constant amplitude stress of 19.95 KSI. This match is illustrated by the points on

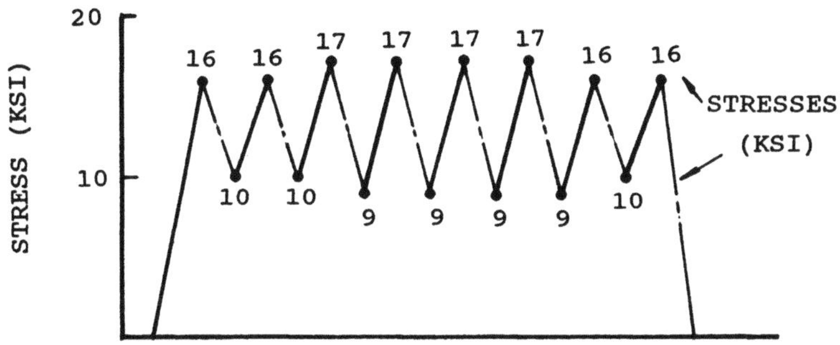

Fig. 37. Simple Example Spectrum For Repeatable Flight

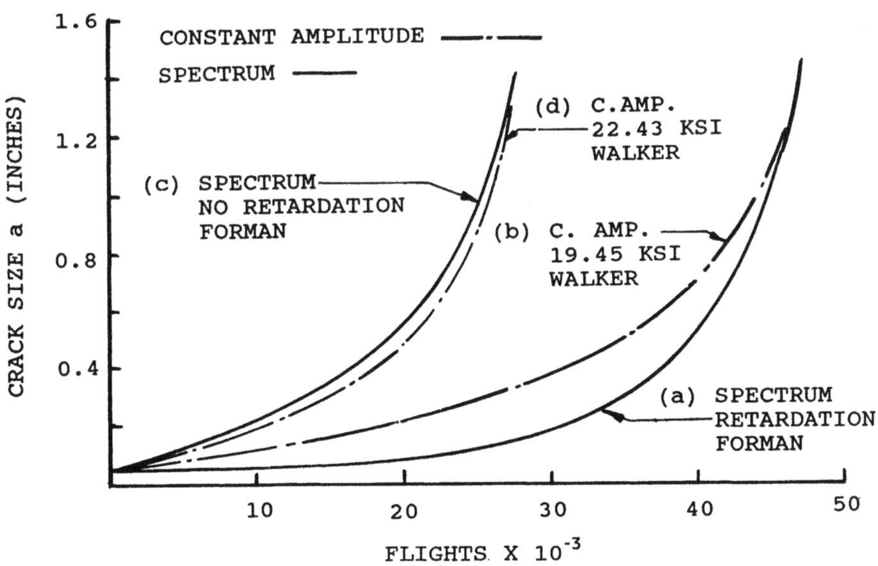

Fig. 38 Crack Growth Comparison Spectrum/Constant Amplitude

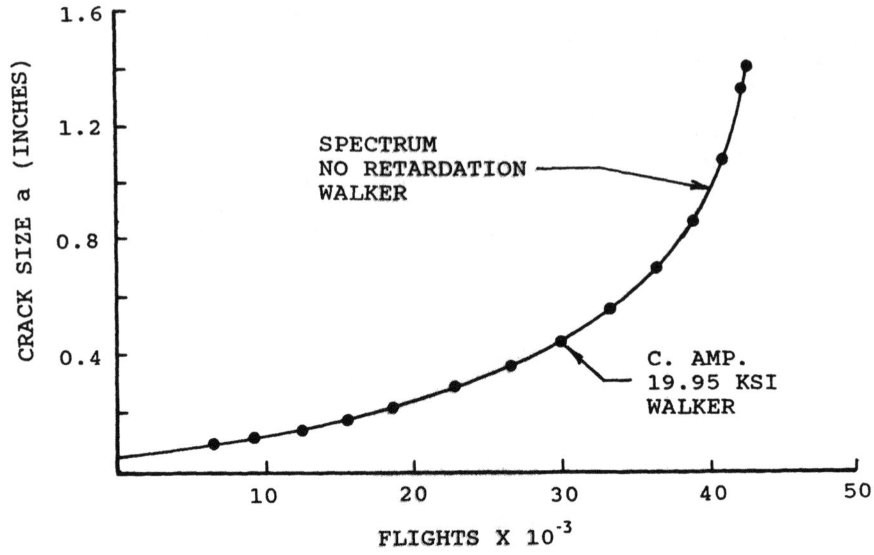

Fig. 39. Crack Growth Comparison Spectrum/Constant Amplitude Using Same Crack Growth Equation

the curve. The purpose of determining the constant amplitude stress level S which matches the original curve is to use this stress in equation 21 to develop a crack growth curve for the modified element. This would be done by calculating G for the modified element accounting for load transfer at the first rivet and the loss in visual detectability when the crack is under the doubler. Equation 14 would be used to develop the new G value based on standard stress intensity expressions found in the literature. This procedure would allow a crack growth curve for the modified structure to be developed without a knowledge of the original stresses but with a knowledge of the original crack growth curve. The purpose of Figure 38 was to illustrate that some caution must be exercised when using this procedure because of the possible mismatched shape illustrated by Figure 38 curves A and D. For example the matching curve B reflects a longer crack growth life from a detectable crack to critical if the detectable size is larger than the initial crack size used to produce the curve. For example, if the detectable size was considered to be 0.15 inches (say) the safe crack growth life from the original curve is 19000 flights but is 35000 flights using the matched curve. Because of this it is better to determine S in equation 21 to fit the curve from the detectable size to the critical size.

Future Possibilities

A method to generate crack growth curves for the modified or repaired structure was discussed using a very simple example case. The approach required a knowledge of the original crack growth curve. Without this curve or a knowledge of the original stress spectrum it is impossible to obtain the required information. One possibility requiring some limited research may be to divide the airplane structure into a number of zones, perhaps six or eight zones for the fuselage, the lower wing surface and the horizontal stabilizer tension surface. Standard conservative stress spectra could be developed, perhaps using the standard TWIST spectrum, for the most critical points in each zone. This could be done using the most conservative 1g and 1P stresses to be found anywhere in the industry. These standard spectra would then be used along with the simplified methods discussed to produce conservative crack growth curves and inspection intervals. The only alternative to this is to persuade the original manufacturer, with all his knowledge of the original stress spectra and loads, to perform the analysis.

Conclusions

- Any repair to an airframe structure has the potential to drastically degrade fatigue life and damage tolerance capability.
- In most cases the critical location for future fatigue initiation is in the basic skin at the first fastener row in the repair doubler.
- Curves are presented illustrating degradation in fatigue life of the basic structure as a function of repair doubler thickness.
- Fatigue life of the basic structure can be calculated as a function of applied gross stress and bearing stress in the skin at the first fastener row.
- Curves are presented, based on displacement compatibility analysis accounting for fastener flexibility, which will enable the critical fastener loads to be determined as a function of skin and doubler thickness for five common repair configurations.
- Data from these curves, together with conservative Sn curves provided, will allow a repairer or modifier to assess the fatigue quality of the repair.
- Some ideas are provided which will enable improved fatigue life and inspectability by laminating the repair doubler.
- Simplified methods which may enable a small modifier or repairer to develop crack growth curves for the basic structure after repair are discussed.

Bottom Line

- All repairs to principal structural elements on aircraft certified to FAR 25.571 amendment 45 or on aircraft assessed to AC 91-56, where a SID document exists for the airplane, must be evaluated for damage tolerance.

- New inspection methods/thresholds/frequencies must be established for the repaired structure.
- The repaired airplane should be inspected at the determined frequencies after the established threshold.

Note

The opinions expressed in this paper are those of the author and not necessarily those of the FAA. The author is an advisor to The Aircraft Certification Service who have final authority.

References

[1]. Swift, T., "Damage Tolerance Analysis of Redundant Structures," Fracture Mechanics Design Methodology, AGARD-LS-97, North Atlantic Treaty Organization, London, England, Jan. 1979, pp. 5-1.

[2] Computer programs LICAFF, GEOFAC and RECYCL, FractuREsearch Inc. 9049 Cupstone Drive, Galena, Ohio 43021, USA.

[3] Jarfall, Lars., "Optimum Design of Joints: The Stress Severity Factor Concept," Aircraft Fatigue Proceedings of the 5th ICAF Symposium, Melbourne, Australia, 22-24 May, 1979.

[4] Jarfall, Lars., "Fatigue Cycling of Riveted or Bolted Joints," FFA report HF-1239. The Aeronautical Research Institute of Sweden, June 1969.

[5] Craig, L., and Goranson, U.G., "Airworthiness Assessment of Boeing Jet Transport Structures." Presented at the 10th ICAF Symposium, Brussels, Belgium, May 16-18, 1979.

[6] Goranson, U.G., and Hall, J., "Airworthiness of Long-Life Jet Transport Structures." Presented to the Royal Aeronautical Society Spring Convention on Long-Life Aircraft Structures, London, England, May 14-15, 1980.

[7] Craig,L., et al "Airworthiness of Long-Life Jet Transport Structures." Presented at the Air Transport Association of America Forum, Miami, Florida, October 1-3, 1980.

[8] Goranson,U.G., et al "Long-Life Damage Tolerant Jet Transport Structures." Presented at ASTM Symposium "Design of Fatigue and Fracture Resistant Structures," Bal Harbor, Florida, Nov. 10-11, 1980.

[9] Gunther, C.K., and Goranson, U.G., "Spectrum Loading Effects on Crack Growth in Major Components of Commercial Transport Aircraft." Presented to the International Conference on Application of Fracture Mechanics to Materials and Structures, Freiburg, Germany, June, 1983.

[10] Rooke, D., and Cartwright, D., "Compendium of Stress Intensity Factors." Printed in England for Her Majesty's Stationary Office by The Hillingdon Press, Uxbridge, Middx. England, Sept 1974.

[11] Feddersen, C.E., "Evaluation and Prediction of the Residual Strength of Center Cracked Tension Panels." Published in "Damage Tolerance in Aircraft Structures," ASTM STP 486, May 1971.

[12] Willenborg, J., Engle, R.M. and Wood, H.A., "A Crack Growth Retardation Model Using An Effective Stress Concept." US Air Force Tech. Memo. 71-1-FBR. Jan. 1971.

28. Case Study and Simulation of Fatigue Damages and DTE of Aging Aircraft - A Review of Researches in Japan

H. TERADA * and K. OHTSUKA **

*National Aerospace Laboratory, 7-44-1 Jindaiji higashi, chofu Tokyo, Japan.

**Association of Air Transport Engineering and Research, Shin Tamachi Bldg, 34-6 Shiba 5-chome, Minato-ku Tokyo, Japan.

Introduction

This is a review paper on the study of aging aircraft structures conducted in Japan.
The paper consists of four (4) parts.
Part 1 is the event report of againg aircraft structural damages.
Part 2 describes fatigue strength evaluation technique and the evaluation of non-destructive inspection method. Investigation for SID of YS-11, turbo-prop transport, is presented in Part 3. Crack propagation of riveted joint with initial damage simulating multi-site damage is presented in Part 4.
The followings are the abbreviation of the papers contributed by Airlines, Manufacturers and National Aerospace Laboratory.

1. Event Report of Aging Aircraft Structure Damages

1.1 Recent Structural Problems on B-747 Wing (Japan Airlines)

As represented by Sec. 41 problems and the skin lap joint design of the pressure vessel, more attention is focused on the B-747 fuselage structure. The trigger for this added attention may have been the result of the special inspection on Sec. 41 structure that JAL performed on some high cycled B-747SR airplanes in 1984. Apart from the fuselage problem, some remarkable problems of B-747 wing structure which were found in the special inspections or recently experienced by JAL are reviewed briefly herein.

Findings:	Cause	Ref.
① Wing Rear Spar to Wing Center Section Upper Splice Fitting and Mating Structure Crack	Bending stress resulting from pressurization cycle and wing flexing	SB 57-2110
② WS 669.5 Rib Crack	Fatigue due to induced loads from engine support fitting	SIIA 747-49

Springer Series in Computational Mechanics
Atluri, Sampath, Tong (Eds.)
Structural Integrity of Aging Airplanes

③ Wing Rear Spar Web Fuel jettison Pump Cutout Hole Crack	Fatigue due to stress concentration	SIIA 747-63
④ Wing Landing Gear Support Beam Outboard End Fitting Crack	Fatigue from corrosion pits	SB 57-2244
⑤ WBL 743.83 Rib Lower Chord Crack	Cyclic flap loading and stress concentration	
⑥ WS 323.82 Rib Attach Stiffener crack	Under investigation	
⑦ WS 586 Rib Aft Back-Up Fitting Crack	Preload due to inadequte shimming	
⑧ WS 697 Rib Tie Crack	Same as ②	

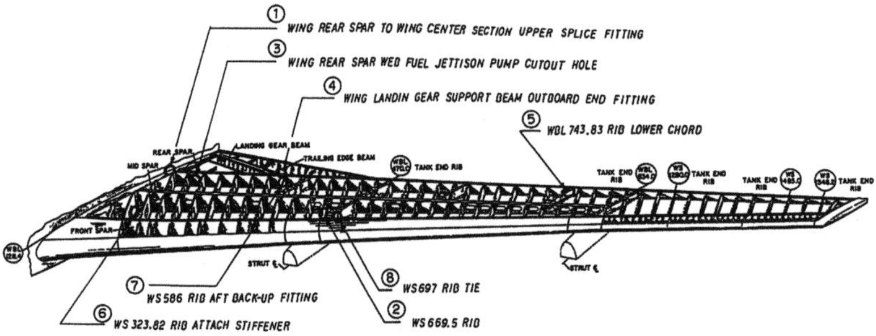

Fig. 1.1. Wing Structure (B-747)

Notable feature of findings is that most cracks were found around the joining points of external load carrying members, like the pylon support, flap track and landing gear beam.

Conclusion:

Knowledge and technique of fatigue and faracture analysis is making steady progress. However, we still have the fatigue cracks in the service aircraft. Most cracks (except stress corrosion cracks) occur due to secondary loads that possibly can not be expected by the present state-of-the-art in the design stage. On this point, the full scale fatigue test plays the important role, and the recent movement to request a two-lifetime fatigue test for certification is reasonable.

1.2 L-1011 MLG Upper Jury Brace Universal Failure (All Nippon Airways)

ANA has experienced two(2) broken and two(2) cracked MLG upper jury brace universal on L-1011. (see Fig. 1.2)

NO.	AC S/N	HOURS	CYCLES	REMARK
1	1134, L/H	15865	13036	FAILURE
2	1128, R/H	18418	14923	CRACK
3	1129, R/H	17695	14653	CRACK
4	1154, L/H	23510	19172	FAILURE

As the countermeasure for the failure, the universal assembly was replaced.

Inspection of Failed Parts

Metallurgical analysis of parts returned to Lockheed showed that the failures or cracks were due to fatigue. All fatigue initiation and initial growth was subsurface. Initiation involved multiple origins which spanned a crack length of over 0.5 inch. It was estimated that the crack was at least 0.03 inch deep. The crack could not be found by sensitive florescent penetrant techniques.

Inspection Interval Revision

The universal should have an initial eddy current inspection by 12,000 flight cycles and subsequent repetitive inspections at 900 flight cycle intervals.

Reference

Lockheed Service Bulletin NO. 093-32-236 (MAR, 1989)

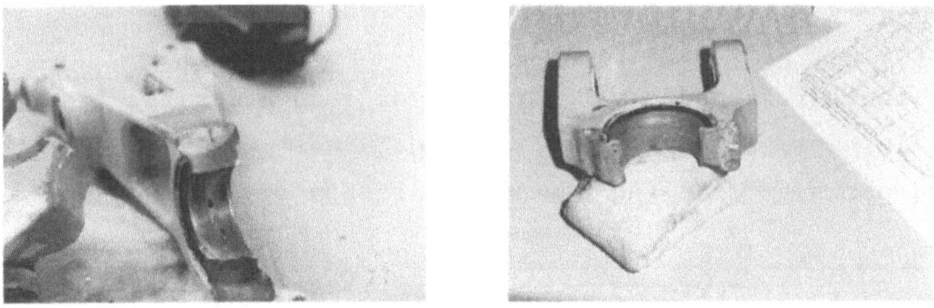

Fig. 1.2

1.3 Fatigue Crack at Frame Fitting of A300 Aircraft (Japan Air System)

The crack was found at FWD fitting (2075T73) of Frame 47 shown in

the following figure, by NDI in accordance with advanced SSIP. The crack occured on aircraft having logged 18,850 flts. and 9 years from delivery as an initial case in service and were reported in the full-scale fatigue test report at 48,600 cycle, due to fatigue by wing bending deflection.
Periodical inspection will be required to propagate to limited crack length for repair.

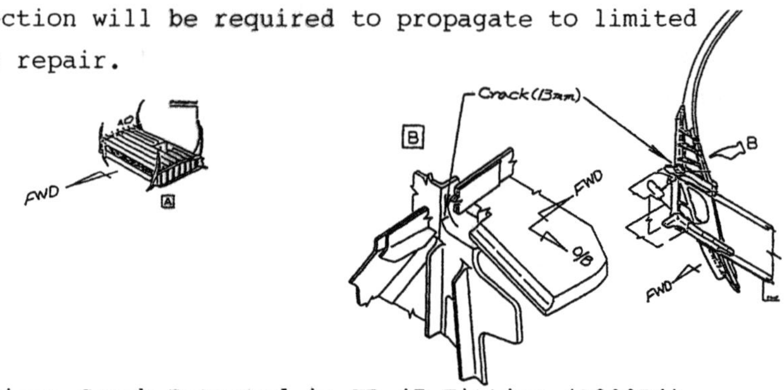

Fig 1.3 Fatigue Crack Detected in FR.47 Fitting (A300B4)

1.4 FWD Cargo Doorjamb Members Damage of DC-9 Aircraft (Japan Air System)

Heavy corrosion were found at 'PAN' mating surface (7075T6) in cavities of the lower cargo doorjamb. The crack of 'JAMB' was found in repair works of the pan by SRM. The cracks occured on aircraft having logged 22,070 flts. and 12 years from delivery and were reported in the information of aircraft manufacturer.
The cracks of 'Jamb' were attributed to stress-corrosion and the corrosion of 'Pan' was due to environmental salty fluid of cargo loadings.
After initial found, several cracks of each aircraft were found in special inspection by X-ray. Damaged 'Jambs' were replaced to improved parts to changing the material from 17-7PH to A-286 CRES.

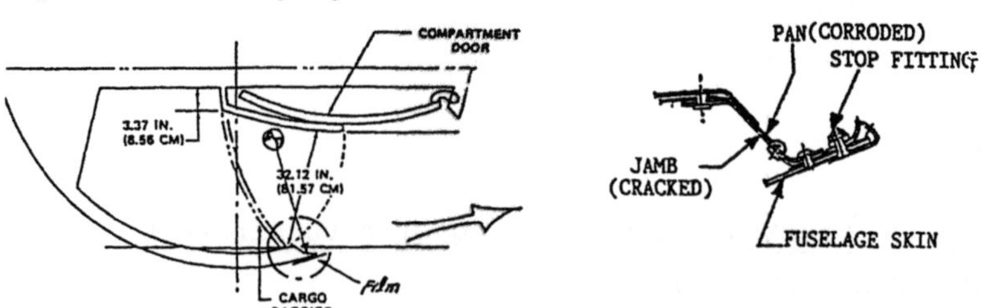

Fig. 1.4 FWD Cargo Doorjamb Members Damage (DC-9-41)

2. Fatigue Strength Evaluation of Lug Joint and Reliability of NDI Technique for Field Application

2.1 Fatigue Strength of Lug Joint (Kawasaki Heavy Industries)

With regard to fatigue strength of the lug joint, many factors such as stiffness of pin, clearance between lug hole and pin and friction of lug surface are considered to affect. In order to clarify these problems, fatigue tests of lug joints were performed.

Test specimen (Fig. 2.1)
Load is transferred from lever to clevis through the pin only.
Treatments such as lubrication to prevent friction are not conducted.

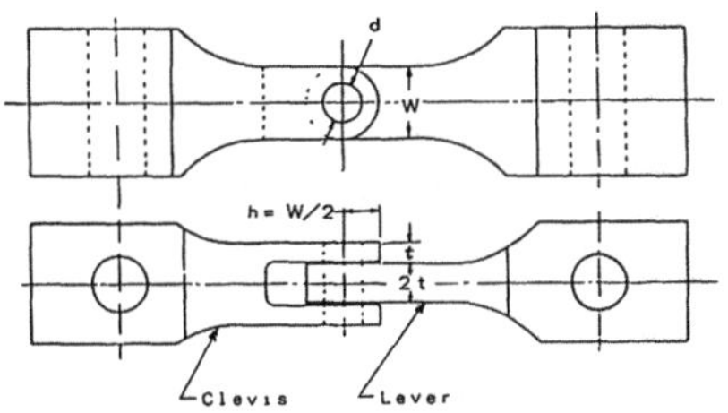

2024-T351 bar
7075-T651 bar
4130 bar

Repeated load
R=0
Constant Amplitude

fitness	clearance*
Positive clearance	-0. 059~-0. 085
Transition	0. 0 ~-0. 004
Tight interference	+0. 011~+0 003

*pin dia-lug hole / pin dia

W/d	W	d	t	Kt
3. 0	38 10	12 70	6 35	3 5
2 0	25 40	12. 70	6. 35	2 75
1 5	19 05	12. 70	6 35	

Fig. 2.1 Test Specimen

Test Result

(1)S-N curve (Fig. 2.2)
This figure shows sharp reduction of S_{max} in $N>10^5$, as compared to the notched specimen S-N curves.
This phenomenon is considered to be induced by fretting effect.

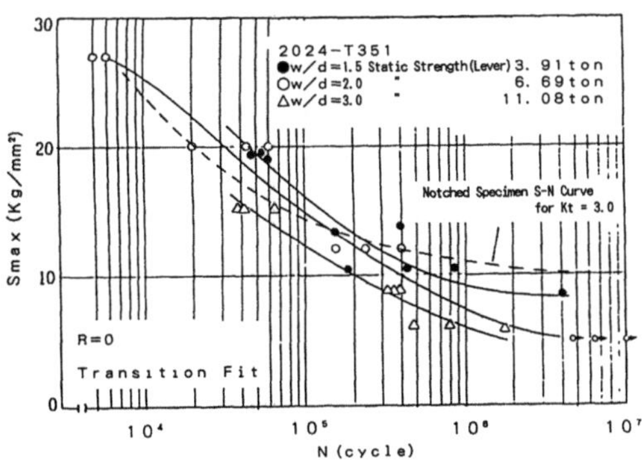

Fig. 2.2 Example of S-N Curve

(2)Fretting factor (Fig. 2.3)
In order to account for this fretting effect on the strength of the lug joint, the fretting factor K_{FRET} was defined as

$$K_{FRET} = \left(\frac{S_{max}{*}}{S_{max}{**}} \right)_N$$

where $S_{max}*$ is obtained from the S-N curve of the notched specimen corresponding to Kt. $S_{max}**$ is obtained from the S-N curve of the lug joint.

It is considered that K_{FRET} depends on the load cycles (N) only and is not influenced by the ratio of w/d.

Procedure for Rough Estimation of Lug Joint S-N curve

(1) Obtain the Kt, Stress concentration factor, of the lugs.
It is the function of the lug configuration (ex. w/d, h/w).

(2) Choose the S-N curve corresponding to the above Kt.

(3) Obtain S_{max}* from the above S-N curve, and K_{FRET} from the K_{FRET}-N curve for the given loading cycle.
Calculate
$S_{max}^{**} = S_{max}^{*}/K_{FRET}$

(4) Draw the S_{max}**-N curve, the lug-Joint S-N curve.

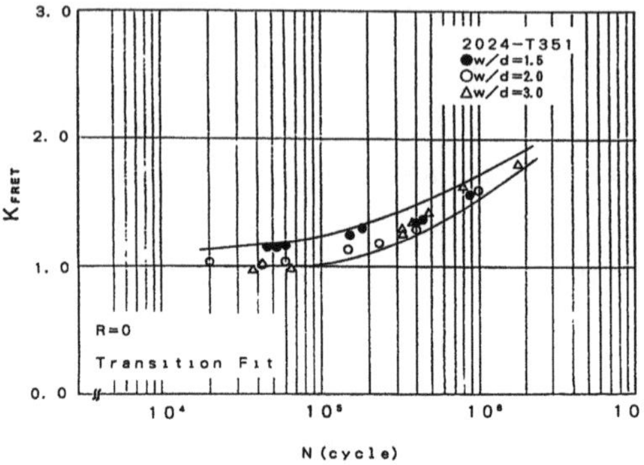

Fig. 2.3 Example of Fretting Factor of Lugs

2.2 Investigation of The Current NDI Technology for Aircraft Structures (Fuji Heavy Industries)

The objective of this study is to investigate the features, limitations and problems of the current NDI technology in the operating field.

The following typical NDI methods were selected from the view point of portability of the equipment.

*To Cracks in Aluminum;

Visual, X-Ray, Eddy Current (Ring Probe, Sliding Probe)

*To Delaminations in Composite;

Tapping, Ultrasonic, Laser Holography, Thermography

Aluminum panels with the Lap Joint and the Butt Joint were selected to simulate one of the focused locations of the current crack problems. Repeated loads were applied to the aluminum test specimens and multiple site cracks were introduced. Honeycomb sandwich panels and laminated plates with various impact damages were selected as the typical damaged composite structures. (See Fig 2.4 for the test specimens.)

After the NDIs were conducted, the specimens were cut for the fractography and the accurate crack/delamination sizes were measured.

TEST ITEM	PANEL	OUTLINE OF THE TEST SPECIMEN	SPECIFICATION	NUMBER OF SPECIMEN	INSPECTION AREA
CRACK IN ALUMINUM	LAP JOINT PANEL	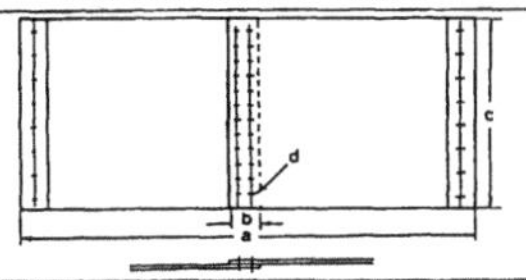	SHEET MATERIAL , 2024C-T3 TYPE I THICKNESS, t=.08in a=500. mm, b=40 mm, c=260. mm RIVET, NAS1097DD6 at d TYPE II·THICKNESS, t= 05in a=502. mm, b=32 mm, c=208. mm RIVET NAS1097DD5 at d RIVET PITCH .4D (2D to EDGE)	TYPE I : 6 TYPE II: 6	REVET HOLES (VISIBLE EXCEPT UNDER THE RIVET HEAD)
	BUTT JOINT PANEL	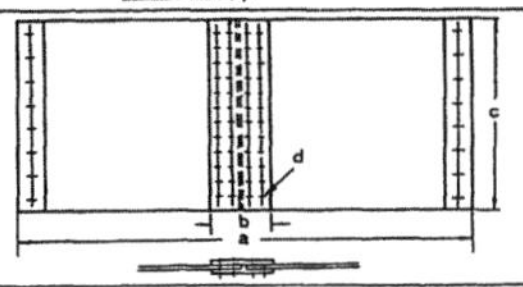	SHEET MATERIAL ; 2024C-T3 TYPE I:THICKNESS; t=.08in a=502. mm, b=82 mm, c=260 mm RIVET, NAS1097DD6 at d TYPE II THICKNESS, t=.05in a=502 mm, b=66 mm, c=208. mm RIVET, NAS1097DD5 at d RIVET PITCH .4D (2D to EDGE)	TYPE I : 6 TYPE II: 6	RIVET HOLES ON THE INNER PLATE (NON -VISIBLE)
DELAMINATION IN COMPOSITE	HONEYCOMB SANDWICH PANEL	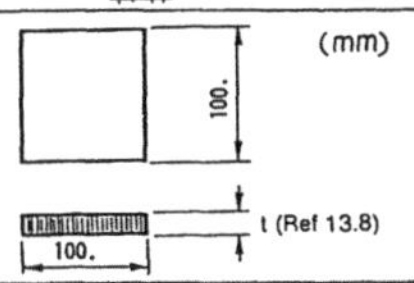	MATERIAL FACE SHEET, AS4/938 5 HARNESS SATIN 2-PLY $(\pm 45)_2$ CORE , NORMEX HONEYCOMB 3 0 LB/FT³	54 9 KINDS OF IMPACT	FACE SHEET, CORE AND BONDING OF DAMAGE-INDUCED AREA
	LAMINATED PLATE	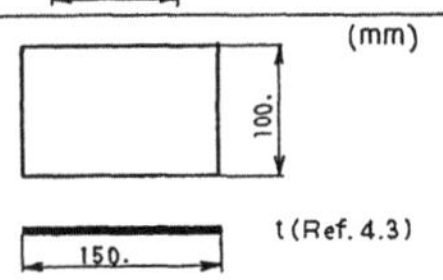	MATERIAL AS4/938 TAPE QUASI-ISOTROPIC LAMINATE 32-PLY $[(45/0/-45/90)_{SYM}]_4$	69 9 KINDS OF IMPACT	DAMAGE-INDUCED AREA

Fig 2.4 Test Specimens

<Test Results>

1. Cracks in Aliminum

The highest detectability has been obtained in Visual inspection as long as the cracks appear on the visible surface. To the inclined cracks in thicknesswise or the closed cracks, it has been hard to detect by X-Ray inspection. The most preferable results have been obtained by Eddy Current inspection. Regarding the probe of eddy current, the sliding type has shown more beneficial ability than the widely used ring type.

An attention must be paid that the eccentric growth on the non-visible side has often occurred in case of the lap joint specimen. The crack detectability to the lap joint has been clearly worse, because of this tendency.

2. Delaminations in Composite

The most preferable ability has been shown in ultrasonic inspection regarding both the detectability and the size measurement. Good detectability has been shown in most NDIs except Thermography for honeycomb sandwich panels. A tendency has been shown that all methods are more effective to the laminated plates than the honeycomb sandwich panels.

3. Development of SID on YS-11 Turbo-prop Transport (Mitsubishi Heavy Industries)

A total of 182 YS-11 were manufactured by 1974 since its maiden flight in 1962, and 151 airplanes are still in service all over the world (see Fig. 3.1)

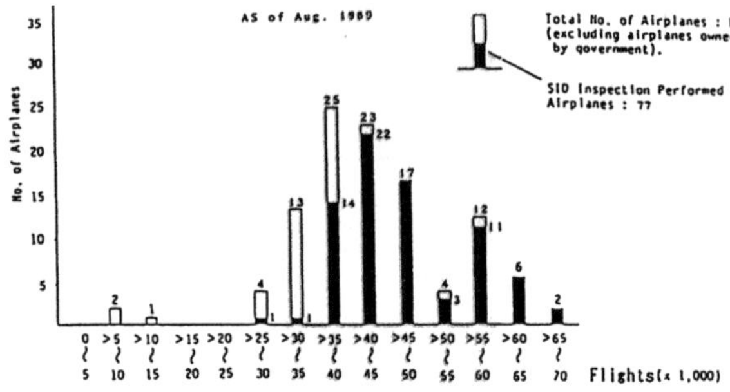

Fig 3.1 Fleet Distribution of YS-11

During this period, no occurrence has been reported of the discrepancies such as fatigue cracks which endanger structural integrity. There are mainly following two factors contributing to the absence of the major discrepancies.

(i) The design criteria was carefully selected during the initial design stage by considering fatigue factors; such as operating stress level, geometric design detail and material choice.

(ii) The full-scale fatigue tests [1] were performed more than 200,000 flights and the proper countermeasures were taken for any deficiencies found by the tests.

The Supplemental Inspection Document (SID) of YS-11 was developed by T/C holder Mitsubishi Heavy Industries (MHI) from 1980 and issued in 1983 with the approval of the regulatory authorities. The SID program on aging aircraft has been continued by MHI since that time. An overview of the SID development processes [2] is given in Fig. 3.2. It is assumed that the corrosion prevention specified on the airplane is properly performed. Therefore, the Corrosion Control Manual comprising the contents shown in Table 3.1 was also prepared in order to minimize the corrosion problems.

A list of the eight items to be inspected per SID is shown in Table 3.2 and the adopted detection limits are shown in Table 3.3.

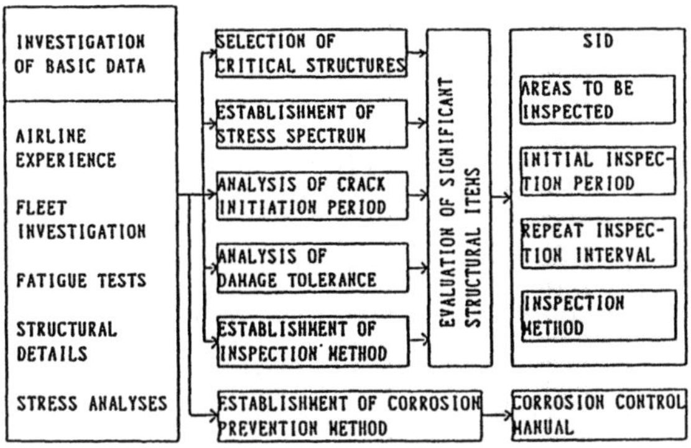

Figure 3.2 Outline of development

Table 3.1 Content of Corrosion Control Manual

1. INSPECTION AND REMOVAL OF CORROSION
   (1) PROCEDURE FOR INSPECTION
      - INSPECTION AREAS
      - INSPECTION METHOD
      - INSPECTION INTERVAL
   (2) PROCEDURE FOR CORROSION REMOVAL
2. CORROSION PREVENTIVE TECHNIQUE
   - AREAS TO BE APPLIED
   - APPLICATION INTERVAL
3. REPAIR PROCEDURE FOR CORRODED AREAS

Table 3.2 List of SID inspection items

L/E : Low Frequency Eddy Current Test
U/T : Ultrasonic Test

TITLE	Nint ORIGINAL	Nint PRESENT	Nrpt	INSP. METHOD
CORNER ANGLE - AFT PRESSURE DOME	35,800 FLT	45,000 FLT	8,000 FLT	L/E
NACELLE SUBFRAME	35,000	46,000	8,000	U/T
FRONT SPAR CAP - OUTER WING, LWR	35,900	46,000	8,000	VISUAL
WING JOINTS - CENTER WING TO OUTER WING				
· SPAR FITTINGS, CENTER WING	39,000	46,000	4,400	L/E
· STRINGER FITTINGS, CENTER WING	63,700	63,700	8,000	L/E
· SPAR FITTINGS, OUTER WING	37,100	46,000	6,000	L/E
· STRINGER FITTINGS, OUTER WING	40,200	45,000	8,000	L/E
MANHOLE DOUBLER - WING LOWER PANEL				
· GROUP A (MANHOLE AT ACCESS PLATES NO. 521~523, 541, 551~555)	36,100	45,000	8,000	VISUAL
· GROUP B (MANHOLE AT ACCESS PLATES NO. 531)	17,100	22,000 FOR YS-11	8,000	VISUAL
	33,800	41,000 FOR YS-11A		
SKIN STRINGER END - WING LOWER PANEL				
· W. STA. 1800, STRINGER #23	36,100	46,000	8,000	VISUAL
· W. STA. 5600, STRINGER #17	36,100	46,000	4,000	VISUAL
· W. STA. 5600, STRINGER #21	36,100	46,000	4,000	VISUAL
· W. STA. 7200, STRINGER #15	36,100	46,000	4,000	VISUAL
· W. STA. 7700, STRINGER #22	36,100	46,000	8,000	VISUAL
SKIN - M.L.C.WHEEL WELL	37,200	45,000	8,000	L/E
SKIN SPLICE - WING LOWER PANEL W. STA. 4000	36,100	46,000	6,000	L/E

Table 3.3 List of detection limits

INSPECTION METHOD	INSPECTION AREAS	DETECTION LIMIT
ULTRASONIC	NACELLE FRAME	2.5mm
HIGH FREQ. EDDY CURRENT	OUTER WING LWR PANEL & SPAR CAP	2.5mm
LOW FREQ. EDDY CURRENT	STRINGER LWR SPLICE FTC-C.WING	6.0mm
	WING FRONT & REAR SPAR LWR CAP	5.0mm
	OUTER WING LWR PANEL	3.5mm
VISUAL	FWD PRESSURE BULKHEAD	25.4mm
	OUTER WING LWR SPAR CAP	12.7mm *

* IN CASE OF VERY GOOD ACCESSIBILITY AND INSPECTABILITY

The multiple site damage (MSD) was not considered when the SID was established. Though the MSD has not been found on the operating fleet yet, the damage tolerance analysis for MSD is recently re-examined with improved accuracy in analysis and its effect on the initial mode The typical result of crack growth is shown in Fig. 3.3. The inspection interval is not changed because the upper limit was 8,000 FLT in consideration of practical inspection frequency when the SID was established. MHI thinks that the inspection interval in SID for YS-11 does not need to be revised from the result of the analysis.

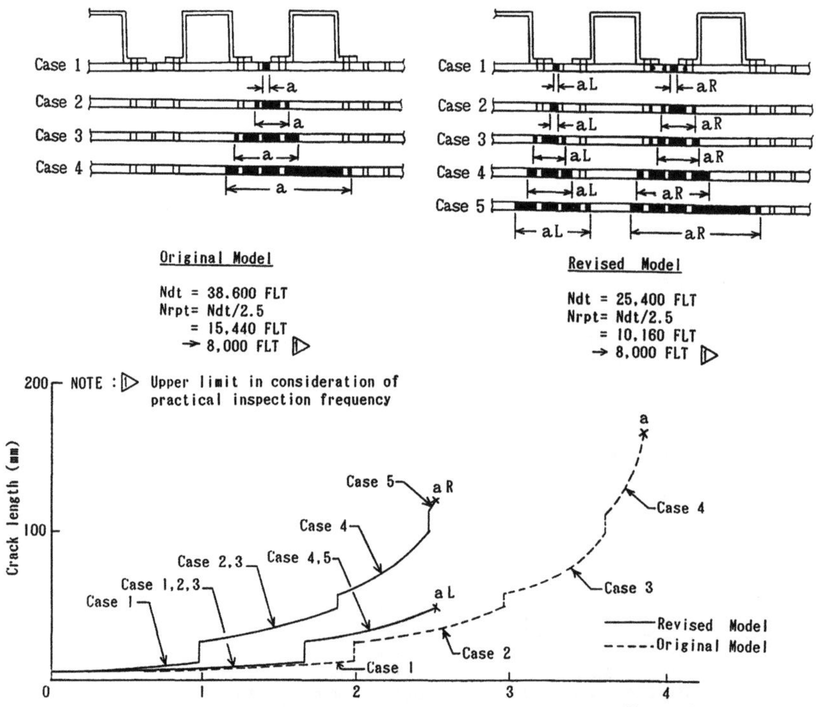

Fig. 3.3 Crack Growth Analysis of MLG-Wheel Well

References

[1] Kamiyama, T., Takeuchi, K., Shima, F., and Kitatani, K. "Full-Scale Fatigue Test of Twin-turboprop Transport", Experimental Mechanics, SESA, April, 1966, p194.

[2] Shiohara, T., Uchimoto, T. and Koga, K., "Development of SID on YS-11 Twin-Turboprop Transport for Continuing Structural Integrity of High Time Airplanes", Proceeding of ICOSSAR '85, IASSAR, May, 1985, VOL. 11, p117.

4. Fatigue Crack Propagation Behavior of 2024-T3 Riveted Plate with Initial Damages (National Aerospace Laboratory)

Fatigue crack propagation of riveted joint specimen of 2024-T3 with several types of notch at rivet holes, which simulates the damage by fretting or corrosion cracking, was studied.
Following five types of specimen were tested.

(a) Center notched sheet specimen. The geometry of the notch is 4mm hole with 1mm saw cut.

(b) Sheet specimen with 15 holes, each of which has 1mm long saw cut at each side of the hole.

(c) Rap joint specimen of single row rivets with 15 rivets. Diameter of the rivet hole was 4mm. Each rivet hole of one plate has 1mm long saw cut at each side. The distance between rivet hole center was 20mm.

(d) The same specimen as (c), but has saw cut only at the center (8th) hole.

(e) Rap joint specimen with double row rivets. Saw cuts were machined before riveted at both sides of all rivet holes of inner row.

Geometry of the specimen tested was 750mm long, 300mm wide and 1mm thick. (See Fig. 4.1) Since the diameter of the rivet head is 8mm, saw cuts are completely hidden by the rivets. Stress applied to the specimen (a) to (d) was (54$\pm$ 42) MPa, which approximately corresponds to the stress applied to the pressure bulkhead of B-747 induced by 0.6 atm. difference. The fuselage structure is subjected to 0.6 atmospheric pressure difference when the altitude of aircraft is 35,000 feet with cabin pressure 0.9 ata..

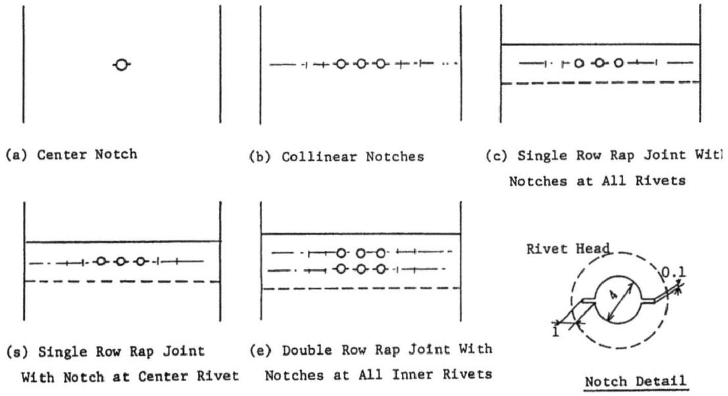

Fig. 4.1 Configuration of the Specimens

As no crack was observed on the double row rivet specimen (type (e)) after $3*10^5$ cycles of above mentioned stress, maximum stress was increased up to 150%.

Testing results and discussions

Crack growth rate is approximately presented by eq. (1) as a function of ΔK.

$$\frac{da}{dn} = \frac{C\Delta K^{\alpha}}{(1-R)K_c - \Delta K} \qquad (1)$$

where, $K_c = 74.4$ MPa $\sqrt{m}$, $C=2.14E-8$, and $\alpha = 3.7$

Fig. 4.2 shows fatigue crack extension behavior of (a) to (e) specimens. For single row rivet joint specimen, equivalent stress intensity factor Keq was used in order to account for the effect of concentrated force by the rivet.

$$Keq = \sqrt{Ka^2 + Kb^2} \qquad (2)$$

$$K_a = \sigma \sqrt{\pi a} F$$

$$K_b = \frac{P}{t\sqrt{\pi a}} f$$

F, f: correction factor
t: thickness

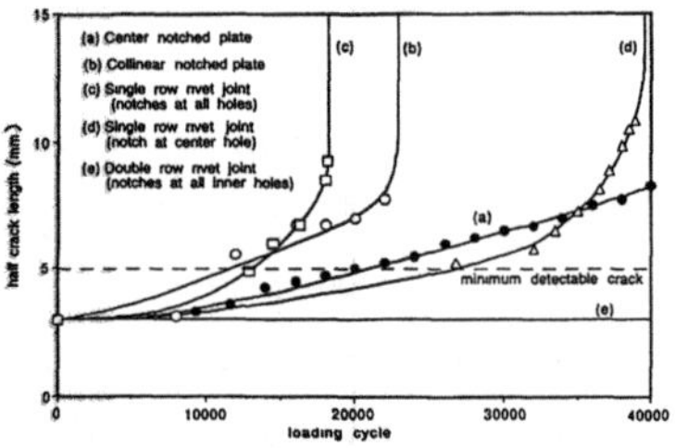

Fig. 4.2 Crack Extension Behavior of (a) to (e) Type of Specimen

The result suggests structural integrity may not be maintained by ordinary inspection period for multiple site damages of single row riveted joint as total fracture occured within 2,000 stress cycles after cracks were detectable at each rivet fastener.

It is noteworthy that out-of-plane deformation of double row rivet joint was very small compared with those of single row specimen. Tightness or load sustaining by surface friction seems to be the most important factor to suppress the crack initiation and extension. It is therefore anticipated that substantial thickness reduction by corrosion lessen surface friction and accelerate crack initiation and propagation.

It was observed both micro- and macroscopically that the crack initiated and propagated with higher rate at inner corner of rivet hole because of bending effect of loose rivet. This phenomenon is not noticeable for double row rivet joint specimen.

MIX
Papier aus verantwortungsvollen Quellen
Paper from responsible sources
FSC® C105338

If you have any concerns about our products,
you can contact us on
ProductSafety@springernature.com

In case Publisher is established outside the EU,
the EU authorized representative is:
Springer Nature Customer Service Center GmbH
Europaplatz 3, 69115 Heidelberg, Germany

Printed by Libri Plureos GmbH
in Hamburg, Germany